QUANTITATIVE METHODS FOR BUSINESS DECISIONS

QUANTITATIVE METHODS FOR BUSINESS DECISIONS

THIRD EDITION

Jon Curwin and Roger Slater

CHAPMAN & HALL

University and Professional Division

London · Glasgow · New York · Tokyo · Melbourne · Madras

Published by Chapman & Hall, 2-6 Boundary Row, London SE1 8HN, UK

Chapman & Hall, 2-6 Boundary Row, London SE1 8HN, UK

Blackie Academic & Professional, Wester Cleddens Road, Bishopbriggs, Glasgow G64 2NZ, UK

Chapman & Hall Inc., One Penn Plaza, 41st Floor, New York NY 10119, USA

Chapman & Hall Japan, Thomson Publishing Japan, Hirakawacho Nemoto Building, 6F, 1-7-11 Hirakawa-cho, Chiyoda-ku, Tokyo 102, Japan

Chapman & Hall Australia, Thomas Nelson Australia, 102 Dodds Street, South Melbourne, Victoria 3205, Australia

Chapman & Hall India, R. Seshadri, 32 Second Main Road, CIT East, Madras 600 035, India

---

First edition 1985
Reprinted 1986
Second edition 1988
Reprinted 1989
Third edition 1991
Reprinted 1993, 1994

Typeset in 10/12pt Times by Best-set Typesetters Ltd, Hong Kong
Printed in Great Britain by Clays Ltd, St Ives plc

ISBN 0 412 40240 8

A catalogue record for this book is available from the British Library
Library of Congress Cataloging-in-Publication Data available

CONTENTS

PREFACE

The third edition of *Quantitative Methods for Business Decisions* has been updated to reflect the changing approach to quantitative information. The ability to use computer based methods and the ability to present the outcome of computer based analysis are seen as increasingly important.

Most chapters have been substantially revised. The coverage of matrices has been extended to form a new chapter, and the construction of decision trees, non-parametric statistical tests and modelling of queues added to existing chapters. There is a new chapter on simulation to reflect the business importance of this modelling technique. A complementary $5\frac{1}{4}''$ disk for use with IBM compatible micro-computers is included, and gives the reader easy access to a quality statistical package, MICROSTATS, and practice data sets. These changes are a response to the many welcome suggestions from readers of the first and second editions. It has not been possible to incorporate all the constructive suggestions and we have conscientiously continued to limit the mathematical rigour of the book. We see the book as presenting a selective view of mathematics and statistics for students new to a business-related course on quantitative methods.

The book is now recommended reading for a wide range of courses. On BTEC programmes, for example, the book has been successfully used as a resource for students individually or in groups to complete activities. On some degree courses the book has been followed chapter by chapter as a means of supporting students with different mathematical backgrounds and rate of progression. On MBA courses the book has been used to illustrate business applications to those already familiar with mathematics at this level. An element common to all those new to quantitative methods, is the need to develop a confidence to solve problems. The book has been designed to give a step by step development of the use of mathematics and statistics (see introductory section 'How to use this book').

The inclusion in the second edition of typical examination papers has been welcomed. This final part of the book has now been extended to include three typical in-course assignments.

We have recognized that increasing access to micro-computers is changing the needs of students and academic staff. For many students a course in computer applications will run in parallel with a course in quantitative methods. We would see such courses as complementary. In this edition we outline how many problems are best solved using a spreadsheet or a package like MICROSTATS. However, many end of year examinations require students to perform calculations by hand and this edition still includes full details of such procedures.

HOW TO USE THIS BOOK

The book has been arranged in eight parts, the first seven of which include two, three or four chapters. These groups of chapters all develop a theme within your quantitative methods course.

Part	*Theme*
One	Quantitative information
Two	Descriptive statistics
Three	Mathematical methods
Four	Measures of uncertainty
Five	Statistical inference
Six	Relating two or more variables
Seven	Mathematical models
Eight	Review and revision

There is a general introduction to each part which should provide a business context for the chapters that follow. At the end of each part and at the end of each chapter there are exercises for practice, revision and discussion.

As with all books of this kind, some chapters may seem familiar and others new and difficult. You may, for example, understand descriptive statistics from your lecture notes but not statistical inference. We suggest you skim read the chapters you think you understand just to make sure. It is worth looking at the worked examples and asking the question 'could I do that?'. The chapters that seem new and difficult you may need to read several times. If you don't understand a particular topic, read through the chapter relatively quickly to get an overall view and then read again with more concentration on the detail. **What you will need to do is establish a way to use this book to your own maximum advantage**. For some it will be a matter of using the book for supportive reading only. You will find plenty of worked examples to complement those given in lectures and your seminars. For others, the book may need to be a self-contained course. You should find a description of all the methods used, problems included at the end of each chapter and answers at the end of the book.

The development of the subject is progressional and you will need to understand some chapters completely before you move on. It is unlikely that you will be able to make sense of the standard deviation covered in Chapter 4, unless you fully understand the calculation of a mean covered in Chapter 3. Numerical difficulties have often impeded the development of understanding in this subject, but the availability of computer packages speeds up calculation and therefore allows data to be analyzed more quickly. The difficulty is no longer getting numerically correct answers but rather being able to interpret these answers.

THE USE OF COMPUTERS

Computers are now widely available to those studying quantitative methods, and recognizing this, we have included a copy of MICROSTATS by Mike Hart with this third edition of the book. This package is similar to MINITAB which may be available where you are studying, and it will be a matter of personal preference which package you choose to use. For those with their own computer, MICROSTATS offers an opportunity to carry out statistical analysis away from their place of study. A user manual for MICROSTATS is included as appendix K.

Other packages are mentioned in the text, such as MARQUIS and SPSS–X. Such packages will help with some of the calculations necessary, especially where survey data is to be analyzed. Each computer package has advantages and disadvantages and different levels of availability from college to college.

It will be an advantage if you have access to a spreadsheet package and that you are familiar with using it. This book does not propose to teach you how to use a spreadsheet! Spreadsheet packages all work in basically the same way and all of the exercises and problems within the book which use spreadsheets can be completed on any such package. What we would suggest is that when you build a spreadsheet model, even a very simple one, you always use formulae wherever possible so that recalculation can be completed quickly and automatically when a single number is changed.

It is not essential to use a computer package to understand the contents of this book, but certain exercises will become rather long and tedious without such a package, and the problems in Chapter 19 cannot be solved by the methods shown unless a package is available.

PART ONE
QUANTITATIVE INFORMATION

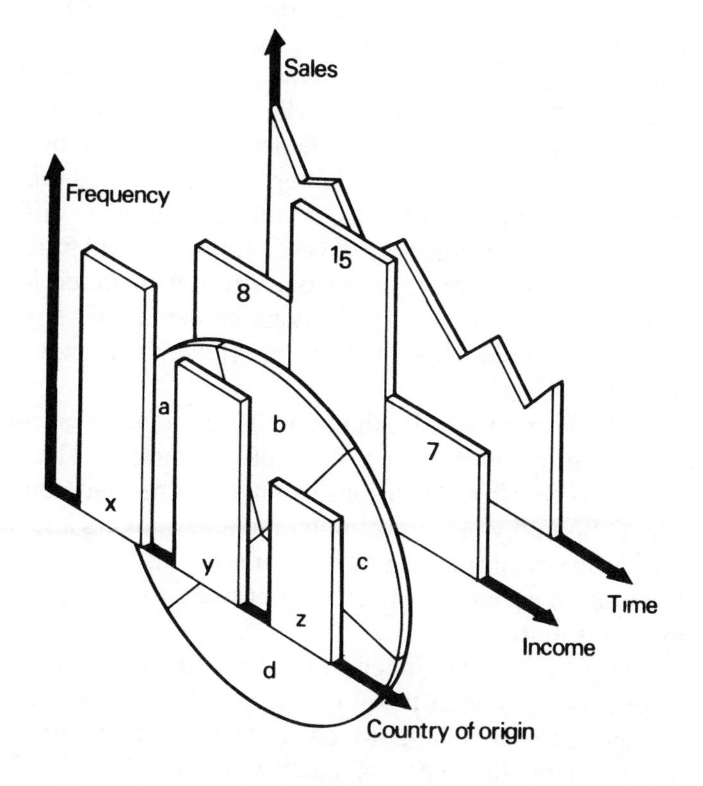

In this part of the book we are concerned with the ways data can be collected and how such data can be effectively presented. Data is a very general term and can include just a few records of time taken to complete a fairly simple task, to the outcomes of a national survey with several thousand respondents all answering 20 or more questions. In all aspects of business we are likely to encounter increasing quantities of data. New technologies literally put data at our fingertips with, for example, share prices in New York or stock levels in a warehouse some distance away being known in minutes. The magnitude and the complexity of the business environment demands more than a verbal description. How many business problems can you describe without numbers? We need to improve our understanding of the numbers; how many, how large and how soon.

Data may have been collected for other reasons, e.g. externally by a government department or internally for accounting reasons, but can find a wide range of new uses. It is always worth checking whether the data you need already exists and in what form. This type of **desk research** will generally provide figures on the number of people by age, by sex, by income or by region. What will often be lacking is particular detail. If you want attitudinal information on products or services, or wish to relate behaviour to other factors you will need to develop the methods to collect this data.

Consider, for example, employment. You will need to define what is meant by the term and who should be included. You will need to recognize that some aspects can be described numerically e.g. the **quantitative** factors and some aspects cannot be described numerically e.g. the **qualitative** factors. Data is available. A single figure, such as the total number employed in the previous year in the UK is precise, needs little in the way of presentation, and will be easily available from published sources. It will, however, communicate little about the nature of employment. If you investigate further you will find the numbers by region, by industry and a range of other factors. As you add to your data, it will become more difficult to present and communicate. As you continue your investigation, you will find that there are a number of factors relating to employment that are not included in the published statistics. If you want to know how some workers perceive their employment, their training opportunities, chances of promotion or overtime pay, you will need methods to collect this data.

Quantitative methods involve more than obtaining the numbers and working out a few statistics. A **statistic** is merely a descriptive number. The purpose of the enquiry will need to be clarified. Decisions will need to be made on who to include and who to exclude. Should you, for example, include smokers and non-smokers in equal numbers in a survey on the health risks of smoking? Should you include car owners in a survey about public transport? The answers to such questions are not as obvious as they might first seem.

The methods for collecting data need to be decided. It has often been said that the answers can only be as good as the questions. It is not just a matter of obtaining the answer a respondent is willing to give. Answers need to be adequate in numerical terms to allow the types of analysis required. It is of limited value knowing that a respondent likes or dislikes a product, if your investigation requires a measure of how much a respondent likes or dislikes a product and a measure of how this attitude relates to other factors. Liking may not be sufficient, if a company needs to know whether liking can or will be turned into the action of purchasing the product or service. You also need to ensure that you are able to collate, analyze and present your data. Familiarity with a computer software package, such as MICROSTATS, will allow you to manage large sets of data, to calculate statistics that would otherwise require considerable time, and to experiment with the data. The last point, is an important one. Data should not just reveal the obvious but should be explored for new patterns and ideas. Looking at the data in a different way may uncover some new insight in your area of interest.

DATA COLLECTION 1

Many business, economic and social questions are not amenable to a simple yes or no answer. They may need clarification and discussion. Possible solutions may need to be presented and criteria agreed for their acceptance or rejection. To consider the arguments and indeed the 'facts' presented, the completeness of current information and the requirements for new information need to be assessed. Decision makers not only need **data** but also need to evaluate the quality of the data.

One dictionary definition of data is 'things known and from which inferences may be deduced'. In this book we are essentially concerned with numerical data. Numerically we can describe, for example, the size of a business, its profitability, its product range, the characteristics of its workforce and a host of other factors. However, numbers alone are unlikely to give us the understanding of the business problem that we require. We also need to take account of the people involved, the culture of the enterprise, the legal and economic environment. In general, most significant business problems are likely to require a multi-disciplinary approach. It is also true that few problems are purely qualitative in nature. If we consider the personnel problem of staff recruitment for example, we soon begin to describe job requirements in terms of age, income and other measureable factors. If we consider another personnel problem of assessing training needs, then we can become involved in a major statistical exercise.

The **completeness** of data is always a problem for the decision maker. It is always possible to collect more and more data. The decision maker will need to decide whether the current data is sufficient for the purpose or whether additional data should be acquired. Data collection takes time and can be costly. The other issue for the decision maker is the quality of the data. Data that has **bias** or is misleading can damage any effective decision-making process. The results of a survey of married women could not, for example, be taken to represent the views of all women. If this survey did not give a fair chance of inclusion to younger married women, then a further source of bias would exist. Whenever we look at data or consider data collection, we need to ask 'what is the problem?' or 'what is the question?'. A prerequisite of any statistical enquiry is an understanding of purpose or more formally, a statement of objectives.

Questions about data fall into broad groups, and in this chapter we will consider each of these groupings:

1. What is the relevant population?
2. What are the sources of data?
3. How many people were asked and how were they selected?
4. How was the information collected from these respondents?
5. Who did not respond?
6. What type of data was collected?

1.1 POPULATION

Identification of the relevant population is essential since data collection can be a costly exercise and contacting large numbers of people who could have nothing to do with the survey will only waste these valuable resources. For example, if you were concerned with the acceptability to women of a new contraceptive pill it would be pointless contacting a group of people, half of whom were men. A similar problem can arise if the group you have identified as the relevant population does not include everyone for whom the survey is relevant, since a range of views or information will be totally missed. If you were interested in why people bought foreign-built cars, but failed to contact purchasers of certain Ford models, for example, then you might fail to identify the fact that some buyers do not realize that their car is foreign-built. The term 'population' can also be used to describe all the items or organizations of interest. An audit, for example, is concerned with the correctness of financial statements. The population of interest to the auditor could be the accounting records, invoices or wage sheets. If we were concerned with job opportunities, the population could be all the local businesses or organizations employing one or more persons.

EXERCISES

1. What is the relevant population to contact regarding a new nappy?
2. Who would be interested in a new metal paint to be sold in large quantities?

Having considered the type of people who would fall into the relevant population, the next problem is to try to identify who these people are; and perhaps even to get a list of their names and addresses. If this list can be obtained, it is called a **sampling frame**. Many surveys, particularly in market research, need a general population of adults, and make use of the Electoral Register. This contains a list of most people in the UK over 18 years old and is updated annually; of course, as about 10% of the population of the UK move each year, others emigrate, some die and there is some immigration, the list cannot be 100% accurate, but it is easily available and very widely used. A study paper from the Central Statistical Office suggests that the Electoral Register is even more accurate in identifying addresses rather than people. (Houses move less frequently than people.) Other groups in the general population may appear on separate lists; for example, all those who are members of the RAC, or all those who are on a credit blacklist. These lists may not be generally available.

When a list does not exist or is not generally available, then those collecting the information may either try to compile a list, or use a method of collection which does not require a sampling frame (see next section). A method used by the Central Statistical Office to identify the population of disabled people, when a new Act of Parliament came into force, was to contact a large number of people in the general population, asking a few simple questions on whether they themselves were disabled or if they knew anyone who was, and if so who they were. This gave a fairly comprehensive list of the disabled, who could then be contacted and asked for more detailed information.

1.2 SOURCES OF DATA

Having considered the purpose of a statistical enquiry and defined the relevant population, we could argue that the most important steps have already been taken. In a market research survey, a statement of objectives and the definition of the target population are seen as critical by most practitioners. If you are not sure about the purpose of the enquiry, and you are not selective about the information collected, what is the likely value of any subsequent, complex statistical analysis? This is surely akin to the computing saying GIGO — 'garbage in, garbage out'.

The next step is to obtain data on the population of interest. A statistical enquiry may require the collection of new data, referred to as **primary** data, or be able to use existing data, referred to as **secondary** data, or may require some combination of both sources. Sources of primary data include observation, group discussions and the use of questionnaires. The distinguishing feature of primary data is its collection for a specific project. As a result, primary data can take a long time to collect and be expensive. Secondary data, in contrast, has been collected for some other purpose. It is usually available at low cost but may be inadequate for the purposes of the enquiry. Where the data requirements are fairly complex, it is generally seen as good practice to first collect the lower cost secondary data, usually of high quality (it has been published) and add to it as necessary. If for example, we were considering the impact of a new shopping centre on the local community, from available statistics we should be able to describe the demographic characteristics of the surrounding area (but would need to define the population first) but not the attitudes of the local residents.

Secondary data can come from within the organisation, internal secondary data, or from outside the organisation, external secondary data. Secondary data may also be called **by-product** statistics, for example, the collection of the number of unemployed is a by-product of paying out benefits and recording the total. The main purpose is to record money paid out, but the by-product is to monitor the number of unemployed in the region or country. Internal data sources include employee records, payroll information and customer orders. The most important source of external secondary data are the official statistics supplied by the Central Statistical Office and other government departments. A useful introduction to the vast range of official statistics is the free booklet *Government Statistics — a brief guide to sources* available from the Information Service Division, Cabinet Office, Great George Street, London, SW1P 3AL. Your local library should be able to direct you to the more extensive *Guide to Official Statistics*. The publications listed below give some indication of the statistics readily available.

The *Annual Abstract of Statistics* is generally regarded as the main reference book for statistics on economic and social life. It includes statistical information, generally for the last 10 years, on climate, population, social services, education, employment, defence, production, energy, transport, trade and a host of other topics.

The *Monthly Digest of Statistics*, published monthly, provides statistical information on some 20 subjects including population, employment, prices, production, energy and transport. The statistics are presented mostly as runs of monthly and quarterly estimates for the last two years. Annual figures go back several more years.

Financial Statistics is published monthly and gives a useful range of financial and monetary statistics including the borrowing of local and national government, profits of companies, interest rate and the balance of payments.

Economic Trends, also published monthly, presents a range of economic measures including selected monthly economic indicators, income, capital formation, production and prices. Articles relating to economic performance are also included. An annual supplement gives data over longer time periods.

Regional Trends provides annually a selection of the main statistical series on a regional basis.

Social Trends is an annual publication that presents a broad description of British society. It includes statistical information on population, the family, employment, income, health, housing, transport, leisure and law enforcement.

The *Employment Gazette* provides monthly information on employment, unemployment, hours worked, earnings, retail prices, industrial disputes and other statistics related to the labour market. Articles and items of news relating to the labour market are also included.

1.3 NUMBERS AND SELECTION

1.3.1 A census

One type of data collection does not require a selection procedure, and this is a census, or complete enumeration of the identified population.

The best example of this type of survey is the Population Census which has been carried out in the UK once every 10 years since 1801 (with the exception of 1941 when rather more urgent matters were being dealt with). While this type of exercise should give highly detailed information and reflect data from all parts of the relevant population, it does take a long time to analyse the data and is very costly. A census is of limited use to the majority of business, social or economic applications, unless the identified population is small. For example, a census of all homes would be an expensive way of estimating the population with television sets. In contrast, if you were representing a manufacturer who sold only to a small number of wholesalers and wanted their views on a new credit-ordering system, then a census would be a suitable method to use.

1.3.2 Selection: random and non-random

Where the relevant identified population is too large for a cost-effective census to be conducted a **sample** of that population must be selected, and individual responses generalized to represent the facts about, or the views of, the entire population. However, the method of selection will have implications for the validity of this generalization procedure: if you were to ask the next five people you see how they would vote at a general election, it is very unlikely that the answers given would be a guide to a general election result. Sampling procedures can be divided into two broad categories: those where individuals are selected by some random method prior to the collection stage, and those where the individuals are non-randomly selected at the collection stage.

Random

Random does not mean haphazard selection, but that each member of the population has some calculable chance of being selected. There is no one in the identified population who *could not* be selected when the sample is set up. A simple random sample gives every individual an *equal* chance of selection. To select a random sample a list or sampling frame is required, where each member is given a number and a series of random numbers (usually generated on a computer) are used to select the individuals to take part in the survey. There is thus no human interference in the selection of the sample, and samples selected in this way will, in the long run, be representative of the population. This is the simplest form of random sampling (see section 1.3.4 for more complex designs).

Non-random

Non-random is a catch-all for other methods of selecting the sample, where there is some judgment made in the selection procedure, and this may lead to some sections of the population being excluded from the sample, for good or bad reasons. For example, if the interviewer is asked to select who will take part in the survey and has a particular aversion to say, tall people, then this group may be excluded. If, then, tall people have different views on the subject of the survey from everyone else, this view will not be represented in the results of the survey. However, a well-conducted non-random survey will produce results more quickly, and at a lower cost, than a random sample; for this reason it is often preferred for market research surveys and political opinion polls.

The most usual form of non-random sampling is the selection of a **quota sample**. In this case various characteristics of the population are noted, for example the divisions on sex, age and job type; and the sample aims to include similar proportions of people with these characteristics. This suggests that if people are representative in terms of known identifiable characteristics they will also be representative in terms of the information being sought by the survey. Having identified the proportions of each type to be included in the sample, each interviewer is then given a set number, or quota, of people with these characteristics to contact. The final selection of the individuals is left up to the interviewer. Interviewers you may have seen or met in shopping precincts are usually carrying out quota sample surveys.

EXERCISE

Apart from groups specifically avoided by a poor interviewer, which groups in the population would be excluded from such a survey?

Setting up a quota survey with a few quotas is relatively simple. The results from the Census of Population will give the proportions of men and women in the population, and also their age distribution.

From the component tables of Table 1.1 we see that the general population over 15 years old consists of 46% men and 54% women, and thus our sample should also exhibit this division of sex. We can obtain similar information on the age distribution of the whole population, but this may be further analyzed to show the separate age distributions of the sexes (part iii) so that we do not get the situation that all of the women in the sample are under 50 years old. Having combined the information from parts (i) and (ii) of Table 1.1 we can take each percentage of the sample size using part (iii), here 1000, to find the number of each type of person to contact.

For a quota sample, it is important that the characteristics on which the quotas are based are easily identified (or at least estimated) by the interviewer, or else a lot of time will be wasted trying to identify the people who will be eligible to take part in the survey. If the number of quotas is large, some of the groups will be very small, even with an overall sample size of 1000.

Table 1.1

(i) Age distribution of population over 15 years

Age (years)	*Percentage*
15–20	19
20–30	25
30–50	26
50+	30

(ii) Sex distribution of population over 15 years

Sex	*Percentage*
Men	46
Women	54

(iii) Population (percentages)

Age (years)	*Men*	*Women*
15–20	10	9
20–30	12	13
30–50	12	14
50+	12	18

(iv) Number for samples of 1000

Age (years)	*Men*	*Women*
15–20	100	90
20–30	120	130
30–50	120	140
50+	120	180

1.3.3 Numbers

So far we have suggested that a census will be too costly for most subjects of business surveys, but that just asking five people would be unlikely to give a full representation of the views of the general public. The variability of the population will influence the sample size required; in the extreme case where everyone held exactly the same opinion, then it would only be necessary to ask one person. (This is a highly unlikely situation!) If everyone in the population held distinct views, then a census would be the only way to elicit the full range of views. (Again, this is very, very unlikely!) Sample size will also be related to how precise the results required from the survey are to be; and if the proportions are to be based on some subgroup of the sample; it is the size of these subgroups that must first be determined.

Since most surveys do not aim to find out a single piece of information, but the answers to a whole range of questions, the determination of sample size can become extremely complex. It has been found that samples of about 1000 give results that are acceptable when surveying the general population. (See section 12.2.3 for the calculation of sample sizes.)

1.3.4 More complex random samples

The simple system outlined above for a random sample will work well with a relatively small population that is concentrated geographically, but would become impractical if it were used on a national scale; you would need the complete Electoral Register for the whole of the UK and might end up visiting one person on Sark, another in the Shetlands and none in the South East. Travel costs would be phenomenal! It would not be impossible, although it would be unlikely, that the whole sample might consist of people living in Wales: this would not be too important if the people of Wales were wholly representative of *all* UK citizens, but on certain issues their views will tend to differ from those of, say England or Scotland, for example. To overcome these types of problem, various other sampling schemes have been developed, but they still retain the basic element of random sampling: that each member of the population has *some chance* of being selected.

Stratification

If there are distinct groups or **strata** within the population that can be identified *before* sample selection takes place, it will be desirable to make sure that each of these groups is represented in the final sample; thus a sample is selected from each group. The numbers from each group may be proportional to the size of the strata, but if there is a small group, it is often wise to select a rather larger proportion of this group to make sure that the variety of their views is represented. In the latter case it will be necessary to weight the results as one group is 'over-represented'.

When it is known that there are subgroups in the population, but it is not possible to identify them before sample selection, it is usual to ask a question which helps to categorize the respondent, such as 'At the last general election which party did you vote for?', and then to divide the results into groups or strata.

This is **post-stratification**. The results from these constructed strata can be weighted to provide more accurate results for the population as a whole.

Clustering

Some populations have groups which within themselves represent all of the views of the general population, for example a town, a college or a file of invoices. If this is the case, it will be much more convenient, and much more cost-effective, to select one or more of these clusters at random and then to select a sample, or carry out a census within the selected clusters.

Multi-stage designs

Even when the designs outlined above are used, there may well be problems over representativeness and costs. To overcome this, many national samples use a series of sampling stages, and at each stage select either all of the subgroups or a random sample. For a national sample in the UK, one may start by noting that the country is split into administrative regions for gas, electricity, civil defence, etc., and that each of these needs to be represented. Each region consists of a number of parliamentary constituencies, which can usually be classified on an urban–rural scale. A random sample of such constituencies may be selected for each region. Parliamentary constituencies are split into wards, and the wards into polling districts, for which the Electoral Register is available, from which a sample of individuals or addresses may be selected. This type of selection procedure will mean that all regions are represented and yet the travelling costs will be kept to a minimum, since interviewing will be concentrated

Table 1.2

Stage	*Sampling unit*	*Number selected*
1	Region	12 (all)
2	Constituency	4 (samples)
3	Ward	3 (samples)
4	Polling district	2 (samples)
5	Individuals	10 (samples)
Sample size = $12 \times 4 \times 3 \times 2 \times 10 = 2880$		

in a few, specific, polling districts. An example of a possible design is given in Table 1.2. To try to ensure that the resultant sample was fully representative, further stages could be added, or stratification could take place at some or all of the stages.

EXERCISES

1. What form of stratification would you use for a sample of car insurance claims?
2. For which type of information would a college be a suitable cluster to use?

1.4 ASKING QUESTIONS

Having identified the relevant population for a survey, and used an appropriate method of selecting a sample of people to give the information that is required, we now need to decide exactly what questions will be used, and how these questions will be given to the people in the sample. No matter how well the first two stages of an investigation are carried out, if biased questions are used, or an interviewer incorrectly records a series of answers, the results of the survey will be worthless.

1.4.1 Questionnaire design

To be successful, a questionnaire needs both a logical structure and well thought-out questions. The structure of the questionnaire should ensure that there is a flow from question to question and topic to topic, as would usually occur in a conversation. Any radical jumps between topics will tend to disorientate the respondent, and will influence the answers given. It is often suggested that a useful technique is to move from general to specific questions on any particular issue.

The Gallup organization has suggested that there are five possible objectives for a question:

1. To find if the respondent is aware of the issue, for example:

   Do you know of any plans to build a motorway between Cambridge and Norwich?

   YES/NO

   The answers that can be expected from a respondent will depend on the information already available and the source of that information (information available can vary from source to source). If the answer to the above question were YES we would then need to ask further questions to ascertain the extent of the respondent's knowledge.
2. To get general feelings on an issue, for example:

   Do you think a motorway should be built?

   YES/NO

   It is one thing to know whether respondents are informed about plans to build a motorway or indeed the merits of a new product but it is another to know whether they agree or disagree. In constructing such a question, the respondent can be asked to provide an answer on a **rating scale** such as:

Strongly agree	Agree	Uncertain	Disagree	Strongly disagree

   A scale of this kind is less restrictive than a YES/NO response and does provide rather more information.
3. To get answers on specific parts of the issue, for example:

   Do you think a motorway will affect the local environment? YES/NO

   In designing a questionnaire we need to decide exactly what issues are to be included. This can be done by using a simple checklist.

If the environment is an issue we need then to decide whether it is the environment in general or a number of factors that make up the environment, such as noise levels and scenic beauty.

4. To get reasons for a respondent's views, for example:
   If against, are you against the building of this motorway because:
   (a) there is an adequate main road already;
   (b) there is insufficient traffic between Cambridge and Norwich;
   (c) the motorway would spoil beautiful countryside;
   (d) the route would mean demolishing a house of national interest;
   (e) other, please specify . . . .

   To find the reasons for a respondent's views is going to require questions of a more complex structure. You will first need to know what his or her views are and then provide the respondent with an opportunity to give reasons why. The conditional statement 'if against' is referred to as a **filter**. The above question is **precoded** (see below); in contrast an **open-ended** question could be used; for example:
   Why are you against the motorway being built? . . . .
5. To find how strongly these views are held, for example:
   Which of the following would you be prepared to do to support your view?
   (a) write to your local councillor;
   (b) write to your MP;
   (c) sign a petition;
   (d) speak at a public enquiry;
   (e) go on a demonstration;
   (f) actively disrupt the work of construction.

   or

   How important is the Hall which would be demolished if the motorway is built? (Place a √ on the grid below)

Should be saved at any cost	1 2 3 4 5 6 7	Of no importance

   In many cases we would want to know not only whether something was considered good or bad but how good or how bad, and to do this we could give a series of possible attitudes or actions, or we could use a **rating scale**. A position on a rating scale provides some measure of attitude. The number of points on the rating scale will depend on the context of the question and method of analysis but in general four point and five point scales are far more common than the seven point scale shown above.

These categories reflect that much of Gallup's work is collecting information on attitudes and issues, but we could add that most surveys will also be asking questions which aim to find factual information about the respondents, their dependants or their possessions. To meet these objectives we can use either open or precoded questions.

An **open** question will allow the respondent to say whatever he or she wishes, for example: Why did you choose to live in Kensington?

This type of question will tend to favour the articulate and educated sections of the community, as they are able to organize and express their thoughts and ideas quickly. If a respondent is finding difficulty in answering, an interviewer may be tempted to probe, or help, and unless this is done carefully, the survey may just reflect the interviewer's views. A further problem with open questions is that, since few interviews are tape recorded, the response that is recorded is that written by the interviewer, who may be forced to edit and abbreviate what is said, and again this can lead to bias. Open questions do, however, often help to put people at their ease and to make sure it is their view which they are giving, rather than one of some pre-arranged group of responses, no one of which is *exactly* their view. Further, at the early development stage of a questionnaire, open questions may be used to identify common responses.

Precoded questions give the respondent a series of possible answers, from which one may be chosen, or an alternative specified. These are particularly useful for factual questions, for example:

How many children do you have?

0 1 2 3 4 5 6 more (circle answer)

When this type of response is used for opinion questions, some respondents will want to give a response between two of the opinions represented by the precoded answers.

Do you agree with the deployment of nuclear weapons in Britain?

Agree ☐
Disagree ☐
Don't know ☐ (tick box)

Some respondents will say 'Yes, but only of a certain type' or 'No, but there is no alternative' or 'Yes, provided there is dual control of their operation'. One reaction to this type of answer is to try to expand the range of precoded answers given, but this does not necessarily solve the problem. An open question may be better. In the example given above, it may be better to ask a series of questions, building up through the objectives suggested by Gallup.

Question wording is also important in eliciting representative responses, as a biased, or leading, question will bias the answers given. Sources of bias in question design identified by the Survey Research Centre are given below.

1. Two or more questions presented as one, for example:

   Do you use self-service garages because they are easy to use and clean? YES/NO

   Here the respondent may use the garages because they are easy to use, but feel that they are dirty and disorganized; or may find them clean but have difficulty in using the petrol pumps.

2. Questions that contain difficult or unfamiliar words, for example:

   Where do you usually shop?

   The difficult word here is 'usually' since there is no clarification of its meaning. An immediate response could be 'usually shop for what?' or 'How often is usually?' People's shopping habits vary with the type of item being purchased, the day of the week the shopping is being done, and often the time of year as well.

   Did you suffer from rubella as a child? YES/NO

   Many people will not know what rubella is, unless the questionnaire is aimed purely at members of the medical profession; it would be much better to ask if the respondent suffered from german measles as a child. This problem will also be apparent if jargon phrases are used in questions.

3. Questions which start with words meant to soften hardness or directness, for example:

   I hope you don't mind me asking this, but are you a virgin? YES/NO

   In this case, the respondent is put on their guard immediately, and may want to use the opening phrase as an excuse for not answering.

   Do you, like most people, feel that Britain should be represented in NATO? YES/NO

   Two possible reactions to this type of leading question are:

   (a) to tend to agree with the statement in order to appear normal, the same as most people; or, in a few cases,
   (b) to disagree purely for the sake of disagreeing.

   In either case, the response does not necessarily reflect the views held by the respondent.

4. Questions which contain conditional or hypothetical clauses, for example:

   How do you think your life would change if you had nine children?

   This is a situation that few people will have considered, and so have never thought about the way in which various aspects of their life would change.

5. Questions which contain one or more instructions to respondents, for example:

   If you take your weekly income, after tax, and when you have made allowances for all of the regular bills, how much do you have left to spend or save?

   This question is fairly long and this may serve to confuse the respondent, but there are also a series of instructions to follow before an answer may be given. Other problems that will arise here are that many incomes are not weekly, and most bills, for gas, electricity, loans, etc., are monthly or quarterly; many people will not make regular savings for bills, but just pay them when they become due, while others will scrupulously save a set amount each week.

The completed questionnaire needs to follow a logical flow, and where questions are used to **filter** respondents, for example, if YES go to question 10 and if NO go to question 18, then all routes through the questionnaire must be consistent with the instructions. A computer package, such as MARQUIS, allows you to type in the questions, specifying the flow from one to another and then checks for flow and consistency. Using such a package, it is possible to develop a questionnaire and then print copies directly from the program.

EXERCISE

Write a series of questions seeking information on topics of your choice to illustrate the problems posed in this section.

1.4.2 Interviews

Once the respondents have been selected and the questionnaire prepared, then the two must be brought together, most frequently by an interviewer. (In a quota sample where the interviewer selects the respondents, this is the only feasible approach.) The interviewer has a key rôle to play in a survey, where only about a third of his or her time is spent in interviewing, the rest being used for travel and locating respondents (40%), editing and clerical work (15%), and preparatory and administrative work (10%).

It is unlikely that a person could just go out and conduct successful interviews; a certain amount of training is necessary to help in recording answers correctly and, in the case of open questions, succinctly. An interviewer's attitude is also important, since if it is not neutral or unbiased, it may influence the respondent. Unbiased questions can be turned into biased ones when a bad interviewer lays stress on one of the alternatives, or 'explains' to the respondent what the question really wants to find out. This explanation, or probing, can be turned into an advantage if the interviewer is fully aware of the aims of the survey and can probe without biasing the response. Complex and intimate topics can be covered by a sympathetic interviewer.

Increasingly, questionnaires are being administered by telephone with the interviewer recording responses, often directly on to a computer. A package, such as MARQUIS, will print the questions onto the screen and the interviewer enters the data directly. The data is automatically stored and can be used for subsequent analysis. Use of direct data entry will reduce the errors introduced in transferring responses from written questionnaires, but errors made by the interviewer cannot be checked. Telephone interviewing is often quicker to organize and complete, but removes any possibility of observing the respondent, or even checking who is being interviewed.

1.4.3 Postal questionnaires

An alternative to interviewing the selected respondents is to post the questionnaire to them, with a reply-paid envelope. This method will yield a considerable saving in time and cost over an interviewer survey, and will allow time for the replies to be considered, documents consulted or a discussion of the answers with other members of the household. (This may be an advantage or disadvantage depending upon the type of survey being conducted.) Since the interviewer is not present, there is no possibility of observing the respondent or probing for more depth in the answers. The method will thus generally be more suitable for surveys looking for mostly factual answers. In general, the questionnaire should be relatively short, to maintain interest and encourage responses, but even so, postal questionnaires tend to discriminate against the less literate members of society, and have a higher response rate from the middle classes.

1.5 NON-RESPONSE

It is almost inevitable that when surveying a human population there will be some non-response, but the researcher's approach should aim at reducing this non-response to a minimum and to find at least some information about those who do not respond. The type of non-response and its recognition will depend upon the type of survey being conducted.

For a preselected (random) sample, some of the individuals or addresses that were selected from the sampling frame may no longer exist, e.g. demolished houses, since few sampling frames are completely up to date. Once the individuals are identified there may be no response for one or more of the following reasons.

Unsuitable for interview
The individual may be infirm or inarticulate in English, and whilst he or she could be interviewed if special arrangements were made, this is rarely done in general surveys.

Those who have moved
These could be traced to their new address, but this adds extra time and expense to the survey; the problem does not exist if addresses rather than names were selected from the sampling frame.

Those out at the time of call
This will often happen but can be minimized by careful consideration of the timing of the call. Further calls can be made, at different times, to try to elicit a response, but the number of recommended recalls varies from one survey organization to another. (The government social survey recommends up to six recalls.)

Those away for the period of the survey
In this case, recalling will not elicit a response, but it is often difficult at first to tell if someone is just out at the time of the call. A shortened form of the questionnaire could be put through the letterbox, to be posted when the respondent returns. Avoiding the summer months will tend to reduce this category of non-response.

Those who refuse to cooperate
There is little that can be done with this group, about 5% of the population, since they will often refuse to cooperate with mandatory surveys such as the Population Census, but the attitude of the interviewer may help to minimize the refusal rate.

Many surveys, particularly the national surveys of complex design, will report the number of non-respondents. In addition, non-respondents may be categorized by reason or cause to indicate whether they differ in any important way from respondents. In a quota sample, there is rarely any recording of non-response, since if one person refuses to answer the questions someone else can be selected almost immediately.

Postal questionnaires have a history of high levels of non-response, starting perhaps with the *Literary Digest* survey of 1936 which posted 10 000 000 questionnaires asking how people would vote in the forthcoming United States presidential election; they received only a 20% response rate, and also made an incorrect prediction of the result of the election. More recent surveys have had response rates of more than 90%, and this change deserves some explanation. Successful postal surveys tend to have relatively few questions, which are usually precoded and mostly factual. Inducements of money or gifts are used to encourage response, but often a well-known sponsoring organization is sufficient to encourage responses. Follow-up letters are also used to those who have not responded after three to four weeks, and in a survey of 14–20 year olds, this helped to increase the response rate from 70% after three weeks to a final figure of 93.3%.

EXERCISE

Why is it desirable to know something about the characteristics of non-respondents?

1.6 TYPES OF DATA

Measurement is about assigning a value or score to an observation. To count the number of home owners in a survey or record precisely the dimension of a car part involves measurement. How a value is determined depends on the *level* of measurement. These levels are distinguished on the basis of ordering and distance properties. Essentially there are four levels of measurement to consider.

If responses are merely classified into a number of distinct categories, where no order or value is implied, only a **categorical** or **nominal** level of measurement has been achieved. The classification of survey respondents on the basis of religious affinity, voting behaviour, or car ownership are all examples of nominal measurement. For data processing convenience, we may code the respondents 0 or 1 (e.g. YES or NO) or 1, 2, 3 (Party X, Party Y or Party Z) but cannot then calculate statistics like the mean and the standard deviation which require measurement made on scales with order and distance. We can make percentage comparisons (e.g. 30% will vote for Party X), present the data using bar charts (see Chapter 2) or use more advanced statistical methods (see Chapter 16). An **ordinal** level of measurement has been achieved when it is possible to rank order all the categories according to some criteria. The preferences indicated on a rating scale ranging from 'strongly agree' to 'strongly disagree' or the classification of respondents by social class (A, B, C1, C2, D, E) are both common examples where ranking is implied. In these examples we can position a response or a respondent but cannot give weight to numerical differences. It is as meaningful to code a five point rating scale 7, 9, 12, 17, 21 as 1, 2, 3, 4, 5; though the latter is generally expected. Only statistics based on order, like the median, really apply. You will, however, find that in market research and other business applications, that the obvious codings are made (e.g. 1 to 5) and then a host of computer-derived statistics calculated. Many of these statistics can be useful for descriptive purposes but you must always be sure about the type of measurement achieved and its statistical limitations. An **interval** scale is an ordered scale where the differences between numerical values are meaningful. Temperature is a classic example of an interval scale, the increase on the centigrade scale between 30 and 40 is the same as the increase between 70 and 80. However, heat cannot be measured in absolute terms (0°C does not mean no heat) and it is not possible to say that 40°C is twice as hot as 20°C. In practice, there are few business-related measurements where the subtlety of the interval scale is of consequence. The highest level of measurement is the **ratio** scale which has all the distance properties of the interval scale and in addition, zero represents the absence of the characteristic being measured. Distance and time are good examples of measurement on a ratio scale. It is meaningful, for example, to refer to 0 time or 0 distance and refer to one journey taking twice as long as another journey or one distance being twice as long as another distance. Measurement that is either interval or ratio is often referred to as **cardinal** measurement. In terms of describing data and calculating statistics, the distinction between categorical, ordinal and cardinal is sufficient.

Data can also be classified as either **qualitative** or **quantitative**. Qualitative data is produced by nominal scales and quantitative data by ordinal, interval or ratio scales. If the qualitative data is a result of observation or group discussions, any form of numerical coding is likely to be a complex activity. A further classification of data that will prove essential in our study of statistics is that of **discrete** and **continuous** data. Data is discrete if the numerical value is the consequence of counting. The number of respondents who vote for Party X or the number of respondents who own a car provide discrete data (on the nominal scale). Continuous data can take any value within a continuum, limited only by the precision of the measurement instrument. Time taken to complete a task could be quoted as 5 seconds or 5.17 seconds or 5.16892 seconds. Time in this case is being measured as continuous on a ratio scale. The difference between types of data and the importance of these differences will become more apparent as you continue your course in statistics. It is worth noting however, that the statistics developed for one level of measurement can always be used at a higher level but not (with validity) at a lower level.

1.7 ALTERNATIVE METHODS

Observation can be more effective than questioning. The use of car seat-belts and the way customers select items from supermarket shelves are two recent examples where observation is claimed to have given more valid results. The success of the method depends critically on the rôle of the observer. If the observer is merely counting events at a distance, the data can be taken as a factual statement. However, if the observer needs to make a judgement or becomes involved with the events, the results can become highly subjective. Is it possible to investigate the safety record of a stretch of motorway objectively by going to the scene of every crash?

Panel surveys are generally concerned with changes over time. The same respondents are asked a series of questions on different occasions. This method is useful in terms of assessing the effectiveness of an advertising campaign e.g. the Christmas drink-drive campaign, or to monitor the political situation e.g. the opinion polls before an election. There are two major problems with panel research. Firstly, respondents can become involved in the nature of the enquiry and as a result change their behaviour (known as panel conditioning). Secondly, as panel members leave the panel (known as panel mortality), those remaining become less representative of the population of interest.

Longitudinal studies follow a group of people, or cohort, over a long period of time. This method tends to require a large initial group and the resources to sustain such a study. It has been used effectively to investigate sociological issues and physical development.

These are many variants of the basic questionnaire method. Some surveys ask the respondents to keep a diary of the events over a short period of time, but this does rely on the accurate recording of all events and a low rate of non-response. The Family Expenditure Survey asks respondents to keep a diary of all purchases over a two-week period. In other cases, the interviewer may use just a checklist to ensure a consistent coverage of topics. Interviews where the respondent has some in-depth knowledge, e.g. with medical experts or the victims of crime, would be limited by a structured questionnaire.

In any market research investigation, the full range of possible methods should be considered. Alternatives to those described in this section include retail audits, group discusions, other qualitative research techniques and projective techniques.

1.8 CONCLUSIONS

The collection of data should be supportive of the decision-making process. Improved knowledge of what people are doing or thinking should lead to a more effective response in providing the competitive products or services that these people are prepared to pay for. The measurement of industrial processes should ensure that safety and quality levels can be maintained and improved. However, the collection of data involves more than obtaining answers or recording dimensions. The purpose of any statistical investigation needs to be clear. A statement that we wish to investigate the management of change within the organization will mean different things to different people, so we need to be clear about our meaning of change or changes, management and the general context. It will have to be decided who to include and exclude. If the workforce is referred to, for example, do we mean only full-time employees, those at a particular location or those doing a particular job? Once the 'who' is decided, the 'how' needs to be considered. It is important to recognize that original data can be costly and time-consuming, and yet still be of limited, short-term value.

It is always worth checking whether data collected for some other purpose is sufficient for present needs. It is a frequently reported experience, that 'desk research' yields some of the information required but also yields other pertinent data and a wealth of new ideas. It is also worth considering how much research is genuinely original! If the purpose of the statistical investigation requires the collection of original data, then the sample survey is probably the most widely used method in business and economics.

Once collected, data needs to be collated and presented. Available computer hardware and software now allows data to be stored, manipulated and analyzed with relative ease. MICROSTATS for example allows the user to enter data to columns. Typically, each column would represent a question, for example age, income, opinion as measured by a rating scale or a numerical coding of a classification, and each row a respondent. To enter data on a single item or question the SET command can be used. If data is being entered simultaneously on a range of items or questions the READ command is more appropriate. Remember to use the SAVE command if you want to store your data for future reference (see Appendix K). However, the output can only be as good as the input, and this depends on the validity of each stage of your statistical enquiry.

1.9 PROBLEMS

1. Why is it essential to define clearly the population when conducting a survey? Illustrate your answer with reference to a survey on driving.
2. Define the population for:
   (a) a survey on the attitudes to smoking in the workplace;
   (b) a survey on the attitudes to parking in residential areas near a new leisure centre;
   (c) sampling public vehicles to check maintenance standards;
   (d) sampling components manufactured using new equipment.
3. Search out information (secondary data) on:
   (a) the number of marriages annually over the last 10 years;
   (b) the annual numbers of medical discharges from United Kingdom service personnel;
   (c) the number of petrol filling stations;
   (d) the quarterly output from the 'clothing and footwear' industries;
   (e) the quarterly numbers of road casualties;
   (f) the notes and coins in circulation with the public;
   (g) the index of retail prices;
   (h) consumer expenditure on beer;
   (i) the number of new dwellings completed by region;
   (j) employment in manufacturing by region;
   (k) the number of Aids cases and deaths;
   (l) air pollution;
   (m) money denoted to charities.
4. Why is the Electoral Register the most widely used sampling frame in the United Kingdom? What are the potential problems with using this sampling frame?
5. How would you construct a sampling frame for a survey of:
   (a) clothes shops in the West Midlands;
   (b) people who regularly eat chocolate;
   (c) customers of a local bakery;
   (d) students on degree courses in economics.
6. What are the differences between a stratified sample and a quota sample? Illustrate your answer by describing the selection of a national sample of the general population.
7. What are the sources of bias in questionnaire design? Write ten biased questions to illustrate your answer.
8. Non-response rates of 5–15% are often quoted for random sample interviewer surveys; why are figures not quoted for quota surveys? Which of the five types of non-response are relevant for quota surveys?
9. Surveys of political opinion typically use a quota sample of a thousand voters, a new sample being selected each month. In the period before a general election, some organizations use panels. Suggest reasons for using a panel in preference to the normal practice.
10. Construct a questionnaire to find the reasons why other students on your course selected your particular college.

11. Describe how you would obtain a sample of ethnic small businesses in the South East. (Note that for our purposes a small business is one with a turnover below £50 000; there is no list of all businesses in the South East.)
12. If you were responsible for briefing interviewers about to conduct a survey to find public awareness of an advertising campaign, which points would you stress?
13. Obtain a copy of a recent Family Expenditure Survey report from the library and analyze the type of sample that is used. Write a brief report on the sample selection methods used by FES.
14. What types of data would a questionnaire yield, if designed to find the reasons students selected a particular college?
15. What sort of data collection problems would you expect, if your investigation involved observing people at work?
16. Using a computer package, such as MICROSTATS, construct your own database of social and economic statistics of the last 10 years.

PRESENTATION OF DATA 2

The coding of questionnaire responses is likely to be the main source of business-related data. These numeric values can come from the counting process or some higher level of measurement, each individual value giving some detail on a single respondent or item. When presenting data we are concerned with the overall picture rather than all the bits. We need to put all the individual details together, like a jigsaw, so that we can see the general pattern emerge. In a survey on drinking and driving behaviour it is the pattern of response that is of interest rather than the answers given by say, a Mrs. Singh or a Mr. Smith. The presentation of data is more than a technical competence to produce results; it requires a thoughtful exploration of what the numbers actually mean. Your selected diagrams should tell a story, and give an insight to the findings of your statistical investigations.

Table 2.1 The number of cases of breakfast bran sold during September by Sellmore PLC

21	50	28	39	41	25	48	35	22	36	55	47	39	27	40	37	51	51	31	23
57	37	46	42	34	32	52	42	34	46	42	26	38	35	29	42	26	45	35	56
40	26	32	43	39	27	43	39	30	36	43	40	37	45	21	38	42	50	28	37
39	45	42	46	20	35	44	26	42	25	50	32	31	54	39	46	37	51	43	35
46	34	52	31	44	49	51	36	38	38	37	47	27	49	44	33	26	44	42	39
30	38	35	24	28	38	21	41	30	34	44	25	44	34	36	35	28	34	46	32
38	22	43	35	56	36	45	32	55	27	49	28	56	19	42	39	44	39	44	25
49	32	39	47	43	38	25	28	31	37	34	30	43	50	30	22	37	46	43	40
40	30	45	36	22	31	50	35	47	49	45	29	40	31	43	39	34	55	44	36
26	47	29	38	39	49	52	38	41	23	26	54	39	33	42	25	36	48	35	42
39	42	37	35	25	43	39	31	34	30	55	38	50	37	43	37	44	29	47	31
35	31	47	35	26	34	40	42	57	21	29	36	41	24	24	46	35	43	52	38
58	39	20	31	42	30	27	38	39	27	37	42	52	38	61	28	45	37	40	31
29	42	49	36	38	25	42	48	46	48	31	56	36	43	40	54	44	32	44	26
36	34	30	33	39	35	27	51	35	37	34	31	40	50	30	34	40	51	43	47
52	48	40	47	32	46	39	24	36	60	32	20	28	36	42	37	39	31	38	40
28	55	36	40	21	52	50	37	39	35	40	36	55	45	32	25	48	30	42	59
46	37	42	51	28	40	47	41	31	30	45	40	48	39	34	36	33	38	27	34
36	32	49	46	36	37	40	37	48	28	39	51	36	30	22	47	31	40	32	37
56	24	45	37	30	50	31	27	36	33	40	49	29	35	48	37	29	53	26	50

Table 2.2 Number of cases of breakfast bran sold during September by Sellmore PLC

Sales	*Frequency*	*Sales*	*Frequency*	*Sales*	*Frequency*
19	1	33	5	47	11
20	3	34	15	48	10
21	5	35	18	49	8
22	5	36	21	50	10
23	2	37	22	51	8
24	5	38	17	52	7
25	9	39	23	53	1
26	10	40	20	54	3
27	9	41	5	55	6
28	11	42	20	56	5
29	8	43	14	57	2
30	14	44	12	58	1
31	17	45	10	59	1
32	12	46	12	60	1
				61	1
					400

Table 2.3

Sales	*Frequency*
20 or less	4
21 up to 25	26
26 up to 30	52
31 up to 35	67
36 up to 40	103
41 up to 45	61
46 up to 50	51
51 up to 55	25
56 up to 60	10
61 or more	1
	400

2.1 TABULATION OF DATA

Whether your collected data includes several hundred or several thousand values, these will need to be grouped together so that the magnitude of the numbers can be understood. The values could be listed as shown in Table 2.1.

It is not immediately obvious from this table how much was sold, how many salespersons were involved or what were the lowest and highest sales. As a first step the data could be arranged in numerical order and the lowest and highest values seen at a glance. However, the list could be very long and the focus on extreme values rather than typical sales. To reduce the scale of the list, a count of how many times a particular value occurs could be made and presented in the form of a **frequency distribution**. The result of arranging the data in this way is shown in Table 2.2.

If there are only a few individual values that occur, as there may well be in the answers to precoded questions on a questionnaire, then this rearrangement will be sufficient to render the data both manageable and readable. However, if there are a large number of individual values, even the rearrangement in Table 2.2 will not suffice to achieve clarity. To reduce the size of the table produced we can amalgamate certain values into groups as shown in Tables 2.3, 2.4 and 2.5.

The tabulations could each be argued to be technically valid, but they do not all succeed in conveying adequate information about sales. Both Tables 2.3 and 2.4 provide some information on the variation of sales, but Table 2.5 tells us very little about the pattern of sales (see section 3.1.3).

Table 2.4

Sales	*Frequency*
25 or less	30
26 up to 35	119
36 up to 45	164
46 up to 55	76
56 or more	11
	400

Table 2.5

Sales	*Frequency*
40 or less	252
41 or more	148
	400

EXERCISE

Imagine you were a prospective employee of Sellmore and were told that your income would be totally dependent on commission from sales; which of the five tables would you prefer to be given as an indication of the performance of current sales staff?

As we move away from Table 2.2 to any of Tables 2.3, 2.4 or 2.5 we lose detailed, individual information. No longer do we know that the most successful sales person in September sold 61 units, we only know that one sold 61 or more (Table 2.3), or that 11 sold 56 or more (Table 2.4), or that 148 sold 41 or more (Table 2.5). The loss of some detail can be worthwhile if the tables can be more easily understood and retain the basic information regarding the differences in our data. In constructing a table, we are no longer concerned with a single observation or measurement but with what is generally true. As a guide most tables are constructed with between 4 and 10 groups depending on the complexity of the data.

In collecting data we are often confronted with a multitude of measurements or counts, on different aspects or different characteristics. A survey questionnaire, for example, will ask a series of related questions. To provide marketing information we may require the detail given in Table 2.6.

A single tabulation of the number of each model sold or sales by region would hide the regional differences between the number of models sold. If counts are made jointly with respect to two or more methods of classification, we obtain a **cross-tabulation** of the data.

The data given in Table 2.1 is on your disk as file DATA1. The data can be read into MICROSTATS using the command READ. Try the commands COUNt, MAXImum and MINImum. You should be able to reproduce the Tables 2.2 to 2.5 using the command HISTogram.

2.2 VISUAL PRESENTATION

One of the most effective ways of presenting numerical information is to construct a chart or diagram. The choice depends on the type of data.

Table 2.6 Numbers sold of four types of industrial trolley by region during the last financial year

		Regions			
Model	*North*	*South*	*East*	*West*	*Total*
Tug	675	60	35	20	790
Conveyor	30	490	30	20	570
Lifter	150	180	235	15	580
Mover	5	20	0	35	60
	860	750	300	90	2000

As a guide we shall make one basic distinction: whether the data set is **discrete** or **continuous**. A data set is discrete if we just make a count, for example the number of people in a room or the number of cars sold last month. A data set is continuous if measurement is made on a continuous scale, for example the time taken to travel to work or the yield in kilograms of a manufacturing process. There are, of course, some exceptions. Technically, money can be seen as discrete, since it changes hands in increments (pence) but is usually treated as continuous because the increments can be relatively small. Age is continuous (we all grow older) but when quoted as age last birthday becomes discrete.

2.2.1 Presentation of discrete data

The counts of industrial trolleys sold by model and region in Table 2.6 is discrete data. To make an easy comparison between regions we can present the same information as percentages (Table 2.7). It can be immediately seen that Tug accounts for 78.49% of unit sales in the North and only 8.00% of unit sales in the South. To calculate these percentages we can first find the fraction and then multiply by 100. Tug, for example, accounts for 675 sales out of the 860 achieved in the northern sales region. The ratio $675 \div 860 = 0.7849$ multiplied by 100 gives the percentage 78.49%. The original totals are often given as base figures.

Pie charts

If we wish to show market shares for a product grouping (Table 2.8) or something similar, we can divide a circle into sectors, where each sector represents size. The size of each sector can be found by taking the appropriate proportion of 360° (Figure 2.1). This type of presentation is only easily understood if a few categories are to be used.

Bar charts

The numbers sold by model or by region can be represented as vertical bars. The height of each bar is drawn in proportion to the number by using

Table 2.7 Percentage unit sales of four types of industrial trolley by region during the last financial year

Model	*Regions* *North*	*South*	*East*	*West*
Tug	78.49	8.00	11.67	22.22
Conveyor	3.49	65.33	10.00	22.22
Lifter	17.44	24.00	78.33	16.67
Mover	0.58	2.67	0.00	38.89
	100.00	100.00	100.00	100.00

Table 2.8 Sales of four types of industrial trolley during the last financial year

Model	*Sales*	*Proportion of total*	*Proportion of 360°*
Tug	790	0.395	142.2
Conveyor	570	0.285	102.6
Lifter	580	0.290	104.4
Mover	60	0.030	10.8
	2000	1.000	360.0

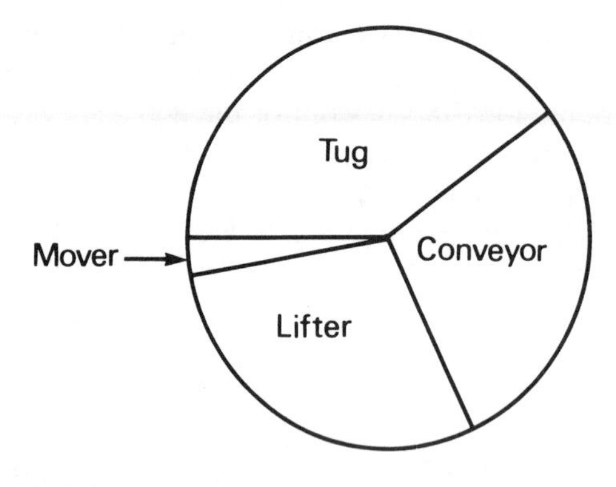

Figure 2.1

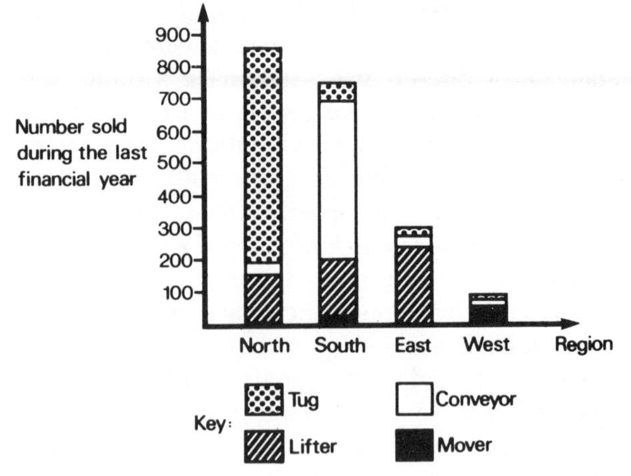

Figure 2.3 Sales of four types of industrial trolley by region during the last financial year

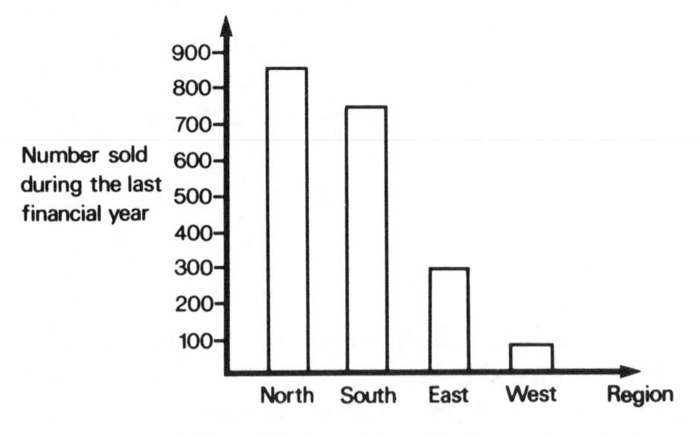

Figure 2.2 Sales of industrial trolley by region during the last financial year

a vertical ruler scale (Figure 2.2). We can also show the number of each model sold by region using a **component bar chart** (Figure 2.3).

Once we begin to examine the composition of totals it can become difficult to see the relative size of some of the components. To overcome this problem, it is often convenient to change the absolute figures into percentages, thus giving bars all of the same length and making direct comparisons possible (Figure 2.4).

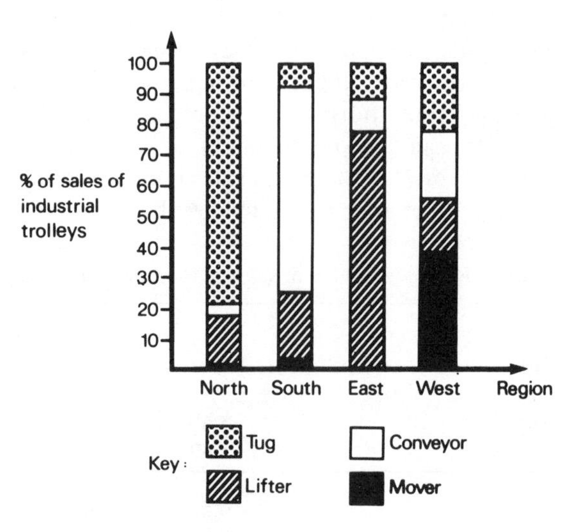

Figure 2.4 Percentage sales by region

Pictograms

Bars can be replaced by appropriate pictures to show comparisons. Whilst this is more eye-catching, it is considerably less accurate and may be misleading. (In Figure 2.5, how many does a

Figure 2.5

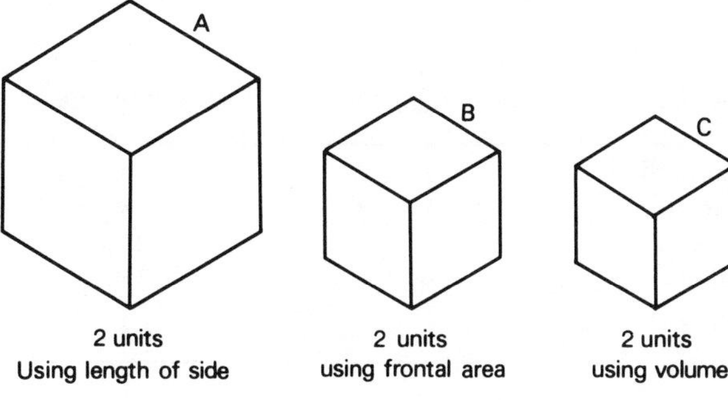

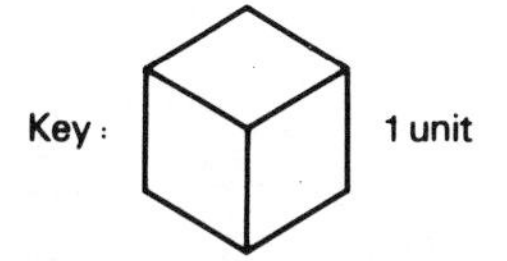

Figure 2.6 Pictogram to show a doubling of the sales of Happy Packs

fraction of a person represent?) Even more confusing are pictograms which use different sized figures to represent different values. An increase could be shown as an increase in height but the visual impression could be in terms of the increase of surface area or volume.

If we are making a single measurement or count, for example, presenting sales by region or turnover by company, a one-dimensional representation is generally clearer. We must try to avoid the confusion possible with the pictograms shown in Figure 2.6. Picture A would almost certainly leave the impression of a more rapid increase in sales than picture B or picture C.

All the charts and diagrams we have considered so far are used to represent discrete data.

2.2.2 Presentation of continuous data

We can treat as continuous any measurement that can be represented by a line rather than just points on a line. Time and length are two good examples of continuous measurement.

Histograms

The distribution of measurement on a continuous scale is presented by means of a histogram. As an

Table 2.9 Income of a particular group of workers

Income	*Number of workers*
less than £200	10
£200 but less than £300	28
£300 but less than £400	42
£400 but less than £600	50
£600 or more	20

example we can use the information on income given in Table 2.9. The data presents us with two problems. Firstly, there are two open-ended groups. We do not know how low the income can be of the workers in the first group or how high the income of the workers can be in the last group. All we can do with open-ended groups of this kind is to assume reasonable lower or upper boundaries on the basis of our own knowledge of the data and the apparent distribution of the data. In this case it may be reasonable to assume a lower boundary of £100 for the first group and an upper boundary of £1000 for the last group. Whatever the decision, it is still a matter of **judgement**. Secondly, if we were to use bars to show the number of workers in each income group, it would appear that there were fewer in the range '£300 but less than £400' than in the range '£400 but less then £600'. However, the range has doubled in size and we need to take account of the increased chance of inclusion. To represent the distribution we plot **frequencies** *in proportion to*

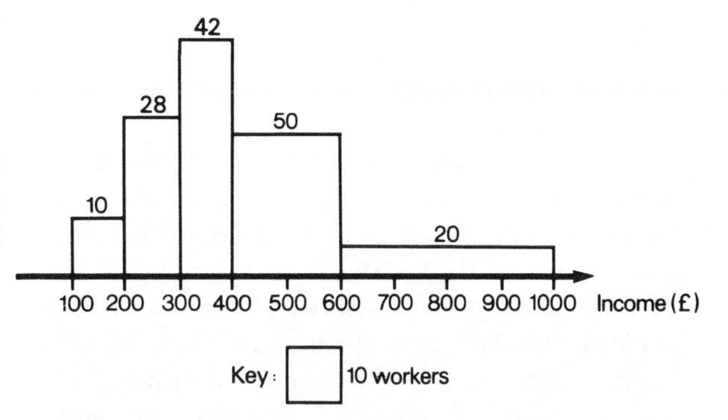

Figure 2.7 A histogram showing the distribution of income

Table 2.10 Method of constructing a histogram

Income	*Frequency*	*Width*	*Scaling factor*	*Height of block*
£100* but less than £200	10	standard	1	10
£200 but less than £300	28	standard	1	28
£300 but less than £400	42	standard	1	42
£400 but less than £600	50	2 times	$\frac{1}{2}$	25
£600 but less than £1000*	20	4 times	$\frac{1}{4}$	5

*Assumed boundary

area. Income is shown as a horizontal ruler scale and frequencies as a series of blocks. These blocks are constructed with reference to a key as show in Figure 2.7. The key provides a standard and all blocks are constructed with reference to this. If we consider the income range '£400 and less than £600' we can note that the doubling of the standard width has had the effect of halving the height.

In practice, we choose one of the groups as our standard and scale as shown in Table 2.10.

2.3 GRAPHICAL PRESENTATION

The way in which sales relate to advertising or the way in which sales change over time can effectively be represented by a graph. To show the relationship between two variables, typically we would construct a graph (also section 6.4).

2.3.1 Graph plotting

In presenting business information, exploring research data or when trying to make predictions, the choice of axes for variables is important since a direction in the relationship is implied. The variable thought to be responsible for the change is plotted on the x-axis (the horizontal axis) and is often referred to as the independent or predictor variable. The variable whose change we are seeking to explain is plotted on the y-axis (the vertical axis) and is referred to as the dependent variable.

Table 2.11

No. of weekly breakdowns	*Operating speed (r.p.m.)*
3	100
3	80
5	120
3	60
1	50
4	90

Consider, for example, the machine data given in Table 2.11.

In this case we would want to know whether the number of machine breakdowns depends on the operating speed. A graph (Figure 2.8) suggests that a relationship does exist (see Chapter 18 for an introduction to the calculation of these relationships). To produce this graph using MIC-

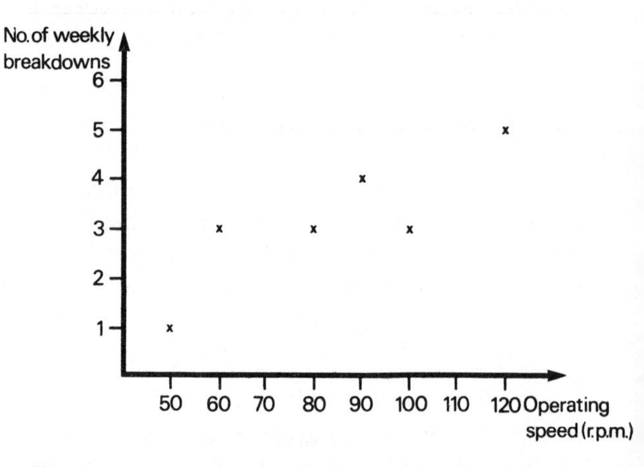

Figure 2.8

ROSTATS with number of breakdowns in column 1 and operating speed in column 2, the command PLOT C2 C1 would be used.

2.3.2 Plotting a time series

Suppose the sales of a domestic appliance over the last five years were recorded as in Table 2.12. In the special case of data recorded against time, time is always plotted on the x-axis as shown in Figure 2.9 (see Chapter 20).

However, a change in scale can affect the impression made graphically as shown in Figures 2.9 and 2.10. It is clear that the sales of this domestic appliance have increased but it is hard to judge from the graphs whether or not the increase is significant. In the same way, we could plot sales against advertising and seek a relationship between the two but would still lack an objective measure of this relationship (see Chapter 17). We would advise caution when interpreting any graphical representations of this kind.

Table 2.12 Sales of a domestic appliance

Year	*Number sola*
1	20 000
2	26 000
3	33 800
4	43 940
5	57 122

Figure 2.9

Figure 2.10

2.3.3 Logarithmic graphs

If we are interested in the rate of change over time we can plot the log of values against time (Table 2.13). In this case we can see that the log values increase at a constant rate once an allowance has been made for the rounding of numbers. If we graph the log of values against time (Figure 2.11), we get a different impression from that given in Figures 2.9 and 2.10.

Sales can be seen to be not only rising, but also rising at a constant rate each year. If we check the sales figure, we find the increase to be 30% each year, e.g.

$$\frac{26\,000 - 20\,000}{20\,000} \times 100 = 30\%$$

Alternatively, we could antilog the increase in log values (0.1140 or 0.1139) to find the multiplication factor 1.30. To estimate the sales in the next year we multiply by 1.30, that is, increase values by 30%.

To produce a graph like Figure 2.11 using MICROSTATS, given year in column 1 and the number sold in column 2, the log transformation would need to be made first, with the log values entered to a new column, say column 3. Then the graph of column 3 against column 1 could be plotted.

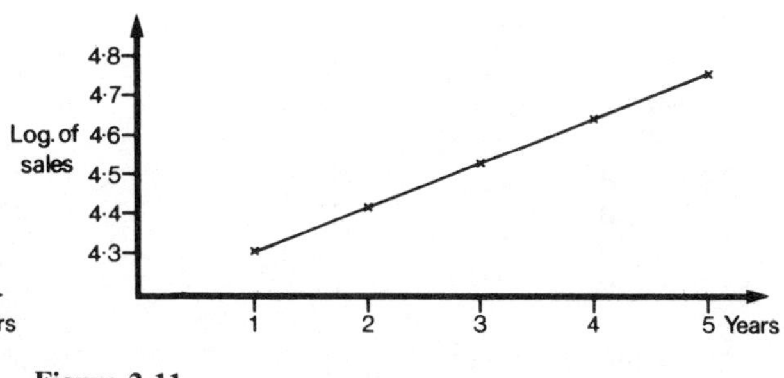

Figure 2.11

Table 2.13 The log values of sales

Year	*Number sold*	*Log of number sold*	*Increase in log values*
1	20 000	4.3010	
2	26 000	4.4150	0.1140
3	33 800	4.5289	0.1139
4	43 940	4.6429	0.1140
5	57 122	4.7568	0.1139

2.3.4 Lorenz curve

One particular application of the graphical method is the Lorenz curve. It is often used with income data or with wealth data to show the distribution or, more specifically, the extent to which the distribution is equal or unequal. This does not imply a value-judgement that there should be equality but only represents what is currently true. To construct a Lorenz curve each distribution needs to be arranged in order of size and then the percentages for each distribution calculated. The percentages then need to be added together to form cumulative distributions which are plotted on the graph.

Let us consider first the information given in Table 2.14. The percentage columns give a direct comparison between population and wealth. It can be seen that the poorest 50% can claim only 10% of total wealth. The cumulative percentage columns allow a continuing comparison between the two. It can also be seen that the poorest 75% of the population can claim 30% of the wealth and the poorest 85% of the population 40% and so on.

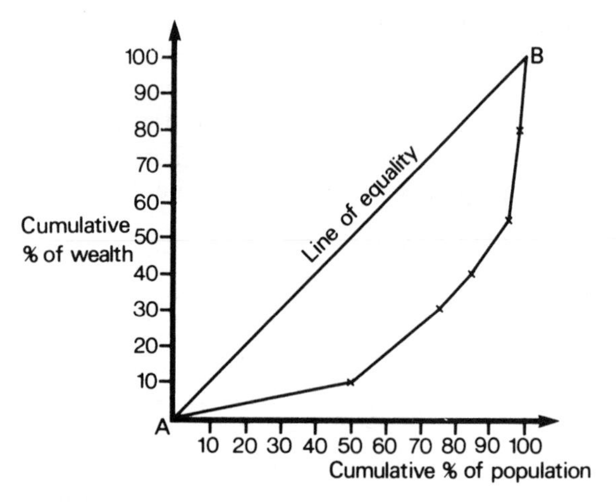

Figure 2.12 The relationship between wealth and population

Note that in Figure 2.12 the point representing zero population and zero wealth (point A) is joined to that representing all of the population and all of the wealth (point B) to show the 'line of equality'. If the points were on this line then there would be an equal distribution of wealth: the further the curve is away from the straight line the

Table 2.14

Group	*Percentage of population*	*Cumulative percentage*	*Percentage of total wealth*	*Cumulative percentage*
Poorest A	50	50	10	10
B	25	(+50) 75	20	(+10) 30
C	10	(+75) 85	10	(+30) 40
D	10	(+85) 95	15	(+40) 55
E	3	(+95) 98	25	(+55) 80
Richest F	2	(+98) 100	20	(+80) 100

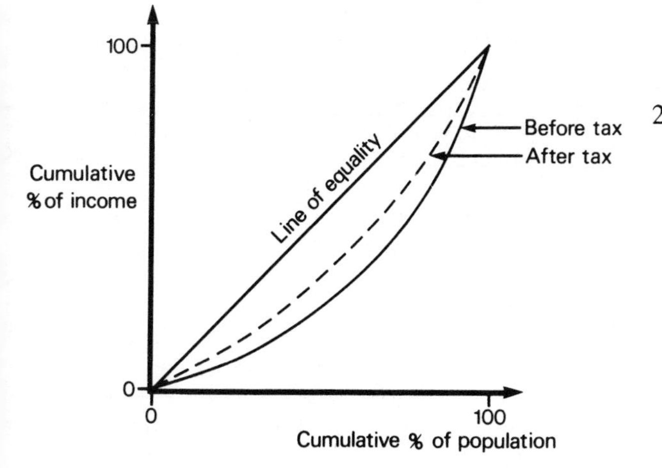

Figure 2.13 The effects of tax on the distribution of income

less equality there is. The curve can also be used to show how the income distribution changes as a result of taxation. Figure 2.13 shows a progressive tax system where the post-tax income distribution is closer to equality than the pre-tax income distribution.

2.4 CONCLUSIONS

The quantity of data that a business needs to manage can be immense. There can literally be hundreds of figures relating to sales, production and other business activities. Data needs to be summarized and presented so that people, not computers, can understand what is happening. You only need to look at the business press to see the importance of clear, concise presentation.

2.5 PROBLEMS

1. Obtain a number of charts and diagrams used to describe quantitative information. Sources could include, for example, newspaper cuttings, building society pamphlets or textbooks. Classify each as being discrete or continuous data and state reasons why you consider them to be informative or misleading.
2. The number of new orders received by a company over the last 25 working days were recorded as follows:

3	0	1	4	4
4	2	5	3	6
4	5	1	4	2
3	0	2	0	5
4	2	3	3	1

   (a) Tabulate the number of new orders in the form of a frequency distribution.
   (b) Construct a bar chart.
   (c) Comment on the distribution.
3. The work required on two types of machine has been categorized as routine maintenance, part replacement and specialist repair. Records kept for the last 12 months provide the following information:

	Frequency	
Work required	*Type X*	*Type Y*
Routine maintenance	11	15
Part replacement	5	2
Specialist repair	4	3

   Present this information using:
   (a) pie charts;
   (b) appropriate bar charts.
4. The mileages recorded for a sample of company vehicles during a given week yielded the following data:

138	164	150	132	144	125	149	157
146	158	140	147	136	148	152	144
168	126	138	176	163	119	154	165
146	173	142	147	135	153	140	135
161	145	135	142	150	156	145	128

   (a) Using the data tabulate a grouped frequency distribution starting with '110 but under 120'.
   (b) Construct a histogram from your frequency distribution.
5. The average weekly household expenditure on a particular range of products has been

recorded from a sample of 20 households as follows:

£8.52	£ 7.49	£ 4.50	£ 9.28	£ 9.98
£9.10	£10.12	£13.12	£ 7.89	£ 7.90
£7.11	£ 5.12	£ 8.62	£10.59	£14.61
£9.63	£11.12	£15.92	£ 5.80	£ 8.31

Tabulate as a frequency distribution and construct a suitable diagram.

6. Information was collected on weekly travel expenditure of residents living in zones A, B and C.

Weekly travel expenditure (£)	*Zones*		
	A	*B*	*C*
under 1	6	4	2
1 but under 3	3	3	3
3 but under 5	1	3	5

Describe this data using charts and diagrams as appropriate.

7. Construct a histogram from the data given in the following table:

Income (£)	*Number*
under 40	175
40 but under 80	229
80 but under 120	241
120 but under 200	269
200 or more	86

8. Construct a histogram from the data given in the following table:
Journey distance to and from work

Miles	*Percentage*
under 1	16
1 and under 3	30
3 and under 10	37
10 and under 15	7
15 and over	9

9. Construct a histogram from the information given in the following table:

Error (£)	*Frequency*
under − 15	20
− 15 but less than − 10	38
− 10 but less than − 5	178
− 5 but less than 0	580
0 but less than 10	360
10 but less than 20	114
20 or more	14

10. Use graphical methods to explore the following data for possible causal relationships.

Year	*Sales (units)*	*Research (£000s)*	*Advertising (£000s)*
1	590	88	142
2	645	99	118
3	495	50	80
4	575	78	42
5	665	97	150
6	810	118	40

11. The sales within an industry have been recorded as follows:

Year	*Quarter 1*	*Quarter 2*	*Quarter 3*	*Quarter 4*
1	40	60	80	35
2	30	50	60	30
3	35	60	80	40
4	50	70	100	50

Graph this data and discuss the relationship between sales and time.

12. A company's advertising expenditure has been monitored for 3 years, giving the following information:

Year	*Quarter 1*	*Quarter 2*	*Quarter 3*	*Quarter 4*
1	10	15	18	20
2	14	16	19	23
3	16	18	20	25

Graph this data and write a short report describing the main features.

13. The results of a company were reported as follows:

Year	*Turnover (£000)*	*Pre-tax Profit (£000)*	*Exports (£000)*
1	7 572	987	2 900
2	14 651	1 682	6 958
3	17 168	2 229	7 580
4	21 024	3 165	9 306
5	25 718	4 273	10 393
6	37 378	6 247	18 280
7	53 988	9 559	28 229
8	79 258	19 646	48 770
9	122 258	32 714	74 410
10	183 338	49 832	95 029

(a) Graph the three sets of data against time.
(b) Graph the log values for the three sets of data against time.
(c) Comment on your graphs outlining the relative merits of those produced in parts (a) and (b).

14. Construct a Lorenz curve for the following data on income.

Income group	*Percentage of people in group*	*Percentage of income*
Poorest paid	10	5
	15	8
	20	17
	20	18
	20	20
	10	15
Highest paid	5	17

15. Construct a Lorenz curve for the following data on wealth-holding in the UK.

Group	*No. of wealth holders (thousands)*	*Amount of wealth (£m)*
Poorest	2 099	1 049.5
	3 530	7 060.0
	2 133	8 532.0
	4 414	33 105.0
	2 588	32 350.0
	1 167	20 422.5
	694	15 615.0
	1 018	38 175.0
	320	24 000.0
	94	14 100.0
Wealthiest	31	34 100.0
	18 088	228 509.0

PART ONE CONCLUDING EXERCISES

1. On the completion of a survey the following tabulations were presented (with question wording) for questions one and two:

   Q1. 'How many years have you been living in this (house/flat)? . . . . '

Number of years	*Frequency*
0–1	137
2–4	209
5–9	186
10–19	229
20+	205

   Q2. For each item below ask 'Do you have . . . ?'

   (a) A fixed bath or shower with a hot water supply:

	Frequency
None	67
Shared	27
Exclusive	871
No answer	1

   (b) A flush toilet inside the house:

	Frequency
None	83
Shared	30
Exclusive	850
No answer	3

   (c) A kitchen separate from living rooms:

	Frequency
None	20
Shared	19
Exclusive	922
No answer	5

   Report on the tabulations given using charts and diagrams when appropriate.

2. Design a questionnaire to ascertain:
   (a) student views on college catering; or
   (b) your colleagues' views on bus travel; or
   (c) your colleagues' views on the use of computers.

   Collect data from at least 30 respondents and use MICROSTATS or a similar package to prepare a brief report.

PART TWO DESCRIPTIVE STATISTICS

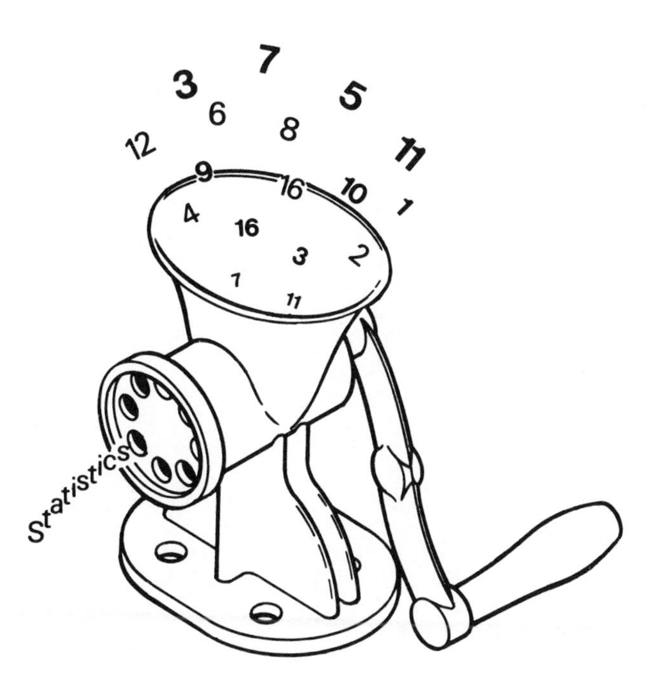

In the first part of this book we have looked at ways of collecting data and illustrating the findings diagrammatically. This may be sufficient for a short presentation of the findings of, say, a survey, but if we wish, or need, to go further, then a numerical description of that data offers a number of benefits. Such a description will often entail the use of averages which may be interpreted as showing a typical result or response, but, as we will see, there are several averages to choose from. We have already seen that data has some variation within it (for example, people vary in their views on 'green' issues), and we will need to describe this variation. Index numbers will allow us to amalgamate data on a wide range of items into a single set of figures, for example, the Retail Price Index.

All of the statistics in this part of the book aim to summarize the mass of detail contained in the individual results of a survey (say a piece of market research), or other data collection exercise (say from a set of management accounts or a production process). These summary statistics aim to describe fully the data by the use of a very small number of figures.

The use of computer packages such as MICROSTATS, MINITAB or SPSSX has made the calculation of such descriptive statistics a very simple matter. Even spreadsheets will calculate descriptive statistics with ease. This removal of the burden of calculation, or number crunching, however, makes the selection of appropriate statistics and their interpretation more important, both in the work situation and within an academic course.

MEASURES OF LOCATION 3

Measures of location are another name for averages, and as such, are familiar to most people. Using the methods shown in Chapter 2 we will be able to illustrate sets of data, but now we want to describe them numerically. Averages will allow us to summarize a data set and then to go on and make comparisons between different data sets separated by either geography or time, for example, the average sales in different regions or the average number of days lost through mechanical breakdown at all plants over a number of years, or average weekly expenditure on public transport by car and non-car owners.

Computer packages will make such calculations fairly simple, but we will still show the 'hand' methods of calculation as we feel that this will help you in choosing an appropriate average to calculate in various circumstances by highlighting the methods used in each case. It is far too easy to calculate all possible averages and confuse, rather than communicate with your audience. Our aim is to find a measure of location which succeeds in describing the average, or typical, value of the data.

3.1 THE MEAN, MEDIAN AND MODE

These three are the most commonly used forms of average for most business data. Each has its own characteristics, and whilst it will be possible to use them interchangeably with some data sets, for others there will be a single average which will be most appropriate. One consideration will be the type of data with which we are dealing; is it categorical, ordinal or cardinal (see Chapter 2)? Secondly we must ask if the data is discrete or continuous; for example, is an average number of children of 1.8 a meaningful answer? A third consideration will be the amount and type of variation in the data; is it all bunched closely together or are there a few extreme values?

The **arithmetic mean** (usually just shortened to the **mean**) is the name used for the simple average which you can already calculate. Almost everyone understands this average and thus it will succeed in communicating the concept of the location of the data to a wide range of people. As we will see, it can also give a very misleading impression with certain types of data. It does not apply to categorical data and its interpretation when used with ordinal data is open to considerable doubt. When used with discrete data it may give an answer which cannot occur, for example a fractional number of people.

The **median** represents the value of the middle item of an ordered list of data. It is becoming more widely used and more generally accepted and will thus communicate to a relatively wide range of people. Where the mean gives misleading results, the median may prove to be more appropriate. Again, it is only really useful for cardinal data. When used with discrete data it will generally give an answer which has actually occurred in the data set.

The **mode** represents the most frequently occurring value or item in the data set. It may not be unique for certain data sets but it does apply to all types of data. Common examples of its use are found when looking for the most frequent response to a particular question in a survey or most popular shoe size.

3.1.1 Untabulated data

The mean

For simple data we have a series of numbers which record the outcome of a measurement, for example, the time in minutes taken to complete a certain job. Suppose that these times were 7, 5, 6, 7, and 8 minutes. To calculate the **mean**, we would add all of the numbers together to find the total time taken, and then divide by the number of values included. Here the mean would be:

$$\bar{x} = \frac{7+5+6+7+8}{5} = \frac{33}{5} = 6.6 \text{ minutes}$$

where $\bar{x}$ (pronounced x bar) is the symbol used to represent the mean. Since the data we have used is continuous, cardinal data, this is a realistic answer. Had the data represented the number of people waiting in a queue (discrete data), then the answer would not represent an actual, feasible occurrence.

As all summary statistics require some form of calculation, a shorthand has developed to describe the necessary steps. Using this shorthand, or notation the calculation of the mean would be written thus:

$$\bar{x} = \frac{\Sigma x}{n}$$

where x represents the individual values, Σ (pronounced sigma) is an instruction to sum values and n is the number of values.

Using MICROSTATS we would put all of the values into a column, say C1 and, at the Command? line, would type in AVER C1 to get the mean. When a large data set exists, using such a package will considerably ease the burden of calculation.

The median

Applying the **median** to the same data, the first step must be to re-arrange the data so that it is in numerical order. Doing this, we have:

5 6 7 7 8

and since there are five numbers, the middle one must be the third. Counting from either the left or the right, this gives us the number 7 as the median. Whether the data represents the time taken to complete a task (continuous data) or the number of people in a queue (discrete data), this is a value which has actually occurred.

With small amounts of data, this procedure is very easy, but when the number of numbers is large, getting them into order can be a very tedious process. Using MICROSTATS with data in column 1, we would type MEDIAN C1 (or MEDI C1).

Where there is an even number of numbers in the data set, there is no middle value, and we have to add the two central values together and divide by two. For example, given the following data:

4 5 6 6 7 8 9 11 14 15

there are ten numbers, and so we take the fifth and sixth values (here 7 and 8), and they give a median value of 7.5. (Note that this represents a situation where the median does not give a value which actually exists in the data set, but since the two numbers we have averaged are close together, the result is not misleading.) This is not a wholly satisfactory solution, but we are normally dealing with data sets which are much larger than this, and thus where averaging the middle two values is unlikely to have any misleading effects. As with all descriptive statistics, it is the interpretation of the statistic, given a particular data set, that is the important challenge.

The mode

The use of the **mode** for simple data is only a question of observation; looking for the most frequently occurring value. Given our original data set:

7 5 6 7 8

we can easily see that 7 occurs twice and is, therefore, the mode. This gives a value which does actually occur in the data and will apply whether the data is discrete or continuous, and whether it is cardinal, ordinal or categorical.

A possible problem with the mode is illustrated by the following data set:

2 3 4 4 5 5 5 6 6 6 7 7 8 9 9

Here there are *two* modes, 5 and 6, since they both occur three times. If an extra data value now becomes available, say another 9, there would then be three modes. If the extra value was a 6, there would only be one mode. Some data will not have a mode that is meaningful and thus may need to be explained in terms of the data and its distribution.

Suppose the salaries of five people over the previous 12 months were recorded as:

£6 000, £6 000, £6 400, £6 500, £10 500

The summary statistics could be given as follows:

mean = £7 080
median = £6 400
mode = £6 000
highest salary = £10 500
lowest salary = £6 000

EXERCISE

If you were thinking of joining these five people and did not have access to the actual figures, which of the summary statistics would you consider the best guide to your future income?

We can note that the median has not been affected by the extreme value, £10 500 in this case. For this reason, the median is often preferred to the mean when analyzing income or wealth data. There are generally a few individuals with very high levels of income or wealth and these high values tend to raise the value of the mean. In general, we would want all these statistics, as together they provide a more complete picture of the data.

EXERCISE

The errors in seven invoices were recorded as follows: −£120, £30, £40, −£8, −£5, £20 and £25. The use of negative and positive signs can be taken to indicate your loss and gain respectively. Calculate appropriate descriptive statistics.
(Answers: mean = −£2.57, median = £20, mode is undefined, lowest value = −£120, highest value = £40.)

3.1.2 Tabulated (ungrouped) discrete data

The mean

Suppose we had just completed a survey of 1440 new cars and wanted to describe the number of faults found. The data set is clearly discrete; the number of faults taking only positive integer values. It could be treated as untabulated (see Section 3.1.1) with the number of faults recorded for each car listed. However, it is likely that at some stage during analysis the data would be tabulated in the form of a frequency distribution as shown in Table 3.1.

Table 3.1 The number of faults recorded in a sample of new cars

Number of faults (x)	*Number of cars (f)*
0	410
1	430
2	290
3	180
4	110
5	20
	1440

Table 3.2 The calculation of the mean from a frequency distribution

x	f	fx
0	410	0
1	430	430
2	290	580
3	180	540
4	110	440
5	20	100
	1440	2090

The mean is $\bar{x} = \frac{2\,090}{1\,440} = 1.451$ faults

To calculate the total number of faults we first multiply the number of faults x, by the number of cars, f for frequency, to obtain a column of sub-totals, fx. We then sum this column, Σfx, to obtain the total number of faults. To calculate the mean (Table 3.2) we divide by the number of values or observations, which is the sum of the frequencies, n $(= \Sigma f)$.

We can note that the first 410 cars contribute 0 faults to the total, the next 430 cars contribute 430 faults to the total (1 each), the next 290 cars contribute 580 faults to the total (2 each) and so on. A modified formula will remind us what we have done:

$$\bar{x} = \frac{\Sigma fx}{n}$$

Note that whilst this is a *correct* answer, there will be no car with 1.451 faults. This is the penalty we pay for using the mean with discrete data.

Where there is a large amount of data, it may be more convenient to use a spreadsheet package to calculate the mean. If the data values (xs) are put into column A and the frequencies (f) are put into column B we can multiply them together (typically + A2*B2) and put the result in column C. Summing columns B and C gives the sum of f and the sum of fx. Reference to these two cells (or summations) allows the simple determination of the mean.

Table 3.3 The calculation of cumulative frequency

x	f	*Cumulative frequency*
0	410	410
1	430	840 = 410 + 430
2	290	1 130 = 840 + 290
3	180	1 310 = 1 130 + 180
4	110	1 420 = 1 310 + 110
5	20	1 440 = 1 420 + 20
	1440	

The median

The tabulation of the number of faults in new cars has ordered the set of data (the first 410 cars have 0 faults, etc.) The median, which is an order statistic, can be found if we cumulate frequencies. The **cumulative frequency** is the number of items with a given value or less. To calculate cumulative frequency we just add the next frequency to the running total (Table 3.3).

To determine the median observation for listed or discrete data we can use the formula

$$(n + 1)/2$$

to locate which item in the data set we require. In this case the median will be the $720\frac{1}{2}$th ordered observation, i.e. it will lie between the 720th and 721st car. From the cumulative frequency it can be seen that 410 new cars have 0 (or less) faults and 840 new cars have 1 fault or less. By deduction, the 720th and 721st new cars both have 1 fault, and hence the median is 1 fault.

The mode

The mode corresponds to the highest frequency count and in this example is 1 fault.

The distribution of faults (a few new cars with a high number of faults) explains differences between the median and the mean.

EXERCISE

An automatic cash dispenser allows any transaction up to the value of £30 in £5 multiples. Recent transactions have been tabulated as follows:

Value of transaction (£)	Number
5	7
10	15
15	12
20	23
25	21
30	22

Determine the mean, median and mode.
(Answers: mean = £20.10, median = £20, mode = £20.)

3.1.3 Tabulated (grouped) continuous data

The mean

If we have been presented with a frequency distribution in which the values have been grouped together, we will no longer know the exact value of each observation. However, we can still *estimate* the descriptive statistics.

Consider the income distribution given in Table 3.4. We do not know the exact weekly income of any of the workers, only the numbers with an income in a given range. We know, for example, that 28 workers have an income of between £200 but less than £300. If we assume that all the values within a group are evenly spread (the larger values tending to cancel the smaller values) then we can represent the group by its **mid-point value**. This is the only realistic assumption that we can make when using continuous data in these circumstances. If we were to use the lower limit value, we would under-estimate the mean, if we were to use the upper limit value we would over-estimate the mean.

Looking at much of the published data, we often find that the first and last groups are left as open-ended, for example, 'under £1 500' or 'over £100 000'. In these cases it will be necessary to make assumptions about the upper or lower limits before we can calculate the mean. There are no specific rules for estimating such end-points, but you should consider the data you are trying to describe. If, for example, we were given data on the 'Age of first driving conviction' with a first group labelled as 'under 17 years' it would hardly be realistic to use a lower limit of zero! (It is quite difficult to drive at a few months old.) Looking at this data, we might decide to use the minimum age at which a driving licence can be obtained (16

Table 3.4 Income of a particular group of workers

Weekly income	*Number of workers*
less than £200	10
£200 but less than £300	28
£300 but less than £400	42
£400 but less than £600	50
£600 or more	20

Table 3.5 The estimation of the mean using mid-points

Weekly income	*Mid-point (x)*	*Number of workers (f)*	*fx*
£100* but less than £200	150	10	1 500
£200 but less than £300	250	28	7 000
£300 but less than £400	350	42	14 700
£400 but less than £600	500	50	25 000
£600 but less than £1 000*	800	20	16 000
		150	64 200

The mean is $\bar{x} = \dfrac{64\,200}{150} = £428$

and the formula $\bar{x} = \dfrac{\Sigma fx}{n}$ where x is the value of mid-points

* Assumed

years), but being caught driving one's parents' Porsche around the M25 at the age of 15 would be likely to lead to some form of conviction. There is no correct answer, it is a question of knowing, or at least thinking about, the data.

Once we have established the limits for each group we can then find the mid-points and use these as the x values in our calculations. An example is shown in Table 3.5.

Using a spreadsheet package will entail putting the mid-points into column A and the frequencies into column B and proceeding as before. If you have a whole series of tabulations to analyze, it may be worth setting up the spreadsheet with sufficient space to hold the largest number of rows of data and putting formulae into the other cells. Now all that will be necessary to find the mean for a new tabulation will be to put in the data and the answer will appear automatically. You might try this for the following exercise.

EXERCISE

The amount of detail in the tabulation will have some effect on the estimates of the mean, as can be seen from the following data:

Amount spent on food per week per household	*No. of households*
Under £10	5
£10 but under £15	13
£15 but under £20	26
£20 but under £25	29
£25 but under £30	32
£30 but under £35	34
£35 but under £40	38
£40 but under £45	45
£45 but under £50	43
£50 but under £55	41
£55 but under £60	39
£60 but under £65	47
£65 but under £70	42
£70 but under £75	28
£75 but under £80	20
£80 and more	18

Amount spent on food per week per household	*No. of households*
Under £20	44
£20 but under £30	61
£30 but under £40	72
£40 but under £50	88
£50 but under £60	80
£60 but under £70	89
£70 but under £80	48
£80 and more	18

Amount spent on food per week per household	*No. of households*
Under £50	265
£50 and more	235

Find the arithmetic mean amount spent on food per household from each of the tabulations. (Answers, using a lower limit of £5 and an upper limit of £100 are: £48.08; £47.92; and £49.825)

Were we to be dealing with discrete data which had been tabulated in this way, we would need to look carefully at the actual end-points given. For example, if we had data on the number of people visiting a shop and a group labelled '10 but under 20', this would represent 10 or 11 or 12 or 13 or 14 or 15 or 16 or 17 or 18 or 19 people. The mid-point would thus be 14.5 and *not* 15. Care must always be taken in identifying limits of groups, for example, the published data on 'Age' is given as '10–14' and '15–20', etc., but since age is continuous, the appropriate mid-points are 12.5 and 17.5.

The median

We can determine the median either by calculation or graphically. The first step in both cases is to find the cumulative frequencies (Table 3.6). We can note from the cumulative frequency col-

Table 3.6 The determination of the median

Weekly income	*Number of workers*	*Cumulative frequency*
less than £200	10	10
£200 but less than £300	28	38
£300 but less than £400	42	80
£400 but less than £600	50	130
£600 or more	20	150

umn that 10 workers have a weekly income of less than £200, 38 workers have a weekly income of less than £300 and so on. In each case, cumulative frequency refers to the upper boundary of the corresponding income range.

Graphical method

To find the median graphically, we plot cumulative frequency against the upper boundary of the corresponding income range and join the points with straight lines (this is the graphical representation of the assumption that values are evenly spread within groups). The resultant cumulative frequency graph or **ogive** is shown in Figure 3.1.

To identify the median value from grouped, continuous data we use the formula $n/2$. The corresponding observation divides the histogram into two equal areas. In this example, the median is the 75th value, which can be read from the ogive as £388.

Calculation of the median

To calculate the median, we must first locate the group that contains the 75th observation, i.e. the median group. Looking down the cumulative frequency column of Table 3.6, we can see that 38 workers have an income of less than £300 and 80 workers have an income of less than £400. The

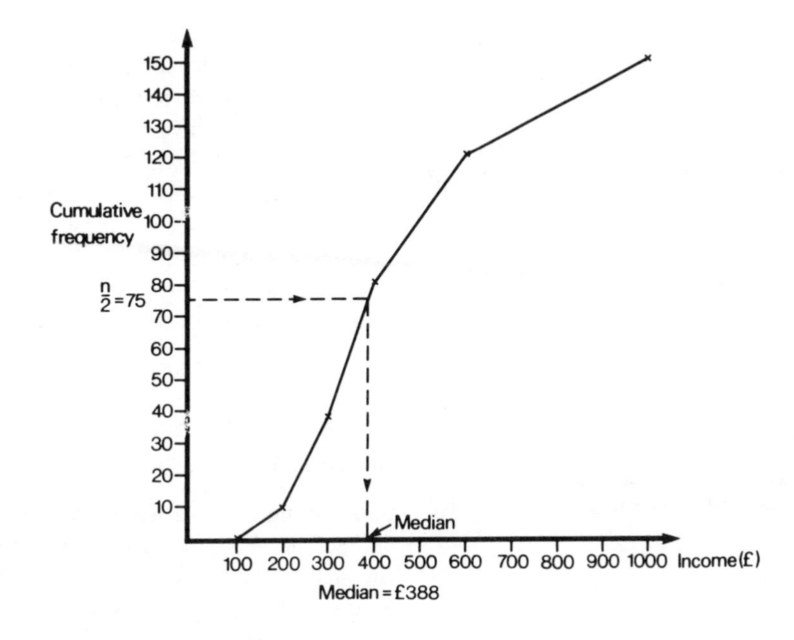

Figure 3.1

75th worker, therefore, must have an income between £300 but less than £400. The median must be £300 plus some fraction of this interval of £100. The median observation lies 37 workers inside the median group; 75 ($n/2$) minus the 38 workers whose weekly income is less than £300. There are 42 workers in the median group so the median lies 37/42th of the way through the interval. The median is equal to:

$$300 + \frac{37}{42} \times 100 = £388.10$$

In terms of a formula we can write:

$$\text{median} = l + i\left(\frac{n/2 - F}{f}\right)$$

where l is the lower boundary of the median group, i is the width of the median group, F is the cumulative frequency up to the median group and f is the frequency in the median group.

Using the figures from the example above:

$$\text{median} = 300 + 100\left(\frac{75 - 38}{42}\right)$$
$$= £388.10$$

The mode

The mode can be defined as the point of greatest density and to estimate this value we refer to the histogram shown in Figure 3.2.

To estimate the mode, we first identify the highest block on the histogram (scaling has already taken place) and join corner points as shown above. In this example the mode is £345.

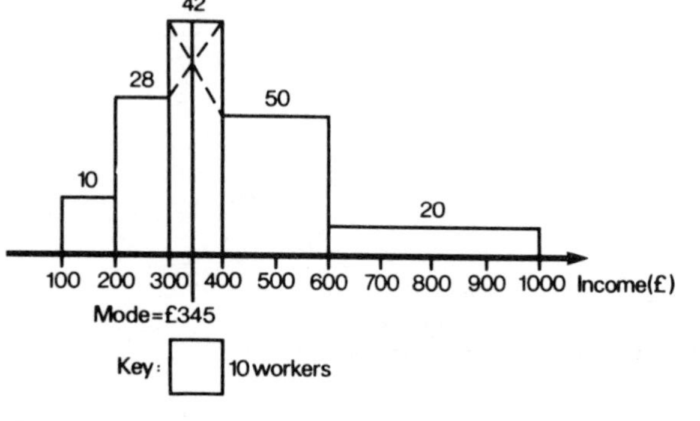

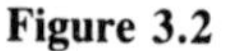
Figure 3.2

EXERCISE

The results of a travel survey were presented as follows:

Journey distance to and from work (miles)	*% of journeys*
0 but under 3	46
3 but under 10	38
10 but under 20	16

Determine the mean, median and mode. (*Hint*: we can use percentages in the same way we used frequencies to determine the mean, median and mode.)

(Answers: mean = 5.56 miles, median = 3.74 miles, mode = 2.56 miles.)

3.2 OTHER MEANS

Two other means are used to describe business data from time to time. The most important of these is the **geometric mean.** This is defined to be 'the nth root of the product of n numbers' and is particularly useful when we are trying to average percentages. (It is also used with index numbers.) For a small number of numbers we can find the geometric mean as follows.

Given the percentage of time spent on a certain task, we have the following data:

30% 20% 25% 31% 25%

Multiplying the five numbers together gives

$$30 \times 20 \times 25 \times 31 \times 25 = 11\,625\,000$$

and, taking the fifth root, the geometric mean is 25.8868%. A simple arithmetic mean would give the answer 26.2% which is an over-estimate of the amount of time spent on the task.

The **harmonic mean** is also used where we are

looking at ratio data, for example miles per gallon or output per shift. It is defined as 'the reciprocal of the arithmetic mean of the reciprocals of the data'. For simple data it is not too difficult to calculate.

For example, if we have data on the number of miles per gallon achieved by five of the representatives of a company:

23 25 26 29 23

then, to calculate the harmonic mean, we find the reciprocals of each number:

0.043478 0.04 0.038461 0.034483 0.043478

find their average:

$$\frac{0.043478 + 0.04 + 0.038461 + 0.034483 + 0.043478}{5}$$

$= 0.03998$

and then take the reciprocal of the answer: 25.0125.

Thus the average fuel consumption of the five representatives is 25.0125 miles per gallon. (The arithmetic mean would be 25.2 m.p.g.)

3.3 THE USE OF MEASURES OF LOCATION

The relative position of each of the statistics from section 3.1 will depend upon the shape of the distribution of the data. Using continuous data, for convenience, we can draw typical frequency distributions to illustrate the three measures.

We have shown the features of the measures on a negatively skewed distribution (i.e. one with a few small values and a larger number of high values), but the measures will vary in their position if the frequency distribution takes a different shape. This is shown in Figure 3.3. (See Chapter 4 for measures of skewness.)

Selection of an appropriate average will depend partly on the nature of the data which you are using, and partly upon the information you wish to convey. Looking at the weekly pay rates for an industry given in Table 3.7 and using the methods shown in section 3.1, we can find the three averages to be:

Arithmetic Mean = £124.10
Median = £102.27
Mode = £ 92.00

Table 3.7 Weekly pay

Weekly pay	*Number of workers*
Under £50	130
£ 50 but under £ 75	156
£ 75 but under £100	198
£100 but under £125	176
£125 but under £150	118
£150 but under £200	76
£200 but under £250	72
£250 but under £300	49
£300 but under £500	19
£500 and more	6

If, however, we look at a breakdown of the figures into the pay of men and the pay of women, a rather different picture emerges (Table 3.8).

EXERCISE

Calculate the mean, median and mode for men and women from the data given in Table 3.8. (Answer: men: mean £148.25, median £120, mode £104; women: mean £88.31, median £79.19, mode £49.)

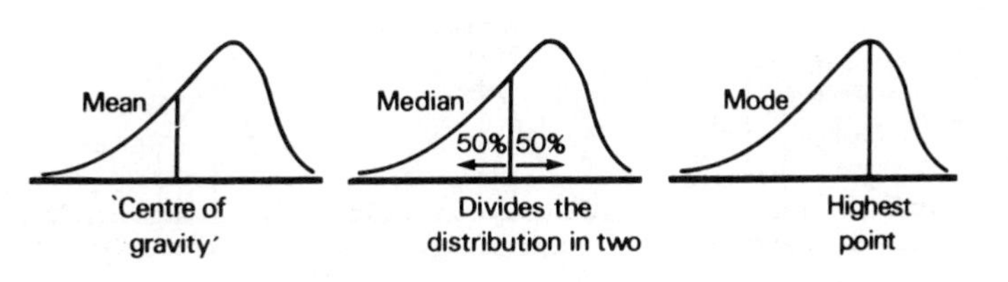

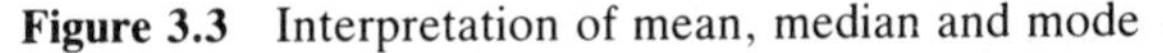

Figure 3.3 Interpretation of mean, median and mode

Table 3.8 Weekly pay

Weekly pay	*Number of men*	*Number of women*
Under £50	30	100
£ 50 but under £ 75	60	96
£ 75 but under £100	114	84
£100 but under £125	120	56
£125 but under £150	90	28
£150 but under £200	60	16
£200 but under £250	60	12
£250 but under £300	42	7
£300 but under £500	18	1
£500 and more	6	0

Thus we have two distinct distributions.

3.4 WEIGHTED MEANS

Suppose that we were given a grouped frequency distribution of weekly income for a particular group of workers and in addition the average income within each of the categories. In this case we could use the set of averages rather than mid-points to calculate an overall mean. In terms of our notation we would need to write the formula as:

$$\bar{x} = \frac{\Sigma \bar{x}_i f_i}{n}$$

where $\bar{x}$ remains the overall mean, $\bar{x}_i$ is the mean in category i and f_i is the frequency in category i.

The calculation using a set of averages is shown in Table 3.9. We would note from the table, for example, that the first 10 workers have an average weekly income of £170, and together earn £1 700. The overall mean is

$$\bar{x} = \frac{68\,180}{150} = £454.53$$

The same result could have been obtained as follows:

$$\bar{x} = \left(170 \times \frac{10}{150}\right) + \left(260 \times \frac{28}{150}\right) + \left(350 \times \frac{42}{150}\right) + \left(590 \times \frac{50}{150}\right) + \left(750 \times \frac{20}{150}\right) = £454.53$$

In terms of describing this procedure the formula can be rewritten as:

$$\bar{x} = \Sigma\left[\bar{x}_i \times \left(\frac{f_i}{n}\right)\right]$$

where f_i/n are the weighting factors.

These weighting factors can be thought of as a measure of size or importance. They can be used

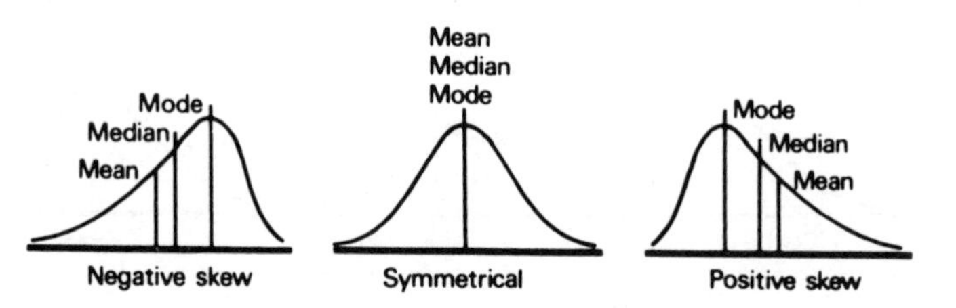

Figure 3.4 The relationship between the mean, median and mode

Table 3.9 The weighting of means

Weekly income	*Category average* ($\bar{x}_i$)	*Number of workers* (f_i)	$\bar{x}_i f_i$
less than £200	£170	10	1 700
£200 but less than £300	£260	28	7 280
£300 but less than £400	£350	42	14 700
£400 but less than £600	£590	50	29 500
£600 or more	£750	20	15 000
		150	68 180

to correct inadequacies in data or to collate results from a survey which was not completely representative.

The method in section 3.1.3 gave a mean of £428 which underestimated the average weekly income. Knowing the means of each of the subgroups in the distribution can be thought of as extra information which helps us to gain a more accurate answer. In the absence of this information we used mid-points of groups, as stated in section 3.1.3.

EXAMPLE

Suppose that two industries, steel manufacture and mining, account for all the economic activity in a small community, and that steel manufacture traditionally accounts for 80% of employment. If 2% become unemployed from steel manufacture and 20% from mining, calculate the overall unemployment rate.

In this example, percentage unemployment from mining is going to have less impact than percentage unemployment from steel manufacture as the latter is four times larger. To average 2% and 20% to find 11% does not allow for this difference in size. To calculate the overall rate of unemployment (weighted average) we can multiply the unemployment rate in each industry by its weighting factor (f_i/n) or proportional size. The weighting factors for steel manufacture and mining are 0.80 and 0.20 respectively.

Rate of unemployment (%)
$= 2 \times 0.8 + 20 \times 0.2 = 5.6\%$.

The overall rate of unemployment is nearer to the 2% for steel manufacture than the 20% for mining owing to the weighting of results.

3.5 CONCLUSIONS

In this chapter we have seen that there are several types of measure of location, or average. By far the most commonly used is the arithmetic mean which is easily understood by almost everyone, but can give misleading results if the data is heavily skewed. Similarly, if the data is made up of different groups, the mean may hide, rather than reveal, the differences between them. Other averages have been developed to combat some of the limitations of the mean, and you should now be in a position to choose the most appropriate average in a particular set of circumstances. The statistics illustrated in the chapter aim to communicate the position, or location, of the data by the use of a typical value, but as we will see in Chapter 4, two data sets which are located in the same place are not necessarily the same.

3.6 PROBLEMS

1. Which descriptive statistics would you use to describe:
   (a) earnings of manual workers in the UK;
   (b) sales turnover of a retail shop;

(c) defective parts in a production process;
(d) strikes.

2. 'The arithmetic mean is the only average ever required'. Write a brief, critical assessment of this statement.
3. The number of new orders received by a company over the last 25 working days were recorded as follows:

3	0	1	4	4
4	2	5	3	6
4	5	1	4	2
3	0	2	0	5
4	2	3	3	1

Determine the mean, median and mode.

4. The number of faults in a sample of new cars has been listed as follows:

0	0	1	3	1	0	0	2	3
1	1	0	1	0	0	4	4	1
1	2	3	0	0	1	1	4	3
0	4	2	2	1	0	0	0	3

Determine the mean, median and mode.

5. The mileages recorded for a sample of company vehicles during a given week yielded the following data:

138	164	150	132	144	125	149	157
146	158	140	147	136	148	152	144
168	126	138	176	163	119	154	165
146	173	142	147	135	153	140	135
161	145	135	142	150	156	145	128

Determine the mean, median and mode. What do these descriptive statistics tell you about the distribution of the data?

Data now becomes available on the remaining ten cars owned by the company and is shown below. How does this new data change the measures of location which you have calculated?

234	204	267	198	179	210	260	290	198	199

6. The average weekly household expenditure on a particular range of products has been recorded from a sample of 20 households as follows:

£8.52	£ 7.49	£ 4.50	£ 9.28	£ 9.98
£9.10	£10.12	£13.12	£ 7.89	£ 7.90
£7.11	£ 5.12	£ 8.62	£10.59	£14.61
£9.63	£11.12	£15.92	£ 5.80	£ 8.31

(a) Determine the mean and median directly from the figures given.
(b) Tabulate the data as a frequency distribution and estimate the mean and median from your table.
(c) Explain any differences in your results from part (a) and part (b).

7. Comparisons are being made between the percentage wage rises given in a group of industries. Each group of workers want to know if they have been given a wage rise which is greater than the average for all of the industries. Given the data below, use both the arithmetic and the geometric means to determine the average rise in wages. What would you report to the various workers?

Industry	*Wage Rise (%)*
A	3.9
B	5.2
C	7.2
D	6.1
E	13.9

8. The number of breakdowns each day on a section of road were recorded for a sample of 250 days as follows:

Number of breakdowns	*Number of days*
0	100
1	70
2	45
3	20
4	10
5	5
	250

Determine the mean, median and mode. Which statistic do you think best describes this data and explain why.

9. Determine the mean, median and mode from the data given in the following table:

Income (£)	*Number*
under 40	175
40 but under 80	229
80 but under 120	241
120 but under 200	269
200 or more	86

10. Determine the mean, median and mode from the data given in the following table: Journey distance to and from work

Miles	*Percentage*
under 1	16
1 and under 3	30
3 and under 10	37
10 and under 15	7
15 and over	9

11. Determine the mean and median from the information given in the following table:

Error (£)	*Frequency*
under −15	20
−15 but less than −10	38
−10 but less than −5	178
−5 but less than 0	580
0 but less than 10	360
10 but less than 20	114
20 or more	15

12. Collect data on the Personal Income Distribution from the *Annual Abstract of Statistics* for the current year. Calculate the mean and median personal income levels. (NB, We suggest that you use a spreadsheet or a computer package to answer this question.)

13. Included in the following table is the average income for those in each income group:

Income (£)	*Average income*	*Number*
under 40	35	175
40 but under 80	64	229
80 but under 120	101	241
120 but under 200	158	269
200 or more	240	86

(a) Determine the mean using the average income figures given.

(b) Is the method used in part (a) as precise as working with a listing of original data?

14. A company files its sales vouchers according to their value so that they are effectively in four strata. A sample of 200 is selected and the strata means calculated.

Stratum	*Number of vouchers*	*Sample size*	*Sample mean (£)*
above £1 000	100	50	1 800
£800 but under £1 000	200	60	890
£400 but under £800	500	50	560
less than £400	1 000	40	180
		200	

Estimate the mean value and total value of the sales vouchers.

MEASURES OF DISPERSION 4

In Chapter 3 we looked at several measures which aimed to show the location of a data set by calculating a typical, or average, value. Such statistics are useful in showing where the data is located, for example that average income in a country was £50 per week ten years ago and is now £175 per week. It is not just the location of the data which is important, but also the **amount of variation** in the data. It may have been that ten years ago, most people in the country had incomes very close to £50, and thus there was little variability; now there may be wide variability such that some people are still earning about £50 per week, whilst others earn £700 per week. In this chapter we introduce ways of measuring this variability, or **dispersion**, in the data and go on to consider ways of comparing different distributions of data. This will enable us to say if there is more dispersion now than at some point in the past, or more dispersion in one region in comparison with another, or between one defined group and another. Considering the income example above, we will be able to describe whether the income distribution is more clustered or not by using statistics, but it is a political and economic question as to whether this is desirable or not.

Measures of dispersion can be divided into two groups, **absolute measures** which consider just the one data set and give an answer in pounds, minutes, age, etc.; and **relative measures** which give an answer as a percentage or proportion, thus allowing direct comparison with other data sets.

4.1 STANDARD DEVIATION

The standard deviation is the most widely used measure of dispersion since it is directly linked to the arithmetic mean. If you are asked to describe a data set and decide that the mean is the most appropriate average, then you would use the standard deviation to describe the variability in that data set. Unlike the mean, however, this statistic is not widely known amongst those who have not studied statistics and it does not have the same, intuitive meaning. It can, therefore, be a less effective tool of communication.

The basic objective of the statistic is to measure the variability of the data, large values, in general, denoting more variation than small values, but in all cases we must remember that it is variability from a given arithmetic mean, and thus it is *not* valid to make comparisons directly between different data sets. (Strictly speaking, if two data sets have the same mean, then such a comparison can be made, but this is rather unlikely in practice.) Remember also, that at this stage, we are merely calculating a measure of description for the data.

Standard deviations are most useful with continuous, cardinal data, although they can be used with discrete data. In the latter case, it is unlikely that the answer will be one of the original values, and it is likely to be a fractional value.

Statistical packages, such as MICROSTATS or MINITAB, make the determination of the standard deviation simple for many data sets, or we could, as an alternative, use a spreadsheet and show the stages in the calculation. The use of such packages will enhance your understanding of this statistic, since you can see the effects of changing one of the data values, or quickly find the standard deviation from a series of different frequency distributions. Calculations by hand are again shown in this chapter since they will reinforce the point that the measure is closely related to the mean, and many examiners still expect you to be able to perform the calculations on simple data sets. Beware of being blinded to the main purpose of the calculations by the detail of the steps to be taken.

The standard deviation is also important in the development of statistical theory, since most traditional theory is based on distributions described by their mean and standard deviation. We will use the standard deviation extensively in Chapters 13, 14, and 15 as we begin to analyze results from sample surveys, and again in Chapters 17 and 18 as we look at the interpretation of correlation and regression analysis.

4.1.1 Untabulated data

We have already seen, in section 3.1.1, how to calculate the mean from simple data, and this will be a necessary first step before we can calculate the standard deviation. Consider, for example, the times recorded to complete a routine job:

7, 5, 6, 7 and 8 minutes

The mean for this data is 6.6 minutes. The differences about this mean are shown diagrammatically in Figure 4.1. To the left of the mean the differences are negative and to the right of the mean the differences are positive. The sum of these differences is zero. In terms of a physical interpretation the mean can be viewed as the centre of gravity and the observations as a number of weights which balance.

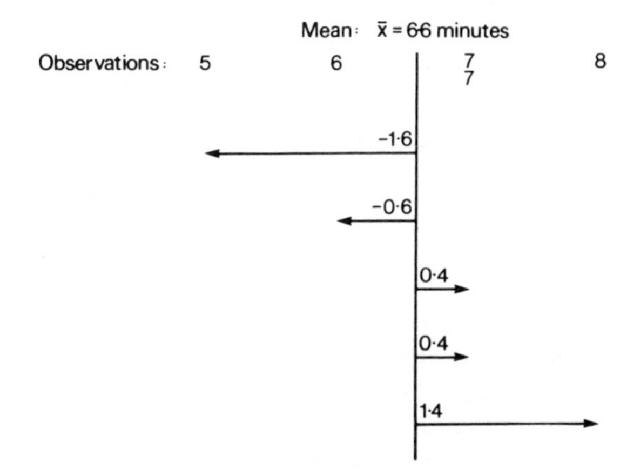

Figure 4.1 The differences about the mean

To calculate the standard deviation we follow six steps.

1. Compute the mean $\bar{x}$.
2. Calculate the differences from the mean $(x - \bar{x})$.
3. Square these differences $(x - \bar{x})^2$.
4. Sum the squared differences $\Sigma(x - \bar{x})^2$.
5. Average the squared differences to find variance:

Table 4.1 The calculation of the standard deviation

x	$(x - \bar{x})$	$(x - \bar{x})^2$
7	0.4	0.16
5	−1.6	2.56
6	−0.6	0.36
7	0.4	0.16
8	1.4	1.96
33		5.20

where $\bar{x} = \frac{33}{5} = 6.6$ minutes

$$s = \sqrt{\left[\frac{\Sigma(x - \bar{x})^2}{n}\right]} = \sqrt{\left[\frac{5.20}{5}\right]}$$

$$= \sqrt{1.04} = 1.02 \text{ minutes}$$

$$\frac{\Sigma(x-\bar{x})^2}{n}$$

6. Square root variance to find standard deviation:

$$\sqrt{\left[\frac{\Sigma(x-\bar{x})^2}{n}\right]}$$

These are illustrated in Table 4.1.

These calculations can be made rather difficult if the mean does not work out to be a convenient number, for example if it were 6.5439287561. To avoid this we could set up a spreadsheet where all of the data values are put into column A and then calculate the mean. The other two columns of Table 4.1 can be constructed by the use of formulae within the spreadsheet, and the standard deviation found once column C is summed. We suggest that you try this for the data in Table 4.1, and then change one of the data values, say the 8, to another value, say 10 to see the effect on the standard deviation. Notice how the standard deviation is affected as you change the 8 to numbers further from the mean, and closer to the mean. (NB, make sure that your spreadsheet re-calculates the mean each time that you change a data value.) An alternative would be to put the data values into a column in MICROSTATS, say C1, and then type in STANDARD C1 (or STAN C1) at the command line.

EXERCISES

1. The errors in seven invoices were recorded as follows: −£120, £30, £40, −£8, −£5, £20 and £25. Calculate the mean and standard deviation.
(Answer: $\bar{x} = -£2.57$, $s = £50.66$.)
2. To check the working consistency of a new machine, the time taken to complete a specific task was recorded on five occasions. On each occasion the recorded time was 30 seconds. Calculate the mean and standard deviation.
(Answer: $\bar{x} = 30$ seconds, $s = 0$ seconds.)

Table 4.2 The number of faults recorded in a sample of new cars

Number of faults	*Number of cars*
0	410
1	430
2	290
3	180
4	110
5	20

4.1.2 Tabulated discrete data

Suppose the results of a survey were given in a tabulated form as in Table 4.2. To calculate standard deviation, we need to allow for the **frequency of occurrence**. In the first group, for example, there are 410 cars with 0 faults, so we would expect 410 differences of 0 from the mean. This requires an additional step of multiplying squared differences by frequency. The formula for standard deviation now becomes

$$s = \sqrt{\left[\frac{\Sigma f(x-\bar{x})^2}{n}\right]}$$

The calculations are shown in Table 4.3.

We could set up a spreadsheet to do these calculations by putting the x values in column A and the frequencies in column B; the rest of the figures being calculated by the use of formulae within the spreadsheet. We suggest that you try this, and then change the frequencies to see the effect on the standard deviation. (NB, make sure that your spreadsheet re-calculates the mean each time, and then uses the updated value in the further calculations.)

4.1.3 Tabulated (grouped) data

When the data is presented as a grouped frequency distribution we must determine initially whether it is discrete or continuous (see section 3.1.3). In either case we will use the mid-points of

Table 4.3 Standard deviation from tabulated discrete data: the number of faults recorded in a sample of new cars

x	f	fx	$(x-\bar{x})$	$(x-\bar{x})^2$	$f(x-\bar{x})^2$
0	410	0	−1.451	2.1054	863.214
1	430	430	−0.451	0.2034	87.462
2	290	580	0.549	0.3014	87.406
3	180	540	1.549	2.3994	431.892
4	110	440	2.549	6.4974	714.714
5	20	100	3.549	12.5954	251.908
	1440	2090			2436.596

$$\bar{x} = \frac{2090}{1440} = 1.451 \text{ faults}$$

$$s = \sqrt{\left[\frac{\Sigma f(x-\bar{x})^2}{n}\right]} = \sqrt{\left[\frac{2436.596}{1440}\right]} = 1.301 \text{ faults}$$

Table 4.4 The estimation of the standard deviation using mid-points

Weekly income	*Frequency* (f)	*Mid-point* (x)	fx	$(x-\bar{x})$	$(x-\bar{x})^2$	$f(x-\bar{x})^2$
less than £200	10	150*	1 500	−278	77 284	772 840
£200 but less than £300	28	250	7 000	−178	31 684	887 152
£300 but less than £400	42	350	14 700	−78	6 084	255 528
£400 but less than £600	50	500	25 000	72	5 184	259 200
£600 or more	20	800*	16 000	372	138 384	2 767 680
	150		64 200			4 942 400

$$\bar{x} = \frac{64\,200}{150} = £428$$

$$s = \sqrt{\left[\frac{\Sigma f(x-\bar{x})^2}{n}\right]} = \sqrt{\left[\frac{4\,942\,400}{150}\right]} = £181.52$$

*Assumed.

the groups in our calculations. Having determined the mid-points, we then continue as shown in the previous section, using the same formula. An example is shown in Table 4.4.

The methodology shown in Table 4.4 will work quite satisfactorily on a spreadsheet, but would be extremely tedious to perform by hand. A little algebraic manipulation of the formula given in section 4.1.2 will give a formula which is much easier to use if the calculations are to be performed by hand (see section 4.7 for a proof of this relationship). This is:

$$s = \sqrt{\left[\frac{\Sigma fx^2}{\Sigma f} - \left(\frac{\Sigma fx}{\Sigma f}\right)^2\right]}$$

An example is shown in Table 4.5.

Even if the calculations are to be performed using a spreadsheet, it will be advantageous to use this formula since it is not necessary to calculate the mean each time the data changes. We suggest that you should set up a spreadsheet to perform the calculations in Table 4.5 and then, again, vary the frequencies slightly to see the effect on the standard deviation.

Table 4.5 The estimation of standard deviation

x	f	fx	x^2	fx^2
150	10	1 500	22 500	225 000
250	28	7 000	62 500	1 750 000
350	42	14 700	122 500	5 145 000
500	50	25 000	250 000	12 500 000
800	20	16 000	640 000	12 800 000
	150	64 200		32 420 000

$$s = \sqrt{\left[\frac{\Sigma fx^2}{n} - \left(\frac{\Sigma fx}{n}\right)^2\right]} = \sqrt{\left[\frac{32\,420\,000}{150} - \left(\frac{64\,200}{150}\right)^2\right]}$$
$$= £181.52$$

4.1.4 The variance

This is defined to be the square of the standard deviation, and is thus easily calculated once the standard deviation is known. It is sometimes used as a descriptive measure of dispersion or variability rather than the standard deviation, but is of more importance as we develop the ideas of statistical theory in Chapters 12, 13 and 14. As we will see in later chapters, you can add variances together, but you cannot add standard deviations. Variance is mentioned here for completeness.

4.2 OTHER MEASURES OF DISPERSION

Whilst the standard deviation is the most widely used measure of dispersion, it is not the only one. As we saw when looking at averages, different averages are appropriate for different situations and the same will be true for measures of dispersion. Furthermore, some of the measures of dispersion are specifically linked to certain measures of location and it would not make sense to mix and match the statistics.

4.2.1 The range

The range is the most easily understood measure of dispersion as it is the difference between the highest and lowest values. If we were concerned with five recorded times of 7, 5, 6, 7 and 8 minutes the range would be 3 minutes (8 minutes − 5 minutes). It is, however, a rather crude measure, which is highly unstable as new data is added, and is naturally affected by a few extreme values. If this type of measure is to be used, it may well be better to quote the highest and lowest values rather than the difference between them. When dealing with data presented as a frequency distribution we will not always know the exact highest and lowest values, and if the groups are open-ended (e.g. 60 and more) then any values used will merely be from assumptions that we have made about the widths of the groups. In such cases there seems little point in quoting either the range or the extreme values.

4.2.2 The quartile deviation

If we are able to quote a half-way value, the median, then we can also quote quarter-way values, the **quartiles**. These are order statistics like the median and can be determined in the same way. With untabulated data or tabulated discrete data it will merely be a case of counting through the ordered data set until we are a quarter of the way through and noting the value. This will be the **first quartile**. Continuing counting until we get three quarters of the way through, and noting the value will give us the **third quartile**.

Table 4.6 Income distribution

Weekly income	*Number of workers*	*Cumulative frequency*
less than £200	10	10
£200 but less than £300	28	38
£300 but less than £400	42	80
£400 but less than £600	50	130
£600 or more	20	150

Where we have continuous data presented as a grouped frequency distribution, further calculations will be necessary, and these are illustrated below. Consider, for example, the income distribution and cumulative frequency given in Table 4.6. The lower quartile (referred to as Q_1), will correspond to the value one-quarter of the way through the data, the 37.5 ordered value:

$$\left(\frac{n}{4} = \frac{150}{4} = 37.5\right)$$

and the upper quartile (referred to as Q_3) to the value three-quarters of the way through the data, the 112.5 ordered value:

$$\left(\frac{3n}{4} = \frac{3}{4} \times 150 = 112.5\right)$$

Graphical method

To estimate any of the order statistics graphically, we plot cumulative frequency against the values to which it refers, as shown in Figure 4.2. The value of the lower quartile can be read from the ogive to be £298 and the value of the upper quartile to be £530.

Calculation of the quartiles

We can adapt the median formula (see section 3.1.3) as follows:

$$\text{Order value} = l + i\left(\frac{O - F}{f}\right)$$

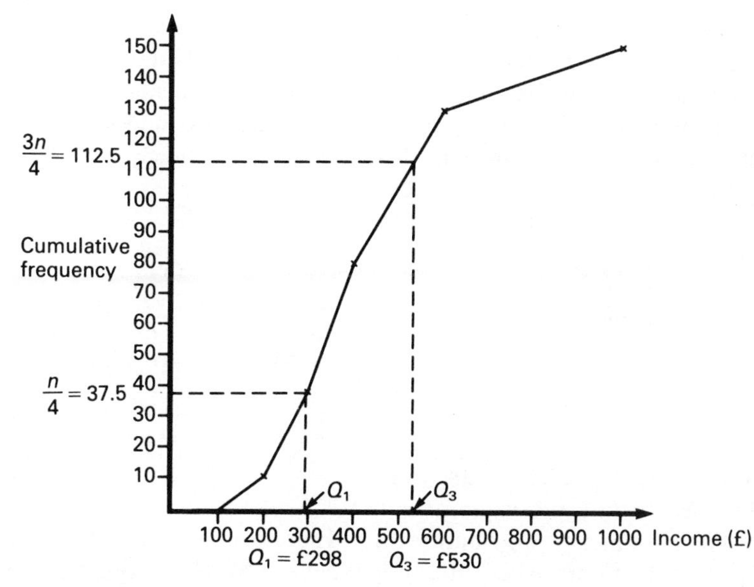

Figure 4.2

where O is the order value of interest, l is the lower boundary of corresponding group, i is the width of this group, F is the cumulative frequency up to this group, and f is the frequency in this group.

The lower quartile will lie in the group '£200 but less than £300' and can be calculated thus:

$$Q_1 = 200 + 100\left(\frac{37.5 - 10}{28}\right) = £298.21$$

The upper quartile will lie in the group '£400 but less than £600' and can be calculated thus:

$$Q_3 = 400 + 200\left(\frac{112.5 - 80}{50}\right) = £530.00$$

The quartile range is the difference between the quartiles:

$$\begin{aligned}\text{Quartile range} &= Q_3 - Q_1\\ &= £530.00 - £298.21\\ &= £231.79\end{aligned}$$

and the quartile deviation (or semi-interquartile range) the average difference:

$$\begin{aligned}\text{Quartile deviation} &= \frac{Q_3 - Q_1}{2}\\ &= \frac{£231.79}{2} = £115.90 \text{ (rounded from } £115.895)\end{aligned}$$

As with the range, the quartile deviation may be misleading. If the majority of the data is towards the lower end of the range, then the third quartile will be considerably further above the median than the first quartile is below it, and when we average the two numbers we will disguise this difference. This is likely to be the case with a country's personal income distribution. In such circumstances, it would be preferable to quote the actual values of the two quartiles, rather than the quartile deviation.

4.2.3 Percentiles

The formula given in section 4.2.2 for an order value, O, can be used to find the value at *any* position in a grouped frequency distribution of continuous data. For data sets which are not skewed to one side or the other, the statistics we have calculated so far will usually be sufficient, but heavily skewed data sets will need further statistics to fully describe them. Examples would include income distribution, wealth distribution and time taken to complete a complex task. In such cases, we may want to use the 95th percentile, i.e. the value below which 95% of the data lies. Any other value between 1 and 99 could also be calculated. An example of such a calculation is shown in Table 4.7.

Table 4.7 Wealth distribution

Wealth	*Number (f)*	*Cumulative frequency*
zero	15 000	15 000
under £1 000	3 100	18 100
under £5 000	2 300	20 400
under £10 000	2 300	22 700
under £25 000	1 600	24 300
under £50 000	1 000	25 300
under £100 000	800	26 100
under £250 000	300	26 400
under £500 000	170	26 570
under £1 000 000	80	26 650
over £1 000 000	50	26 700

For this wealth distribution, the first quartile and the median are both zero. The third quartile is £4 347.83. None of these statistics adequately describes the distribution.

To calculate the 95th percentile, we find 95% of the total frequency, here

$$0.95 \times 26\,700 = 25\,365$$

and this is the item whose value we require. It will be in the group labelled 'under £100 000' which has a frequency of 800 and a width of 50 000 (i.e. 100 000 − 50 000). Using the formula, we have:

$$\begin{aligned}\text{95th Percentile} &= £50\,000 + 50\,000\left(\frac{£25\,365 - £25\,300}{800}\right)\\ &= £54\,062.50\end{aligned}$$

Therefore, 95% of this sample have wealth of below £54 062.50; or alternatively, 5% of the sample have wealth above £54 062.50.

EXERCISE

Using the same data, calculate the 90th percentile and the 99th percentile for wealth.
(Answers: £22 468.75, and £298 529.41)

4.3 RELATIVE MEASURES OF DISPERSION

All of the measures of dispersion described earlier in this chapter have dealt with a single set of data. In practice, it is often important to compare two or more sets of data, maybe from different areas, or data collected at different times. In Part Five we look at formal methods of comparing the difference between sample observations, but the measures described in this section will enable some initial comparisons to be made. The advantage of using relative measures is that they do not depend upon the units of measurement of the data.

4.3.1 Coefficient of variation

This measure calculates the standard deviation from a set of observations as a percentage of the arithmetic mean:

$$\text{coefficient of variation} = \frac{s}{\bar{x}} \times 100$$

Thus the higher the result, the more variability there is in the set of observations. If, for example, we collected data on personal incomes for two different years, and the results showed a coefficient of variation of 89.4% for the first year, and 94.2% for the second year, then we could say that the amount of dispersion in personal income data had increased between the two years. Even if there has been a high level of inflation between the two years, this will not affect the coefficient of variation, although it will have meant that the average and standard deviation for the second year are much higher, in absolute terms, than the first year.

4.3.2 Coefficient of skewness

Skewness of a set of data relates to the shape of the histogram which could be drawn from the data. The type of skewness present in the data can be described by just looking at the histogram, but it is also possible to calculate a measure of skewness so that different sets of data can be compared. Three basic histogram shapes are shown in Figure 4.3, and a formula for calculating skewness is shown below.

$$\text{coefficient of skewness} = \frac{3(\text{mean} - \text{median})}{\text{standard deviation}}$$

A typical example of the use of the coefficient of skewness is in the analysis of income data. If the coefficient is calculated for gross income before tax, then the coefficient gives a large positive result since the majority of income earners receive relatively low incomes, whilst a small proportion of income earners receive high incomes. When the coefficient is calculated for the same group of earners using their after tax income, then, although a positive result is still obtained, its size has decreased. These results are typical of a progressive tax system, such as that in the UK. Using such calculations it is possible to show that the distribution of personal incomes in the UK

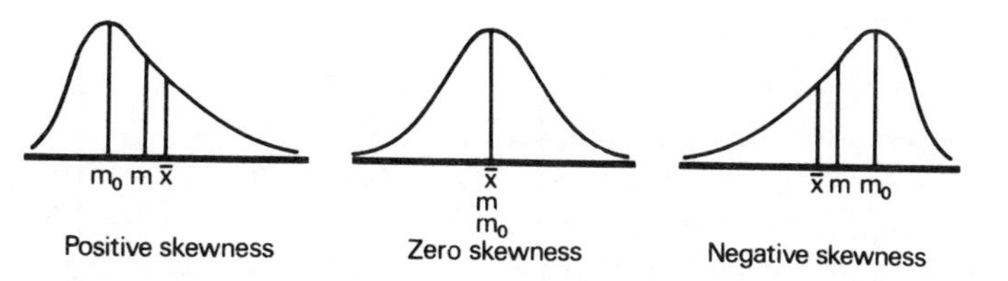

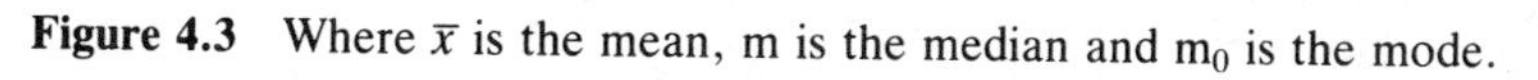
Figure 4.3 Where $\bar{x}$ is the mean, m is the median and m_0 is the mode.

has changed over time. A discussion of whether or not this change in the distribution of personal incomes is good or bad will depend upon your economic and political views; the statistics highlight that the change has occurred.

EXERCISE

Using the income data given in Table 4.4 determine the coefficient of variation and measure of skewness. (Answer: coefficient of variation = 42.42%; skewness = 0.66

4.4 VARIABILITY IN SAMPLE DATA

We would expect the results of a survey to identify differences in opinions, income and a range of other factors. The extent of these differences can be summarized by an appropriate measure of dispersion (standard deviation, quartile deviation, range). Market researchers, in particular, seek to explain differences in attitudes and actions of distinct groups within a population. It is known, for example, that the propensity to buy frozen foods varies between different groups of people. As a producer of frozen foods you might be particularly interested in those most likely to buy your products. Supermarkets of the same size can have very different turnover figures and a manager of a supermarket may wish to identify those factors most likely to explain the differences in turnover. A number of **clustering algorithms** have developed in recent years that seek to explain differences in sample data.

As an example, consider the following algorithm or procedure that seeks to explain the differences in the selling prices of houses:

1. Calculate the mean and a measure of dispersion for all the observations in your sample. In this example we could calculate the average price and the range of prices (Figure 4.4).
   It can be seen from the range that there is considerable variability in price relative to the average price. Usually the standard deviation would be preferred to the range as a measure of dispersion for this type of data.

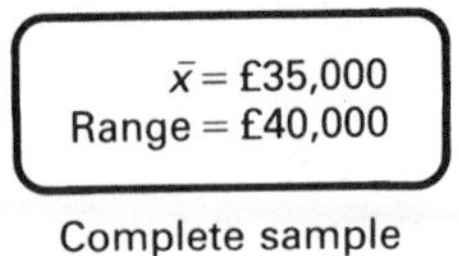

Figure 4.4

2. Describe which factors explain most of the difference (range) in price, e.g. location, house-type, number of bedrooms. If location is considered particularly important, we can divide the sample on that basis and calculate the chosen descriptive statistics (Figure 4.5).
   In this case we have chosen to segment the sample by location, areas X and Y. The smaller range within the two new groups indicates that there is less variability of house prices within areas. We could have divided the sample by some other factor and compared the reduction in the range.

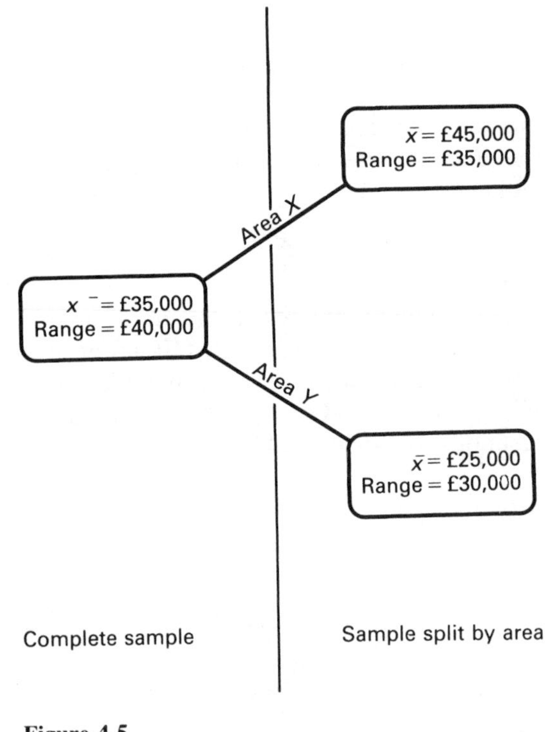

Figure 4.5

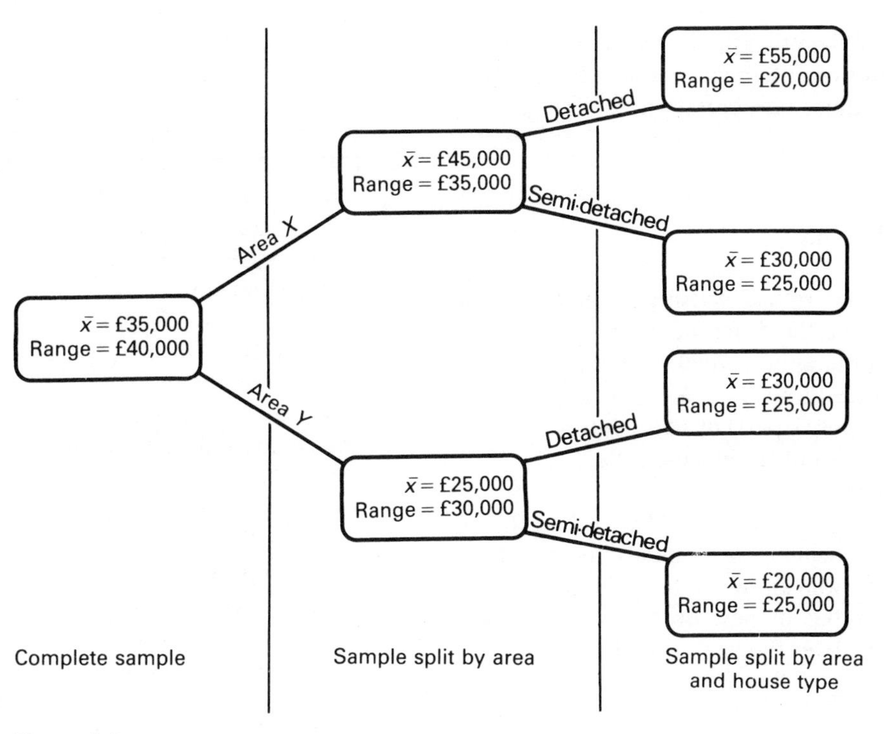

Figure 4.6

3. Divide the new groups and again calculate the descriptive statistics. We could divide the sample a second time on the basis of house-type (Figure 4.6).
4. The procedure can be continued in many ways with many splitting criteria.

A more sophisticated version of this procedure is known as the automatic interactive detection technique. It has been used as a basis for market segmentation and as a guide for multiple regression (see Chapter 19).

4.5 CONCLUSIONS

Describing the variability of a data set or a sample is more complex than describing its location or average value since there is less intuitive meaning to the statistics which we use. Where the data are closely packed and near to the average, then we will obtain relatively low values for the measure of dispersion; where the data are widely spread, the larger values will be obtained. Comparisons between samples are valid only if we use relative measures of dispersion.

Description of variability will be important when we are trying to make business decisions. If there is little variation, we could use the average value from the sample and expect to describe most responses fairly. When variability is relatively large, it is quite likely that any particular member of the sample will be quite different from the average value.

The question then arises of why such variability should exist. In market research, for example, relatively large overall variability may result from the clustering of certain groups of respondents. In this case we may consider market segmentation and target groups within the population. (Variability within such target groups may be much less than variability in the population in general.) Where political opinion is widely spread on an attitudinal scale, we may wish to relate this finding to other explanatory factors. Standard deviation and the other measures of variation are only measures of difference, and it is these differences that require interpretation.

4.6 PROBLEMS

1. Which descriptive statistics would you use to describe the differences in:
   (a) earnings of manual workers in the UK;
   (b) sales turnover of a retail shop;
   (c) defective parts in a production process;
   (d) wealth of the richest 10%;
   (e) strikes.
2. Write a brief statement (200 words) explaining the meaning of standard deviation to someone who knows nothing about statistics.
3. The number of new orders received by a company over the last 25 working days were recorded as follows:

3	0	1	4	4
4	2	5	3	6
4	5	1	4	2
3	0	2	0	5
4	2	3	3	1

   Determine the range, quartile deviation and standard deviation.
4. The number of faults in a sample of new cars has been listed as follows:

0	0	1	3	1	0	0	2	3
1	1	0	1	0	0	4	4	1
1	2	3	0	0	1	1	4	3
0	4	2	2	1	0	0	0	3

   Determine the range, quartile deviation and standard deviation. Which of these measures of dispersion would you consider most appropriate for the data given? Explain the reasons for your choice.
5. The mileages recorded for a sample of company vehicles during a given week yielded the following data:

138	164	150	132	144	125	149	157
146	158	140	147	136	148	152	144
168	126	138	176	163	119	154	165
146	173	142	147	135	153	140	135
161	145	135	142	150	156	145	128

   Determine the range, quartile deviation and standard deviation from these figures.

   The data below now becomes available on the mileages of the other ten cars belonging to the company.

234	204	267	198	179	210	260	290	198	199

   Recalculate the range, quartile deviation and standard deviation, and comment on the changes to these statistics.
6. The average weekly household expenditure on a particular range of products has been recorded from a sample of 20 households as follows:

£8.52	£ 7.49	£ 4.50	£ 9.28	£ 9.98
£9.10	£10.12	£13.12	£ 7.89	£ 7.90
£7.11	£ 5.12	£ 8.62	£10.59	£14.61
£9.63	£11.12	£15.92	£ 5.80	£ 8.31

   (a) Determine the mean and standard deviation directly from the figures given.
   (b) Tabulate the data as a frequency distribution and estimate the mean and standard deviation from your table.
   (c) Explain any differences in your results from part (a) and part (b).
7. The number of breakdowns each day on a section of road were recorded for a sample of 250 days as follows:

Number of breakdowns	*Number of days*
0	100
1	70
2	45
3	20
4	10
5	5
	250

   Calculate the range, quartile deviation and standard deviation.

8. Determine the quartile deviation and standard deviation of incomes from the data given in the following table:

Income (£)	*Number*
under 40	175
40 but under 80	229
80 but under 120	241
120 but under 200	269
200 or more	86

9. Determine the quartile deviation and standard deviation from the data given in the following table.

Journey distance to and from work in miles	*Percentage*
under 1	16
1 and under 3	30
3 and under 10	37
10 and under 15	7
15 and over	9

10. Determine the mean and standard deviation from the data given in the following table.

Error(£)	*Frequency*
under −15	20
−15 but less than −10	38
−10 but less than −5	178
−5 but less than 0	580
0 but less than 10	360
10 but less than 20	114
20 or more	14

11. Given the following distribution of weekly household income:

Household income (£)	*% of all households*
under 30	13.7
30 but under 40	7.6
40 but under 60	11.6
60 but under 80	13.4
80 but under 100	14.2
100 but under 120	13.0
120 but under 150	12.3
150 or more	14.2

(a) calculate the mean and standard deviation;
(b) estimate the percentage of households with a weekly income below the mean;
(c) determine the median and quartile deviation;
(d) contrast the values you have determined in parts (a), (b) and (c) and comment on the skewness (if any) of the distribution.

12. The following annual salary data has been collected from two distinct groups of skilled workers within a company.

Annual salary (£)	*No. from group A*	*No. from group B*
8 000 but under 10 000	5	0
10 000 but under 12 000	17	19
12 000 but under 14 000	21	25
14 000 but under 16 000	3	4
16 000 but under 18 000	1	0
18 000 but under 20 000	1	0

(a) Determine the mean and standard deviation for each group of skilled workers.
(b) Determine the coefficient of variation and a measure of skew for each group of skilled workers.
(c) Discuss the results obtained in parts (a) and (b).

13. The time taken to complete a particularly complex task has been measured for 250 individuals and the results are shown below.

Time taken	*No. of people*
under 5 minutes	2
under 10 minutes	2
under 15 minutes	3
under 20 minutes	5
under 25 minutes	5
under 30 minutes	18
under 40 minutes	85
under 50 minutes	92
under 60 minutes	37
over 60 minutes	1

Estimate the maximum time taken by someone in the quickest:
(a) 1%
(b) 5%
(c) 10%

14. You have been given the following data from a sample of 20 individuals:

Code number	*Sex*	*Age*	*Employment*	*Amount spent weekly on alcoholic drink (£)*
1	0	20	0	8.83
2	1	33	0	4.90
3	1	50	1	0.71
4	0	48	0	5.70
5	0	47	0	6.20
6	0	19	0	7.40
7	1	21	1	3.58
8	0	64	0	4.80
9	1	32	0	4.50
10	1	57	1	2.80
11	0	49	0	4.60
12	0	18	0	5.30
13	1	39	1	3.42
14	0	28	0	10.15
15	0	51	0	6.20
16	1	43	0	4.80
17	1	40	0	3.82
18	0	22	1	7.70
19	1	30	0	6.20
20	0	60	0	4.45

Age: number of years
Sex: male 0, female 1
Employment: working 0, not working 1

Measure and explain the variation in the amount spent weekly on alcoholic drinks with reference to the other factors given.

4.7 APPENDIX

Proof that $\sqrt{\left[\frac{\Sigma f(x-\bar{x})^2}{n}\right]} = \sqrt{\left[\frac{\Sigma fx^2}{n} - \left(\frac{\Sigma fx}{n}\right)^2\right]}$.

$$\frac{1}{n}\Sigma f(x-\bar{x})^2 = \frac{1}{n}\Sigma f(x^2 - 2x\bar{x} + \bar{x}^2)$$

$$= \frac{1}{n}(\Sigma fx^2 - 2\bar{x}\Sigma fx + \bar{x}^2\Sigma f)$$

$$= \frac{1}{n}(\Sigma fx^2 - 2n\bar{x}^2 + n\bar{x}^2)$$

$$= \frac{1}{n}(\Sigma fx^2 - n\bar{x}^2)$$

$$= \frac{1}{n}\left[\Sigma fx^2 - n\left(\frac{\Sigma fx}{n}\right)^2\right]$$

$$= \frac{\Sigma fx^2}{n} - \left(\frac{\Sigma fx}{n}\right)^2$$

Notes: $\bar{x}$ is a constant as far as Σ is concerned; $\Sigma f = n$; $\Sigma fx = n\bar{x}$ since $\bar{x} = \Sigma fx/n$.

INDEX NUMBERS 5

We are often concerned with describing changes in economic, business and social variables over time. Where the changes can be measured on a cardinal scale, then it can be effectively represented by an **index number**. Many of the situations we seek to represent involve very large numbers, for example, the Gross Domestic Product of a country; the size of such numbers are almost meaningless and also disguise the amount of change taking place. In such situations, an index number will be much easier to understand and will highlight the changes which are taking place.

Index numbers were originally developed by economists to allow the changing value of a range of goods to be incorporated into a single number, an **aggregated** index number. This concept is still seen today in their use to represent the changes in the retail prices of goods as represented by the general index of retail prices, from which we may calculate measures of inflation. Businesses may use index series to compare their performance or prices with published series, or the performance of other companies.

5.1 THE INTERPRETATION OF AN INDEX NUMBER

An index is a scaling of numbers so that a start is made from a base figure of 100. If, for example, we had the price of bread for four years given as:

Year	*Price*
1	0.25
2	0.30
3	0.40
4	0.47

then to construct a simple index, we need to scale the £0.25 to the base figure of 100. To do this, we divide by £0.25 and multiply by 100. We now take each of the other prices, divide by £0.25 and multiply by 100 to get the index for the price of bread, as shown below:

Year	*Price*		*Index*
1	0.25	$(0.25/0.25) \times 100$	100
2	0.30	$(0.30/0.25) \times 100$	120
3	0.40	$(0.40/0.25) \times 100$	160
4	0.47	$(0.47/0.25) \times 100$	188

Therefore, all of the subsequent index numbers relate to the base year chosen. Consider the example in Table 5.1. In this case, year 1 is the **base year**. The subsequent index numbers all measure the change from the base year. The index number 135 is taken to mean that there has been a *35% increase* in the measured quantity from the base year (year 1) to year 4. Similarly, the index number 150 is taken to measure a *50% increase* from the base year. To calculate a percentage increase we first find the difference between the two figures, divide by the base figure and then multi-

Table 5.1 An index

Year	*Index 1*
1	100
2	110
3	120
4	135
5	150

ply by 100. The percentage increase from 100 to 150 can be calculated as

$$\frac{150 - 100}{100} \times 100 = 50\%.$$

In the same way an increase in hourly pay from £1.80 to £2.70 is 50%:

$$\frac{2.70 - 1.80}{1.80} \times 100 = 50\%.$$

One convenience of index numbers is that by starting from 100, the percentage increase from the base year is found just by subtraction. However, the differences thereafter are referred to as **percentage points**. It can be seen from Table 5.1 that there has been a 15 percentage point increase from year 4 to year 5. The percentage increase, however, is

$$\frac{150 - 135}{135} \times 100 = 11.11\%.$$

There are no 'hard and fast' rules for the choice of a base year and, as shown in Table 5.2, any year can be the base year.

Each of the indices measures the same change over time (to two decimal places). The percentage increase from year 1 to year 2 using index 2, for example, is

$$\frac{100.00 - 90.91}{90.91} \times 100 = 10.00\%.$$

To change the base year (move the 100) requires only a scaling of the index up or down. If we want index 1 to have year 2 as the base year (construct index 2), we can use the equivalence between 110 and 100 and multiply index 1 by this scaling factor 100/110.

EXERCISE

Scale index 2 in such a way that the base year becomes year 5 (index 5).

In terms of the mathematics, the choice of a base year really does not matter but in practice, there are a number of important considerations. As the index gets larger the same percentage change is represented by a larger difference. A change from 100 to 120 is the same as a change from 300 to 360 but the impression can be very different. If, for example, our index were used as a measure of inflation, as with the Retail Price Index, we would not want the index to move very far from 100. When, as is usually the case, an index measures the change in an aggregate of heterogeneous items, we may need periodically to revise the items included in the index. Suppose a manufacturer constructed a productivity index using as a measure of productivity the times taken to make the most popular products. As new products appear and established products disappear, the manufacturer would need to reconsider the basis of the index. Also, we tend to choose a typical

Table 5.2 Indices that all measure the same change

Year	*Index 1*	*Index 2*	*Index 3*	*Index 4*	*Index 5*
1	100	90.91	83.33	74.07	66.67
2	110	100.00	91.67	81.48	73.33
3	120	109.09	100.00	88.89	80.00
4	135	122.73	112.50	100.00	90.00
5	150	136.36	125.00	111.11	100.00

month, quarter or year as our starting point. If we were to start an index of car sales in August, when new letter registrations are introduced, the index of 1 month later would give the misleading impression of car sales falling. Index numbers are constructed to show generally what is happening and may be adjusted to allow for predictable changes at certain times of the year. Unemployment figures, for example, are adjusted each July, to allow for the number of school leavers, and the corresponding index can also be adjusted to allow for this disturbance in the general trend (for seasonal adjustment refer to Chapter 20). In practice, we may be confronted with a change of base year as shown in Table 5.3.

Table 5.3 A change of base year

Year	*'Old' index*	*'New index*
3	120	
4	135	
5	150	100
6		115
7		125

We can use the equivalence of 150 in the 'old' index with 100 in the 'new' index at year 5. We either scale down the 'old' index using a multiplication factor of 100/150 as shown in Table 5.4 or scale up the 'new' index using 150/100 as shown in Table 5.5.

Table 5.4 Scaling down the 'old' index

Year		*'New' index*
3	120 × 100/150 =	80
4	135 × 100/150 =	90
5		100
6		115
7		125

A change in base year is often accompanied by a change in definition and the user must be aware of the effect this may have, particularly on the composition of the index (i.e. what items are included in it and the importance attached to them).

Table 5.5 Scaling up the 'new' index

Year		*'Old' index*
3		120
4		135
5		150
6	115 × 150/100 =	172.5
7	125 × 150/100 =	187.5

EXERCISE

A company has constructed an efficiency index to monitor the performance of its major production plant. After 3 years it was decided that a new index should be started owing to the major changes in the production process. The indices are given below:

Year	Existing index	New index
1	140	
2	155	
3	185	100
4		105
5		107

1. Calculate the percentage increase in efficiency from year 1 to year 5.
   (Answer: 41.4%.)
2. Construct another index using year 2 as the base year.
   (Answer: 90.3, 100.0, 119.4, 125.3, 127.7.)

Note that no allowance has been made for changes in the methods used to construct the indices.

5.2 THE CONSTRUCTION OF INDEX NUMBERS

Index numbers are perhaps best known for measuring the change in price or prices over time. For our examples we will use the information given in Table 5.6.

Table 5.6 The prices and consumption of tea, coffee and chocolate drinks by a representative individual in a typical week

	Year 0		*Year 1*		*Year 2*	
Drinks	*Price*	*Quantity*	*Price*	*Quantity*	*Price*	*Quantity*
Tea	8	15	12	12	16	10
Coffee	15	3	17	3	18	4
Chocolate	22	1	23	3	24	5

The price can be taken as the average amount paid in pence for a cup and the quantity as the average number of cups drunk per person per week.

5.2.1 The simple price index

If we want to construct an index for the price of one item only we first calculate that ratio of the 'new' price to the base year price, the **price relative**, and then multiply by 100. In terms of a notation

$$\frac{P_n}{P_0} \times 100$$

where P_0 is the base year price and P_n is the 'new' price.

A simple index for the price of tea, taking year 0 as the base year can be calculated as in Table 5.7. The doubling of the price of tea from 8p to 16p gives a 100% increase in the index; 100 to 200. The increase from 12p to 16p is 50 percentage points or $33\frac{1}{3}\%$.

In reality we are likely to drink more than just tea. In constructing, say, an index of beverage prices, we may wish to include coffee and chocolate drinks.

Table 5.7 A simple price index

Year	*Price*	P_n/P_0	*Simple price index*
0	8	1.0	100
1	12	1.5	150
2	16	2.0	200

EXERCISE

Calculate a simple price index for coffee with year 0 as the base year.
(Answer: 100, 113, 120.)

5.2.2 The simple aggregate price index

To include all items, we could sum the prices year by year and construct an index from this sum. If the sum of the prices in the base year is Σp_0 and the sum of the prices in year n is ΣP_n then the simple aggregate price index is

$$\frac{\Sigma P_n}{\Sigma P_0} \times 100$$

The calculations are shown in Table 5.8.

This particular index ignores the amounts consumed of tea, coffee and chocolate drinks. In particular, the construction of this index ignores both consumption patterns and the units to which price refers. If, for example, we were given the price of tea for a pot rather than a cup, the index would be different.

5.2.3 The average price relatives index

To overcome the problem of units, we could consider price ratios of individual commodities instead of their absolute prices and treat all price movements as equally important. In many cases,

Table 5.8 The simple aggregate price index

Drinks	P_0	P_1	P_2
Tea	8	12	16
Coffee	15	17	18
Chocolate	22	23	24
	$\Sigma P_0 = 45$	$\Sigma P_1 = 52$	$\Sigma P_2 = 58$

Year	$\Sigma P_n/\Sigma P_0$	*Simple aggregate price index*
0	45/45 = 1.00	100
1	52/45 = 1.15	115
2	58/45 = 1.29	129

the goods we wish to include will be measured in very different units. Breakfast cereal could be in price per packet, potatoes price per pound and milk price per pint bottle. As an alternative to the simple aggregate price index we can use the average price relatives index:

$$\frac{1}{k}\Sigma(P_n/P_0) \times 100$$

where k is the number of goods. Here the price relative, P_n/P_0, for a stated commodity will have the same value whatever the unit for which the price is quoted.

Table 5.9 The average price relatives index

Drinks	P_0	P_1	P_2	P_1/P_0	P_2/P_0
Tea	8	12	16	1.50	2.00
Coffe	15	17	18	1.13	1.20
Chocolate	22	23	24	1.05	1.09
				$\Sigma(P_1/P_0) = 3.68$	$\Sigma(P_2/P_0) = 4.29$

Year	$(1/k)\Sigma(P_n/P_0)$	*Average price relatives index*
0	1.00	100
1	(1/3)(3.68) = 1.23	123
2	(1/3)(4.29) = 1.43	143

In comparing Tables 5.9 and 5.8 we can see that the average price relatives index, in this case, shows larger increases than the simple aggregate price index. To explain this difference we could consider just one of the items, tea. The *value* of tea is low in comparison to other drinks so it has a smaller impact on the totals in Table 5.8. In contrast, the *changes* in the price of tea are larger than any of the other drinks and this makes a greater impact on the totals in Table 5.9. To construct a price index for all goods and sections of the community we need to take account of the quantities bought.

It is not just a matter of comparing what is spent year by year on drinks, food, transport or housing. If prices and quantities are *both* allowed to vary, an index for the amount spent could be constructed but not an index for prices. If we want a price index we need to control quantities. In practice, we consider a **typical basket** of goods in which the quantity of goods of each kind is fixed and find how the cost of that basket has changed over time. To construct an index for the price of beverages we need the quantity information for a selected year as given in Table 5.6.

5.2.4 The Laspeyre index

This index uses the quantities bought in the base year to define the typical basket. It is referred to as a **base-weighted index** and compares the cost of this basket of goods over time. This index is calculated as

$$\frac{\Sigma P_n Q_0}{\Sigma P_0 Q_0} \times 100$$

where $\Sigma P_0 Q_0$ is the cost of the base year basket of goods in the base year and $\Sigma P_n Q_0$ is the cost of the base year basket of goods in any year (thereafter) n.

It can be seen from Table 5.10 that we only require the quantities from the chosen base year (Q_0 in this case). The index implicitly assumes that whatever the price changes, the quantities purchased will remain the same. In terms of economic theory, no substitution is allowed to take place. Even if goods become relatively more ex-

Table 5.10 The Laspeyre index

Drinks	P_0	Q_0	P_1	P_2
Tea	8	15	12	16
Coffee	15	3	17	18
Chocolate	22	1	23	24

P_0Q_0	P_1Q_0	P_2Q_0
120	180	240
45	51	54
22	23	24
187	254	318

Year	$\Sigma P_nQ_0/\Sigma P_0Q_0$	*Laspeyre index*
0	1.00	100
1	254/187 = 1.36	136
2	318/187 = 1.70	170

Table 5.11 The Paasche index

Drinks	P_0	P_1	Q_1	P_2	Q_2
Tea	8	12	12	16	10
Coffee	15	17	3	18	4
Chocolate	22	23	3	24	5

P_1Q_1	P_0Q_1	P_2Q_2	P_0Q_2
144	96	160	80
51	45	72	60
69	66	120	110
264	207	352	250

Year	$\Sigma P_nQ_n/\Sigma P_0Q_n$	*Paasche index*
0	1.00	100
1	264/207 = 1.28	128
2	352/250 = 1.41	141

pensive it assumes that the same quantities are bought. As a result, this index tends to *overstate* inflation.

5.2.5 The Paasche index

This index uses the quantities bought in the current year for the typical basket. This **current year weighting** compares what a basket of goods bought now (in the current year) would cost with what the same basket of goods would have cost in the base year. This index is calculated as

$$\frac{\Sigma P_nQ_n}{\Sigma P_0Q_n} \times 100$$

where ΣP_nQ_n is the cost of the basket of goods bought in the year n at year n prices and ΣP_0Q_n is the cost of the year n basket of goods at base year prices.

As the basket of goods is allowed to change year by year, the Paasche index is not strictly a price index and as such has a number of disadvantages. Firstly, the effects of substitution would mean that greater importance is placed on goods that are relatively cheaper now than they were in the base year. As a consequence, the Paasche index tends to *understate* inflation.

Secondly, the comparison between years is difficult because the index reflects both changes in price and the basket of goods. Lastly, the index requires information on the current quantities and this may be difficult or expensive to obtain.

5.2.6 Other indices

The Laspeyre and Paasche methods of index construction can also be used to measure quantity movements with prices as the weights.

Laspeyre quantity index using base year prices as weights:

$$\frac{\Sigma P_0Q_n}{\Sigma P_0Q_0} \times 100.$$

Paasche quantity index using current year prices as weights:

$$\frac{\Sigma P_nQ_n}{\Sigma P_nQ_0} \times 100.$$

To measure the change in value the following 'value' index can be used:

$$\frac{\Sigma P_n Q_n}{\Sigma P_0 Q_0} \times 100.$$

5.2.7 Calculations using spreadsheets

Looking at the methodology used in the construction of index numbers we see that we are continually multiplying one column of numbers by another. This is exactly the sort of calculation which can be quickly and easily put into a spreadsheet. If the items within the index are put into the first column (A) and two columns are assigned to each year, one for price and one for quantity, then any of the summations that are required can be found.

Table 5.12 Extract from a spreadsheet

	A	B	C	D	E	F
1	Items	Year 0		Year 1		
2		Price	Quantity	Price	Quantity	
3	Tea	8	15	12	12	
4	Coffee	15	3	17	3	
5	Chocolate	22	1	23	3	
6						
7						
8						
9						

Looking at Table 5.12, we see a section of a typical spreadsheet. If we now multiply B3 by C3 and put the result in F3 we are beginning to find P_0Q_0. Doing this for all of the items, and summing the column gives the required total. For such a small example, it is probably quicker to perform the calculations by hand, but where there are a large number of items, the spreadsheet will have considerable advantages of speed and accuracy. It will also allow us to see the effect of changes to an individual price or quantity on the overall index number.

EXERCISE

Using the data given below, set up a spreadsheet to calculate the following all items index numbers:

1. Laspeyres price index;
2. Paasche price index;
3. Laspeyres quantity index;
4. Paasche quantity index; and
5. Value index

for years 1 and 2 with year 0 as the base year.

	Year 0		*Year 1*		*Year 2*	
Items	*Price*	*Quantity*	*Price*	*Quantity*	*Price*	*Quantity*
Bread	25	4	27	5	29	4
Potatoes	8	6	9	7	10	9
Carrots	21	2	22	2	22	3
Swede	35	1	35	1	40	1
Cabbage	45	1	46	2	55	1
Soup	18	5	23	4	25	5
Cake	53	2	62	2	78	1
Jam	40	1	45	2	56	1
Tea	23	3	25	3	29	4
Coffee	58	2	60	3	60	3

Answers:

		Price	Quantity	
Laspeyres	1	110.85	122.87	
	2	124.31	110.56	
Paasche	1	109.54	121.41	
	2	120.68	107.33	
Value	1			134.59
	2			133.43

Check the working of your spreadsheet by seeing the effects of changing the details on potatoes in Year 2 to Price = 15 and Quantity = 6 and for coffee, to Price = 70 and Quantity = 1.

5.3 THE WEIGHTING OF INDEX NUMBERS

Weights can be considered as a measure of importance (see section 3.3). The Laspeyre index and the Paasche index both refer to a typical basket of goods. The prices are weighted by the quantities in these baskets. In measuring a diverse range of items, it is often more convenient to use **amount spent** as a weight rather than a quantity. If we consider travel, for example, it could be more meaningful to define expenditure on public transport than the number of journeys. In the same way, we could enquire about the expenditure on meals bought and consumed outside the home rather than the number of meals and their price. Expenditure on public transport, meals outside the home and other items are **additive** since money units are homogeneous; the number of journeys, number of meals and number of shirts are not.

In constructing a base-weighted index we can use

$$\frac{\Sigma w P_n/P_0}{\Sigma w} \times 100$$

where P_n/P_0 are the price relatives (see section 5.2.1) and w are the weights. Each weight is the amount spent on the item in the base year.

Consider again our example from section 5.2 (Table 5.13).

Table 5.13 Weighted price index

Goods	P_0	Q_0	w	P_1	P_1/P_0	P_2	P_2/P_0
Tea	8	15	120	12	1.50	16	2.00
Coffee	15	3	45	17	1.13	18	1.20
Chocolate	22	1	22	23	1.05	24	1.09

Goods	w	$w \times P_1/P_0$	$w \times P_2/P_0$
Tea	120	180.00	240.00
Coffee	45	50.85	54.00
Chocolate	22	23.10	23.98
	187	253.95	317.98

Year	$(\Sigma w P_n/P_0)/\Sigma w$	*Base-weighted index*
0	1.00	100
1	253.95/187 = 1.36	136
2	317.98/187 = 1.70	170

It is no coincidence that this base-weighted index is identical to the Laspeyre index of Table 5.10. The identity is proven below:

$$\text{Laspeyre index} = \frac{\Sigma P_n Q_0}{\Sigma P_0 Q_0} \times 100.$$

Let the weights, w, equal the amount spent on items in the base year, e.g. $w = P_0 Q_0$:

$$\text{Laspeyre index} = \frac{\Sigma P_n Q_0}{\Sigma w} \times 100.$$

If we note that $Q_0 = w/P_0$ then

$$\text{Laspeyre index} = \frac{\Sigma w \times P_n/P_0}{\Sigma w} \times 100.$$

The weights only need to represent the relative order of magnitude and in practice are scaled to sum to 1000. If we were to multiply each of the weights in Table 5.13 by 1000/187, to obtain the sum of weights of 1000, the value of the index would not change. The items included in the Retail Price Index are assigned weights in this way.

5.4 THE GENERAL INDEX OF RETAIL PRICES

Of all the indices constructed to measure economic and social change in recent times, the Retail Price Index (RPI) is probably the most prominent. It can be seen as a measure of economic or political performance and provides an accepted measure for the cost of living. Increases in wage rates have been justified in terms of the increases in the index. Recent or anticipated changes often form the basis of a wage claim. The use of the RPI extends beyond wage bargaining. Savings and pensions, in many cases, have been **index-linked**; they increase in proportion to the index. All forms of economic planning take some account of inflation. Economists, for example, will distinguish between **real** and **nominal** disposable income, nominal being in terms of the currency (pounds in the pocket) and real referring to the purchasing power. If the index has doubled over a specified period, purchasing power will have halved if income remains the same.

The RPI includes a range of goods and services bought by the *typical* household. The index excludes those households with the highest incomes and those consisting of retired people mainly dependent on state pensions and benefits. Separate price indices are available for pensioners, one and two person households.

Coverage includes housing and travel but excludes saving, investment, life insurance, income tax and national insurance. The relative importance of the various categories is shown by weights in Table 5.14. The weights are revised annually using the most recent results from the Family Expenditure Survey' but excluding the top 4% of income earners and pensioner households. Each group is excluded to try to ensure that the spending patterns represented by the weights are typical of the majority of households. Separate index series and weighting structures are constructed for one-person and two-person pensioner households. The weights for 1990 are shown in Table 5.14.

The RPI has been rebased to start with January 1987 = 100 (as an exception to our usual algebraic rules we are able to write base year = 100). To obtain now a measure of inflation over recent years the indices will need to be scaled as shown in section 5.1.

As far as the RPI is concerned, using the weights given in Table 5.14, food accounts for 15.8% of expenditure and tobacco for 3.4%. For a one-person pensioner household, food is 32% of expenditure whilst tobacco is only 2.8%. In the case of a two-person pensioner household, the respective figures are 33% and 3.6%.

As an example of the effect of a price change in one category on the overall RPI, consider what will happen if the cost of fuel and light increases by 10%. From the table, this category represented 5% of expenditure (a weight of 50), so an increase of 10% will raise this to 5.5% (a weight of 55). The sum of the weights would now be 1005. The increase in the index is therefore:

$$\frac{1005-1000}{1000} \times 100 = 0.5\%$$

As fuel and light accounts for 5% of expenditure, a 10% increase will only add 0.5% to the overall index. In practice, the monthly change in the Retail Price Index reflects many changes in the prices of the basket of goods.

The typical basket of goods and services and their weightings shown in Table 5.14 will exactly match the expenditure patterns of only a small group of households. Some families will spend more on some items and less on others, but like all averages, the basic rises in prices can be reflected through such a typical basket.

Table 5.14 Weights for 1990

Groups	RPI	Pensioners 1 person	2 persons
Food	158	320	330
Catering	47	31	26
Alcoholic drink	77	28	38
Tobacco	34	28	36
Housing	185	0	0
Fuel and light	50	173	124
Household goods	71	90	89
Household services	40	82	53
Clothing and footwear	69	61	67
Personal goods and services	39	58	56
Motoring expenditure	131	22	90
Fares and travel costs	21	21	16
Leisure goods	48	49	49
Leisure services	30	37	26
Total	1000	1000	1000

5.5 CONCLUSIONS

Index numbers play an increasingly important rôle in describing the economy in which we live and the performance of businesses. Their great advantages are that percentage increases from the base year can be seen at a glance, and that the numbers are of a manageable, and understandable size. As we have seen, we are able to aggregate a wide range of heterogeneous items into a single index series, which will enhance our comprehension of an overall situation, for example, the level of inflation in an economy. In the presentation of accounting information allowance needs to be made for inflation. Historic cost accounting (with no allowance for inflation) works well only in periods of stable prices. In current cost accounting (CCA), adjustments are made in proportion to relevant indices. The Government Statistical Service publishes *Price Index Numbers for Cur-*

rent Cost Accounting specifically for this purpose.

Index numbers can be misleading if care is not taken. When an index is re-based it is important to note the last value of the previous series. When items are excluded, or new items included in an index, there may be drastic movements in the series which do not reflect major changes in prices or quantities, but merely the changed composition of the index.

5.6 PROBLEMS

1.

Year	*Index*
1	100
2	115
3	120
4	125
5	130
6	145

(a) Change the base year for this index to year 4.
(b) Find the percentage rises from year 3 to year 4 and year 5 to year 6.

2.

Year	*'Old' index*	*'New' index*
1	100	
2	120	
3	160	
4	190	100
5		130
6		140
7		150
8		165

(a) Scale down the 'old' index for years 1 to 3.
(b) Scale up the 'new' index for years 5 to 8.
(c) Explain the reasons for being cautious when merging indices of this kind.

3. The following information was recorded for a range of DIY items.

	No. of items bought			*Price per item (£)*		
Items	*Year 0*	*Year 1*	*Year 2*	*Year 0*	*Year 1*	*Year 2*
W	4	5	7	3	5	10
X	3	3	4	4	6	15
Y	3	2	2	4	7	19
Z	2	2	1	5	9	25

(a) Construct a simple price index for item Y using year 0 as the base year.
(b) Determine the simple aggregate price index using year 0 as the base year.
(c) Determine the price relatives index using year 0 as the base year.
(d) Calculate the Laspeyre price index using year 0 as the base year.
(e) Calculate the Paasche price index using year 0 as the base year.
(f) Calculate the Laspeyre quantity index using year 0 as the base year (note: here we wish to keep prices fixed, and thus the appropriate formula is

$$[\Sigma(P_0Q_n)/\Sigma(P_0Q_0)] \times 100).$$

(g) Calculate the Paasche quantity index using year 0 as the base year (note: here we are using the current year prices as the fixed weights, and thus the appropriate formula is

$$[\Sigma(P_nQ_n)/\Sigma(P_nQ_0)] \times 100).$$

(h) Why do the Laspeyre and Paasche indices give such different answers for year 2 in parts (d) to (g)?

4. The following data gives the wages paid to four groups of workers and the number of workers in each group.

	Year 0		*Year 1*		*Year 2*	
Groups	*Wage*	*No. of workers*	*Wage*	*No. of workers*	*Wage*	*No. of workers*
Managerial	300	40	330	50	390	70
Skilled	255	60	270	70	270	70
Semi-skilled	195	60	240	60	270	70
Labourer	90	100	150	100	240	80

(a) Construct a simple index of wages for skilled workers using year 0 as the base year.

(b) Calculate the simple aggregate index of wages using year 0 as the base year.
(c) Calculate the Laspeyre index of wages using year 0 as the base year.
(d) Calculate the Passche index of wages using year 0 as the base year.

5. The average prices of four commodities are given in the following table:

Commodity	*Average price per unit (£)* Year 0	*Year 1*	*Year 2*
A	101	105	109
B	103	106	107
C	79	93	108
D	83	89	86

The number of units used annually by a certain company is approximately 400, 200, 600 and 100 for the commodities *A*, *B*, *C* and *D* respectively. Calculate a weighted price index for years 1 and 2 using year 0 as the base year.

6. With reference to published government statistics, e.g. *Monthly Digest of Statistics* or *Economic Trends*:
   (a) Extract indices showing the change in prices and average wages since 1980. In particular note how these indices are defined and their consistency over time.
   (b) Draw comparisons between the changes in the two indices.

7. Use the *Monthly Digest of Statistics* to find the weights given to each group in the Retail Price Index for 1962, 1974 and the current year. Use these weights to comment on the relative importance of each group over this period.

8. Over the last fifteen years data has been recorded on prices and wages in a certain country. Index numbers for this data are shown in the table below.

Year	*Prices (old)*	*Prices (new)*	*Wages (old)*	*Wages (new)*
0	145.3		100.0	
1	148.5		103.2	
2	154.7		107.5	
3	163.6		112.4	
4	172.3		123.4	
5	194.1	100.0	136.3	
6		109.5	145.1	
7		114.9	146.3	
8		119.4	150.5	
9		125.6	158.1	100.0
10		130.6		105.4
11		136.6		111.9
12		144.2		121.1
13		150.3		130.0
14		158.9		141.3
15		171.3		154.6

As can be seen, both index series have been rebased during this time period.
   (a) Create a single index series for prices with year 0 as the base year.
   (b) Create a single index series for wages with year 0 as the base year.
   (c) Construct a graph showing these two index series.
   (d) Find the year-by-year percentage change in prices and the year-by-year percentage changes in wages and graph your results.
   (e) Comment on the implications of your results for the standard of living within the country. What reservations might you have about using this data to infer conclusions about the standard of living? What extra information would you require to make more positive statements about the standard of living?

9. Use a spreadsheet with the data overleaf to find the all items index numbers listed below for years 1 and 2 with year 0 as the base year.
   (a) Laspeyres price index
   (b) Laspeyres quantity index
   (c) Paasche price index
   (d) Paasche quantity index
   (e) a value index

Item			*Prices*		*Quantities*	
	Year 0	*Year 1*	*Year 2*	*Year 0*	*Year 1*	*Year 2*
A	3.00	3.25	3.40	198	237	287
B	2.30	2.45	2.55	300	307	296
C	6.10	6.10	6.50	800	755	789
D	4.20	4.33	4.44	200	290	300
E	5.70	5.89	5.99	351	427	389
F	12.50	12.60	12.89	107	110	104
G	0.56	0.76	0.79	1106	1473	1145
H	1.60	1.66	1.89	852	841	773
I	13.60	13.99	14.99	390	409	400
J	29.99	33.99	49.99	17	29	50

Further information now comes to light which changes various price and quantity data in the table. Use your spreadsheet to find the effect on the ten index numbers you have already constructed in each of the following:

(f) Year 2, quantities change to:
A = 207, B = 400, C = 1545;

(g) Year 2, prices change to:
A = 3.70, B = 2.75, D = 4.54,
F = 12.99, G = 0.60, J = 59.99;

(h) Year 2, quantities change to:
B = 196, C = 710, E = 289
F = 85, G = 70.

PART TWO CONCLUDING EXERCISE

Collect data on income distribution in the United Kingdom for the whole population and for men and women separately, obtaining data for 1970 and for the most recent year. Using this data obtain the following statistics:

1. the average income in 1970 for the whole population, for men and for women;
2. a measure of dispersion for each group in 1970;
3. the average income now for the whole population, for men and for women;
4. a measure of dispersion for each group now;
5. an index series (constructed with 1970 as the base year) of the average income of each group for each year between 1970 and the present day.

Write a brief report on the basis of your results and explain why there is still a difference in average earnings between men and women.

[We suggest that you use the data provided by the Inland Revenue and available in the *Annual Abstract of Statistics*.]

PART THREE
MATHEMATICAL METHODS

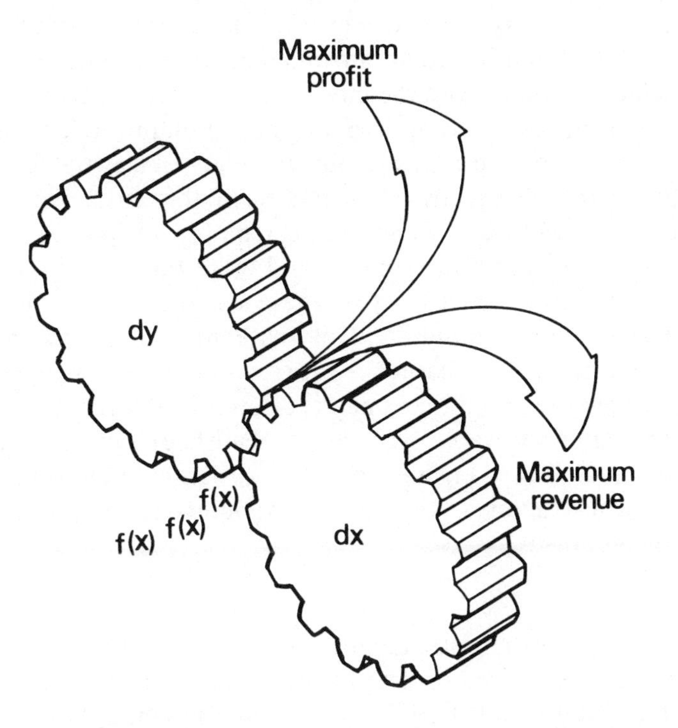

The charts and diagrams constructed and the statistics calculated in the previous chapters have been concerned with describing available data. However we can extend our understanding of business, economic and social situations by attempting to describe various factors and relationships mathematically. Mathematics should make our lives easier! Mathematically we can show how a company's profits are likely to change with increases in output, for example. Mathematics allows precise statements of how we see a business and allows the development and application of concepts like marginal cost and marginal revenue. Logical arguments can be developed effectively and presented mathematically e.g. if A = B and B = C then A = C.

Mathematics provides a powerful link between the real world of data (which needs interpretation) and the abstract world of thinking (that may need business application). Indeed, there are many business facts and theories that can only be presented with ease mathematically. A word of warning though: challenge the mathematics and statistics presented to you and ask the question 'What does this mean?'. The statistics may not lie but they may conceal the truth. A mathematical derivation may be correct but the underlying assumptions may not be realistic in business terms. When tackling a range of business problems it is often useful to formulate your ideas mathematically and then compare the outcomes with actual data collected on the problem. If there are major differences between your mathematical outcomes and collected data you can explore the reasons why. Typically we can ask questions about the correctness of our ideas, the assumptions we are making and the completeness of the data.

In this section of the book, we consider how the algebraic notation of x perhaps extended to x, y and z can improve our understanding of business. Any variable quantity, or **variable**, can be given the label x, whether it is travel distance to work, hourly wage rate or the level of output giving maximum profit. To understand more fully a particular situation we may attempt to relate one variable, perhaps travel time, to another variable say travel distance. In this case, we could label the variable of interest, travel time, y and the explanatory variable, travel distance, x. We can refer to y as being a **function** of x and write this as $y = f(x)$. If it takes 10 minutes to travel one mile, then $y = 10x$ where x is the number of miles travelled and y is the travel time in minutes. This mathematical description of travel is referred to as a **model**. In Chapter 9, for example, we will model (describe mathematically) how the value of money is related to interest rates.

The use of computer software, especially spreadsheets, has become particularly important in the modelling of business problems. Many of the instructions required have a mathematical structure. We may, for example need to interpret an expression of the following form:

(A2 * B5)/(D20 − D21).

The use of brackets and other mathematical signs remains the same. In this case we would need to multiply the values from cells A2 and B5 to obtain a first number, then subtract the value in cell D21 from the value in cell D20 to obtain a second number and finally divide the first number by the second number. Available computer software can plot the graphs you will see in the next chapter and take the work out of many of the calculations. As with all computing applications you will need to use selectively the available software and interpret carefully the output.

MATHEMATICAL RELATIONSHIPS

6

The use of mathematics provides a precise way of describing a range of situations. Whether we are looking at production possibilities or constructing an economic model of some kind, mathematics will effectively communicate our ideas and solutions. Reference to x or y should not take us back to the mysteries of school algebra but should give us a systematic way of modelling and solving some problems. Most things we measure will take a range of values, for example, people's height, and mathematically they are described by variables. A variable z, for example could measure temperature throughout the day, the working speed of a drilling machine or a share price. The important point here, is that the variable z is what we want it to be. The value that z actually takes will depend on a number of other factors. Temperature throughout the day will depend on location; the working speed of a drilling machine on the materials being used; and share price on economic forecasts.

Suppose you wish to calculate your weekly pay. If you work 40 hours each week and you are paid £10 per hour, then your weekly pay will be 40 × £10 = £400, which is a specific result. However, if the number of hours you work each week and the hourly rate of pay can both vary, you will need to go through the same calculation each week to determine your weekly pay. To describe this process, you can use mathematical notation. If the letter h represents the number of hours worked and another letter p represents the hourly rate of pay in pounds per hour, then the general expression hp describes the calculation of weekly pay. If $h = 33$ and $p = 11$ then weekly pay will equal £363. It should be noted at this point that the symbol for multiplication is only used when necessary; hp is used in preference to $h \times p$ although they mean exactly the same. To show the calculation of the £363 we would need to write 33 hours × £11 per hour.

6.1 INTRODUCTION TO ALGEBRA

The use of algebra provides one powerful way of dealing with a range of business problems. As we shall see, a single answer is rarely the result of one correct method, or likely to provide a complete solution to a business-related problem. Consider **breakeven analysis** as an example. Problems are often specified in such a way that a single value of output can be found where the total cost of production is equal to total revenue i.e. no profit is being made. This single figure may well be useful. A manager could be told, for example, that the division will need to breakeven within the next six months. This single figure will not, however, tell the manager the level of output necessary to achieve an acceptable level of profits by the end of the year. Even where a single figure is sufficient, it will still be the result of a number of simplifying assumptions. If, for example, we relax the assumption that price remains the same regardless of output the modelling of this business problem will become more complex.

Suppose a company is making a single product and selling it for £390. Whether the company is successful or not will depend on both the control of production costs and the ability to sell. If we recognize only two types of cost, fixed cost and variable cost, we are already making a simplifying assumption. Suppose also that fixed costs, those that do not vary directly with the level of output (e.g. rent, lighting, administration) equal £6 500 and variable costs, those costs that do vary directly with the level of output (e.g. direct labour, materials) equal £340 per unit.

The cost and revenue position of this company could be described or modelled using a spreadsheet format. Output levels could be shown in the first column (A) of the spreadsheet. The corresponding levels of revenue and costs could then be calculated in further columns (assuming that all output is sold). Finally a profit or loss figure could be determined as shown in Table 6.1.

Algebra provides an ideal method to describe the steps taken to construct this spreadsheet. If we use x to represent output and p to represent price then revenue r is equal to px. Finally, total cost c is made up of two components; a fixed cost, say a, and a variable cost which is the product of variable cost per unit, say b, and the level of output. To summarize:

$$r = px$$

and

$$c = a + bx.$$

Profit, usually written as π, (it would be confusing to use p for price and profit) is the difference between revenue and total cost:

$$\begin{aligned}\pi &= r - c\\ &= px - (a + bx).\end{aligned}$$

It should be noted here that the brackets signify that all the enclosed terms should be taken together. The expression for profit can be simplified by collecting the x terms together:

$$\pi = px - bx - a.$$

The minus sign before the bracket was an instruction to subtract each of the terms included within the bracket. As x is now common to the first two terms we can take x outside new brackets to obtain:

$$\pi = (p - b)x - a.$$

or using the numbers from the example above:

$$\begin{aligned}\pi &= (390 - 340)x - 6\,500\\ &= 50x - 6\,500.\end{aligned}$$

These few steps of algebra have two important consequences in this example. Firstly, we can see that profit has a fairly simple relationship to output, x. If we would like to know the profit corresponding to an output level of 200 units, we merely let $x = 200$ to find that profit is equal to $50 \times 200 - 6\,500$ or £3 500. Secondly, algebra allows us to develop new ideas or concepts. The

Table 6.1 Format of spreadsheet to determine breakeven point for single-product company

	A *Output*	*B* *Revenue*	*C* *Fixed cost*	*D* *Var cost*	*E* *Tot cost*	*F* *Profit/loss*
1	100	39 000	6 500	34 000	40 500	−1 500
2	110	42 900	6 500	37 400	43 900	−1 000
3	120	46 800	6 500	40 800	47 300	−500
4	130	50 700	6 500	44 200	50 700	0
5	140	54 600	6 500	47 600	54 100	500
6	150	58 500	6 500	51 000	57 500	1 000

difference shown between price and variable cost per unit, $(p - b)$, or £50 in this example, is known as **contribution to profit**. Each unit sold represents a gain of £50 for the company; whether a loss or a profit is being made depends on whether the fixed costs have been covered or not.

Consider now the breakeven position. If profit is equal to 0, then

$$0 = 50x - 6\,500.$$

To obtain an expression with x on one side and numbers on the other, we can subtract $50x$ from both sides:

$$-50x = -6\,500.$$

To determine the value of x, divide both sides by -50:

$$x = -6\,500/-50 = 130.$$

You need to remember here that if you divide a minus by a minus the answer is a plus.

The purpose of the above example is not to show the detail of breakeven analysis but rather the power of algebra. If we consider the above division, the following interpretation is possible: the breakeven level is the number of £50 gains needed to cover the fixed cost of £6 500. You are likely to meet the concept of contribution to profit in studies of both accountancy and marketing.

In reality, few companies deal with a single product. Suppose a company produces two products, x and y, selling for £390 and £365 and with variable cost per unit of £340 and £305 respectively. Fixed costs are £13 700. A spreadsheet model could be developed as shown in Table 6.2 (again assuming all output is sold).

The spreadsheet reveals that a two-product company can breakeven in more than one way. The notation used so far can be extended to cover companies producing two or more products. Using x and y subscripts on price and variable costs, for a two-product company we could write

$$r = p_x x + p_y y$$
$$c = a + b_x x + b_y y$$

Profit is the difference between revenue and total costs:

$$\pi = (p_x x + p_y y) - (a + b_x x + b_y y)$$
$$(p_x - b_x)x + (p_y - b_y)y - a.$$

It can be seen in this case that profit is made up of the contribution to profit from two products less fixed cost. Substituting the numbers from the example we have

$$\pi = (390 - 340)x + (365 - 305)y - 13\,700$$
$$= 50x + 60y - 13\,700.$$

You may recall that one equation with one variable or indeed two independent equations with two variables can be uniquely solved. As we have shown, one equation with two variables does not have a unique solution but rather a range of possible solutions. In a business context, a range of solutions may be preferable as these offer management a choice.

Table 6.2 Format of spreadsheet to determine breakeven points for a two-product company

A Output X	*B* Output Y	*C* Revenue	*D* Fixed cost	*E* Variable cost	*F* Total cost	*G* Profit/loss
120	110	86 950	13 700	74 350	88 050	−1 100
130	110	90 850	13 700	77 750	91 450	−600
140	110	94 750	13 700	81 150	94 850	−100
142	110	95 530	13 700	81 830	95 530	0
120	120	90 600	13 700	77 400	91 100	−500
130	120	94 500	13 700	80 800	94 500	0
140	120	98 400	13 700	84 200	97 900	500

EXAMPLE

For the two-product company described, what level of output would be required from x to achieve a profit of £1 000 if the level of output for y was fixed at 150?

Solution: Substituting known values into the profit equation:

$$1\,000 = (390 - 340)x + (365 - 305)\times 150 - 13\,700$$
$$1\,000 = 50x + 9\,000 - 13\,700$$
$$50x = 5\,700$$
$$x = 114$$

6.2 POWERS

When multiplying the same quantity together several times, it will be much more convenient to use powers rather than to write that quantity down each time, thus:

$$a \times a = a^2$$
$$a \times a \times a = a^3$$
$$a \times a \times a \times a = a^4, \text{ etc. (where } a \text{ is any number)}$$

Once we begin to manipulate quantities raised to powers, we can develop rules as follows:

$$a^3 \times a^2 = (a \times a \times a) \times (a \times a) = a^5 = a^{3+2}$$

i.e. for multiplication, we *add* the powers.

$$a^3/a^2 = \frac{a \times a \times a}{a \times a} = a^1 = a^{3-2}$$

i.e. for division, we *subtract* the powers. Note also that a on its own can be written as a^1.

Another example:

$$a^4/a^4 = \frac{a \times a \times a \times a}{a \times a \times a \times a} = a^{4-4} = a^0$$

Using the rule for division, we can see that any number (apart from 0) divided by itself will give a result to the power zero. In terms of the number system, any number raised to the power zero will be equal to 1.

If we have a^3/a^6, this is:

$$\frac{a \times a \times a}{a \times a \times a \times a \times a \times a} = \frac{1}{a \times a \times a} = a^{3-6} = a^{-3} = \frac{1}{a^3}$$

i.e. a negative power means that we take the *reciprocal* value.

Since we add powers when multiplying, then:

$$a^{1/2} \times a^{1/2} = a^1 = a$$

and the number which multiplied by itself gives a must be the square root of a, thus:

$a^{1/2} = \sqrt{a}$

$a^{1/3} = \sqrt[3]{a}$ the cube root of a

$a^{3/2} = \sqrt{a^3}$ the square root of a cubed, and so on.

If we are faced by brackets which are raised to powers, we treat them in the same ways; so that

$$x(a + b) = ax + bx$$
$$(a + b)^2 = (a + b)(a + b)$$
$$= a(a + b) + b(a + b)$$
$$= a^2 + ab + ba + b^2$$

and since $ab = ba$

$$= a^2 + 2ab + b^2$$
$$(a - b)^2 = a^2 - 2ab + b^2$$

(Note: $a^3 + a^2$ cannot be simplified by expressing it as a to a single power.)

6.3 ARITHMETIC AND GEOMETRIC PROGRESSIONS

In much of our work we are not just looking at a single number but rather a range of numbers. In developing a solution to a business problem, perhaps using a spreadsheet model, we are likely to generate a list of figures rather than a single figure. A **sequence** of numbers is just an ordered list, for example:

18, 23, 28, 33, 38, 43

and

9, 36, 144, 576, 2304, 9216

If the sequence can be produced using a particular rule or law, it is referred to as a **series** or a **progression**. It is particularly useful if we can recognize a pattern in a list. In the first example, 5 is being added each time and in the second example, the previous term is being multiplied by 4. As we shall see, the first sequence is an arithmetic progression and the second sequence is a geometric progression.

6.3.1 Arithmetic progression

A sequence is said to be an arithmetic progression if the difference between the terms remains the same e.g. +5. This difference is called the **common difference** and is usually denoted by d. If a is the first term, then the successive terms of an arithmetic progression are given by

$a, a+d, a+2d, a+3d, \ldots$

The nth term is given by

$t_n = a + (n-1)d$

EXAMPLE

Given the sequence

18, 23, 28, 33, 38, 43

find the value of the 12th term.
Solution: since $a = 18$ and $d = 5$,

$t_{12} = 18 + 11 \times 5 = 73$

The sum of an arithmetic progression with n terms is given by

$$S_n = \frac{n}{2}[2a + (n-1)d]$$

A proof of this formula is given in section 6.8 for the interested reader.

EXAMPLE

Given the same sequence

18, 23, 28, 33, 38, 43

what is the sum of this sequence:

1. with the 6 terms shown; and
2. continued to 12 terms?

Solution: By substitution:

1. $S_6 = 6/2\ (2 \times 18 + 5 \times 5) = 3(36 + 25) = 183$.
2. $S_{12} = 12/2\ (2 \times 18 + 11 \times 5) = 6(36 + 55) = 546$.

EXAMPLE

You have been offered a new contract. The starting salary is £12 000 per annum and there is an incremental increase of £800 at the end of each year. What will your salary be at the end of year 6 and what will your total earnings be over this period?

We could of course, develop a spreadsheet solution to this problem as shown in Table 6.3.

Table 6.3 Format of spreadsheet to show incremental salary increases

A	*B*
Year	*Salary*
1	12 000
2	12 800
3	13 600
4	14 400
5	15 200
6	16 000
Total	84 000

As an alternative, we could substitute values as necessary:

$t_6 = 12\,000 + 5 \times 800 = 16\,000$
$S_6 = 6/2\ (2 \times 12\,000 + 5 \times 800) = 84\,000$

6.3.2 Geometric progression

A sequence is said to be a geometric progression if the ratio between the terms remains the same e.g. 4. This constant ratio, r, is called the **common ratio**. The successive terms of a geometric progression are given by

$a, ar, ar^2, ar^3, ar^4, \ldots$

The nth term is given by

$t_n = ar^{n-1}$

EXAMPLE

Given the sequence

9, 36, 144, 576, 2304, 9216

find the value of the 12th term.
Solution: Since $a = 9$ and $r = 4$,

$$t_{12} = 9 \times (4)^{11} = 9 \times 4\,194\,304 = 37\,748\,736.$$

It should be noted at this point that geometric progressions have a tendency to grow very quickly when the value of r is greater than 1. This observation has major implications if human populations or the debtors of a company can be modelled using such a progression.

The sum of an geometric progression with n terms is given by

$$S_n = \frac{a(r^n - 1)}{r - 1}$$

A proof of this formula is given in section 6.9 for the interested reader.

EXAMPLE

Given the same sequence

9, 36, 144, 576, 2304, 9216

what is the sum of this sequence:

1. with the 6 terms shown; and
2. continued to 12 terms?

Solution: By substitution,

1. $S_6 = \dfrac{9(4^6 - 1)}{4 - 1} = 9 \times \dfrac{4095}{3} = 12\,285.$

2. $S_{12} = \dfrac{9(4^{12} - 1)}{4 - 1} = 9 \times \dfrac{16\,777\,215}{3} = 50\,331\,645.$

EXAMPLE

You have just been offered a different contract. The starting salary is still £12 000 per annum but the increase will be 6% of the previous year's salary rather than a fixed rate. What will your salary be at the end of year 6 and what will your total earnings be over this period?

We could again develop a spreadsheet solution as shown in Table 6.4.

Table 6.4 Format of spreadsheet to show percentage salary increases

A	*B*
Year	*Salary*
1	12 000.00
2	12 720.00
3	13 483.20
4	14 294.19
5	15 149.72
6	16 058.71
Total	83 703.82

Again, by the substitution of values, using $r = 1.06$ (to give a 6% increase per annum)

$$t_6 = 12\,000(1.06)^6 = 16\,058.71$$

and $S_6 = 12\,000 \dfrac{(1.06^6 - 1)}{1.06 - 1} = 83\,703.82$

6.4 GRAPHS

The organization of a consistent system of representing points on a graph was a major step for-

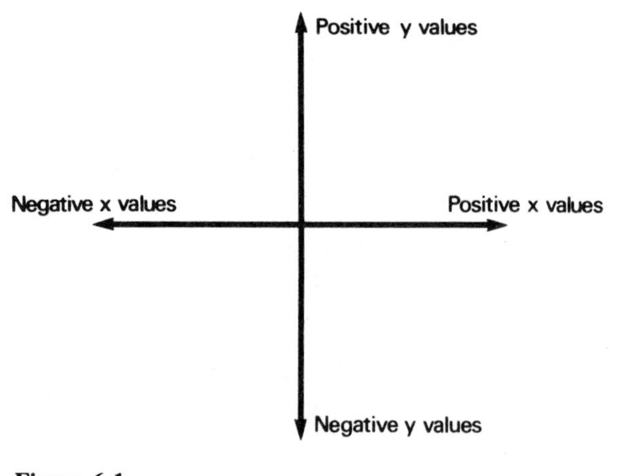

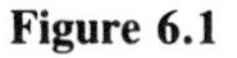
Figure 6.1

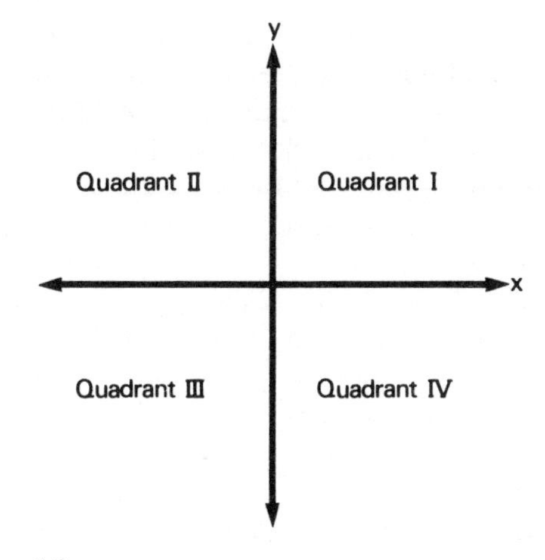

Figure 6.2

ward in the development of mathematics. Two straight lines placed at right angles to each other are referred to as axes (or Cartesian axes), and these divide the page into four sections or quadrants (Figure 6.1). It is usual to refer to the horizontal axis as the x-axis and the vertical axis as the y-axis.

Quadrants are often numbered as in Figure 6.2 and the majority of work in business-related areas will be in quadrant I, where we have positive values of both x and y (sometimes called the positive quadrant), with some work in quadrant IV. The reason for this is that the x-axis is often used to measure output or sales of a product and negative values would have no meaning.

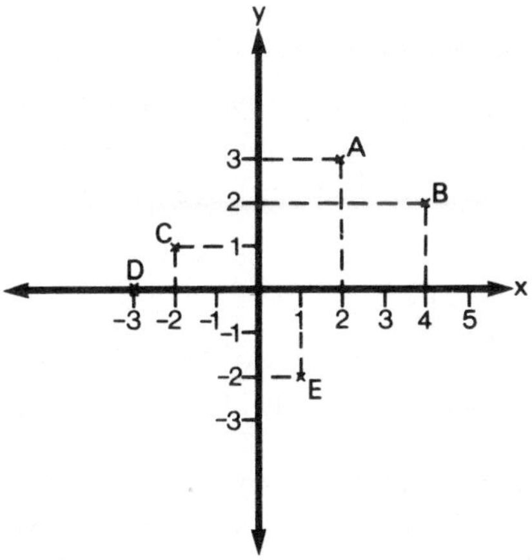

Figure 6.3

In contrast, profits typically plotted on the y axis, can be negative for the unfortunate company.

With these axes we may now identify individual points by two numbers (called **coordinates**), the first of which represents the position in relation to the x-axis and the second the position in relation to the y-axis; the coordinates are usually enclosed in brackets. Thus in Figure 6.3 the point labelled A has an x-value of 2 and a y-value of 3 and so would be quoted as (2, 3).

The point B is at $x = 4$, $y = 2$ and is written as (4, 2). For the other points, C is $(-2, 1)$, D is $(-3, 0)$ and E is $(1, -2)$. Note that for the point D which is on the x-axis, the y-value is 0; points on the y-axis would have a zero x-value and where the axes cross, usually called the **origin**, the coordinates will be (0, 0).

Graphs provide a powerful visual representation of how one variable y relates to another variable x. We can see, for example how in Figure 6.4 the values of y tend to fall as the values of x increase. We would expect the number of industrial accidents to fall as the time allocated to safety training increased.

Of particular importance is how we decide which variable to plot on which axis. Graphs imply a dependency of y on x. In a simple agricultural example, we would expect crop yield to increase with the increased application of fertilizer (at least in the short term). Crop yield would be plotted on the y axis and application of fertilizer on the x axis. Although the observed pattern of points cannot be assumed to show a 'cause and an effect' (see section 17.2) we do think in terms of x being the **independent** variable and y being the **dependent** variable. This terminology can be help-

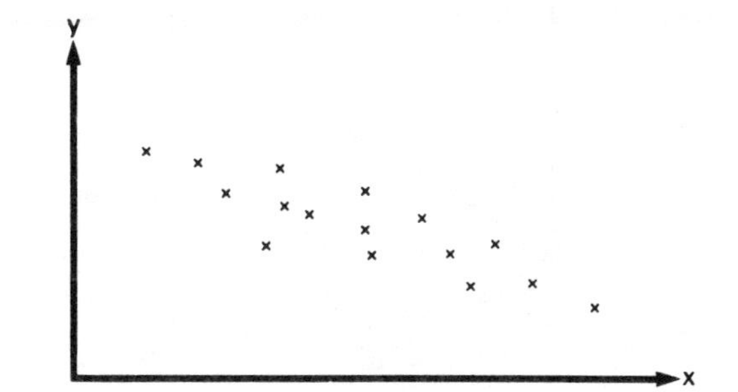

Figure 6.4

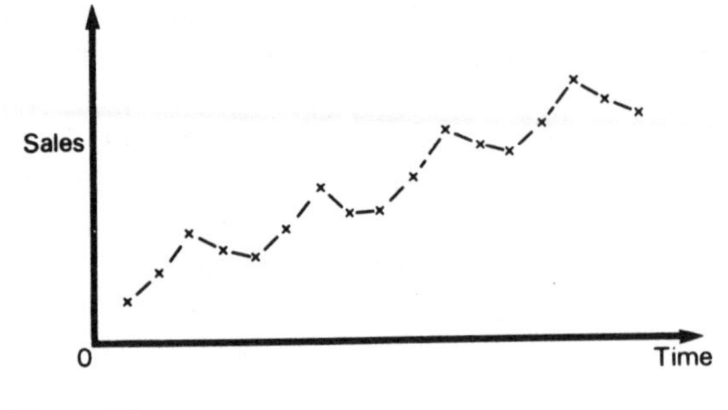

Figure 6.5

ful. The x variable is the one we can control, such as the application of fertilizer or expenditure on advertising, and the y variable measures the consequence, such as increased yield or increased sales. It must always be remembered that the graphs drawn, and certainly any inferences made, also have to make business sense. In the special case of values plotted against time, time series data, time is always shown on the x axis. The variation revealed in such data, see Figure 6.5 for example, is particularly useful in modelling economic changes over time (see Chapter 20).

However the importance of graphs in this chapter is not merely the illustration of collected data (see Chapter 2 and Part Six) but rather the representation of mathematical relationships. We have shown how a single–product and a two–product

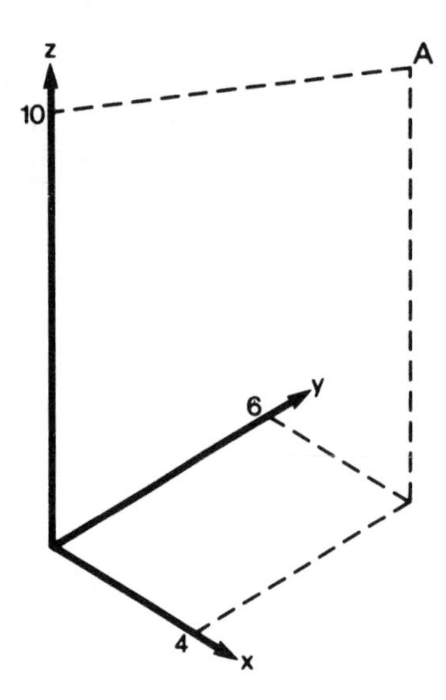

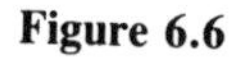

Figure 6.6

company can breakeven using both a spreadsheet format and equations. Graphs can be used very effectively to communicate how breakeven levels can be achieved and where a profit or loss is being made. Graphs are used extensively to explain a range of economic ideas including the theories of supply and demand.

Whilst the majority of the graphs used in this course will only have two axes, i.e. two-dimensional graphs, it is possible to extend the system to deal with three measurements, as in Figure 6.6. Here the point A would be represented by three numbers (4, 6, 10) or (x, y, z). Drawing such three dimensional graphs by hand can be quite difficult, but software available on many computer systems will allow such surfaces to be visualized quickly. The system may be further extended but visual representation is precluded for four or more dimensions.

6.5 FUNCTIONS

A function is the specification of the relationship between a variable y and a variable x. This means that if we are given a value of x, we will also be able to find the related value of y. We will only have a function if each value of x is related to a definite value of y.

An example of a simple function would be if workers were given a 10% pay rise, and the old wage level was x; the new wage level would be

$$\text{new wage level} = x + 0.1x = 1.1x$$

The new wage level y depends only on a single variable x. This is written as $y = f(x)$. If workers are given a further $\frac{1}{4}$% for each year of service, say z, then

$$y = 1.1x + 0.025z$$

The new wage level is now a function of two variables, x and z, and this can be written $y = f(x, z)$.

Below we examine several basic types of function which are used in economic modelling and which form the basis of some solution techniques in business (see Chapters 21 and 23).

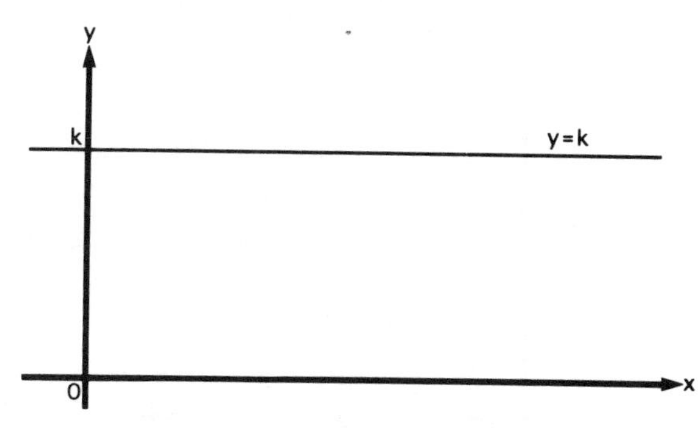

Figure 6.7

6.5.1 Constant functions

In this case, no matter what value x takes, the value of y remains the same, as shown in Figure 6.7. The line representing $y = k$ passes through the y-axis at a value k and goes off, at least in theory, to infinity in both directions. Constant functions of this type will appear in linear programming problems (as in Chapter 21).

If the values of x for which $y = k$ are limited to a particular group (known as the **domain** of the function) then we may use the following symbols:

$< x$ means less than x;
$> x$ means greater than x;
$\leqslant x$ means less than or equal to x;
and $\geqslant x$ means greater than or equal to x.

Now if the constant function only applies between $x = 0$ and $x = 10$, then this will be as illustrated in Figure 6.8 and is written as:

$y = k$ for $0 \leqslant x \leqslant 10$
$= 0$ elsewhere

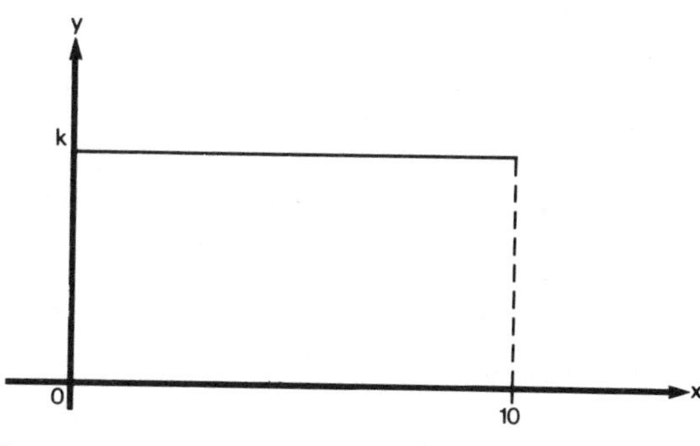

Figure 6.8

We could use this idea of the domain of a function to represent graphically a book price with discounts for quantity purchase, as in Figure 6.9.

6.5.2 Linear functions

A linear function is one that will give a straight line when we draw the graph (at least in two dimensions; it has a similar, but more complex meaning in three, four or more dimensions). This function occurs frequently when we try to apply quantitative techniques to business-related problems, and even where it does not apply exactly, it may well form a close enough approximation for the use to which we are putting it.

If the value of y is always equal to the value of x

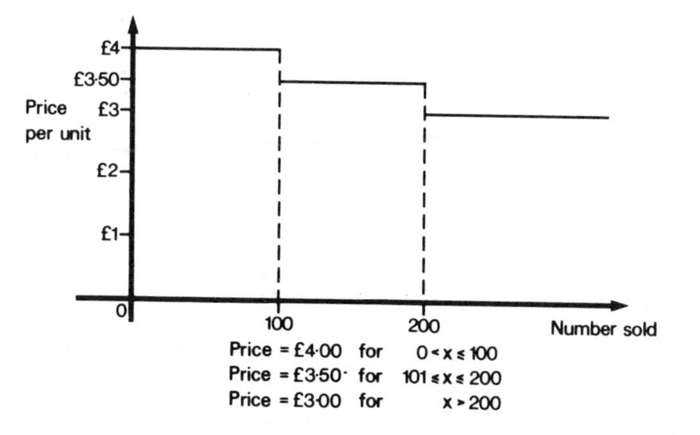

Figure 6.9

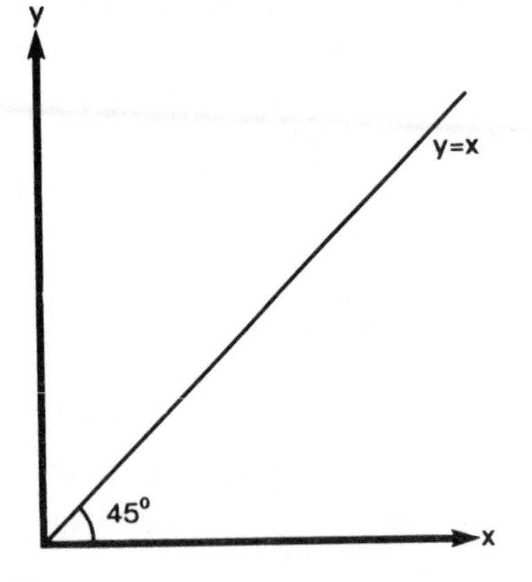

Figure 6.10

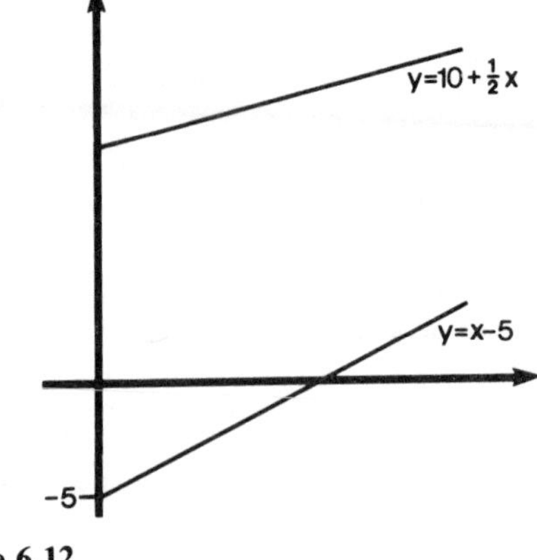

Figure 6.12

then we shall obtain a graph as shown in Figure 6.10. This line will pass through the origin and will be at an angle of 45° to the x-axis, provided that both axes use the same scale. This function was used when we considered Lorenz curves. We may change the angle or slope of the line by multiplying the value of x by some constant, called a **coefficient**. Two examples are given in Figure 6.11. These functions still go through the origin; to make a function go through some other point on the y-axis, we add or subtract a constant. The point where the function crosses the y-axis is called the **intercept**. Two examples are given in Figure 6.12.

The general format of a linear function is $y = a + bx$, where a and b are constants. To draw the graph of a linear function we need to know two points which satisfy the function. If we take the function

$$y = 100 - 2x$$

then we know that the intercept on the y-axis is 100 since this is the value of y if $x = 0$. If we substitute some other, convenient, value for x, we can find another point through which the graph of the function passes. Taking $x = 10$, we have:

$$y = 100 - 2 \times 10 = 100 - 20 = 80$$

so the two points are (0, 100) and (10, 80). Marking these points on a pair of axes, we may

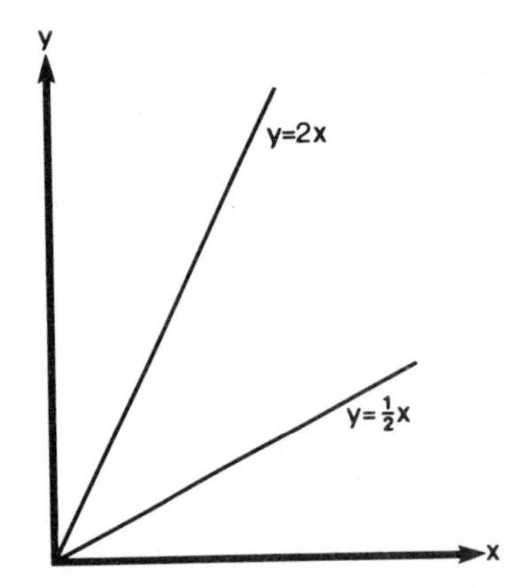

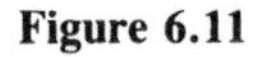
Figure 6.11

Figure 6.13

join them up with a ruler to obtain a graph of the function (see Figure 6.13).

Fixed costs, such as £6 500 and £13 700 in our breakeven examples, do not change with output level and would be represented by a horizontal straight line.

EXAMPLE

Given the following revenue and total cost functions

$r = 390x$
$c = 6\,500 + 340x$

show graphically how the company can breakeven.

When output is 0 ($x = 0$), revenue will equal 0 and total cost will equal £6 500. When output is 2 000 ($x = 200$), revenue will equal £78 000 and total cost will equal £74 500. The breakeven point, and output ranges of profit and loss are shown in Figure 6.14.

EXERCISE

Develop the spreadsheet shown in Table 6.1 and if a graphics facility is available, produce the graph shown in Figure 6.14.

EXAMPLE

Given the following profit function for a two-product company

$\pi = 50x + 60y - 13\,700$

show graphically the possible ways in which this company can breakeven.

To illustrate a technique which will prove useful later, instead of rearranging the formula in terms of y, we first let $x = 0$ and find the value of y and then let $y = 0$ and find the value of x. When $x = 0$, $y = 228$ (rounded down). When $y = 0$, $x = 274$. We can then plot a breakeven line, as in Figure 6.15, showing the various product combinations

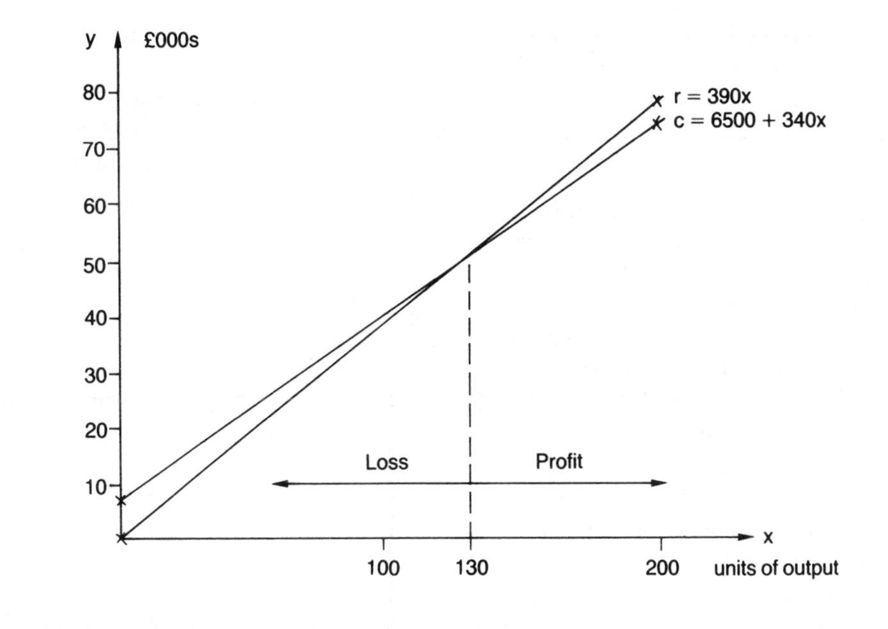

Figure 6.14 Breakeven level of profit

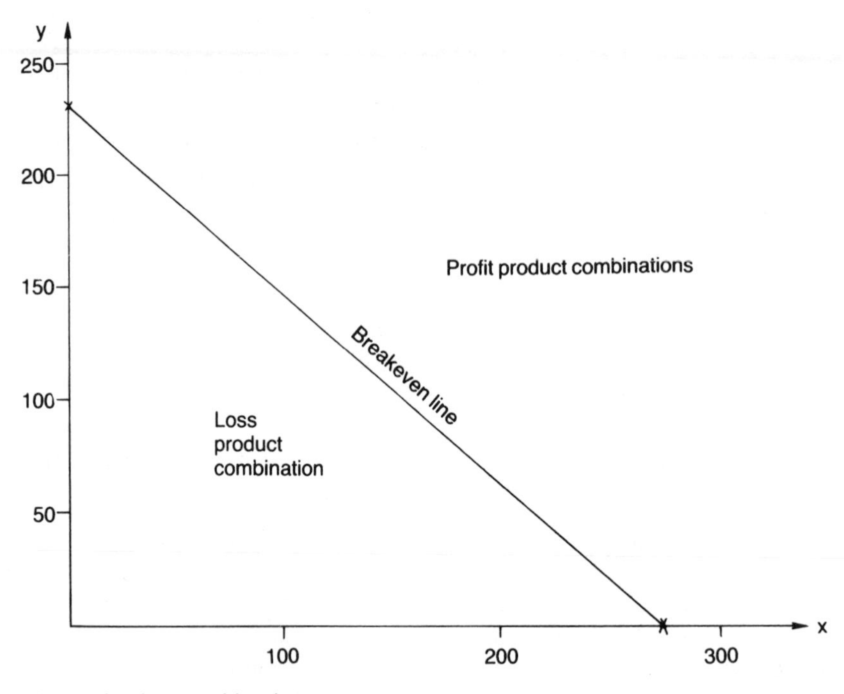

Figure 6.15 Profit production combinations

of output that allow the company to breakeven. On the same graph, we can also see how the company can make a loss or a profit.

We may use two linear functions to illustrate the market situation in economics, allowing one to represent the various quantities demanded over a range of prices, and the other to represent supply conditions. Where these two functions cross is known as the **equilibrium point** (point E in Figure 6.16), since at this price level the quantity demanded by the consumers is exactly equal to the amount that the suppliers are willing to produce, and thus the market is cleared. Note that we follow the tradition of economics texts and place quantity on the x-axis and price on the y-axis. Figure 6.16 might illustrate a situation in which the **demand function** is:

$$P = 100 - 4Q$$

and the **supply function** is:

$$P = 6Q$$

Since we know that the price (P_E) will be the same on both functions at the equilibrium point we can manipulate the functions to find the numerical values of P_E and Q_E:

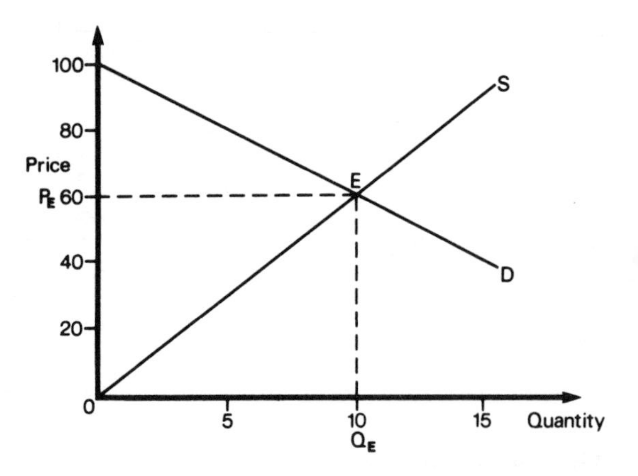

Figure 6.16

$$
\begin{aligned}
\text{(Demand)}\quad P &= P \quad \text{(Supply)}\\
100 - 4Q &= 6Q\\
100 &= 10Q\\
10 &= Q
\end{aligned}
$$

If $Q = 10$, then

$$P = 100 - 4Q = 100 - 40 = 60$$

thus $P_E = 60$ and $Q_E = 10$.

This system could also be used to solve pairs of linear functions which are both true at some point; these are known as **simultaneous equations** (a more general version of the demand and supply relationship above). Taking each equation in turn we may construct a graph of that function; where the two lines cross is the solution to the pair of simultaneous equations, i.e. the values of x and y for which they are both true.

If $\quad 5x + 2y = 34$
and $\quad x + 3y = 25$

then reading from the graph in Figure 6.17, we find that $x = 4$ and $y = 7$ is the point of intersection of the two linear functions.

This system will work well with simple equations, but even then is somewhat time-consuming: there is a simpler method for solving simultaneous equations, which does not involve the use of graphs.

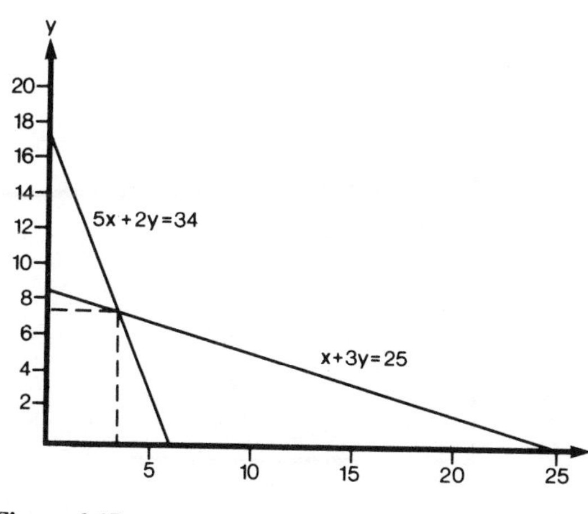

Figure 6.17

EXAMPLES

1.
$$
\begin{aligned}
5x + 2y &= 34\\
x + 3y &= 25
\end{aligned}
$$

If we multiply each item in the second equation by 5 we have

$$5x + 15y = 125$$

and since both equations are true at the same time we may subtract one equation from the other (here, the first from the new second)

$$
\begin{aligned}
5x + 15y &= 125\\
5x + \ \ 2y &= \ \ 34\\
\hline
13y &= \ \ 91
\end{aligned}
$$

Therefore, $\quad y = 7$

Having found the value of y, we can now substitute this into either of the original equations to find the value of x:

$$
\begin{aligned}
5x + 2(7) &= 34\\
5x + 14 &= 34\\
5x &= 20\\
x &= 4
\end{aligned}
$$

2.
$$
\begin{aligned}
6x + 5y &= 27\\
7x - 4y &= \ \ 2
\end{aligned}
$$

Multiply the first question by 4 and the second by 5

$$
\begin{aligned}
24x + 20y &= 108\\
35x - 20y &= \ \ 10\\
\hline
\end{aligned}
$$

Add together $\quad 59x = 118$
Therefore, $\quad x = 2$
substitute $\quad 6 \times 2 + 5y = 27$

$$
\begin{aligned}
12 + 5y &= 27\\
5y &= 15\\
y &= 3
\end{aligned}
$$

and thus the solution is $x = 2$, $y = 3$.

Note that we will return to the solution of simultaneous equations in Chapter 7, using matrix algebra.

Returning now to graphs of linear functions, we may use the method developed above to find the equation of a linear function that passes through two particular points.

EXAMPLES

1. If a linear function goes through the points $x = 2, y = 5$ and $x = 3, y = 7$ we may substitute these values into the general formula for a linear function, $y = a + bx$, to form a pair of simultaneous equations:

for (2, 5) $\quad 5 = a + 2b$
for (3, 7) $\quad 7 = a + 3b$

Subtracting the first equation from the second gives

$$2 = b$$

and substituting back into the first equation gives

$$5 = a + 2 \times 2$$
$$5 = a + 4$$

Therefore, $\quad a = 1$

Now substituting the values of a and b back into the general formula, gives

$$y = 1 + 2x$$

2. A linear function goes through (5, 40) and (25, 20),

thus
$$40 = a + 5b$$
$$20 = a + 25b$$
$$20 = \quad -20b$$

Therefore $\quad b = -1$
$$40 = a - 5$$
Therefore $\quad a = 45$
and thus $\quad y = 45 - x$

An alternative method for finding the equation is to label the points as (x_1, y_1) and (x_2, y_2) and then substitute into

$$\frac{y_1 - y}{y_2 - y_1} = \frac{x_1 - x}{x_2 - x_1}.$$

Taking the last example, we have:

$$\frac{40 - y}{20 - 40} = \frac{5 - x}{25 - 5}$$
$$\frac{40 - y}{-20} = \frac{5 - x}{20}$$

Multiplying both sides by 20 gives:

$$-(40 - y) = (5 - x)$$
$$-40 + y = 5 - x$$
$$45 - x = y$$

6.5.3 Quadratic functions

A quadratic function has the general equation

$$y = ax^2 + bx + c$$

and once the values of a, b and c are given we have a specific function. This function will produce a curve with one bend, or change of direction. (It is usually said to have one **turning point**.) If the value assigned to the coefficient of x^2, a, is negative, then the shape in Figure 6.18 will be produced, while if a is positive, the shape in Figure 6.19 will result.

To construct the graph of a function it is necessary to produce a table of values, plot the pairs of values, and join them together.

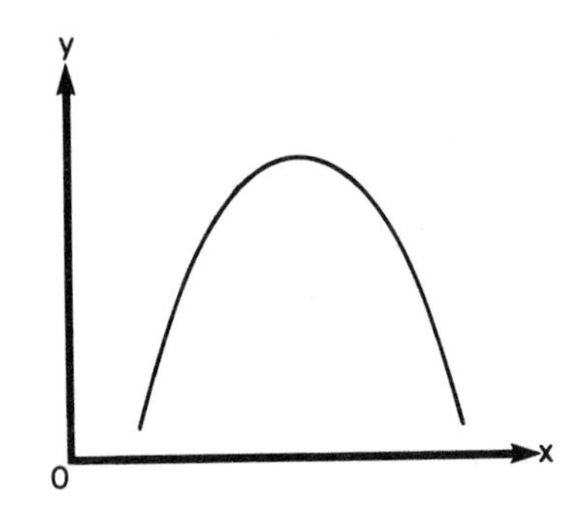

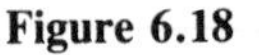
Figure 6.18

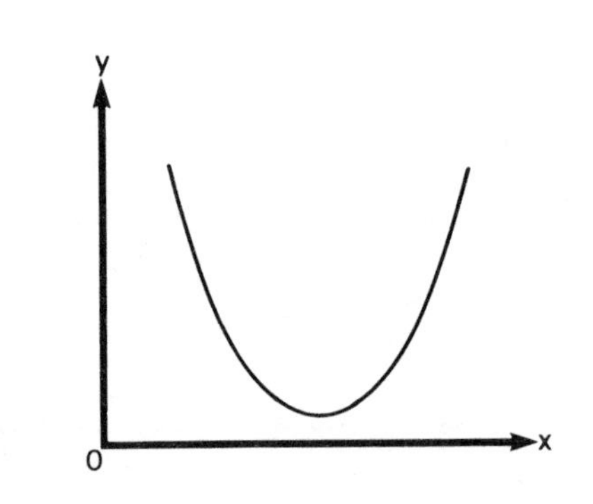

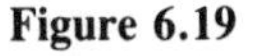
Figure 6.19

EXAMPLE

Consider the function:

$y = x^2 - 2x + 1$

Here we decide on a range of x values and write out each part of the function across the top of the table. We then work out each column for each value of x. Finally we add across each row to give the y-value on the right-hand side.

x	x^2	$-2x$	$+1$	y
−2	4	+4	+1	9
−1	1	+2	+1	4
0	0	0	+1	1
1	1	−2	+1	0
2	4	−4	+1	1
3	9	−6	+1	4

It is not always obvious at this stage for which range of values of x we should calculate the equivalent values of y; but we usually want to show where the curve changes direction and you will notice from the table above that the change in y decreases as we approach the change in direction and then increases again, so that if the changes in y are increasing as x increases, you should try lower values of x (Fig. 6.20).

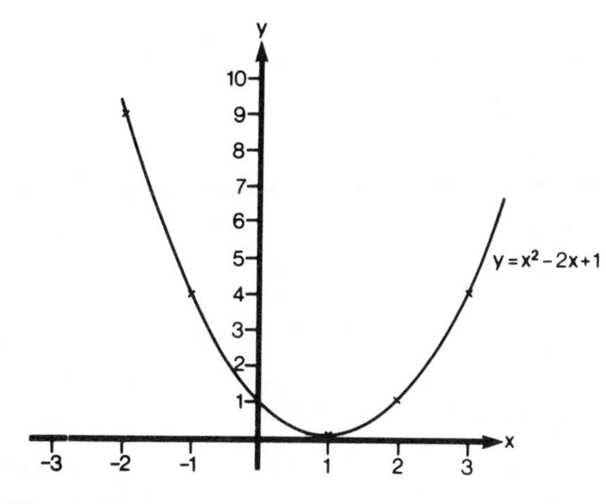

Figure 6.20

EXERCISE

Given the function $y = x^2 - 2x + 1$

1. reproduce the table of x and y values and, if possible graph y against x using an appropriate spreadsheet package.
2. produce the graph shown as Figure 6.20 using a specialist graph plotting package if such a package is available to you. What advantages do such packages offer you as a business user?

A further example where the x^2 coefficient is negative is given below.

EXAMPLE

$y = -x^2 - 2x + 1$ (Figure 6.21).

x	$-x^2$	$-2x$	$+1$	y
−3	−9	+6	+1	−2
−2	−4	+4	+1	1
−1	−1	+2	+1	2
0	0	0	+1	1
1	−1	−2	+1	−2

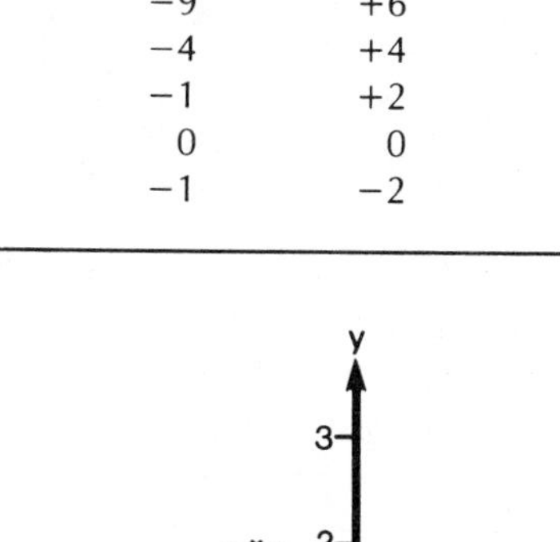

Figure 6.21

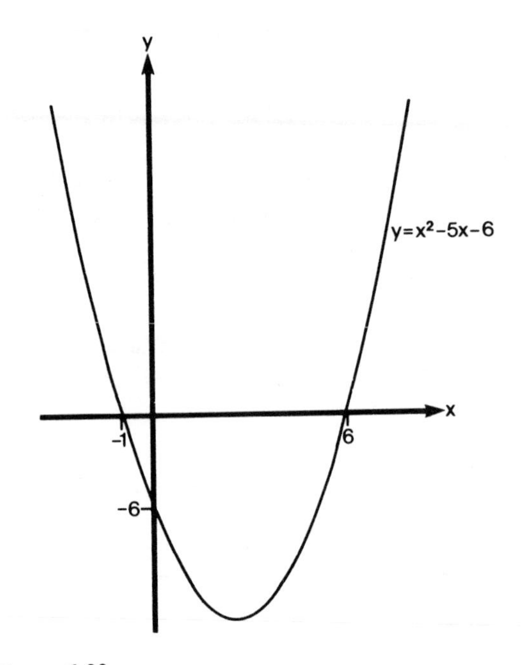

Figure 6.22

We are often interested in finding where a quadratic function crosses the x axis; the values of x when $y = 0$. These points are known as the **roots** of the quadratic equation e.g. the values of x that make $ax^2 + bx + c = 0$. These roots may be found in several ways.

We can graph the function and refer directly to the graph. It can be seen in Figure 6.22, for example, that the roots are -1 and 6. To check the correctness of this result, we can substitute these values back into the original equation and obtain a y value of 0.

A second method is to recognize that an equation such as $x^2 - 5x - 6 = 0$ can often be written as the product of two factors, say $(x + p)$ and $(x + q)$ where p and q are two real numbers:

$$x^2 - 5x - 6 = (x + p)(x + q) = 0$$

In this case it can be shown that $p = -6$ and $q = 1$. So

$$(x - 6)(x + 1) = 0$$

Now for this to be true, either the first factor equals zero or the second factor equals zero, hence

$$x - 6 = 0, \quad \text{i.e. } x = 6$$
$$\text{or} \quad x + 1 = 0, \quad \text{i.e. } x = -1$$

and so the roots are -1 and $+6$. The easiest functions to break down in this way are those with just x^2 as the first term. The two numbers in the factors when multiplied together must give the third term, or constant; and the numbers when multiplied by the other x and added (allowing for signs) must give the coefficient of x (here $-6x + 1x = -5x$).

EXAMPLE

If

$$x - 5x + 4 = 0$$

then

$$(x - 4)(x - 1) = 0$$

and roots are $+1$ and $+4$.

The third method of finding roots is to use a formula. If $ax^2 + bx + c = 0$ then the roots are at

$$x = \frac{-b \pm \sqrt{(b^2 - 4ac)}}{2a}$$

and we substitute the appropriate values of a, b and c to find these roots. A proof of this formula is given in section 6.10 for the interested reader.

EXAMPLE

$$x^2 - 5x + 4 = 0$$

then $a = 1$; $b = -5$; $c = +4$ so the roots are at

$$x = \frac{-(-5) \pm \sqrt{[(-5)^2 - 4(1)(4)]}}{2(1)}$$

$$= \frac{5 \pm \sqrt{[25 - 16]}}{2}$$

$$= \frac{5 \pm \sqrt{9}}{2}$$

$$= \frac{5 + 3}{2} \text{ or } \frac{5 - 2}{2}$$

$$= \frac{8}{2} \text{ or } \frac{2}{2}$$
$$= 4 \text{ or } 1.$$

This method will always give the roots, but beware of *negative* values for the expression under the square root sign ($b^2 - 4ac$). If this is negative then the function is said to have **imaginary roots**, since in normal circumstances we cannot take the square root of a negative number. At this level, these imaginary roots need not concern us.

EXAMPLE

$2x^2 - 4x - 10 = 0$

then $a = 2$, $b = -4$, $c = -10$ so the roots are at

$$x = \frac{4 \pm \sqrt{(16 + 80)}}{4}$$
$$= \frac{4 \pm \sqrt{96}}{4}$$
$$= \frac{4 \pm 9.8}{4}$$
$$= \frac{13.8}{4} \text{ or } \frac{-5.8}{4}$$
$$= 3.45 \text{ or } -1.45.$$

Quadratic functions are often used to represent cost equations, such as marginal cost or average cost, and sometimes profit functions. When profit is represented by a quadratic function, then we can use the idea of roots either to find the range of output for which any profit is made, or we can specify a profit level and find the range of output for which at least this profit is made.

EXAMPLE

If profit $= -x^2 + 8x + 1$ where x represents output, then if the specified profit level is 8, we have:

$$-x^2 + 8x + 1 = 8$$
$$-x^2 + 8x - 7 = 0$$
$$(x - 7)(1 - x) = 0$$
$$x = 7 \text{ or } 1$$

This is illustrated in Figure 6.23.

We have seen how two pairs of points are required to determine the equation of a straight line. To find a quadratic function, three pairs of points are required. One method of doing this is shown below.

EXAMPLE

If a quadratic function goes through the points

$x = 1, \quad y = 7$
$x = 4, \quad y = 4$
$x = 5, \quad y = 7$

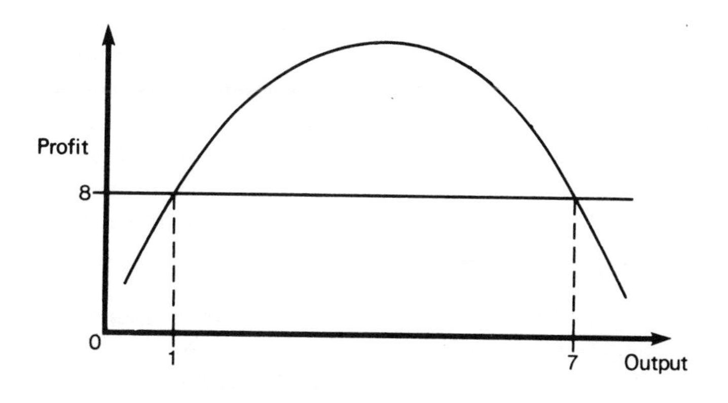

Figure 6.23

then we may take the general equation $y = ax^2 + bx + c$ and substitute:

$$(1,7) \quad 7 = a(1)^2 + b(1) + c = a + b + c \qquad 6.1$$
$$(4,4) \quad 4 = a(4)^2 + b(4) + c = 16a + 4b + c \qquad 6.2$$
$$(5,7) \quad 7 = a(5)^2 + b(5) + c = 25a + 5b + c \qquad 6.3$$

Rearranging the first equation 6.1 gives

$$c = 7 - a - b$$

and substituting for c into equation 6.2 gives

$$\begin{aligned} 4 &= 16a + 4b + (7 - a - b) \\ &= 15a + 3b + 7 \\ -3 &= 15a + 3b. \end{aligned} \qquad 6.4$$

Substituting again for c but into equation 6.3 gives

$$\begin{aligned} 7 &= 25a + 5b + (7 - a - b) \\ &= 24a + 4b + 7 \\ 0 &= 24a + 4b \end{aligned} \qquad 6.5$$

We now have two simultaneous equations (6.4 and 6.5) in two unknowns (a and b).
Multiplying equation 6.4 by 4 and equation 6.5 by 3 to create equal coefficients of b, gives:

$$\begin{aligned} -12 &= 60a + 12b \\ 0 &= 72a + 12b \end{aligned}$$

which, if one is subtracted from the other gives

$$12 = 12a.$$

Therefore, $a = 1$
From equation 6.5, we have:

$$0 = 24 + 4b$$

Therefore,

$$b = -6$$

and from equation 6.1

$$\begin{aligned} c &= 7 - a - b \\ &= 7 - 1 + 6 = 12 \end{aligned}$$

and thus the quadratic function is:

$$y = x^2 - 6x + 12.$$

6.5.4 Cubics and polynomials

A cubic function has the general equation:

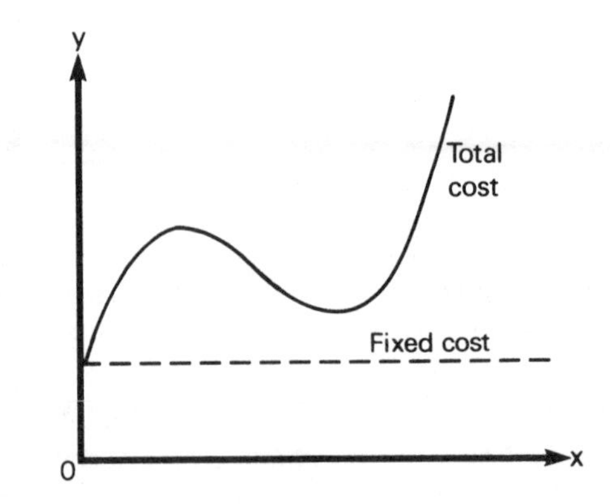

Figure 6.24

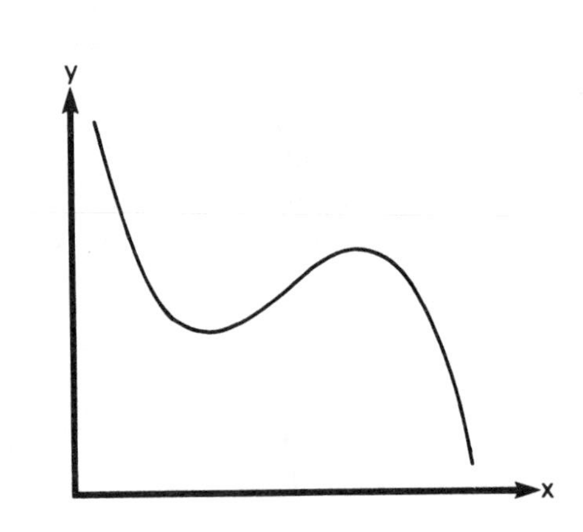

Figure 6.25

$$y = ax^3 + bx^2 + cx + d$$

and will have two turning points when $b^2 > 4ac$. They are often used to represent total cost functions in economics (Figure 6.24 and 6.25). We could go on extending the range of functions by adding a term in x^4, and then one in x^5 and so on. The general name for functions of this type is **polynomials**. For most business and economic purposes we do not need to go beyond cubic functions, but for some areas, the idea that a sufficiently complex polynomial will model any situation will appear.

6.5.5 Exponential functions

Within mathematics, those values that arise in a wide variety of situations and a broad spectrum of

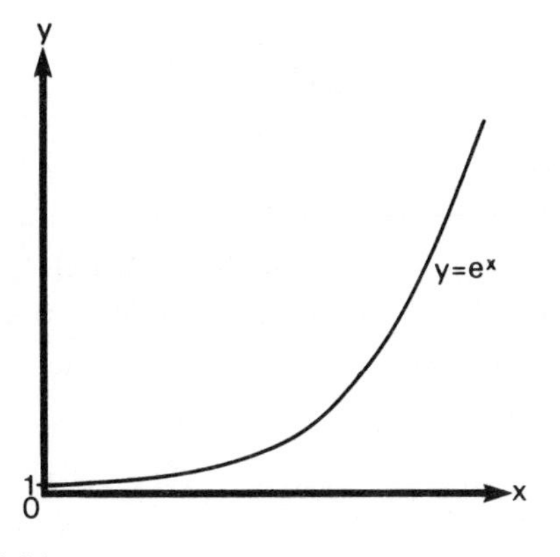

Figure 6.26

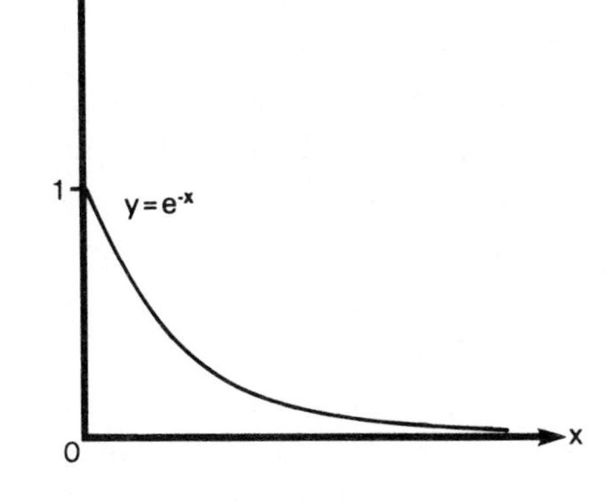

Figure 6.27

applications (and often run to a vast number of decimal places) tend to be given a special letter. One example most people will have met is π (pi). Exponential functions make use of another such number, e (Euler's constant) which is a little over 2.7. Raising this number to a power gives the graph in Figure 6.26, or, if the power is negative, the result is as Figure 6.27.

The former often appears in growth situations, and will be of use in considering money and interest (see Chapter 9). The latter may be incorporated into models of failure rates, market sizes and many probability situations (see Chapter 11).

6.6 CONCLUSIONS

Mathematics provides a useful range of tools for describing, analysing and solving business problems. On some occasions we may need to balance a few numbers and on others, understand the theoretical development of a discipline, like economics, using equation systems. The tools of mathematics support a range of problem solving activities but first we must learn to use the tools.

6.7 PROBLEMS

For each of the following, find x in terms of y:

1. $4x + 3y = 2x + 21y$
2. $2x - 4y = 3(x + y) + 5y$
3. $x + y + 3x = 4y + 2x - 7y - 2x$
4. $3x + 5(x - y) = 2(2x + 3y) + y$
5. $x^2(x^2 + 4x + 3) - 3y(4y + 10) - x^4 - 4(x^3 - 10)$ $= 3x^2 - 2y(6y - 10)$

Simplify each of the following expressions:

6. $a^2 \times a^3$
7. $a \times a^2/a^3$
8. $(a + b)a^2/b$
9. $(a^{1/2} \times a^2 \times a^{3/2})/a^2$
10. $a^2(a^5 + a^8 - a^{10})/a^4$
11. $(1/2)^3$
12. $6x(2 + 3x) - 2(9x^2 - 5x) - 484 = 0$

Expand the brackets and simplify the following expressions:

13. $a(2a + b)$
14. $(a + b)^2 - ab$
15. $(a + 2b)^2$
16. $(a + b)(a - b)$
17. $(a + b)^3 - (a - b)^3$
18. $(3x + 6y)(4x - 2y)$

For each of the following arithmetic progressions, find (a) the 10th term and (b) the sum given that each progression has 12 terms:

19. 5, 8, 11, 14, 17, . . .
20. 112, 126, 140, 154, . . .
21. 57, 49, 41, 33, 25, . . .

22. If the third term and the sixth term of an arithmetic progression are 11 and 23 respectively, determine the ninth term.

For each of the following geometric progressions, find (a) the 7th term and (b) the sum given that the progression has 10 terms:

23. 4, 12, 36, 108, . . .
24. 6, −12, 24, −48, . . .
25. 4, 3.2, 2.56, 2.048, . . .

26. If the third term and the fifth term of a geometric progression are 225 and 5625 respectively, determine the sixth term.

Construct a graph and mark the following points upon it:

27. (a) (1, 1)
 (b) (2, 3)
 (c) (3, 2)
 (d) (−1, 4)
 (e) (2, −3)
 (f) (−2, −1)

28. Shade the area on the graph where x is above 0 but below 4 and y is above 0 but below 2.
29. Construct graphs of the following functions:

 (a) $y = 0.5x \quad 0 < x < 20$
 (b) $y = 2 + x \quad 0 < x < 15$
 (c) $y = 25 - 2x \quad 0 < x < 15$
 (d) $y = 2x + 4 \quad 3 < x < 6$
 (e) $y = 3$
 (f) $x = 4 - 0.5y \quad 0 < x < 5$
 (g) $y = x^2 - 5x + 6 \quad -2 < x < 6$
 (h) $y = x^3 - 2x^2 + x - 2 \quad -2 < x < 6$

30. Solve the following simultaneous equations:

 (a) $4x + 2y = 11$
 $3x + 4y = 9$
 (b) $2x - 3y = 20$
 $x + 5y = 23$
 (c) $6x + 2y = 20$
 $2x - 3y = 14$
 (d) $4x + 5y = 22.75$
 $7x + 6y = 31.15$

31. Find the roots of the following functions:

 (a) $x^2 - 3x + 2 = 0$
 (b) $x^2 - 2x - 8 = 0$
 (c) $x^2 + 13x + 12 = 0$
 (d) $2x^2 - 6x - 40 = 0$
 (e) $3x^2 - 15x + 8 = 0$
 (f) $x^3 - 6x^2 - x + 6 = 0$ (Hint, use a graph)
 (g) $-2x^2 + 5x + 40 = 0$

32. A demand function is known to be linear $[P = f(Q)]$, and to pass through the following points:

 $Q = 10,\ P = 50;\ Q = 20,\ P = 30$

 Find the equation of the demand function.
33. A supply function is known to be linear $[P = f(Q)]$, and to pass through the following points:

 $Q = 15,\ P = 10;\ Q = 40,\ P = 35$

 Find the equation of the supply function.
34. Use your answers to questions 32 and 33 to find the point of equilibrium for a market having the respective demand and supply functions.
35. A demand function is known to be linear $[P = f(Q)]$ and to pass through the following points:

 $Q = 5,\ P = 25;\ Q = 9,\ P = 7$

 An associated supply function is also linear $[P = f(Q)]$ and passes through:

 $Q = 5,\ P = 10,\ Q = 25,\ P = 12$

 Find the equations of each function, and hence find the point of equilibrium.

 If the demand function now shifts to the right, to pass through the points:

 $Q = 10,\ P = 25;\ Q = 14,\ P = 7$

 find the new equilibrium position.
36. A market has been analyzed and the following points estimated on the demand and supply curves:

Output demand	*Price*
1	1902
5	1550
10	1200

Output supplied	*Price*
1	9
5	145
10	540

(a) Determine the equations of the demand and supply functions, assuming that in both cases price $= f(\text{output})$, and that the function is quadratic. (NB. Neither function applies above an output of 20.)
(b) Determine the equilibrium price and quantity in this market.

6.8 PROOF OF AP

Proof that

$$S_n = \frac{n}{2}\left[2a + (n-1)d\right].$$

The sum of an arithmetic progression with n terms may be written

$$S_n = a + (a+d) + (a+2d) + \ldots + [a + (n-1)d]$$

This expression may also be written as

$$S_n = [a + (n-1)d] + [a + (n-1)d - d] + [a + (n-1)d - 2d] + \ldots + (a+d) + a$$

by reversing the order of the terms on the right-hand side of the equation.

Adding the two equations gives

$$\begin{aligned} 2S_n &= [2a + (n-1)d] + [a + d + a + (n-1)d - d] + \ldots + [a + (n-1)d + a] \\ &= [2a + (n-1)d] + [2a + (n-1)d] + \ldots \end{aligned}$$

but as there are n terms

$$2S_n = n[2a + (n-1)d]$$

and thus,

$$S_n = \frac{n}{2}\left[2a + (n-1)d\right]$$

6.9 PROOF OF GP

Proof that

$$S_n = \frac{a(r^n - 1)}{r-1}.$$

The sum of a geometric progression with n terms may be written:

$$S_n = a + ar + ar^2 + ar^3 + \ldots + ar^{n-1}.$$

Multiplying both sides by r gives

$$rS_n = ar + ar^2 + ar^3 + ar^4 + \ldots + ar^n.$$

Subtracting the first equation from the second gives

$$rS_n - S_n = ar^n - a$$
$$S_n(r-1) = a(r^n - 1)$$

So,

$$S_n = \frac{a(r^n - 1)}{r-1}$$

6.10 PROOF OF QUADRATIC FORMULA

Proof that

$$x = \frac{-b \pm \sqrt{b^2 - 4ac}}{2a}$$

The form of a quadratic equation is

$$ax^2 + bx + c = 0$$

Dividing by a and rearranging gives

$$x^2 + \frac{b}{a}x = -\frac{c}{a}$$

Adding $\frac{b^2}{4a^2}$ to both sides gives

$$x^2 + \frac{b}{a}x + \frac{b^2}{4a^2} = \frac{b^2}{4a^2} - \frac{c}{a}$$

This can be written

$$\left(x + \frac{b}{2a}\right)^2 = \frac{b^2 - 4ac}{4a^2}$$

Taking the square root of each side gives

$$x + \frac{b}{2a} = \pm\frac{\sqrt{b^2 - 4ac}}{2a}.$$

Hence

$$x = \frac{-b \pm \sqrt{b^2 - 4ac}}{2a}$$

MATRICES 7

The algebra used in Chapter 6 allows the specification and solution of a range of business problems and provides the framework for the theoretical development of disciplines such as economics. We have seen single equations solved in terms of x and simultaneous equations solved in terms of x and y. However, we must be able to deal effectively with more complex problems. Matrix notation provides a way of describing these more complex problems and the rules of matrices provide a way of manipulating these problems. It must be remembered that matrix algebra only provides a convenient notation and some new methods of solution. Matrices will not solve problems that are not amenable to solution by other methods. As with the application of all quantitative methods, problem specification is the important first step.

Matrices are particularly useful in solving sets of simultaneous equations. The compact form of notation is consistent with the use of arrays in computer programming, and many computer packages will offer the facility of matrix manipulation e.g. MINITAB. To illustrate the application of matrix algebra, an input/output model is developed in section 7.4 where it is shown how output from one industry is the input to another. Clearly, as the number of industries considered increases, the complexity of modelling increases. Matrices are also used extensively in multiple regression (Chapter 19), and the methods of matrix algebra in more advanced work on linear programming (Chapter 21). Further applications include probability modelling, such as Markov chains (Chapter 10).

7.1 WHAT IS A MATRIX?

A marix is a **rectangular array** of numbers arranged in **rows** and **columns** and is characterized by its size (or **order**), written as (no. of rows × no. of columns). The whole matrix is usually referred to by a capital letter, whilst individual numbers, or **elements**, within the matrix are referred to by lower case letters, usually with a suffix to identify in which row and in which column they appear. Note that a matrix does not have a numerical value, it is merely a convenient way of representing an array of numbers.

$$\text{If}\quad \mathbf{A} = \begin{bmatrix} 4 & 8 & 17 & 12 \\ 21 & 3 & 19 & 17 \\ 10 & 21 & 4 & 2 \end{bmatrix}$$

then the order of matrix **A** is (3×4) and the element a_{13} is the 17 since it is in the first row and the third column.

It is often convenient to use the double subscript notation where a_{ij} denotes the element located in the ith row and jth column of matrix **A** and b_{ij} denotes the element located in the ith row and jth column of matrix **B**.

If a matrix has only one row, then it is known as a **row vector**: if it only has one column, then it is a **column vector**. e.g.

$$\mathbf{B} = [4\ 8\ 7] \quad \mathbf{C} = \begin{bmatrix} 10 \\ 12 \\ 28 \\ 49 \\ 102 \end{bmatrix}$$

The order of matrix **B** is (1×3) and the order of matrix **C** is (5×1).

EXAMPLE

A company, Comfy Chairs Ltd, produces three types of chair, the Classic, the Victorian and the Modern in its existing factory. Weekly output of the Classic, the Victorian and the Modern types in mahogany are 30, 60 and 80 respectively, and in teak, 20, 30 and 40 respectively.

This information can easily be represented by a matrix, say **D**, where columns refer to chair type and the rows to the wood used.

$$\mathbf{D} = \begin{bmatrix} 30 & 60 & 80 \\ 20 & 30 & 40 \end{bmatrix}$$

7.2 MATRIX MANIPULATION

We will find that many sets of data and many situations can be described in terms of rows and columns. In this section we consider the rules governing matrix algebra that will allow us to manipulate these rows and columns.

7.2.1 Equality of matrices

The concept of equality is fundamental to all algebra. In matrix algebra, two matrices **A** and **B**, are equal only if they are of the same size *and* their corresponding elements are equal, $a_{ij} = b_{ij}$ for *all* values of i and j.

EXAMPLE

Matrix **E** is only equal to matrix **D** if matrix **E** is of the same size, (2 × 3), and each element of matrix **E** is equal to each element of matrix **D**.

$$\text{If} \quad \mathbf{E} = \begin{bmatrix} a & b & c \\ d & e & f \end{bmatrix}$$

then $a = 30$, $b = 60$, $c = 80$, $d = 20$, $e = 30$ and $f = 40$.

7.2.2 Addition and subtraction of matrices

To add or subtract matrices they must be of the same order; they are then said to be **comformable for addition**. When this is true, for addition the corresponding elements in each matrix are added together and for subtraction, each of the second elements is subtracted from the corresponding elements in the first. If **A** and **B** are two matrices of the same size, then $\mathbf{A} + \mathbf{B} = [a_{ij} + b_{ij}]$ and $\mathbf{A} - \mathbf{B} = [a_{ij} - b_{ij}]$ for all values of i and j.

$$\text{If} \quad \mathbf{A} = \begin{bmatrix} 10 & 15 \\ 20 & 14 \end{bmatrix} \quad \mathbf{B} = \begin{bmatrix} 21 & 13 \\ 12 & 17 \end{bmatrix}$$

$$\text{then } \mathbf{A} + \mathbf{B} = \begin{bmatrix} 10 & 15 \\ 20 & 14 \end{bmatrix} + \begin{bmatrix} 21 & 13 \\ 12 & 17 \end{bmatrix}$$

$$= \begin{bmatrix} (10+21) & (15+13) \\ (20+12) & (14+17) \end{bmatrix} = \begin{bmatrix} 31 & 28 \\ 32 & 31 \end{bmatrix}$$

$$\text{and } \mathbf{A} - \mathbf{B} = \begin{bmatrix} 10 & 15 \\ 20 & 14 \end{bmatrix} - \begin{bmatrix} 21 & 13 \\ 12 & 17 \end{bmatrix}$$

$$= \begin{bmatrix} (10-21)(15-13) \\ (20-12)(14-17) \end{bmatrix} = \begin{bmatrix} -11 & 2 \\ 8 & -3 \end{bmatrix}$$

If we have:

$$\begin{bmatrix} 10 & 20 \\ 40 & 5 \end{bmatrix} - \begin{bmatrix} 10 & 20 \\ 40 & 5 \end{bmatrix} = \begin{bmatrix} 0 & 0 \\ 0 & 0 \end{bmatrix}$$

then the result is a **zero matrix** (which performs the same function as zero in ordinary arithmetic).

Addition of matrices is said to be **commutative** (sequence makes no difference) since **A** + **B** = **B** + **A**.

EXAMPLE

The company, Comfy Chairs Ltd, have presented proposed weekly output from a new factory in matrix **E** shown below.

$$\mathbf{E} = \begin{bmatrix} 30 & 30 & 0 \\ 18 & 12 & 0 \end{bmatrix}$$

Again, columns represent chair types and rows represent wood used. What would the combined weekly output be from the existing and new factories?

$$\mathbf{D}+\mathbf{E}=\begin{bmatrix}30 & 60 & 80\\ 20 & 30 & 40\end{bmatrix}+\begin{bmatrix}30 & 30 & 0\\ 18 & 12 & 0\end{bmatrix}=\begin{bmatrix}60 & 90 & 80\\ 38 & 42 & 40\end{bmatrix}$$

7.2.3 Scalar multiplication of a matrix

A matrix may be multiplied by a single number or **scalar**. To do this we multiply each element of the matrix by the scalar, e.g.

$$5\times\begin{bmatrix}4 & 8 & 3\\ 17 & 2 & 12\end{bmatrix}=\begin{bmatrix}20 & 40 & 15\\ 85 & 10 & 60\end{bmatrix}$$

In terms of matrix notation, if **A** is the matrix and c is the scalar then $c\mathbf{A}=[ca_{ij}]$ for all values of i and j.

EXAMPLE

Suppose Comfy Chairs Ltd plan to increase output across the range from their existing factory by 10%. To increase output across the range by 10%, we need to multiply each element by 1.1.

$$1.1\times\mathbf{D}=\begin{bmatrix}1.1\times 30 & 1.1\times 60 & 1.1\times 80\\ 1.1\times 20 & 1.1\times 30 & 1.1\times 40\end{bmatrix}$$

$$=\begin{bmatrix}33 & 66 & 88\\ 22 & 33 & 44\end{bmatrix}$$

It is often convenient to reverse this argument, and take a common factor out of the matrix (cf. taking a common factor out of a bracket), to make further calculations easier, e.g.

$$\begin{bmatrix}10 & 170 & 100\\ 20 & 90 & 95\\ 140 & 30 & 50\end{bmatrix}=10\times\begin{bmatrix}1 & 17 & 10\\ 2 & 9 & 9.5\\ 14 & 3 & 5\end{bmatrix}$$

When the *same* matrix is to be multiplied by a series of scalars, we have:

$$a\mathbf{A}+b\mathbf{A}+c\mathbf{A}+d\mathbf{A}=(a+b+c+d)\mathbf{A}$$

7.2.4 Multiplication of matrices

When two matrices are to be multiplied together it is first necessary to check that the multiplication is possible; the matrices must be **conformable for multiplication**. This condition is satisfied if the number of columns in the first matrix is equal to the number of rows in the second matrix. The outcome of multiplication will be a matrix with the same number of rows as the first matrix and the same number of columns as the second matrix. If matrix **A** is of order $(a\times b)$ and matrix **B** is of order $(c\times d)$, then for multiplication to be possible, b must equal c and the new matrix produced by the product **AB** will be of order $(a\times d)$.

Thus a (2×3) matrix multiplied by a (3×2) matrix will give a (2×2) result; whereas a (3×2) matrix multiplied by a (2×3) matrix will give a (3×3) result. It is not possible to multiply a matrix of order (4×8) by a matrix of order (5×3). Note that even though the resultant matrix may be of the same order, multiplication is **not commutative** since **AB** does not necessarily equal **BA**.

The process of multiplication involves using a particular row from the first matrix and a particular column from the second matrix; placing the result as a single element in the result matrix, e.g.

$$\mathbf{A}=\begin{bmatrix}1 & 4 & 7\\ 2 & 5 & 8\\ 3 & 6 & 9\end{bmatrix}\quad \mathbf{B}=\begin{bmatrix}10 & 13\\ 11 & 14\\ 12 & 15\end{bmatrix}\quad \mathbf{A}\times\mathbf{B}=\mathbf{C}$$

To find $\mathbf{A}\times\mathbf{B}$, we will work out each element separately. For example, taking the *first* row of **A** and the *first* column of **B** gives the element c_{11} in the first row and column of **C**, i.e.

$$[1\ 4\ 7]\begin{bmatrix}10\\ 11\\ 12\end{bmatrix}=(1\times 10)+(4\times 11)+(7\times 12)=138=c_{11}$$

Note that we have gone along the row of the first matrix and down the column of the second matrix, multiplying the corresponding elements.

To find the *second* element in the *first* row of **C**, we take the *first* row of the matrix **A**, and the *second* column of the matrix **B**.

$$[1\ 4\ 7]\begin{bmatrix}13\\ 14\\ 15\end{bmatrix}=(1\times 13)+(4\times 14)+(7\times 15)=174=c_{12}$$

This process continues, using the second row from **A**, and then the third row. In general, the

mth row of **A** by the nth column of **B** gives the element c_{mn} in the mth row and nth column of **C**.

EXERCISE

Calculate the remaining elements of the matrix **C**. Answer:

$$\mathbf{C} = \begin{bmatrix} 138 & 174 \\ 171 & 216 \\ 204 & 258 \end{bmatrix}$$

Note that it is not possible to find $\mathbf{B} \times \mathbf{A}$ because the number of columns in **B** is not the same as the number of rows in **A**.

EXAMPLE

$$\begin{bmatrix} 1 & 3 \\ 2 & 5 \end{bmatrix} \begin{bmatrix} 3 & 2 \\ 5 & 7 \end{bmatrix} = \begin{bmatrix} 18 & 23 \\ 31 & 39 \end{bmatrix}$$

$$\begin{bmatrix} 3 & 2 \\ 5 & 7 \end{bmatrix} \begin{bmatrix} 1 & 3 \\ 2 & 5 \end{bmatrix} = \begin{bmatrix} 7 & 19 \\ 19 & 50 \end{bmatrix}$$

This confirms that $\mathbf{A} \times \mathbf{B} \neq \mathbf{B} \times \mathbf{A}$. Since the order in which the matrices are multiplied together is crucial to the product obtained, it is useful to have a terminology to show this. In the product $\mathbf{A} \times \mathbf{B}$, **A** is said to be **postmultiplied** by **B**, whilst **B** is said to be **premultiplied** by **A**.

We have seen, above, the zero matrix which performs the same function in matrix algebra as a zero in arithmetic. We also need a matrix which will perform the function that unity (one) plays in arithmetic (e.g. $5 \times 1 = 5$).

For any matrix **A**, what matrix has no effect when **A** is postmultiplied by it? i.e.

$$\mathbf{A} \times ? = \mathbf{A}$$

The matrix which performs this function is known as the **identity matrix** and consists of 1s on the diagonal from top left (a_{11}) to bottom right (a_{nn}) and 0s for all other elements. Note that all identity matrices are *square* and that the 1s are said to be on the leading or **principal diagonal**.

Thus: $$\begin{bmatrix} 5 & 3 \\ 2 & 1 \end{bmatrix} \begin{bmatrix} 1 & 0 \\ 0 & 1 \end{bmatrix} = \begin{bmatrix} 5 & 3 \\ 2 & 1 \end{bmatrix}$$

or $$\begin{bmatrix} 6 & 8 & 10 \\ 12 & 20 & 14 \\ 2 & 75 & 3 \end{bmatrix} \begin{bmatrix} 1 & 0 & 0 \\ 0 & 1 & 0 \\ 0 & 0 & 1 \end{bmatrix} = \begin{bmatrix} 6 & 8 & 10 \\ 12 & 20 & 14 \\ 2 & 75 & 3 \end{bmatrix}$$

If **I** is used for the identity matrix, we have:

$\mathbf{AI} = \mathbf{IA} = \mathbf{A}$.

EXAMPLE

Suppose Comfy Chairs Ltd. decide not to implement an overall increase in output across the range by 10% from their existing factory but rather increase the Classic by 5%, the Victorian by 10% and the Modern by 15% in both mahogany and teak.

This can be done by multiplying matrix **D** by a new matrix **F**:

$$\mathbf{DF} = \begin{bmatrix} 30 & 60 & 80 \\ 20 & 30 & 40 \end{bmatrix} \begin{bmatrix} 1.05 & 0 & 0 \\ 0 & 1.10 & 0 \\ 0 & 0 & 1.15 \end{bmatrix}$$

$$= \begin{bmatrix} 31.5 & 66 & 92 \\ 21 & 33 & 46 \end{bmatrix}$$

EXAMPLE

Suppose we want to know the total number of chairs produced in mahogany or teak from the existing factory before any increase.

We can do this by multiplying matrix **D** by a column vector of 1s:

$$\begin{bmatrix} 30 & 60 & 80 \\ 20 & 30 & 40 \end{bmatrix} \begin{bmatrix} 1 \\ 1 \\ 1 \end{bmatrix} = \begin{bmatrix} 170 \\ 90 \end{bmatrix}$$

7.2.5 Inverse of a matrix

As in arithmetic, division of one matrix by a second matrix is equivalent to multiplying the first by the **inverse** of the second, bearing in mind that the operation is not commutative. To transfer a matrix from one side of an equation to the other, both sides are multiplied by the inverse of the matrix. The inverse of a matrix **A** is denoted by $\mathbf{A}^{-1}$. A matrix multiplied in *either order* by the inverse of itself will always give the identity matrix, i.e.

$$\mathbf{AA}^{-1} = \mathbf{A}^{-1}\mathbf{A} = \mathbf{I}$$

Suppose we have:

$$\mathbf{A} \times \mathbf{B} = \mathbf{C}$$

and want to find **B**. If we *premultiply* both sides by $\mathbf{A}^{-1}$:

$$\begin{aligned} \mathbf{A}^{-1}\mathbf{AB} &= \mathbf{A}^{-1}\mathbf{C} \\ \mathbf{IB} &= \mathbf{A}^{-1}\mathbf{C} \\ \mathbf{B} &= \mathbf{A}^{-1}\mathbf{C} \end{aligned}$$

To find **A**, we can postmultiply both sides by $\mathbf{B}^{-1}$:

$$\begin{aligned} \mathbf{ABB}^{-1} &= \mathbf{CB}^{-1} \\ \mathbf{AI} &= \mathbf{CB}^{-1} \\ \mathbf{A} &= \mathbf{CB}^{-1} \end{aligned}$$

Finding the inverse of a matrix is *only* possible if the matrix is *square*, i.e. has the same number of rows as columns, and we shall initially look at the special case of a (2×2) matrix and then suggest a method for larger matrices.

The inverse of a (2 × 2) matrix

If

$$\mathbf{A} = \begin{bmatrix} a_{11} & a_{12} \\ a_{21} & a_{22} \end{bmatrix}$$

then

$$\mathbf{A}^{-1} = \frac{1}{(a_{11}a_{22} - (a_{12}a_{21})}\begin{bmatrix} a_{22} & -a_{12} \\ -a_{21} & a_{11} \end{bmatrix}$$

Putting this into words, the fraction in front of the matrix is one over the product of the two elements on the principal diagonal minus the product of the other two elements. Within the matrix the two elements on the principal diagonal change places and the other two elements change sign, e.g. if

$$\mathbf{A} = \begin{bmatrix} 1 & 2 \\ 3 & 4 \end{bmatrix}$$

then

$$\begin{aligned} \mathbf{A}^{-1} &= \frac{1}{(1 \times 4) - (2 \times 3)}\begin{bmatrix} 4 & -2 \\ -3 & 1 \end{bmatrix} \\ &= -\frac{1}{2}\begin{bmatrix} 4 & -2 \\ -3 & 1 \end{bmatrix} \\ &= \begin{bmatrix} -2 & 1 \\ 1.5 & -0.5 \end{bmatrix} \end{aligned}$$

To check the answer we can find $\mathbf{AA}^{-1}$:

$$\begin{bmatrix} 1 & 2 \\ 3 & 4 \end{bmatrix}\begin{bmatrix} -2 & 1 \\ 1.5 & -0.5 \end{bmatrix} = \begin{bmatrix} 1 & 0 \\ 0 & 1 \end{bmatrix}$$

The denominator of the fraction we calculated above is called the **determinant** of the matrix, and if this is 0 then the matrix is said to be **singular**, and does not have an inverse.

Larger matrices

To find the inverse for a matrix larger than (2×2) is a somewhat more complex procedure and may be done by using the method of **cofactors** or by **row operations**. We shall give an example of how to find the inverse for a (3×3) matrix. (For larger matrices it is probably advisable to use a computer program.)

To find an inverse using the cofactors, we must first find the determinant of the matrix, denoted by $|\mathbf{A}|$. If

$$\mathbf{A} = \begin{bmatrix} a_{11} & a_{12} & a_{13} \\ a_{21} & a_{22} & a_{23} \\ a_{31} & a_{32} & a_{33} \end{bmatrix}$$

then

$$\begin{aligned} |\mathbf{A}| &= a_{11}\begin{vmatrix} a_{22} & a_{23} \\ a_{32} & a_{33} \end{vmatrix} - a_{21}\begin{vmatrix} a_{12} & a_{13} \\ a_{32} & a_{33} \end{vmatrix} + a_{31}\begin{vmatrix} a_{12} & a_{13} \\ a_{22} & a_{23} \end{vmatrix} \\ &= a_{11}(a_{22}a_{33} - a_{23}a_{32}) - a_{21}(a_{12}a_{33} - a_{13}a_{32}) \\ &\quad + a_{31}(a_{12}a_{23} - a_{13}a_{22}) \end{aligned}$$

A cofactor of an element consists of the determinant of those elements which are not in the same row and not in the same column as that element. Thus the cofactor of a_{11} is:

$$\begin{vmatrix} a_{22} & a_{23} \\ a_{32} & a_{33} \end{vmatrix}$$

and this can be evaluated as $(a_{22}a_{33}) - (a_{23}a_{32})$. To form the inverse matrix, each element of the initial matrix is replaced by its *signed* cofactor. (Note the signs of the cofactors replacing a_{12}, a_{21}, a_{23} and a_{32}.)

$$\begin{bmatrix} \begin{vmatrix} a_{22} & a_{23} \\ a_{23} & a_{33} \end{vmatrix} & -\begin{vmatrix} a_{21} & a_{23} \\ a_{31} & a_{33} \end{vmatrix} & \begin{vmatrix} a_{21} & a_{22} \\ a_{31} & a_{32} \end{vmatrix} \\ -\begin{vmatrix} a_{12} & a_{13} \\ a_{32} & a_{33} \end{vmatrix} & \begin{vmatrix} a_{11} & a_{13} \\ a_{31} & a_{33} \end{vmatrix} & -\begin{vmatrix} a_{11} & a_{12} \\ a_{31} & a_{32} \end{vmatrix} \\ \begin{vmatrix} a_{12} & a_{13} \\ a_{22} & a_{23} \end{vmatrix} & -\begin{vmatrix} a_{11} & a_{13} \\ a_{21} & a_{23} \end{vmatrix} & \begin{vmatrix} a_{11} & a_{12} \\ a_{21} & a_{22} \end{vmatrix} \end{bmatrix}$$

After each element of this new matrix has been evaluated, it is **transposed**; that is each column is written as a row, so that a_{12} becomes a_{21}. The resultant matrix is then multiplied by one over the determinant. This is likely to become considerably clearer as we work through an example.

EXAMPLE

$$\mathbf{A} = \begin{bmatrix} 1 & 2 & 3 \\ 1 & 3 & 5 \\ 1 & 5 & 12 \end{bmatrix}$$

then

$$\begin{aligned} |\mathbf{A}| &= 1 \times (3 \times 12 - 5 \times 5) - 1 \times (2 \times 12 - 3 \times 5) + 1 \\ &\quad \times (2 \times 5 - 3 \times 3) \\ &= 11 - 9 + 1 \\ &= 3 \end{aligned}$$

Replacing elements by cofactors, we have:

$$\begin{bmatrix} \begin{vmatrix} 3 & 5 \\ 5 & 12 \end{vmatrix} & -\begin{vmatrix} 1 & 5 \\ 1 & 12 \end{vmatrix} & \begin{vmatrix} 1 & 3 \\ 1 & 5 \end{vmatrix} \\ -\begin{vmatrix} 2 & 3 \\ 5 & 12 \end{vmatrix} & \begin{vmatrix} 1 & 3 \\ 1 & 12 \end{vmatrix} & -\begin{vmatrix} 1 & 2 \\ 1 & 5 \end{vmatrix} \\ \begin{vmatrix} 2 & 3 \\ 3 & 5 \end{vmatrix} & -\begin{vmatrix} 1 & 3 \\ 1 & 5 \end{vmatrix} & \begin{vmatrix} 1 & 2 \\ 1 & 3 \end{vmatrix} \end{bmatrix}$$

$$= \begin{bmatrix} 11 & -7 & 2 \\ -9 & 9 & -3 \\ 1 & -2 & 1 \end{bmatrix}$$

Transposing this, we have:

$$\begin{bmatrix} 11 & -9 & 1 \\ -7 & 9 & -2 \\ 2 & -3 & 1 \end{bmatrix}$$

This is called the **adjunct matrix**, and $\mathbf{A}^{-1}$ is this matrix multiplied by the reciprocal of the determinant.

$$\mathbf{A}^{-1} = \frac{1}{3}\begin{bmatrix} 11 & -9 & 1 \\ -7 & 9 & -2 \\ 2 & -3 & 1 \end{bmatrix}$$

EXERCISE

Check that $\mathbf{AA}^{-1} = \mathbf{I}$.

To find the inverse using row operations, we create a **partitioned matrix**, by putting an identity matrix alongside the original matrix:

$$\left[\begin{array}{ccc|ccc} 1 & 2 & 3 & 1 & 0 & 0 \\ 1 & 3 & 5 & 0 & 1 & 0 \\ 1 & 5 & 12 & 0 & 0 & 1 \end{array}\right]$$

Our objective now is to multiply and divide each row, or add and subtract rows until the matrix on the *left* of the partition is an identity. At that point, whatever is to the *right* of the partition will be the inverse of the original matrix. We already have a 1 at a_{11}, so to change the 1 at a_{21} to 0 we may subtract row 1 from row 2. (Note that we subtract corresponding elements for the *whole* row.)

$$\left[\begin{array}{ccc|ccc} 1 & 2 & 3 & 1 & 0 & 0 \\ 0 & 1 & 2 & -1 & 1 & 0 \\ 1 & 5 & 12 & 0 & 0 & 1 \end{array}\right]$$

To alter the 1 to a 0 at a_{31}, we again subtract row 1 from row 3.

$$\left[\begin{array}{ccc|ccc} 1 & 2 & 3 & 1 & 0 & 0 \\ 0 & 1 & 2 & -1 & 1 & 0 \\ 0 & 3 & 9 & -1 & 0 & 1 \end{array}\right]$$

To alter the 2 to a 0 at a_{12}, we can subtract *two* times row 2 from row 1.

$$\left[\begin{array}{ccc|ccc} 1 & 0 & -1 & 3 & -2 & 0 \\ 0 & 1 & 2 & -1 & 1 & 0 \\ 0 & 3 & 9 & -1 & 0 & 1 \end{array}\right]$$

To alter the 3 to a 0 at a_{32}, subtract *three* times row 2 from row 3.

$$\left[\begin{array}{ccc|ccc} 1 & 0 & -1 & 3 & -2 & 0 \\ 0 & 1 & 2 & -1 & 1 & 0 \\ 0 & 0 & 3 & 2 & -3 & 1 \end{array}\right]$$

To reduce the 3 to a 1 at a_{33}, divide row 3 by 3.

$$\left[\begin{array}{ccc|ccc} 1 & 0 & -1 & 3 & -2 & 0 \\ 0 & 1 & 2 & -1 & 1 & 0 \\ 0 & 0 & 1 & 2/3 & -1 & 1/3 \end{array}\right]$$

To alter the -1 to a 0 at a_{13}, add row 3 to row 1.

$$\left[\begin{array}{ccc|ccc} 1 & 0 & 0 & 11/3 & -3 & 1/3 \\ 0 & 1 & 2 & -1 & 1 & 0 \\ 0 & 0 & 1 & 2/3 & -1 & 1/3 \end{array}\right]$$

To alter the 2 to a 0 at a_{23}, subtract *two* times row 3 from row 2.

$$\left[\begin{array}{ccc|ccc} 1 & 0 & 0 & 11/3 & -3 & 1/3 \\ 0 & 1 & 0 & -7/3 & 3 & -2/3 \\ 0 & 0 & 1 & 2/3 & -1 & 1/3 \end{array}\right]$$

As you will see, this is the same answer that was achieved by using the cofactors method. Note also that $(\mathbf{AB})^{-1} = \mathbf{B}^{-1}\mathbf{A}^{-1}$ as you can easily prove.

7.3 SOLUTION OF SIMULTANEOUS EQUATIONS

Matrices provide a methodology to manage sets of equations. Most of the equations encountered at this level will give unique solutions e.g. $x = 5$ but this is not always the case. The single equation $x + 2y = 10$ does not have a unique solution but a number of solutions e.g. $x = 0$ and $y = 5$ *or* $x = 2$ and $y = 4$ *or* $x = -10$ and $y = 10$ etc. In general, we need the same number of independent equations as variables to obtain unique solutions.

7.3.1 Solving equations using matrix inverses

Consider the equations:

$$7x + 4y = 80$$
$$5x + 3y = 58$$

This pair of equations may be written in matrix notation as:

$$\begin{bmatrix} 7 & 4 \\ 5 & 3 \end{bmatrix} \begin{bmatrix} x \\ y \end{bmatrix} = \begin{bmatrix} 80 \\ 58 \end{bmatrix}$$

or

$$\mathbf{Ax} = \mathbf{b}$$

If we premultiply by $\mathbf{A}^{-1}$ we have

$$\begin{aligned} \mathbf{A}^{-1}\mathbf{Ax} &= \mathbf{A}^{-1}\mathbf{b} \\ \mathbf{Ix} &= \mathbf{A}^{-1}\mathbf{b} \\ \mathbf{x} &= \mathbf{A}^{-1}\mathbf{b} \end{aligned}$$

The result will hold for *any* set of simultaneous equations where there are as many equations as unknowns; and thus if we premultiply the vector **b** by the inverse of the matrix **A**, we shall be able to find the values of the unknowns that satisfy the equations. Here:

$$\mathbf{A}^{-1} = \frac{1}{21 - 20} \begin{bmatrix} 3 & -4 \\ -5 & 7 \end{bmatrix}$$

Therefore,

$$\begin{bmatrix} x \\ y \end{bmatrix} = \begin{bmatrix} 3 & -4 \\ -5 & 7 \end{bmatrix} \begin{bmatrix} 80 \\ 58 \end{bmatrix}$$

$$= \begin{bmatrix} 240 - 232 \\ -400 + 406 \end{bmatrix}$$

$$= \begin{bmatrix} 8 \\ 6 \end{bmatrix}$$

Thus $x = 8$ and $y = 6$.

If we have three equations:

$$\begin{aligned} 4x_1 + 3x_2 + x_3 &= 8 \\ 2x_1 + x_2 + 4x_3 &= -4 \\ 3x_1 \qquad + x_3 &= 1 \end{aligned}$$

then it can be shown that the inverse of the **A** matrix is:

$$\mathbf{A}^{-1} = \frac{1}{31} \begin{bmatrix} 1 & -3 & 11 \\ 10 & 1 & -14 \\ -3 & 9 & -2 \end{bmatrix}$$

and that the solution to the equations is:

$$\begin{bmatrix} x_1 \\ x_2 \\ x_3 \end{bmatrix} = \frac{1}{31} \begin{bmatrix} 1 & -3 & 11 \\ 10 & 1 & -14 \\ -3 & 9 & -2 \end{bmatrix} \begin{bmatrix} 8 \\ -4 \\ 1 \end{bmatrix}$$

$$= \frac{1}{31} \begin{bmatrix} 31 \\ 62 \\ -62 \end{bmatrix}$$

$$= \begin{bmatrix} 1 \\ 2 \\ -2 \end{bmatrix}$$

thus $x_1 = 1$, $x_2 = 2$ and $x_3 = -2$.

7.3.2 Solving equations using Cramer's rule

Consider the following set of equations:

$$a_1x + b_1y + c_1z = k_1$$
$$a_2x + b_2y + c_2z = k_2$$
$$a_3x + b_3y + c_3z = k_3$$

We need to find the determinant of the matrix of coefficients, **A**, and the determinant of this matrix with each column replaced in turn by the column vector of constant terms.

The determinant of matrix **A** may be written

$$|\mathbf{A}| = \begin{vmatrix} a_1 & b_1 & c_1 \\ a_2 & b_2 & c_2 \\ a_3 & b_3 & c_3 \end{vmatrix}$$

and the determinants of the matrix with column replacement

$$|\mathbf{A}_1| = \begin{vmatrix} k_1 & b_1 & c_1 \\ k_2 & b_2 & c_2 \\ k_3 & b_3 & c_3 \end{vmatrix} \quad |\mathbf{A}_2| = \begin{vmatrix} a_1 & k_1 & c_1 \\ a_2 & k_2 & c_2 \\ a_3 & k_3 & c_3 \end{vmatrix}$$

$$|\mathbf{A}_3| = \begin{vmatrix} a_1 & b_1 & k_1 \\ a_2 & b_2 & k_2 \\ a_3 & b_3 & k_3 \end{vmatrix}$$

If $|\mathbf{A}| \neq 0$, the unique solution is given by:

$$x = \frac{|\mathbf{A}_1|}{|\mathbf{A}|} \quad y = \frac{|\mathbf{A}_2|}{|\mathbf{A}|} \quad z = \frac{|\mathbf{A}_3|}{|\mathbf{A}|}$$

EXAMPLE

Given

$$\begin{bmatrix} 4 & 3 & 1 \\ 2 & 1 & 4 \\ 3 & 0 & 1 \end{bmatrix} \begin{bmatrix} x_1 \\ x_2 \\ x_3 \end{bmatrix} = \begin{bmatrix} 8 \\ -4 \\ 1 \end{bmatrix}$$

$$|\mathbf{A}| = \begin{vmatrix} 4 & 3 & 1 \\ 2 & 1 & 4 \\ 3 & 0 & 1 \end{vmatrix} = 31 \quad |\mathbf{A}_1| = \begin{vmatrix} 8 & 3 & 1 \\ -4 & 1 & 4 \\ 1 & 0 & 1 \end{vmatrix} = 31$$

$$|\mathbf{A}_2| = \begin{vmatrix} 4 & 8 & 1 \\ 2 & -4 & 4 \\ 3 & 1 & 1 \end{vmatrix} = 62 \quad \begin{vmatrix} 4 & 3 & 8 \\ 2 & 1 & -4 \\ 3 & 0 & 1 \end{vmatrix} = -62$$

Therefore

$$x_1 = \frac{|\mathbf{A}_1|}{|\mathbf{A}|} = \frac{31}{31} = 1$$

$$x_2 = \frac{|\mathbf{A}_2|}{|\mathbf{A}|} = \frac{62}{31} = 2$$

$$x_3 = \frac{|\mathbf{A}_3}{|\mathbf{A}|} = \frac{-62}{31} = -2$$

7.3.3 Equations without unique solutions

Not all sets of simultaneous equations will give unique solutions like $x_1 = 1$, $x_2 = 2$ and $x_3 = -2$. Consider the following.

$$2x + y - z = 9$$
$$x - y + 2z = 5$$
$$4x + 2y - 2z = 18$$

This can be written:

$$\begin{bmatrix} 2 & 1 & -1 \\ 1 & -1 & 2 \\ 4 & 2 & -2 \end{bmatrix} \begin{bmatrix} x \\ y \\ z \end{bmatrix} = \begin{bmatrix} 9 \\ 5 \\ 18 \end{bmatrix}$$

Using the notation $\mathbf{Ax} = \mathbf{b}$, we could attempt to find the inverse of **A** by the partitioned matrix method.

$$\left[\begin{array}{ccc|ccc} 2 & 1 & -1 & 1 & 0 & 0 \\ 1 & -1 & 2 & 0 & 1 & 0 \\ 4 & 2 & -2 & 0 & 0 & 1 \end{array}\right]$$

Divide row 1 by 2, so that $a_{11} = 1$

$$\left[\begin{array}{ccc|ccc} 1 & 1/2 & -1/2 & 1/2 & 0 & 0 \\ 1 & -1 & 2 & 0 & 1 & 0 \\ 4 & 2 & -2 & 0 & 0 & 1 \end{array}\right]$$

Take row 1 from row 2, so that $a_{21} = 0$

$$\left[\begin{array}{ccc|ccc} 1 & 1/2 & -1/2 & 1/2 & 0 & 0 \\ 0 & -3/2 & 5/2 & -1/2 & 1 & 0 \\ 4 & 2 & -2 & 0 & 0 & 1 \end{array}\right]$$

Take 4 times row 1 from row 3, so that $a_{31} = 0$

$$\left[\begin{array}{ccc|ccc} 1 & 1/2 & -1/2 & 1/2 & 0 & 0 \\ 0 & -3/2 & 5/2 & -1/2 & 1 & 0 \\ 0 & 0 & 0 & -2 & 0 & 1 \end{array}\right]$$

Divide row 2 by 5 and add to row 1, so that $a_{13} = 0$

$$\left[\begin{array}{ccc|ccc} 1 & 1/5 & 0 & 2/5 & 1/5 & 0 \\ 0 & -3/2 & 5/2 & -1/2 & 1 & 0 \\ 0 & 0 & 0 & -2 & 0 & 1 \end{array}\right]$$

Multiply row 2 by $-2/3$, so that $a_{22} = 1$

$$\left[\begin{array}{ccc|ccc} 1 & 1/5 & 0 & 2/5 & 1/5 & 0 \\ 0 & 1 & -5/3 & -1/3 & -2/3 & 0 \\ 0 & 0 & 0 & -2 & 0 & 1 \end{array}\right]$$

We cannot obtain an identity matrix on the left hand side. In this case we cannot obtain a unique solution, because the equations are not independent. Looking at the original set of equations, it can be seen that the third equation was double the first. We only have two independent equations and three variables.

If we now consider only the first two independent equations, and take twice equation 1 from equation 2 we can eliminate z:

$$5x + y = 23$$

At this stage we can identify combinations of x and y that satisfy this condition, but not unique solutions.

7.4 LEONTIEF INPUT-OUTPUT ANALYSIS

This input-output model was first developed by Leontief in the 1940s. It recognized that industries or economic sectors do not produce in isolation but rely on each other. The output from one sector may be the input to the same sector, or to another productive sector, or may be consumed as final demand. Final demand refers to the non-productive sector, including households and exports. The agricultural sector for example, will retain some of its output as seed or breeding stock, transfer the majority of its output to the food processing industries with the remainder of its output going directly for home consumption or exports.

In a complex economy, the segregation may be fairly arbitrary since many companies produce a range of products, that may fall into several sectors, but in the UK and the rest of Europe, the segregation is still attempted by using a Standard Industrial Classification of about 40 sectors. We will not attempt to model a complete system of 40 sectors (!) but will analyze a hypothetical economy with three productive sectors. The same approach however, could be applied to a larger system.

If the three productive sectors have outputs x_1, x_2 and x_3, part of this output will go to other productive sectors and part to final demand. An important feature of this model is that for each sector, *total output is equal to total input*. An example is given in Table 7.1.

The first three columns show the inputs to sectors X_1, X_2 and X_3. It can be seen in the first column that sector X_1 uses 20 units of its own output, 40 units from sector X_2 and 10 units from sector X_3. The remaining 130 units are referred to as **primary inputs** and include items such as labour and raw materials. The fourth column gives final demand which is the difference between total output and output used by other productive sectors. A **control economy** would attempt to manage these final demand levels

Table 7.1

Outputs from:	*Inputs to:* sector X_1	sector X_2	sector X_3	*Final demand*	*Total output*
X_1	20	10	60	110	200
X_2	40	10	50	150	250
X_3	10	60	30	400	500
Other inputs	130	170	360		
	200	250	500		

through the planning of industry or sectorial output. Clearly if high proportions of output are destined for other industries, the amount available for final demand will be reduced.

To produce 200 units from sector X_1, we need inputs of 20, 40 and 10 from sectors X_1, X_2, and X_3 respectively. Expressed as proportions we get 0.10, 0.20 and 0.05 which are known as the **input-output coefficients** or **technical coefficients**. Another important feature of the model is that these input-output coefficients remain *constant* regardless of changes in final demand. These input-output coefficients are placed into a matrix **A**, and using matrix notation,

$a_{11} = 0.10$, $a_{21} = 0.20$ and $a_{31} = 0.05$.

In general, a_{ij} is the proportion of input from X_i in the output of X_j. The matrix **A** is known as the **input-output matrix** and can be written:

$$\mathbf{A} = \begin{bmatrix} a_{11} & a_{12} & a_{13} \\ a_{21} & a_{22} & a_{23} \\ a_{31} & a_{32} & a_{33} \end{bmatrix} = \begin{bmatrix} 0.10 & 0.04 & 0.12 \\ 0.20 & 0.04 & 0.10 \\ 0.05 & 0.24 & 0.06 \end{bmatrix}$$

The input-output system shown in Table 7.1 can be represented by matrices:

$$\begin{bmatrix} 0.10 & 0.04 & 0.12 \\ 0.20 & 0.04 & 0.10 \\ 0.05 & 0.24 & 0.06 \end{bmatrix} \begin{bmatrix} 200 \\ 250 \\ 500 \end{bmatrix} + \begin{bmatrix} 110 \\ 150 \\ 400 \end{bmatrix} = \begin{bmatrix} 200 \\ 250 \\ 500 \end{bmatrix}$$

In summary:

$$\mathbf{AX} + \mathbf{D} = \mathbf{X}$$

where X is the output matrix and **D** is the demand matrix. This equation is known as the input-output equation. To find outputs to meet any projected final demands we must solve the above equation in terms of **X**. Using matrix algebra:

$$\begin{aligned} \mathbf{X} - \mathbf{AX} &= \mathbf{D} \\ (\mathbf{I} - \mathbf{A})\mathbf{X} &= \mathbf{D} \\ (\mathbf{I} - \mathbf{A})^{-1}(\mathbf{I} - \mathbf{A})\mathbf{X} &= (\mathbf{I} - \mathbf{A})^{-1}\mathbf{D} \\ \mathbf{X} &= (\mathbf{I} - \mathbf{A})^{-1}\mathbf{D} \end{aligned}$$

Taking matrix **A** away from the identity matrix **I** gives

$$(\mathbf{I} - \mathbf{A}) = \begin{bmatrix} 0.90 & -0.04 & -0.12 \\ -0.20 & 0.96 & -0.10 \\ -0.05 & -0.24 & 0.94 \end{bmatrix}$$

The inverse of $(\mathbf{I} - \mathbf{A})$ is

$$\frac{10}{77\,132} \begin{bmatrix} 8\,784 & 664 & 1\,192 \\ 1\,930 & 8\,400 & 1\,140 \\ 960 & 2\,180 & 8\,560 \end{bmatrix}$$

EXERCISE

Using the figures from the example show that

$\mathbf{X} = (\mathbf{I} - \mathbf{A})^{-1}\mathbf{D}$

Having developed the input-output model, we are now told that the final demand levels required from X_1, X_2 and X_3 are 77 132, 231 396 and 385 660 respectively. To achieve these levels we can solve the input-output equation in terms of **X**, to determine industry outputs.

$$\begin{bmatrix} x_1 \\ x_2 \\ x_3 \end{bmatrix} = \frac{10}{77\,132} \begin{bmatrix} 8\,784 & 664 & 1\,192 \\ 1\,930 & 8\,400 & 1\,140 \\ 960 & 2\,180 & 8\,560 \end{bmatrix} \begin{bmatrix} 77\,132 \\ 231\,396 \\ 385\,660 \end{bmatrix}$$

$$= \begin{bmatrix} 167\,360 \\ 328\,300 \\ 503\,000 \end{bmatrix}$$

The outputs required from sectors X_1, X_2 and X_3 are 167 360, 328 300 and 503 000.

The equation system can be used to show the relationships between industries within an economy and the effect on final demand of various output and input changes. It was used in the UK in the early 1970s to look at the effects of a reduction of oil supplies on various industrial sectors. However the difficulty with this system is not only its size for a complex economy but the implicit assumption that the input-output coefficients remain constant. This is unlikely to remain true as technical progress and innovation change the ways in which some industries operate. To take this into account, a new matrix of input-output coefficients must be derived every few years and this can be a costly and time-consuming process.

The measurement of inputs/outputs can also present problems. It is generally assumed that we can express all inputs and outputs in the same units, say millions of pounds. Clearly the outputs from some sectors are more tangible than others. In a model of the economy, how do we measure the output from the public sector?

7.5 CONCLUSIONS

As we have seen, matrices provide a method for solving a range of problems. Once we begin to work with more than two equations, the algebra can become quite tedious. In practice, we would look for a computer-based method of solution for the larger problems. In terms of specifying a problem, matrices provide a convenient form of notation.

7.6 PROBLEMS

$$\mathbf{A} = [4 \quad 2 \quad 1] \quad \mathbf{B} = \begin{bmatrix} 6 \\ 8 \\ 10 \end{bmatrix}$$

$$\mathbf{C} = \begin{bmatrix} 2 & 1 \\ 4 & 5 \end{bmatrix} \quad \mathbf{D} = \begin{bmatrix} 5 & 7 \\ 8 & 10 \end{bmatrix}$$

$$\mathbf{E} = \begin{bmatrix} 4 & 12 & 8 \\ 7 & 2 & 5 \\ 9 & 1 & 3 \end{bmatrix} \quad \mathbf{F} = \begin{bmatrix} 1 & 2 & 10 \\ 1 & 0 & 2 \\ 1 & 7 & 1 \end{bmatrix}$$

$$\mathbf{G} = \frac{1}{58}\begin{bmatrix} -14 & 68 & 4 \\ 1 & -9 & 8 \\ 7 & -5 & -2 \end{bmatrix} \quad \mathbf{H} = [10 \quad 5 \quad 8]$$

Using the matrices given above, find:

1. $\mathbf{C}+\mathbf{D}$
2. $4\mathbf{A}$
3. $\mathbf{AB}$
4. $\mathbf{CD}$
5. $\mathbf{DC}$
6. $\mathbf{BC}$
7. $\mathbf{E}+\mathbf{F}$
8. $\mathbf{AE}$
9. $\mathbf{EF}$
10. $\mathbf{FB}$
11. $\mathbf{C}^{-1}$
12. $\mathbf{E}^{-1}$
13. $(\mathbf{AB})^{-1}$
14. $\mathbf{BA}$
15. $\mathbf{FG}$
16. $\mathbf{GF}$
17. $\mathbf{HF}$
18. $\mathbf{BH}$
19. $\mathbf{HB}$
20. $\mathbf{BHF}$

Solve the following sets of simultaneous equations using matrix algebra:

21. $4x+2y=11$, $3x+4y=9$
22. $6x+3y=20$, $2x-3y=14$
23. $5x-3y=26$, $2x+2y=4$
24. $10x+10y=6$, $3x+7y=11$
25. $4a+3b+c=15$, $2a+5b+7c=47$, $5a+6b-2c=7$
26. $10a+3b+4c=8$, $20a-5b+2c=12$, $25a-b+5c=16$

Using Cramer's rule, solve

27. $4x+2y-z=57$, $x-y+2z=-12$, $2x+2y+z=31$
28. $2a+b+c=7$, $b+3c=9$, $4a-b=1$

Using a computer-based method, solve

29. $a+2b+c+2d=5$, $2a+b+c-d=1$, $3a+b+3c+d=10$, $a-b-c+2d=4$
30. $4x_1+2x_2-x_3+x_4=29$, $x_1+x_2+x_3-x_4=7$, $2x_1+3x_2+x_4=20$, $3x_1+x_2-x_3+2x_4=22$

Using the notation developed in the previous section, analyze the following input/output tables:

31.

Outputs from:	*Inputs to: A*	*B*	*Final demand*	*Total output*
A	10	40	50	100
B	30	20	50	100
Other	60	40		
	100	100		200

(a) Identify **A**, **X**, **D** and verify that the relationship holds.
(b) If final demand changes to 220 for *A* and 170 for *B*, find the level of output required from each sector.
(c) Reconstruct the table above given these new final demands.

32.

Outputs from:	*Inputs to: A*	*B*	*Final demand*	*Total output*
A	5	85	10	100
B	20	120	360	500
Other	75	295		
	100	500		600

(a) Verify that the relationship between **A**, **X**, and **D** holds for this table.
(b) Find the outputs necessary from each sector if the final demand changes to 50 for *A* and 500 for *B*.

33.

Outputs from:	*Inputs to:* *A*	*B*	*C*	*Final demand*	*Total output*
A	20	10	30	40	100
B	20	40	60	80	200
C	20	20	120	140	300
Other	40	130	90		
	100	200	300		600

(a) Find the matrix of technical coefficients.
(b) Find $(\mathbf{I}-\mathbf{A})$.
(c) Show that the inverse of $(\mathbf{I}-\mathbf{A})$ is:

$$\frac{10}{342}\begin{bmatrix} 46 & 4 & 9 \\ 16 & 46 & 18 \\ 18 & 9 & 63 \end{bmatrix}$$

(d) Confirm that $\mathbf{X} = (\mathbf{I}-\mathbf{A})^{-1}\mathbf{D}$.
(e) If final demand now changes to 342 for *A*, 684 for *B* and 1026 for *C*, find the levels of output necessary for each sector.

34.

Outputs from:	*Inputs to:* *A*	*B*	*C*	*Final demand*	*Total output*
A	10	10	10	170	200
B	10	20	20	350	400
C	100	300	70	530	1 000
Other	80	70	900		
	200	400	1 000		1 600

Using the information given above, find the levels of output necessary from each sector to support final demand figures of 1 000 for *A*, 1 000 for *B* and 400 for *C*.

THE USE OF CALCULUS 8

We have seen in Chapter 6 how a range of business situations and economic relationships can be described mathematically. Using equations, for example, we can determine the breakeven point of a company or the price where market demand is equal to market supply. However, we are not always just interested in a particular level of output, or price, or point in time but often how *change* is taking place. We may be interested in the rate at which profits are increasing or decreasing, or the effects of a unit change in price, or sales growth over time.

Calculus is about the measurement of change. It provides the methodology to calculate the rate of change and indeed, whether the rate of change is increasing or decreasing. Calculus also provides a notation to describe change. Given that y is a function of x ($y = f(x)$), the change in y resulting from a change in x is written dy/dx.

8.1 DIFFERENTIATION

In this section we consider an intuitive development of differentiation (8.1.1), the rules of differentiation (8.1.2) and finally, the theoretical development of differentiation (8.1.3). Many courses at the first-year level will not expect you to be familiar with the theoretical aspects of the subject and you can leave section 8.1.3 without a loss of continuity. Indeed, if you are new to the subject, you can choose to leave section 8.1.3 on your first reading of this chapter and come back to it at a later stage.

8.1.1 Differentiation: an intuitive approach

Differentiation measures the change in y resulting from a change in x. If we were looking at a graph plot of y against x, we would think of this change in terms of gradient; and this is helpful to some extent. Suppose we were interested in the cost of car hire. The total cost in pounds, y, could be plotted against the number of miles, x. The gradient would give the cost of each extra mile.

If the total cost was a fixed charge of £200, then $y = 200$. No x term is included as the cost does not depend on the number of miles travelled. The change in y resulting from a change in x is 0, as an extra mile makes no difference to cost. In this case, $dy/dx = 0$.

If the cost of car hire had two components, a fixed charge of £150 and a mileage charge of 20p per mile, then $y = 150 + 0.20x$. Each extra mile would cost 0.20 and $dy/dx = 0.20$. In the case of a linear function, the gradient is constant (the increase in y resulting from a unit increase in x) and the change measured by the differential dy/dx is equal to the gradient.

The cost of car hire could be expressed in a more complex form:

$$y = 150 + 0.30x - 0.0005x^2 \quad \text{for } 0 \leqslant x \leqslant 500$$
$$y = 150 + 0.04x \quad \text{for } x > 500$$

There are two major issues to consider here: continuity and the meaning of change. Firstly, for differentiation to be valid the function must be **continuous** within the relevant range i.e. no gaps or jumps. In this case, we are dealing with two functions both continuous within the ranges given; differentiation is valid except at the boundary points of $x = 0$ and $x = 500$ where the functions are discontinuous. As we will see from the rules of differentiation given in section 8.1.2:

$$dy/dx = 0.30 - 0.001x \quad \text{for } 0 < x < 500$$
and
$$dy/dx = 0.04 \quad \text{for } x > 500$$

Secondly, a quadratic function like $y = 150 + 0.30x - 0.0005x^2$ produces the type of curve shown in Figure 6.21. The rate of change is not constant and will depend on the value of x.

When $x = 100$, $dy/dx = 0.2$, and
when $x = 200$, $dy/dx = 0.1$

But does dy/dx measure gradient? Now the y values corresponding to $x = 100$ and $x = 200$ are 175, $(150 + 0.30 \times 100 - 0.0005 \times 100^2)$ and 190, $(150 + 0.30 \times 200 - 0.0005 \times 200^2)$ respectively. This increase of 15 units in y achieved over an increase of 100 units in x gives an average increase of 0.15 units in y for each unit increase in x, over this range. Differentiation does not give the average increase over a range of x values but rather a measure of change occurring at a point on the curve. Differentiation can be thought of as giving the gradient of a tangent to a curve (see section 8.1.3).

Before moving on to consider the rules of differentiation, let us again consider the equation $y = 150 + 0.30x - 0.0005x^2$. We are often presented with such equations and given little explanation, but it is worth pausing to examine the structure of this equation. It can be rewritten $y = 150 + (0.30 - 0.0005x)x$. The three components of cost are therefore a fixed cost, plus a charge per mile times the number of miles. The charge per mile, $(0.30 - 0.0005x)$, depends on the number of miles, x, and as x increases, the charge made decreases. You will also find this form of equation when you look at revenue functions where price per unit falls as sales increase. It is always worth asking the question 'What does this equation mean?'.

8.1.2 Differentiation: the rules

To obtain a value for dy/dx we can apply a number of rules:

1. Given a function of the form $y = k$ where k is a constant,

   $$\frac{dy}{dx} = 0.$$

   A constant, by its nature does not change (depend on x) and therefore has zero rate of change.

EXAMPLE

If $y = 23$, then $dy/dx = 0$.

---

2. Given a function of the form $y = ax$ where a is a constant,

   $$\frac{dy}{dx} = a.$$

   For each unit increase in x there is a constant a unit increase in y — the rate of change.

EXAMPLE

If $y = 4.5x$, then $dy/dx = 4.5$.

---

3. Given a function of the form $y = ax + k$

   $$\frac{dy}{dx} = a.$$

   When we differentiate this more general linear function, the rate of change is still a constant given by the x coefficient, because for a function consisting of several parts or terms, we differentiate each part separately and then put the results together.

EXAMPLE

If $y = 32.8 + 0.96x$, then $dy/dx = 0.96$.

---

4. Given a function of the form $y = ax^n$

$$\frac{dy}{dx} = nax^{n-1}.$$

EXAMPLE

If $y = 6x^3$, then $a = 6$ and $n = 3$

$$\frac{dy}{dx} = 3 \times 6x^2 = 18x^2$$

EXAMPLE

If $y = 10x^2$, then $dy/dx = 2 \times 10x^1 = 20x$ since $x^1 = x$

EXAMPLE

If

$y = 6x^3 - 4x^2 + 10x - 50 + 3x^{-2}$

then

$$\frac{dy}{dx} = 3 \times 6x^2 - 2 \times 4x^1 + 10x^0 + 0 + (-2) \times 3x^{-3}$$
$$= 18x^2 - 8x + 10 - 6x^{-3}$$

NB $x^0 = 1$.

To complete this section, we need to consider certain other functions which have special **derivatives**. A derivative is the result of differentiation and is also called the differential coefficient.

EXAMPLE

If $y = e^x$ (the exponential function described in section 6.6.5) then $dy/dx = e^x$.

EXAMPLE

If $y = \log_e x$ then $dy/dx = 1/x$.

8.1.3 Differentiation: the theoretical development

Calculus, as discussed, is about the measurement of change. However, the rate of change can be interpreted in two ways: the average rate of change and the instantaneous rate of change. If you consider a car travelling from Birmingham to Manchester, then its **average speed** (i.e. the rate of change of distance in relation to time) may be 50 miles per hour, but its speed at any particular moment may be much less (for example 5 m.p.h. in heavy traffic) or rather more (for example 70 m.p.h. on the motorway). The driver, looking at the speedometer at a particular moment, can tell the **instantaneous speed**. Now it has been argued that there can be no such thing as instantaneous speed, since speed is to do with the distance travelled, but try convincing someone who has been driving and met a wall (which was presumably stationary) that he could not be travelling at a speed at the instant the vehicle and the wall met!

Calculus, then is about the **instantaneous rate of change** and *not* the average rate of change.

If we consider the graph of a function (as in Figure 8.1) we can see that there is an increasing change in y ($\Delta y_1 < \Delta y_2$) for equal changes in x

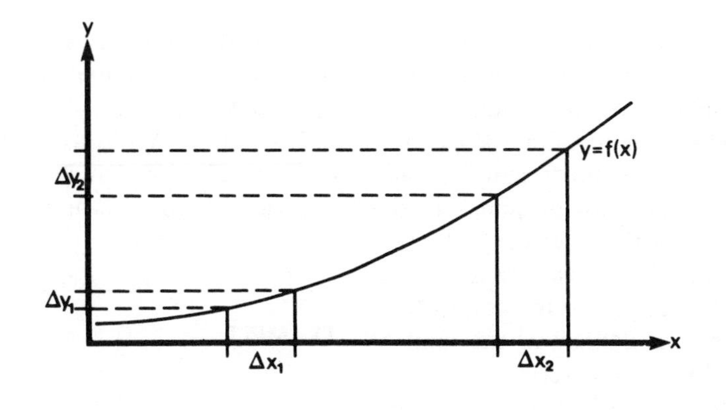

Figure 8.1

($\Delta x_1 = \Delta x_2$) as we move to the right. The symbol Δ (Greek capital delta) indicates an interval in the value of a variable, e.g. Δx represents the difference between two values of x. Thus there are different rates of change in y for different values of x. To find the average rate rate of change between two values of x, we would divide the change in y by the change in x. Here,

$\frac{\Delta y_1}{\Delta x_1}$ is less than $\frac{\Delta y_2}{\Delta x_2}$.

If a small section of the graph of the function is magnified (as in Figure 8.2), we see that the ratio $\Delta y/\Delta x$ not only measures the average rate of change of y between two values of x, but also the gradient or slope of a straight line (known as a **chord**) drawn between two points on the graph. As we reduce the change in x the straight line gets shorter and shorter and hence its slope or rate of change gets nearer and nearer to the slope of the curve at a single point; but if we have no change in x, then there is no change in y, although the function is still changing (cf. the car mentioned above). Now if the straight line between two points can represent the average slope of the graph between the points, is there a straight line at one point whose slope is that of the graph at that point? The answer is yes!

Figure 8.3 shows a straight line which touches the curve at just one point (known as a **tangent**) and the slope of this line will be the slope of the curve at the point where the two touch. Since

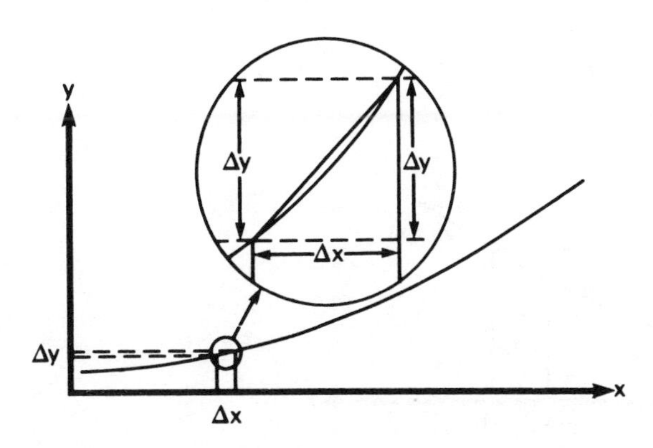

Figure 8.2

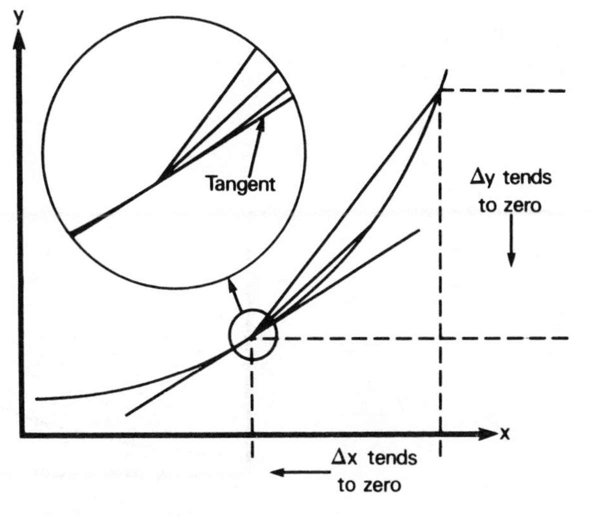

Figure 8.3

there is only one point now in question, the slope of the curve at that point measures not the average but the instantaneous rate of change.

Formally, the measurement of the slope of a curve at a point is valid if the function is **continuous**, i.e. there are no gaps or jumps in the function; thus if $y = f(x)$ then both x and y are continuous variables in the relevant range or domain. The slope of the tangent at the point where it touches the function is defined as the limit of the ratio $\Delta y/\Delta x$ as Δx tends to zero.

A **limit** is a number to which the ratio gets nearer and nearer, as the **interval** Δx gets smaller. For example the ratio

$$\frac{x + \Delta x}{x}$$

will get close and closer to 1 as the interval Δx gets smaller. Thus the limit of this expression will be 1.

Formally, the change in x is thought of as a distance, h, so that the interval Δx is from x to $x + h$. Since $y = f(x)$, the interval Δy extends from $f(x)$ to $f(x + h)$ (see Figure 8.4).

We saw above that an average slope of a function was

$$\frac{\Delta y}{\Delta x} = \frac{y_2 - y_1}{x_2 - x_1}$$

or, in Figure 8.4,

$$\frac{f(x+h) - f(x)}{(x+h) - x} = \frac{f(x+h) - f(x)}{h}$$

and it is this ratio, as h gets smaller and smaller, that gives the instantaneous rate of change of $f(x)$ at a single value x.

The limit of $\Delta y/\Delta x$ as $h \to 0$ is denoted by dy/dx or $f'(x)$. Here d signifies an infinitesimal change in a variable (cf. Δ which signifies finite change).

EXAMPLE

If $f(x) = x^2$, then $f(x + h) = (x + h)^2 = x^2 + 2xh + h^2$ so the ratio $\Delta y/\Delta x$ is given by:

$$\frac{\Delta y}{\Delta x} = \frac{(x^2 + 2xh + h^2) - x^2}{h}$$

$$= \frac{2xh + h^2}{h}$$

$$= 2x + h$$

and as $h \to 0$, this gets closer and closer to $2x$. Thus, for

$$y = x^2$$

$$\frac{dy}{dx} = 2x$$

So the rate of change of the function $y = x^2$ at $x = 2$, for example, is $2x = 2 \times 2 = 4$.

Note that the rate of change of a function is often referred to as the **gradient** of the function.

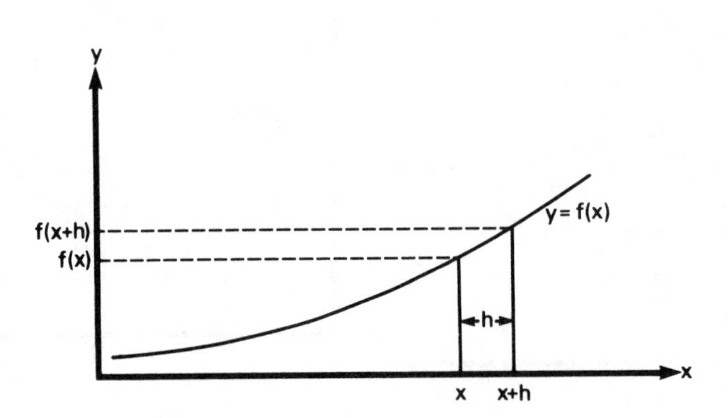

Figure 8.4

8.2 ECONOMIC APPLICATIONS I

Within economics, there are several functions which are related to each other as function to derivative. For instance, the total cost (TC) represents the cost of producing a particular amount of the product; the marginal cost (MC) is the cost of producing one extra unit and so the marginal cost of a given output is the rate at which total cost is changing at that output. Thus, if we have a total cost function and differentiate it, we will find that the result is a marginal cost function. This relationship will also hold for revenue functions.

Thus if $y = \text{TC}$ then $\dfrac{dy}{dx} = \text{MC}$

if $\quad y = \text{TR}$ then $\dfrac{dy}{dx} = \text{MR}.$

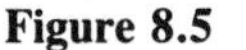

where TR is total revenue and MR is marginal revenue. For example:

if $\quad \text{TC} = 40 + 10x + 2x^2 + x^3$
then $\text{MC} = \quad 10 \ + 4x \ + 3x^2.$

(per unit change in output when output $= x$). For example:

if $\quad \text{TR} = 4x$
then $\text{MR} = 4.$

(per unit change in sales when sales $= x$).

We may also use the idea of averaging to find the average cost of, or revenue from, a given output. If total revenue is £100 from an output of 5, then the average revenue (AR) will be 100/5 = £20 per unit. So if

$$\text{TR} = 100x - 10x^2$$
$$\text{AR} = \frac{1}{x}(100x - 10x^2)$$
$$= \frac{100x}{x} - \frac{10x^2}{x}$$
$$= 100 - 10x$$

(per unit when x units are sold). Similarly if

$$\text{AC} = x^2 - 10x + 38$$
then $\text{TC} = x(x^2 - 10x + 38)$
$$= x^3 - 10x^2 + 38x.$$

If we begin with a cubic total cost function and find the marginal cost function, which will be

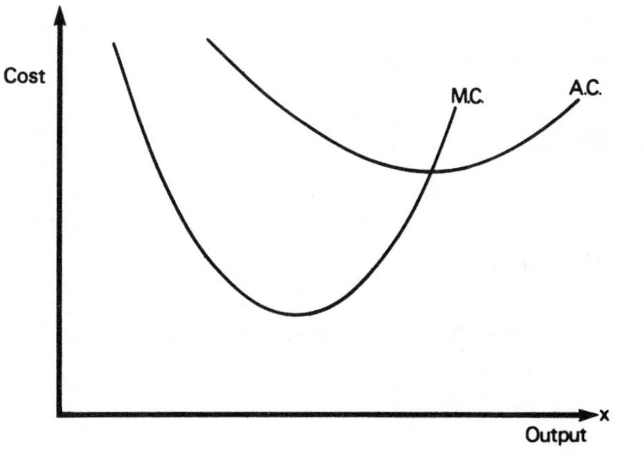

Figure 8.5

quadratic, and the average cost function, which will also be quadratic, then graph these, we will have the typical economic diagram, as in Figure 8.5.

With a linear demand curve (i.e. an AR function) we will obtain the relationship in Figure 8.6, where the marginal revenue function will also be linear, but have a slope twice that of the average revenue function.

Since if $\text{AR} = a + bx$
then $\quad \text{TR} = ax + bx^2$
and $\quad \text{MR} = a + 2bx.$

Elasticity of demand measures the responsiveness of the quantity sold to various factors: initially,

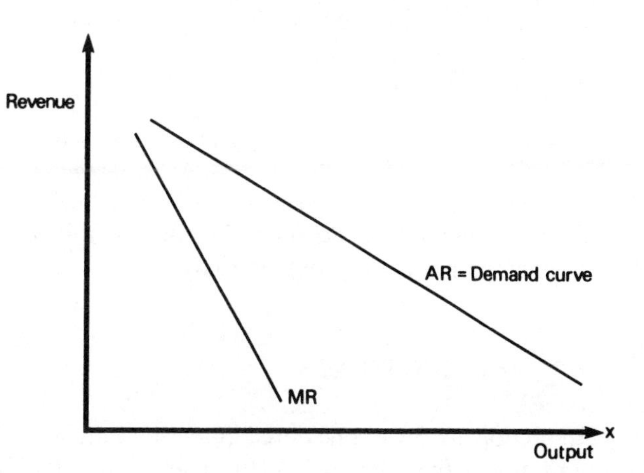

Figure 8.6

the most useful of these is price elasticity (E_D), which is defined as:

$$E_D = \frac{dq}{dp} \times \frac{p}{q}.$$

EXAMPLE

If the demand curve is given by

$p = 100 - 5q$

then

$$\frac{dp}{dq} = -5$$

and since

$$\frac{dq}{dp} = \frac{1}{\frac{dp}{dq}}$$

then

$$\frac{dq}{dp} = -\frac{1}{5}$$

At a quantity of 10, the price is:

$p = 100 - 5(10) = 50$

So, elasticity is:

$$E_D = \frac{1}{-5} \times \frac{50}{10} = -1$$

At a quantity of 8, the price is 60 and elasticity is

$$E_D = \frac{1}{-5} \times \frac{60}{8} = -1.5$$

So for a linear demand function, there is a *constant* slope but a *changing* elasticity.

8.3 TURNING POINTS

In section 6.5.3 we considered quadratic functions and stated that the sign of the coefficient of x^2 determined the shape of the function. With func-

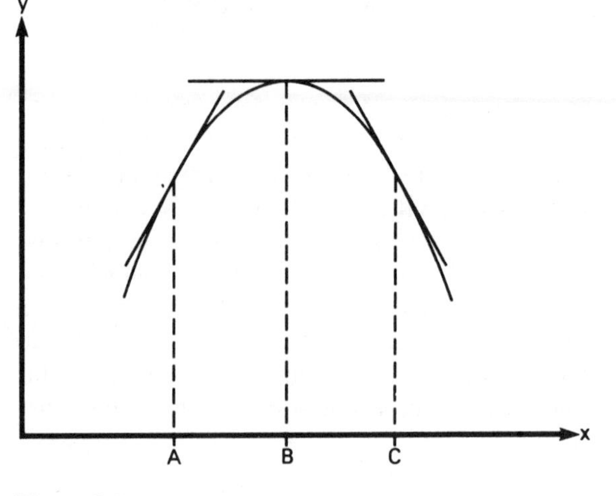

Figure 8.7

tions that include higher powers of x the graph will have several changes of direction. We can use the method of differentiation developed above to locate both *where* these changes of direction occur and *which way* the change affects the function. This is the method of locating the **maximum** and **minimum** values of a function. By a 'maximum' value we do not necessarily mean one that is greater than all other values (a **global** maximum), it could simply be greater than all neighbouring values only (a **local** maximum). Similar definitions apply to minimum values. For the economic applications at this stage, this distinction between local and global maxima is not of great importance.

If you consider the function represented in Figure 8.7 you can see that the slopes of the function at points A, B and C are quite different. At point A the function is increasing and thus the slope is positive. At point C the function is decreasing, and thus the slope is negative. However, at point B, it is just changing direction and is neither going up nor going down; therefore the slope is zero. Now the slope of a function can be found by differentiation.

In Figure 8.8, the first graph represents a function which has a maximum at the point A and a minimum at the point B, and the second graph shows dy/dx against x for this function. This shows that to the left of the point A there is a

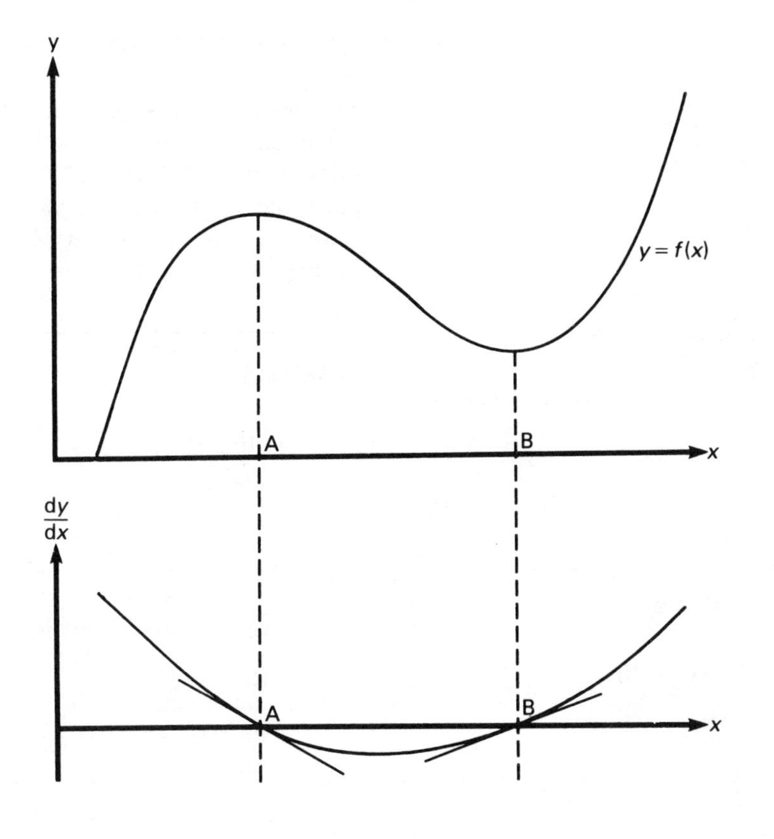

Figure 8.8

positive slope, but that slope is decreasing. At A there is zero slope, and so $dy/dx = 0$. After A the original function is decreasing and so dy/dx is negative, but when the point B is reached, where the original function begins to go up again, there will be zero slope, and dy/dx will again be 0.

To the right of point B, the original function is increasing, and hence the second graph is in the positive region.

From this we see that at both **turning points** the value of $dy/dx = 0$.

There is more information that we can gain from Figure 8.8. Looking at the second graph we find that at the point A, the value of dy/dx is decreasing as x increases, and thus has a negative slope, while at the point B, the value of dy/dx is increasing, and thus has a positive slope. Thus, if we can find the slope of the dy/dx function, we can distinguish between maximum points like A and minimum points like B.

In effect, this means differentiating the original function a second time, and we denote this second differential by a slightly different symbol:

$$\frac{d^2y}{dx^2}$$

The process of differentiating is still the same, only it is now applied to the function dy/dx.

The rules for distinguishing a maximum from a minimum value are thus:

for a maximum, $\frac{dy}{dx} = 0$ and $\frac{d^2y}{dx^2} < 0$;

for a minimum, $\frac{dy}{dx} = 0$ and $\frac{d^2y}{dx^2} > 0$.

EXAMPLE

1. If

$$y = 2x^2 - 8x + 50$$

then

$$\frac{dy}{dx} = 4x - 8 = 0$$

therefore $4x = 8$
and hence

$$x = \frac{8}{4} = +2$$

so the turning point is at $x = 2$.

$\frac{d^2y}{dx^2} = 4$, which is *positive*, and so the turning point is a minimum.

The value of y is $2(2)^2 - 8(2) + 50 = 8 - 16 + 50 = 42$
Thus the function $y = 2x^2 - 8x + 5$ has a minimum point at (2, 42).

2. $y = \frac{1}{3}x^3 - 4x^2 + 15x + 10$

then

$$\frac{dy}{dx} = x^2 - 8x + 15 = 0$$

factorizing gives $(x - 3)(x - 5) = 0$ so $x = 3$ and 5 and there is a turning point for each of these values.

$$\frac{d^2y}{dx^2} = 2x - 8.$$

Now at $x = 3$, $2x - 8 = 2(3) - 8 = -2$, (negative $\Rightarrow$ maximum)
and at $x = 5$, $2x - 8 = 2(5) - 8 = +2$, (positive $\Rightarrow$ minimum)

Also, at $x = 3$, $y = \frac{1}{3}(3)^3 - 4(3)^2 + 15(3) + 10 = 28$

and, at $x = 5$, $y = \frac{1}{3}(5)^3 - 4(5)^2 + 15(5) + 10 = 26.67$.

Thus the function

$$y = \frac{1}{3}x^3 - 4x^2 + 15x + 10$$

has a maximum at (3, 28) and a minimum at (5, 26.67).

8.4 ECONOMIC APPLICATIONS II

The economist is often interested in finding the maximum value of certain relationships for profit, sales revenue, welfare, etc., and minimum values, particularly for cost functions. The methods developed above will allow us to specify where these turning points occur.

EXAMPLES

1. If a firm has

$$TR = 40x - 8x^2$$
and
$$TC = 8 + 16x - x^2$$

where x = thousands of units of product, then its profit function (π) will be the difference between its total revenue and total cost:

$$\begin{aligned}\pi &= TR - TC \\ &= (40x - 8x^2) - (8 + 16x^2 - x^2) \\ &= 40x - 8x^2 - 8 - 16x + x^2 \\ &= -8 + 24x - 7x^2.\end{aligned}$$

If we wish to find where maximum profit occurs, we differentiate to give

$$\frac{d\pi}{dx} = 24 - 14x$$

and find that $d\pi/dx$ is zero when $x = 24/14 = 1.714$, and $d^2\pi/dx^2 = -14$, which is negative, thus representing a maximum.

So the firm will achieve maximum profit at an output of $x = 1.714$, and since x is measured in thousands, this is 1714 units. Profit here is 12.571. If the firm wished to maximize its sales revenue (i.e. TR), we have:

$$TR = 40x - 8x^2$$
$$\frac{dTR}{dx} = 40 - 16x$$

and when $dTR/dx = 0$, $x = 40/16 = 2.5$ and $d^2TR/dx^2 = -16$, which is negative, thus representing a maximum.

So, maximum sales revenue will be at 2 500 units, but profit here would only be 8.25.

2. If a firm's total cost function is

$$TC = 200 + 10x - 6x^2 + x^3$$

then the 200 represents fixed cost, since it does not vary with the level of output. If we remove this fixed cost, we will be left with total variable cost (TVC) with

$$TVC = 10x - 6x^2 + x^3$$

and average variable cost (AVC), can be found by dividing by x:

$$AVC = 10 - 6x + x^2$$

We often want to know the output to give minimum AVC, and to find this, we differentiate:

$$\frac{dAVC}{dx} = -6 + 2x = 0$$

therefore $x = 3$ and $\frac{d^2AVC}{dx^2} = +2$, positive, thus minimum.
So, minimum average variable cost is at an output of 3.

8.5 FURTHER NOTES

1. If the function to be differentiated is the product of two functions, then there is a method of differentiating without having to multiply the two functions out, e.g.

$$y = (2x^2 + 10x + 5)(6x^3 + 12x^2)$$

let $u = 2x^2 + 10x + 5$
and $v = 6x^3 + 12x^2$
so if $y = uv$

then, in general,

$$\frac{dy}{dx} = v \times \frac{du}{dx} + u \times \frac{dv}{dx}.$$

Now if $u = 2x^2 + 10x + 5$ then $\frac{du}{dx} = 4x + 10$

and if $v = 6x^3 + 12x^2$ then $\frac{dv}{dx} = 18x^2 + 24x$

and

$$\frac{dy}{dx} = (6x^3 + 12x^2)(4x + 10) + (2x^2 + 10x + 5)(18x^2 + 24x)$$

$$= 24x^4 + 60x^3 + 48x^3 + 120x^2 + 36x^4 + 48x^3 + 180x^3 + 240x^2 + 90x^2 + 120x.$$

$$= 60x^4 + 336x^3 + 450x^2 + 120x.$$

2. If the function to be differentiated consists of one function divided by another, then the following method is appropriate.

$$y = \frac{(10x^2 + 6x + 5)}{(12x^3 + 15x^2)}$$

so let $u = 10x^2 + 6x + 5$
and $v = 12x^3 + 15x^2$.

Then, if

$$y = \frac{u}{v}$$

then, in general,

$$\frac{dy}{dx} = \frac{v \times \frac{du}{dx} - u \times \frac{dv}{dx}}{v^2}.$$

If $u = 10x^2 + 6x + 5$ then $\frac{du}{dx} = 20x + 6$

and if $v = 12x^3 + 15x^2$ then $\frac{dv}{dx} = 36x^2 + 30x$

$$\frac{dy}{dx}$$

$$= \frac{(12x^3 + 15x^2)(20x + 6) - (10x^2 + 6x + 5)(36x^2 + 30x)}{(12x^2 + 15x^2)^2}$$

$$= \frac{240x^4 + 300x^3 + 72x^3 + 90x^2 - (360x^4 + 300x^3 + 216x^3 + 180x^2 + 180x^2 + 150x)}{144x^6 + 360x^5 + 225x^4}$$

$$= \frac{-120x^4 - 144x^3 - 270x^2 - 150x}{144x^6 + 360x^5 + 225x^4}$$

$$= \frac{-6x(20x^3 + 24x^2 + 45x + 25)}{x^4(144x^2 + 360x + 225)}$$

Note that it is not always necessary to carry out all of the multiplications and simplifications can help if made earlier in the calculation.

3. If the function to be differentiated is a function of another function, then the following method is appropriate.

If $y = (10x^2 + 6x)^3$

let $u = 10x^2 + 6x$ thus $\frac{du}{dx} = 20x + 6$

and $y = u^3$ thus $\frac{dy}{du} = 3u^2$

and, in general,

$$\frac{dy}{dx} = \frac{dy}{du} \times \frac{du}{dx}$$

$$= 3u^2(20x + 6)$$
$$= 3(10x^2 + 6x)^2(20x + 6)$$
$$= 3(100x^4 + 120x^3 + 36x^2)(20x + 6).$$

This could be further simplified.

4. If the function is exponential then if

$$y = e^x$$
$$\frac{dy}{dx} = e^x$$

If $y = e^{ax}$ then $\frac{dy}{dx} = ae^{ax}$

e.g. $y = e^{2x}$ and $\frac{dy}{dx} = 2e^{2x}$

If $y = e^{(x^2+6)}$ let $u = x^2 + 6$.

Then $y = e^u$ and $\frac{du}{dx} = 2x$

so $\frac{dy}{dx} = \frac{dy}{du} \times \frac{du}{dx}$

$$= e^u \times 2x = 2xe^{(x^2+6)}.$$

5. If $y = \log_e x$ then $\frac{dy}{dx} = \frac{1}{x}$

If $y = \log_e(x^3 + 2x + 3)$
let $u = x^3 + 2x + 3$
then $y = \log_e(u)$
and $\frac{du}{dx} = 3x^2 + 2$
so $\frac{dy}{dx} = \frac{dy}{du} \times \frac{du}{dx}$

$$= \frac{1}{u}(3x^2 + 2)$$

$$= \frac{3x^2 + 2}{x^3 + 2x + 3}.$$

8.6 INTEGRATION

Differentiation has allowed us to find an expression for the rate of change of a function; what we need now is some method of reversing the process, i.e. obtaining the original function when the rate of change is known. This process is **integration**.

If $y = x^2$, we know that the rate of change is $2x$; thus if we integrate $2x$ we must get x^2. However, if $y = x^2 + 10$, the rate of change is still $2x$, and integrating will give us x^2 and *not* $x^2 + 10$. This is because the constant in the initial expression has a zero rate of change, and therefore disappears when it is differentiated. When we integrate, then, we should add a constant c to the expression we obtain, and we will need some further information if we are to find the specific value of this constant. For example, if

$$\frac{dy}{dx} = 2x$$

integrating gives

$$y = x^2 + c$$

If we also know that $y = 10$ when $x = 0$ we have

$$10 = 0^2 + c$$

therefore $c = 10$ and hence $y = x^2 + 10$.

The symbol used for integration is an 'old style' S, i.e. $\int$, and it is usual to put dx after the expression to show that we are integrating with respect to x. Thus

$$\int(2x)dx = x^2 + c.$$

As with differentiation, there is a general formula for integration. If

$$y = ax^n$$

where a is a constant then

$$\int y dx = \frac{ax^{n+1}}{n+1} + c$$

for all values of n except $n = -1$. In that special case we have

$$\int \frac{1}{x} dx = \log_e x + c.$$

Reverting to the more usual case, if

$y = 15x^4$
$a = 15$ and $n = 4$ so

$$\int y dx = \frac{15x^{4+1}}{4+1} + c$$

$$= \frac{15x^5}{5} + c$$
$$= 3x^5 + c$$

When there are several terms in the function we may treat each separately, e.g.

$$y = 10x^3 + 6x^2 - 4x + 10$$
$$(a = 10,\ n = 3) \quad (a = 6,\ n = 2) \quad (a = -4,\ n = 1) \quad (a = 10,\ n = 0)$$

$$\int y\,dx = \frac{10x^{3+1}}{3+1} + \frac{6x^{2+1}}{2+1} - \frac{4x^{1+1}}{1+1} + 10x + c$$
$$= \frac{10x^4}{4} + \frac{6x^3}{3} - \frac{4x^2}{2} + 10x + c$$
$$= 2.5x^4 + 2x^3 - 2x^2 + 10x + c$$

Note that integrating the constant, 10, gives $10x$.

Integration can be viewed as a summation process; for example, if you sum all of the marginal costs (the cost of producing one more) up to some point, then you will obtain total cost. This idea can also be used to find a sum of areas bounded by a curve between two points (known as definite integration), provided the function is positive. For example

$$\int_2^5 (10x + 5)dx$$

means find the integral of $10x + 5$ i.e. find the sum of all products $(10x + 5)dx$, between the values $x = 2$ and $x = 5$.

Integrating gives the indefinite integral $5x^2 + 5x + c$, but the definite integral is usually written as

$$\left| 5x^2 + 5x \right|_2^5$$

omiting the constant, because in the evaluation (below) it cancels itself out.

Now we evaluate this at $x = 5$, and at $x = 2$ and subtract the second from the first:

$$(5x^2 + 5x)_{(x=5)} - (5x^2 + 5x)_{(x=2)}$$
$$= (125 - 25) \quad - (20 + 10)$$
$$= \quad 150 \quad - \quad 30$$
$$= 120$$

This problem is illustrated in Figure 8.9, the shaded area being the 120 found above.

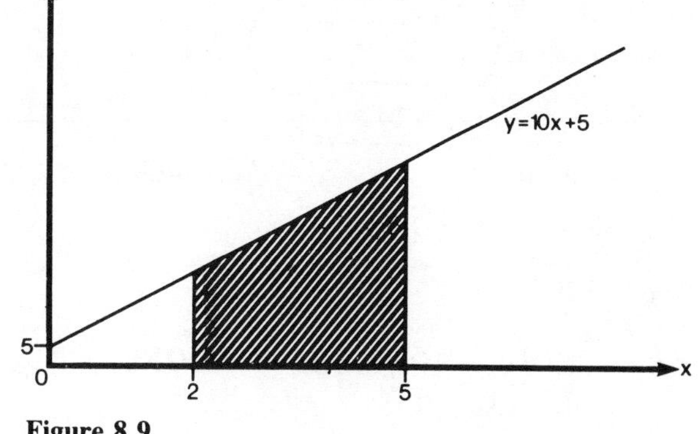

Figure 8.9

Within economics, the process of integration will allow us to go from marginal functions to total functions. Thus

$$\int \text{MR}dx = \text{TR} + c$$
$$\int \text{MC}dx = \text{TC} + c'$$

But with TR, there is rarely any revenue if output is zero, so in general the constant will also be zero. With TC there is a cost to the firm even if no production takes place, and so c' will be non-zero and positive. It will represent fixed cost.

If the marginal cost function for a firm has been identified as:

$$\text{MC} = 2x^2 - 8x + 10$$

and fixed costs are known to be 100, then the total cost function is given by:

$$\text{TC} = \int \text{MC}dx$$
$$= \int (2x^2 - 8x + 10)dx$$
$$= \tfrac{2}{3}x^3 - 4x^2 + 10x + c.$$

TC = 100 when $x = 0$, therefore $c = 100$, and thus the total cost function is:

$$\text{TC} = \tfrac{2}{3}x^3 - 4x^2 + 10x + 100.$$

8.7 ECONOMIC SUMMARY

Figure 8.10 summarizes the application of calculus to simple economic models, usually of one firm or one market.

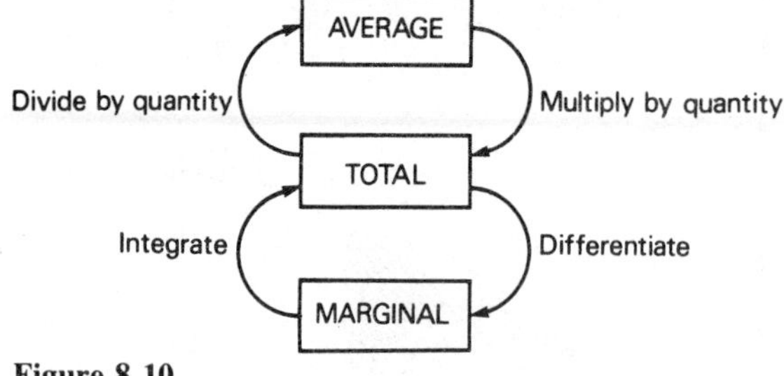

Figure 8.10

8.8 FUNCTIONS OF MORE THAN ONE VARIABLE

The simple economic models discussed so far embody functions of only *one variable.* As we move on to consider more complex economic and business situations, we find that there are *several* factors which will affect the outcome. Where the exact relationship is unknown, it must be estimated by statistical means (see Chapter 19), but for many examples in economics an exact function can be specified. For a duopoly (two sellers), the sales of one company's product may be a function of their own price, the price charged by the competitor and the respective amounts spent on advertising. In production theory, it is the *combination* of both capital and labour which determines the amount produced and not just one factor of production; while in welfare economics we consider the combination of goods which is available to the consumer or community. For all of these situations, while we are interested in the final outcome of particular decisions, we are also interested in the individual contribution of each factor, and also in ways of making optimal decisions on each factor in order to optimize the final outcome.

Consider a function such that y is determined by x and z; for example:

$$y = 10x^2 + 6xz + 15z^2.$$

We can find the rate of change of y with respect to x if we *temporarily* hold the value of z constant when differentiating. (To distinguish this from differentiation where $y = f(x)$ we now use a new symbol $\delta y/\delta x$.)

Differentiating, we have:

$$\frac{\delta y}{\delta x} = 20x + 6z.$$

Note that $10x^2$ was treated as before, giving $20x$, $6xz$ was treated as $(6z)x$, giving $6z$ and $15z^2$ was treated as a constant.

We can also find the rate of change of y with respect to z by temporarily holding x constant. Thus

$$\frac{\delta y}{\delta z} = 6x + 30z.$$

The process described above is known as **partial differentiation** since we are finding a rate of change while part of the function is held constant. As with the function of single variable, each of the functions we have obtained may be differentiated again, but note the number of outcomes:

$\frac{\delta^2 y}{\delta x^2} = 20$, i.e. holding z constant again and differentiating $20x + 6z$.

$\frac{\delta^2 y}{\delta z^2} = 30$, i.e. holding x constant again and differentiating $6x + 30z$.

$\frac{\delta^2 y}{\partial x/\partial z} = 6$, i.e. holding x constant and differentiating $20x + 6z$.

$\frac{\delta^2 y}{\delta z \delta x} = 6$, i.e. holding z constant and differentiating $6x + 30z$.

Note, however, that $\frac{\delta^2 y}{\delta x \delta z} = \frac{\delta^2 y}{\delta x \delta x}$ in general.

We may now use these results to find the maxima and minima for a particular function. For a **maximum**:

$$\frac{\delta y}{\delta x} = 0 \quad \text{and} \quad \frac{\delta y}{\delta z} = 0$$

$$\frac{\delta^2 y}{\delta x^2} < 0 \quad \text{and} \quad \frac{\delta^2 y}{\delta z^2} < 0$$

$$\text{and} \quad \left(\frac{\delta^2 y}{\delta x^2}\right)\left(\frac{\delta^2 y}{\delta z^2}\right) \geqslant \left(\frac{\delta^2 y}{\delta x \delta z}\right)^2$$

For a **minimum**:

$$\frac{\delta y}{\delta x} = 0 \quad \text{and} \quad \frac{\delta y}{\delta z} = 0$$

$$\frac{\delta^2 y}{\delta x^2} > 0 \quad \text{and} \quad \frac{\delta^2 y}{\delta z^2} > 0$$

$$\text{and} \quad \left(\frac{\delta^2 y}{\delta x^2}\right)\left(\frac{\delta^2 y}{\delta z^2}\right) \geqslant \left(\frac{\delta^2 y}{\delta x \delta z}\right)^2$$

EXAMPLES

1. If $y = 5x^2 + 4xz - 60x - 40z + 4z^2 + 1\,000$

then $\frac{\delta y}{\delta x} = 10x + 4z - 60 = 0$

$$\frac{\delta y}{\delta z} = 4x - 40 + 8z = 0$$

Rearranging gives two simultaneous equations:

$$10x + 4z = 60$$
$$4x + 8z = 40$$

multiplying up the first by 2 gives

$$20x + 8z = 120.$$

Subtracting the second gives

$$16x = +80.$$

Therefore, $x = 5$ and substituting gives $z = 2.5$

$$\frac{\delta^2 y}{\delta x^2} = 10, \quad \frac{\delta^2 y}{\delta z^2} = 8;$$

both positive, and

$$\frac{\delta^2 y}{\delta x \delta z} = 4$$

$$\left(\frac{\delta^2 y}{\delta x^2}\right)\left(\frac{\delta^2 y}{\delta z^2}\right) = 10 \times 8 = 80$$

which is greater than

$$\left(\frac{\delta^2 y}{\delta x \delta z}\right)^2 = 4^2 = 16$$

Thus, the function has a minimum when $x = 5$ and $z = 2.5$ (the y value will be 800).

2. Consider a monopolist with two products, A and B, who wishes to maximize total profit. The two demand functions are:

$$P_A = 80 - q_A$$
$$P_B = 50 - 2q_B$$

and the total cost function is

$$TC = 100 + 8q_A + 6q_B + 14q_A^2 + 4q_B^2 + 4q_Aq_B.$$

The total revenue function will consist of two parts:

$TR_A = 80q_A - q_A^2$ and $TR_B = 50q_B - 2q_B^2$

thus the total profit function is

$$\pi = TP_A + TR_B - TC$$
$$= -100 + 72q_A + 44q_B - 15q_A^2 - 6q_B^2 - 4q_Aq_B.$$

$$\frac{\partial \pi}{\partial q_A} = 72 - 30q_A - 4q_B = 0.$$

$$\frac{\partial \pi}{\partial q_B} = 44 - 12q_B - 4q_A = 0.$$

Rearranging these equations gives two simultaneous equations:

$$30q_A + 4q_B = 72$$
$$4q_A + 12q_B = 44$$

multiplying the first by 3 gives

$$90q_A + 12q_B = 216$$

Then subtracting the second gives

$$86q_A = 172$$

Therefore $q_A = 2$ and substituting gives $q_B = 3$.

$$\frac{\partial^2 \pi}{\partial q_A^2} = -30, \quad \frac{\partial^2 \pi}{\partial q_B^2} = -12;$$

both negative and,

$$\frac{\partial^2 \pi}{\partial q_A \partial q_B} = -4$$

but $(-30)(-12) > (-4)^2$.

Thus maximum profit is where the quantity of A sold is 2, and the quantity of B sold is 3. By substitution into the profit function, we find that the level of profit will be 38. The prices charged for the two products are found from the demand functions; the price of A is 78 and the price of B is 44.

8.9 MAXIMIZATION AND MINIMIZATION SUBJECT TO CONSTRAINTS

Many economics problems have **limitations** upon the solution that may be obtained: production may be limited by the available supply of raw materials or by the capacity of the current factory; cost may have to be kept to a minimum subject to certain, predetermined, output levels; or a consumer's utility maximized subject to the budget. In each of these cases, if we simply maximize or minimize the function it is unlikely that this will also satisfy the constraint and thus we need to take the constraint into consideration during the optimization process. For simple functions it may be possible to substitute the constraint into the function, but the more general method is that of **Lagrangian multipliers**.

8.9.1 Substitution

For functions of two variables, we may be able to substitute the constraint into the function. Consider the following example:

$$z = 10 + 3x - 4x^2 + 10y - 2y^2 + xy$$

to be maximized subject to the constraint that $x = 2y$.

Substituting, we have:

$$\begin{aligned} z &= 10 + 6y - 16y^2 + 10y - 2y^2 + 2y^2 \\ &= 10 + 16y - 16y^2 \end{aligned}$$

$\frac{dz}{dy} = 16 - 32y = 0$ therefore $y = \frac{1}{2}$ and hence $x = 1$.

$\frac{d^2z}{dy^2} = -32$ therefore it is a maximum.

EXERCISE

Find the values of x and y to maximize the function without the constraint. (Answer: $x = 22/31; y = 83/31$.)

8.9.2 Lagrangian multipliers

For more complex functions the method above will become tedious and thus we now place the constraint as part of the function we are to optimize. Again, we will work through an example. Suppose

$$z = 10x - 4x^2 + 20y - y^2 + 4xy$$

subject to $2x + 4y = 100$. The first step is to rearrange the constraint as follows:

$$100 - 2x - 4y = 0.$$

We then multiply it by a new unknown value, λ (lambda), which is the Lagrangian multiplier:

$$\lambda(100 - 3x - 4y) = 0$$

and this function is added to the function z to give a new function z^*. Therefore

$$z^* = 10x - 4x^2 + 20y - y^2 + 4xy + \lambda(100 - 2x - 4y).$$

This will not change the value of z since, as we have just seen, the term added is equal to 0, and it can be shown that if we find the optimum point or points of the function z^* then these will be the optimum points for the original function z subject to the constraint.

Partially differentiating z^*, we have:

$$\frac{\delta z^*}{\delta x} = 10 - 8x + 4y - 2\lambda = 0$$

$$\frac{\delta z^*}{\delta y} = 20 - 2y + 4x - 4\lambda = 0$$

and

$$\frac{\delta z^*}{\delta \lambda} = 100 - 2x - 4y = 0$$

(thus satisfying the constraint).

As we have seen in Chapter 7, there are several ways of solving three simultaneous equations and these will not be repeated here: solving the three equations gives $x = 10$, $y = 20$, and $\lambda = 5$ (although strictly we do not need to know the value of λ).

Further differentiation shows that $x = 10$, $y = 20$ is a maximum for the function z subject to the constraint.

(Note that although $\left(\frac{\partial^2 z}{\partial x^2}\right)\left(\frac{\partial^2 z}{\partial y^2}\right)$ equals $\left(\frac{\partial^2 z}{\partial y \partial x}\right)^2$, and is not 'greater than', there is sufficient evidence to show a maximum.)

EXAMPLE

Consider a Cobb-Douglas production function

$Q = 10L^{1/2}K^{1/2}$

subject to the constraint that $4L + 10K = 100$, where L is the amount of labour and K is the amount of capital.

Rearranging the constraint, we have

$$100 - 4L - 10K = 0$$

and $\quad \lambda(100 - 4L - 10K) = 0$

We thus wish to maximize the function

$Q^* = 10L^{1/2}K^{1/2} + \lambda(100 - 4L - 10K)$

$$\frac{\partial Q^*}{\partial L} = 5L^{-1/2}K^{-1/2} - 4\lambda = 0 \quad \text{so} \quad \lambda = \frac{5K^{1/2}}{4L^{1/2}}$$

$$\frac{\partial Q^*}{\partial K} = 5L^{1/2}K^{1/2} - 10\lambda = 0 \quad \text{so} \quad \lambda = \frac{5L^{1/2}}{10K^{1/2}}$$

Thus

$$\frac{5K^{1/2}}{4L^{1/2}} = \frac{5L^{1/2}}{10K^{1/2}}$$
$$50K = 20L$$
$$L = 2.5K$$

But

$$\frac{\delta Q^*}{\partial \lambda} = 100 - 4L - 10K = 0$$
$$100 - 10K - 10K = 0$$

so $K = 5$ and hence $L = 12.5$.

Further partial differentiation shows this to be a maximum, and thus production (Q) is maximized subject to the constraint when $L = 12.5$ and $K = 5$. Note that the function Q itself, when unconstrained, has no maximum value.

EXERCISE

Check that the constraint is satisfied and find the value of Q.
(Answer: $Q = 79.06$.)

8.10 CONCLUSIONS

Calculus was developed during the seventeenth century from problems experienced by physical scientists, like Galileo, trying to describe the world. It arose from the need to give an adequate mathematical account of changes in motion. Although the more immediate applications of calculus were found in the physical sciences, calculus has played an important part in the development of other discipines; particularly economics. Whenever we need to consider the effects of change, calculus provides a language and a set of techniques to manage the mathematical descriptions.

8.11 PROBLEMS

Differentiate each of the following functions:

1. $y = 12$
2. $y = 10 + 2x$
3. $y = 3x + 7$
4. $y = 2x^5$
5. $y = 14x^3 - 3x^4$
6. $y = 17x^2 + 14x + 10$
7. $x = 7y + 2y^2 - 4y^3$
8. $p = 2q^2 - 10 + 4q^{-1}$
9. $y = e^x$
10. $r = 2s + 4s^2 + s^{1/2}$
11. $y = 20x^7 + 4x^6 - 3x^5 - 7x^4 + 4x^3 - 11x^2 + 2x + 57 - x^{-1} + 9x^{-2}$
12. $y = \frac{1}{x^2} - \frac{2}{x^3} - \frac{4}{3x^4}$
13. If average revenue for a company is given by:

    $AR = 100 - 2x$ (x is quantity)

    find the marginal revenue function and the price when $MR = 0$.
    Graph the AR, MR, and TR functions.
14. A firm's total costs are given by the following function:

    $TC = (1/3)x^3 - 5x^2 + 30x$ (x is quantity)

    Find the average cost and marginal cost functions. Graph the three functions.
15. Using the information in the previous two questions, find the quantity levels where:
    (a) $AR = AC$; and
    (b) $MR = MC$.

16. A firm's average cost function is given as:

$AC = x^2 - 3x + 25$ (x = quantity)

and its total revenue function as:

$TR = 60x - 2x^2$

Find the quantity levels where:
(a) MC = MR; and
(b) AR = AC.

17. Given the following demand function:

$P = 40 - 2x$

find the price elasticity of demand at quantity levels of:
(a) 8;
(b) 10;
(c) 12.
Construct a graph of the average revenue and marginal revenue functions. What may you determine from this graph and your previous calculations?

18. An oligopolist sells a product at £950 per unit of production, and estimates the average revenue function to consist of two linear segments.

Below 10 units, AR = 990 when output = 2
AR = 965 when output = 7
Above 10 units, AR = 840 when output = 12
AR = 565 when output = 17

(a) Find the equation of the average revenue function below 10 units of output.
(b) Find the equation of the average revenue function above 10 units of output.
(c) Find the equations of the *two* marginal revenue functions, assuming that total revenue is zero if output is zero.
(d) Sketch the average and marginal revenue functions.
(e) Find the price elasticity of demand for a price rise and for a price fall from the current position.
(f) Find the range of marginal cost figures that are consistent with profit maximization and the current output level (N.B. for profit maximization MC = MR).

Find the maximum and/or minimum of each of the following functions, checking the second order conditions (i.e. d^2y/dx^2):

19. $y = x^2 - 10x + 25$
20. $y = -2x^2 + 40x + 1\,000$
21. $y = 3x^2 + 6x + 5$
22. $y = 2x + 1$
23 $y = 3x^2 - 10x + 4$
24. $y = \frac{1}{3}x^3 - 3x^2 + 5x + 10$
25 $a = -4b^2 + 2b + 10$
26. $y = 2x^3 - 12x^2 + 12x + 10$
27 $y = 25$
28. $y = 0.25x^4 - 2x^3 + 5.5x^2 - 6x + 40$

29. A company's profit function is given by:

profit $= -2x^2 + 40x + 10$ (where x = output).

Determine the profit maximizing output level for the company, checking that the second order conditions are met.

30. A firm has the following total revenue and total cost functions:

$TR = 100x - 2x^2$
$TC = (\frac{1}{3})x^3 - 5x^2 + 30x$ (where x = output).

Find the output level to maximize profits, and the level of profit achieved at this output.

31. A firm has an average revenue function of

$AR = 60 - 2x$

and an average cost function of

$AC = x^2 - 3x + 25$ (where x = output).

Find the output level for maximum profit, and the level of profit. (NB check the second order conditions.)

32. Given the following average cost function, show that the marginal cost function cuts the average cost function at the latter's minimum point.

$AC = 2x^2 - 4x + 100$

Use the rules of differentiation outlined in section 8.5 to find the first differentials of the following functions:

33. $y = (x^2 + 6)(2x + 5)$
34. $y = (3x^2 + 4x + 10)(2x^2 + 6x + 5)$
35. $y = (4x^3 + 6x^2)^3$
36. $y = e^{x^2/2}$
37. $y = (4x + 6x^2)/(x^3 + 6x)$
38. $y = 2x/(3x^4 + 10)$

Integrate each of the following functions:

39. $y = 4x$
40. $y = 3x^2 + 4x + 10$
41. $y = 2x^3 + 6x^2 + 10x^{-2}$
42. $y = 25$
43. $y = x^2 + x + 5$
44. $y = x + 10$ between limits of $x = 2$ and $x = 4$.
45. $y = 3x^2 + 4x + 5$ between limits of $x = 0$ and $x = 5$.
46. $y = -2x^2 + 40x + 10$ between limits of $x = 0$ and $x = 10$.
47. A firm's market demand function is

$$\text{AR} = 150 - 2x$$

and its marginal cost function is

$$\text{MC} = \tfrac{1}{3}x^2 - 5x + 30.$$

(a) Find the level of output to maximize profits.
(b) Find the level of profit and the price at this point.
(c) Find the price elasticity of demand at this point.
(d) Sketch the total revenue, total cost and profit functions.
(e) If a tax of 10 per unit is imposed on the firm, what effect will this have on the production level for maximum profit.

48. A company is faced by the following marginal cost and marginal revenue functions:

$$\text{MC} = 16 - 2Q$$
$$\text{MR} = 40 - 16Q$$

It is also known that fixed costs are 8 when production is zero. Find:

(a) the total cost function,
(b) the total revenue function,
(c) the output to give maximum sales revenue,
(d) the output to give maximum profit,
(e) the range of output for which profit is at least 1; and
(f) sketch the total cost and total revenue functions.

Find the first partial derivatives for each of the following:

49. $y = 4xz$
50. $y = 2x^3 + 4x^2z + 2xz^2 - 3z^3$
51. $y = 7x + 3x^2z - x^3z^2 - 2xz + 5xz^2 - 7xz^3 - 10$
52. $y = 10x^4 - 15z^3$
53. $q = 2p_1^2 - 3p_1 + 4p_1p_2 - 5p_2 + p_2^2$
54. $r = 2t(s^2 + s + 5) - 3s(2t^2 - 4t + 7)$

Find the maxima or minima of the following functions:

55. $y = 2x^2 - 2xz + 2.5z^2 - 2x - 11z + 20$
56. $y = 3x^2 + 3xz + 3z^2 - 21x - 33z + 100$
57. $y = -2x^2 - 2xz - 4z^2 + 40x + 90z - 150$
58. $z = 19x - 4x^2 + 16y - 2xy - 4y^2$
59. $y = x^2 + xz + z^2$
60. $y = 100 - 4x$
61. If a firm's total costs are related to its workforce and capital equipment by the function

$$\text{TC} = 10L^2 + 10K^2 - 25L - 50K - 5LK + 2000$$

where L = thousands of employees and K = thousands of pounds invested in capital equipment, find the combination of labour and capital to give minimum total cost. Find this cost and show that it is a minimum.

62. A monopolist has two products, X and Y, which have the following demand functions:

$$P_X = 26.2 - X$$
$$P_Y = 24 - 2Y.$$

The total costs of the monopolist are given by

$$\text{TC} = 5X^2 + XY + 3Y^2.$$

Determine the amounts of X and Y the monopolist should produce to maximize profit, and the amount of profit produced.

63. A company is able to sell two products, X and Y, which have demand functions:

$$P_X = 52 - 2X$$
$$P_Y = 20 - 3Y$$

and has a total cost function

$$\text{TC} = 10 + 3X^2 + 2Y^2 + 2XY$$

(a) Determine the profit maximizing levels of output for X and Y and the level of this profit.
(b) If product Y were not produced, determine the production level of X to maximize profit and the level of this profit.

(c) If product X were not produced, determine the production level of Y to maximize profit and the level of this profit.

Maximize or minimize the following functions subject to their constraints, using either substitution or the method of Lagrangian multipliers.

64. $y = 10 + 20x + 6z - 4x^2 - 2z^2$ subject to $x = 3z$.
65. $y = 4x^2 - 6x + 7z + 3z^2 + 2xz$ subject to $x = 5z$.
66. $z = 10x - x^2 + 14y - 2y^2 + xy$ subject to $2x + 3y = 3100$.
67. $z = 6x - 3x^2 + 40y - 8y^2 + 5xy$ subject to $4x - 5y = 30$.
68. $z = 2xy$ subject to $x + y = 1$.
69. Maximize the production function

$$Q = 4L^{1/2}K^{1/2}$$

subject to the constraint that $3L + 5K = 200$, finding the values for L and K.

70. Find the values of L and K to maximize the production function

$$Q = 8L^{1/4}K^{3/4}$$

subject to constraint $L + 8K = 1\,000$.

71. A firm's profit is given by the function

$$\pi = 600 - 3x^2 - 4x + 2xy - 5y^2 + 48y$$

where π denotes profit, x output and y advertising expenditure.

(a) Find the profit maximizing values of x and y and hence determine the maximum profit. Confirm that second order conditions are satisfied.

(b) If now the firm is subject to a budget constraint $2x + y = 5$ determine the new values of x and y which maximize profit.

(c) Without further calculation determine the effect of a constraint $2x + y = 8$.

THE TIME VALUE OF MONEY

9

In the 'civilized' world we have come to value money for what it can buy us now and for what it can buy us in the future. As individuals we continually decide whether to purchase items now with the aid of a loan or wait until our savings are sufficient. In making these decisions we need to consider how much the loan would cost and how long saving would take. House purchase for many is now the most important single expenditure of a lifetime involving sums many times their annual disposable income. Few would consider saving the complete amount before buying, even if they expected savings to keep abreast of inflation. In general, a deposit is saved and the rest acquired by a mortgage or an endowment policy.

The performance of a business for the most part is assessed in terms of monetary value. It may aim to provide a good service, a good product and achieve a good market share but it will still need to make and continue to make money to survive. Decisions will continually need to be made between competing projects with differing costs and differing returns over time.

In this chapter we consider what makes us save, invest and borrow in the ways we do.

9.1 INTEREST: SIMPLE AND COMPOUND

An interest rate, usually quoted as a percentage, gives the gain we can expect from each £1 saved. If, for example, we were offered 10% per annum we would expect a gain of 10 pence for each £1 saved. The interest received each year will depend on whether the interest is calculated on the basis of simple interest or compound interest.

9.1.1 Simple interest

If money is invested in a saving scheme where the interest earned each year is paid out and not reinvested, then the saving scheme offers simple interest. The simple interest, I, offered at the end of each year is

$$I = A_0 \times \frac{r}{100}$$

where A_0 is the initial sum invested and r is the rate of interest as a percentage.

EXAMPLE

What is the simple interest gained from a investment of £120 for a year at an interest rate of 8% per annum?

$I = 120 \times 0.08 = £9.60.$

As simple interest remains the same over the life of the investment, the interest gained over t years is given by $I \times t$ and the value of the investment (including the initial sum A_0) by $A_0 + It$.

EXAMPLE

A sum of £120 is invested at simple interest for 5 years at an interest rate of 14% per annum. What interest is paid over the 5 years and what is the value of the investment?

Interest over 5 years = £120 × 0.14 × 5 = £84.

Value of the investment = £120 + £84 = £204.

9.1.2 Compound interest

If the interest gained each year is added to the sum saved, we are looking at compound interest. The sum carried forward grows larger and larger and the interest gained each year grows larger and larger; the interest is **compounded** on the initial sum plus interest to date. Given an interest rate of 10%, after 1 year our £1 would be worth £1.10 and after 2 years our £1 would be worth £1.21.

The sum at the end of each year can be calculated using the following formula:

$$A_t = A_0\left(1 + \frac{r}{100}\right)^t$$

where A_0 is the initial sum invested, A_t is the sum after t years and r is the rate of interest as a percentage.

EXAMPLE

What amount would an initial sum of £100 accumulate to in 8 years if it were invested at a compound interest rate of 10% per annum?

By substitution we obtain

$$A_8 = 100\left(1 + \frac{10}{100}\right)^8 = 100(1.1)^8 = 100(2.143588)$$
$$= £214.36$$

After 8 years of investment, at a compound rate of interest of 10%, the initial sum of £100 would be worth £214.36.

A proof of this formula will demonstrate the principle involved. Suppose we invest an initial sum of A_0 at an annual interest rate of $r\%$. After 1 year, this sum would be worth

$$A_1 = A_0 + \frac{r}{100}A_0$$
$$= A_0\left(1 + \frac{r}{100}\right)^1.$$

The term in brackets gives the annual gain of each £1. An interest rate of 10% would give a multiplicative factor of 1.1 in the first year. If the sum at the end of the first year is then carried forward for one more year, the investment of A_0 at the end of the second year would be worth

$$A_2 = A_1 + \frac{r}{100}A_1$$
$$= A_1\left(1 + \frac{r}{100}\right)^1.$$

We can substitute the value we already have for A_1 to obtain

$$A_2 = A_0\left(1 + \frac{r}{100}\right)^1\left(1 + \frac{r}{100}\right)^1$$
$$= A_0\left(1 + \frac{r}{100}\right)^2.$$

If the interest rate is 10%, then the multiplicative factor of 1.1 is used twice to obtain the value of the investment after 2 years.

Continuing in this way, we can obtain the general formula for any number of years t:

$$A_t = A_0\left(1 + \frac{r}{100}\right)^t.$$

The example has assumed that the interest is paid at the end of the year, so that the £1 invested at 10% gains 10p each year. Some investments are of this type, but many others give or charge interest every 6 months (e.g. building societies) or more frequently (e.g. interbank loans interest is charged per day). The concept of compound interest can deal with these situations provided that we can identify how many times per year interest is paid. If, for example, 5% interest is paid every 6 months, during the second 6 months we will be earning interest on more than the initial amount invested. Given an initial sum of £100, after 6

months we have £100 + £5 (interest) = £105 and after 1 year we would have £105 + £5.25 (interest) = £110.25.
Calculation takes place as follows:

$$A_1 = £100\left(1 + \frac{0.1}{2}\right)^{2\times 1} = £110.25$$

i.e. divide the annual rate of interest (10% in this case) by the number of payments each year, and multiply the power of the bracket by the same number of periods.

More generally,

$$A_t = A_0\left(1 + \frac{r}{100 \times m}\right)^{mt}$$

where m is the number of payments per year.

9.1.3 The use of present value tables

The growth of an investment is determined by the interest rate and the time scale of the investment as we can see from the multiplicative factor:

$$\left(1 + \frac{r}{100}\right)^t$$

Any mathematical term of this kind can be tabulated to save repeated calculations. In this particular case, the tabulation takes the form of present value factors (see Appendix F) which can be used indirectly to find a compound interest factor. These factors are the reciprocal of what is required:

$$\text{present value factor} = \frac{1}{\left(1 + \frac{r}{100}\right)^t}$$

hence

$$\left(1 + \frac{r}{100}\right)^t = \frac{1}{\text{present value factor}}$$

If we were interested in the growth over 8 years of an investment made at 10% per annum we could first find the present value factor of 0.4665 from tables and then use the reciprocal value of 2.1436 to calculate a corresponding accumulated sum.

EXAMPLE

What amount would an initial sum of £150 accumulate to in 8 years if it were invested at a compound interest rate of 10% per annum?

Using tables

$$A_8 = 150 \times \frac{1}{0.4665} = 150 \times 2.1436 = £321.54.$$

It should be noted that the use of tables with four significant digits sometimes results in rounding errors.

9.2 DEPRECIATION

In the same way that an investment can increase by a constant percentage each year as given by the interest rate, the book value of an asset can decline by a constant percentage. If we use r as the depreciation rate we can adapt the formula for compound interest to give

$$A_t = A_0\left(1 - \frac{r}{100}\right)^t$$

where A_t becomes the book value after t years.

EXAMPLE

Use the 'declining balance method' of depreciation to find the value after 3 years of an asset initially worth £20 000 and whose value declines at 15% per annum.

By substitution we obtain

$$A_3 = £20\,000\left(1 - \frac{15}{100}\right)^3 = £20\,000\,(0.85)^3$$
$$= £12\,282.50.$$

A manipulation of this formula gives the following expression for the rate of depreciation.

$$r = \left(1 - \sqrt[t]{\frac{A_t}{A_0}}\right) \times 100$$

where A_0 is the original cost and A_t is the salvage (or scrap) value after t years.

EXAMPLE

The cost of a particular asset is known to be £20 000 and its salvage value is estimated at £8 000 after a useful life of 5 years. Determine the book value of this asset at the end of each year.

By substitution

$$r = \left(1 - \sqrt[5]{\frac{£8000}{£20\,000}}\right) \times 100 = 16.74\%$$

We can now depreciate the asset at 16.74% a year:

	£
cost	20 000.00
depreciation year 1 (16.74% of £20 000)	3 348.00
book value (end of year 1)	16 652.00
depreciation year 2 (16.74% of £16 652)	2 787.54
book value (end of year 2)	13 864.46
depreciation year 3 (16.74% of £13 864.46)	2 320.91
book value (end of year 3)	11 543.55
depreciation year 4 (16.74% of £11 543.55)	1 932.39
book value (end of year 4)	9 611.16
depreciation year 5 (16.74% of £9611.16)	1 608.91
book value (end of year 5)	8 002.25

It is worth noting that this procedure is unlikely to provide the scrap value exactly, due to the rounding errors. In this case, it is an accounting practice to make an adjustment in the final year's depreciation to get the exact salvage value.

9.3 PRESENT VALUE

The formula for compound interest can be rearranged to allow the calculation of the amount of money required now to achieve a specific sum at some future point in time given a rate of interest:

$$A_0 = A_t \frac{1}{\left(1 + \frac{r}{100}\right)^t}$$

EXAMPLE

What amount would need to be invested now to provide a sum of £242 in 2 years' time given that the market rate of interest is 10%?

By substitution we obtain

$$A_0 = £242 \times \frac{1}{\left(1 + \frac{10}{100}\right)^2}$$
$$= £242 \times 0.826446 = £200$$

The amount required now to produce a future sum can be taken as a measure of the worth or the **present value** of the future sum. A choice between £200 now and £200 in 2 years' time would be an easy one for most people. Most would prefer the money in their pockets now. Even if the £200 were intended for a holiday in 2 years' time, it presents the owner with opportunities; one of which is to invest the sum for 2 years and gain £42. The choice between £200 now and £242 in 2 years' time, however, would be rather more difficult. If the interest rate were 10%, the present value of £242 in 2 years' time would be £200. Indeed, if one were concerned only with interest rates there would be an indifference between the two choices.

Present value provides a method of comparing monies available at different points in time. As such, it presents these in business with a basis for making decisions. The calculation of the present value of future sums of money is referred to as **discounting**.

EXAMPLE

You need to decide between two business opportunities. The first opportunity will pay £700 in 4 years' time and the second opportunity will pay £850 in 6 years' time. You have been advised to discount these future sums by using the interest rate of 8%.

By substitution we are able to calculate a present value for each of the future sums.
First opportunity:

$$A_0 = £700 \times \frac{1}{\left(1 + \frac{8}{100}\right)^4}$$
$$= £700 \times 0.7350$$
$$= £514.50$$

Second opportunity:

$$A_0 = £850 \times \frac{1}{\left(1 + \frac{8}{100}\right)^6}$$
$$= £850 \times 0.6302$$
$$= £535.67$$

On the basis of present value (or discounting) we would choose the second opportunity as the better business proposition. In practice, we would need to consider a range of other factors.

The present value factors of 0.7350 and 0.6302 calculated in this example could have been obtained directly from tables (see section 9.1.3); look at the present value factors given in Appendix F. We can explain the meaning of these factors in two ways. Firstly, if we invest $73\frac{1}{2}$ pence at 8% per annum it will grow to £1 in 4 years and 63 pence (0.6302) will grow to £1 in 6 years. Secondly, if the rate of interest is 8% per annum, £1 in 4 years' time is worth $73\frac{1}{2}$ pence now and £1 in 6 years' time is worth 63 pence now.

EXAMPLE

A business needs to choose between two investment options. It has been decided to discount future returns at 12%. The expected revenues and initial cost are given as follows:

Year	*Estimated end of year revenue* *option 1*	*option 2*
1	300	350
2	350	350
3	410	350
Cost in year 0	300	250

In calculating the present value from different options we generally refer to a discount rate or the rate of return on capital rather than the interest rate. These rates tend to be higher than the market interest rate and reflect the cost of capital to a particular business. Net present value (NPV) for each option is the sum of the constituent present values.

The present value factors can be obtained directly from tables or calculated using $(1 + 0.12)^{-t}$ for the years $t = 1$ to 3. In this example, the costs are immediate and therefore are not discounted.

Year	*Present value factors*	*Revenues option 1*	*Revenues option 2*	*Present value option 1*	*Present value option 2*
1	0.8929	300	350	267.87	312.515
2	0.7972	350	350	279.02	279.02
3	0.7118	410	350	291.838	249.13
	Present value (PV)			838.728	840.665
	Cost			−300	−250
	Net present value (NPV)			538.728	590.665

Although the total revenue over 3 years is slightly higher with option 1, the value now to a business is higher with option 2. A more immediate revenue presents a business with more immediate opportunities. In can be seen in this example that option 2 offers the business an extra £50 in the first year which can itself be used for additional gain, and is especially useful in maintaining cash flow.

The comparison of these two options depends

crucially on the 'time value of money', that is the discount rate, the estimates given for revenue and the completeness of information.

EXERCISE

Construct a speadsheet model for the above example. Use your model to evaluate the two options using a discount rate of a) 8% and b) 14%.

This type of exercise looks straightforward in a textbook but presents a series of problems when it is to be used in business. Initial costs of each project to be considered will be known, but there may be extra costs involved in the future which cannot be even estimated at the start, e.g. a change in tariffs in a country to which the company exports. All further cash flow information must be estimated, since it is to come in the future, and is thus open to some doubt: if we are dealing with a new product these cash flows are likely to be based on market research (see Chapter 1 on survey methods).

A further practical difficulty is to decide upon which discount rate to use in the calculations. This could be:

1. the market rate of interest;
2. the rate of return gained by the company on other projects;
3. the average rate of return in the industry;
4. a target rate of return set by the board of directors; or
5. one of these plus a factor to allow for the 'riskiness' of the project — high risk projects being discounted at a higher rate of interest.

High risk projects are likely to be discriminated against in two ways: by the discount rate used which is likely to be high, and in the estimated cash flows which are often conservatively estimated.

All attempts to use this type of present value calculation to decide between projects make an implicit assumption that the project adopted will succeed.

Net present value is only an *aid* to management in deciding between projects, as it only considers the monetary factors, and those only as far as they are known, or can be estimated: there are many personal, social and political factors which may be highly relevant to the decision. If a company is considering moving its factory to a new site, several sites may be considered, and the net present value of each assessed for the next 5 years. However, if one site is in an area that the managing director (or spouse) dislikes intensely, then it is not likely to be chosen. The workforce may be unwilling to move far, so a site 500 miles away could present difficulties. There may be further environmental problems, which are not costed by the company, but are a cost to the community, e.g. smoke, river pollution, extra traffic on country roads.

EXERCISES

1. Should a company try to include environmental (or social) costs in its calculations of net present value? How could these be incorporated into the calculations?
2. Evaluate the two investment options given in the example on the preceding page using a discount rate of 9%.
   (Answer: option 1, NPV = £586.42; option 2, NPV = £635.96.)
3. Suppose we are given more complete information on the two options. The business discovers that it will cost £10 a year in each of years 1 to 3 to implement option 1 and cost an additional £20 immediately to implement option 2. In addition, option 1 will have a scrap value (book value) of £30 at the end of year 3 and option 2 a scrap value of £28 at the end of year 4. Use a discount rate of 10% to evaluate these two options and comment on your analysis.
   (Answer: option 1, NPV = £567.67; option 2, NPV = £619.50.)

9.4 THE INTERNAL RATE OF RETURN

The method of calculating the internal rate of return is included in this chapter because it provides a useful alternative to the net present value method of investment appraisal. The internal rate of return (IRR), sometimes referred to as the **yield**, is the discount rate that produces a net present value of 0. This is the value for r, which equates the net present value of cost to the net present value of benefits.

EXAMPLE

Suppose £1 000 is invested now and gives a return of £1 360 in one year. The internal rate of return can be calculated as follows:

net present value of cost = net present value of benefits

$$£1\,000 = \frac{1\,360}{(1 + r)}$$

since cost is incurred now and benefits are received in one year. So,

$$100(1 + r) = 1\,360$$
$$1\,000 + 1\,000r = 1\,360$$
$$r = 360/1\,000$$
$$= 0.36 \text{ or } 36\%.$$

EXAMPLE

Suppose £1 000 is invested now and gives a return of £800 in one year and a further £560 after two years. We can proceed in the same way:

$$1\,000 = \frac{800}{(1 + r)} = \frac{560}{(1 + r)^2}$$

Multiplying by $(1 + r)^2$ gives

$$1\,000(1 + r)^2 = 800(1 + r) + 560$$
$$1\,000(1 + r)^2 - 800r - 800 - 560 = 0$$
$$1\,000 + 2\,000r + 1000r^2 - 800r - 1\,360 = 0$$
$$1\,000r^2 + 1\,200r - 360 = 0$$

The solution to this quadratic equation (see section 6.5.3) is given by $r = -1.4485$ or 0.2485. Only the latter, positive value has a business interpretation, and the internal rate of return in this case is 0.2485 or more simply expressed as 25%.

The internal rate of return cannot be calculated so easily for longer periods of time and is estimated either using a graphical method or by linear interpolation.

9.4.1 The internal rate of return by graphical method

To estimate the internal rate of return graphically we need to determine the net present value of an investment corresponding to two values of the discount rate r. The accuracy of this method is improved if one value of r gives a small positive net present value and the other value of r a small negative value.

EXAMPLE

Suppose an investment of £460 gives a return of £180, £150, £130 and £100 at the end of years 1, 2, 3 and 4.

Using a spreadsheet format we can determine the net present values of this cash flow for two values of r. In this example, $r = 8\%$ and $r = 10\%$ provide reasonable answers.

Year	Cash inflow	PVF $r = 8\%$	PV $r = 8\%$	PVF $r = 10\%$	PV $r = 10\%$
0	−460		−460		−460
1	180	0.9259	166.66	0.9091	163.64
2	150	0.8573	128.60	0.8264	123.96
3	130	0.7938	103.19	0.7513	97.67
4	100	0.7350	73.50	0.6830	68.30
Net present value			11.95		−6.43

where PVF is the present value factor and PV is the present value.

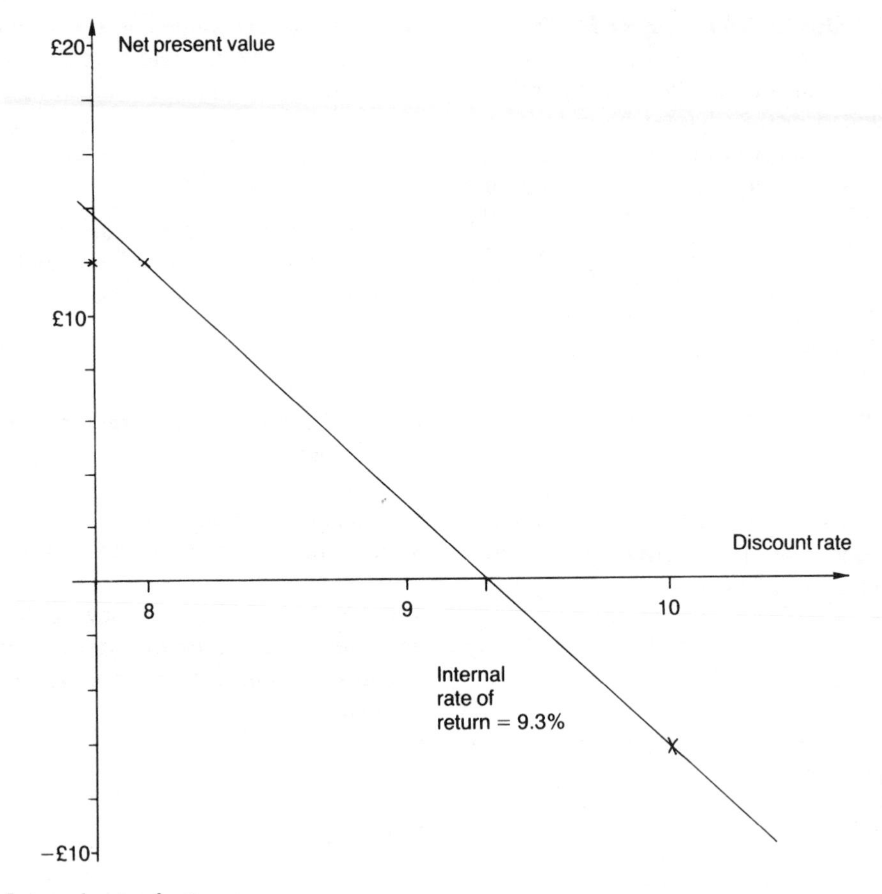

Figure 9.1 Internal rate of return

EXERCISE

Construct the preceding spreadsheet in such a way that it can accommodate different values of *r*.

We are now able to plot net present value against *r*, as shown in Figure 9.1.
By joining the plotted points we can obtain a line showing how net present value decreases as the discount rate increases. When net present value is 0, the corresponding *r* value is the estimated internal rate of return, 9.3% in this case.

9.4.2 The internal rate of return by linear interpolation

Again we need to determine the net present value of an investment corresponding to two values of *r*. Instead of joining points on a graph with a straight line, we calculate a value of *r* between these two values on a proportionate basis. If

r_1 = the lower discount rate,
r_2 = the upper discount rate,
NPV_1 = the net present value corresponding to r_1 and
NPV_2 = the net present value corresponding to r_2

then

$$IRR = r_1 + \frac{NPV_1}{(NPV_1 - NPV_2)} \times (r_2 - r_1)$$

EXAMPLE

Given $r_1 = 8\%$, $r_2 = 10\%$, $NPV_1 = £11.95$ and $NPV_2 = -£6.43$,

$$\begin{aligned} IRR &= 8 + \frac{11.95}{11.95 + 6.43} \times (10 - 8) \\ &= 8 + 0.65 \times 2 \\ &= 9.3\% \end{aligned}$$

Linearity is assumed in these calculations (and the graphical method) but the relationship between net present value and discount rate is non-linear. Provided the two selected values of r are fairly close to the internal rate of return, the estimation procedure produces a result of sufficient accuracy.

The decision of whether to invest or not will depend on how the internal rate of return compares with the discount rate being used by a company (a measure of the cost of capital to the company) or other organisation. If the calculations are specified on a spreadsheet, then changes in parameters are very quick to make and the effect easy to observe.

9.5 INCREMENTAL PAYMENTS

In section 9.1.2 we considered the growth of an initial investment when subject to compound interest. Many saving schemes will involve the same sort of initial investment but will then add or subtract given amounts at regular intervals. If x is an amount added at the end of each year, then the sum receivable, S, at the end of t years is given by

$$S = A_0\left(1 + \frac{r}{100}\right)^t + \frac{x\left(1 + \frac{r}{100}\right)^t - x}{r/100}$$

An outline proof of this particular formula is given in section 9.10.

EXAMPLES

1. A saving scheme involves an initial investment of £100 and an additional £50 at the end of each year for the next 3 years. Calculate the receivable sum at the end of 3 years assuming that the annual rate of interest paid is 10%.

   By substitution we obtain

$$\begin{aligned} S &= £100\left(1 + \frac{10}{100}\right)^3 + \frac{£50\left(1 + \frac{10}{100}\right)^3 - £50}{10/100} \\ &= £133.10 + £165.50 \\ &= £298.60 \end{aligned}$$

   The sum is in two parts, the first being the value of the initial investment (£133.10) and the second being the value of the end of year increments (£165.50).

   An alternative is to calculate the growing sum year by year.

	Value of investment
Initial sum	£100
Value at the end of year 1	£110
+ increment of £50	£160
Value at the end of year 2	£176
+ increment of £50	£226
Value at the end of year 3	£248.60
+ increment of £50	£298.60

   If the rate of interest or the amount added at the end of the year changed from year to year it would no longer be valid to substitute into the formula given.

   In the case of regular withdrawals, we use a negative increment.

2. It has been decided to withdraw £600 at the end of each year for 5 years from an investment of £30 000 made at 8% per annum compound.

   In this example, we have a negative increment of £600. By substitution we obtain:

$$\begin{aligned} S &= £30\,000\,(1 + 0.08)^5 + \\ &\quad \frac{(-£600)(1 + 0.08)^5 - (-£600)}{0.08} \\ &= £44\,079.842 - £3519.9606 \\ &= £40\,559.881 \end{aligned}$$

9.5.1 Sinking funds

A business may wish to set aside a fixed sum of money at regular intervals to achieve a specific sum at some future point in time. This sum, known as a sinking fund, may be in anticipation of some future investment need such as the replacement of vehicles or machines.

EXAMPLE

How much would we need to set aside at the end of each of the following 5 years to accumulate £20 000, given an interest rate of 12% per annum compound?

We can substitute the following values:

S = £20 000
$A_0 = 0$ (no saving is being made immediately)
r = 12%
t = 5 years

to obtain

$$£20\,000 = 0 + \frac{x(1+0.12)^5 - x}{0.12}$$

$$£20\,000 \times 0.12 = 0.7623x$$

$$x = £3148.37$$

where x is the amount we would need to set aside at the end of each year.

9.5.2 Annuities

An annuity is an arrangement whereby a fixed sum is paid in exchange for regular amounts to be received at fixed intervals for a specified time. Such schemes are usually offered by insurance companies and are particularly attractive to retired people.

EXAMPLE

How much is it worth paying for an annuity of £1 000 receivable for the next 5 years and payable at the end of each year, given interest rates of 11% per annum?

We can substitute the following values:

$S = 0$ (final value of investment)
$x = -£1\,000$ (a negative increment)
r = 11%
$t = 5$

to obtain

$$0 = A_0(1+0.11)^5 + \frac{(-£1\,000)(1+0.11)^5 - (-£1000)}{0.11}$$

$$A_0 = £1\,000 \times \frac{(1+0.11)^5 - 1}{0.11 \times (1+0.11)^5}$$

$$= £1\,000 \times \frac{1 - \dfrac{1}{(1+0.11)^5}}{0.11}$$

$$= £1\,000 \times 3.695\,89$$

$$= £3695.89$$

where A_0 is the value of the annuity. (Note that $1/(1+0.11)^5 = 0.5935$ from Appendix F.)

The present value of the annuity is £3695. We could have calculated the value of the annuity by discounting each of £1000 receivable for the next 5 years by present value factors.

9.5.3 Mortgages

The most common form of mortgage is an agreement to make regular repayments in return for the initial sum borrowed. At the end of the repayment period the outstanding debt is zero.

EXAMPLE

What annual repayment at the end of each year will be required to repay a mortgage of £25 000 over 25 years if the annual rate of interest is 14%?

We can substitute the following values

$S = 0$ (final value of mortgage)
$A_0 = -£25\,000$ (a negative saving)
r = 14%
t = 25

to obtain

$$0 = -£25\,000(1+0.14)^{25} + \frac{x(1+0.14)^{25} - x}{0.14}$$

$$x = £25\,000\frac{(1+0.14)^{25} \times 0.14}{(1+0.14)^{25} - 1}$$

$$= £25\,000 \times \frac{0.14}{1 - \frac{1}{(1+0.14)^{25}}}$$

$$= £25\,000 \times 0.1455$$

$$= £3657.46$$

where x is the annual repayment.

The multiplicative factor of 0.1455 is referred to as the **capital recovery factor** and can be found from tabulations.

9.6 ANNUAL PERCENTAGE RATE (APR)

For an interest rate to be meaningful, it must refer to a period of time e.g. per annum or per month. Legally, the annual percentage rate (APR), also known as the actual percentage rate, must be quoted in many financial transactions. The APR represents the true cost of borrowing over the period of a year when the compounding of interest, typically per month, is taken into account.

EXAMPLE

Suppose a credit card scheme charges 3% per month compound on the balance outstanding. What is the amount outstanding at the end of one year if £1 000 has been borrowed and no payments have been made?

To just multiply the monthly rate of 3% by 12 would give the **nominal rate of interest** of 36% which would underestimate the true cost of borrowing. Using the method outlined in section 9.1.2:

balance outstanding = £1 000(1 + 0.03)^{12} = £1 425.76
interest paid = £1 425.76 − £1 000 = £425.76

From this we can calculate the APR:

$$\text{APR} = \frac{\text{total interest paid in one year}}{\text{initial balance}} \times 100$$

$$= \frac{£425.76}{£1000} \times 100 = 42.58\%$$

We can develop a formula for the calculation of the APR. The balance outstanding at the end of one year is given by

$$A_0\left(1 + \frac{r}{100}\right)^m$$

where m is the number of payment periods, usually 12 months. The total interest paid is given by

$$A_0\left(1 + \frac{r}{100}\right)^m - A_0$$

To the calculate the annual increase we compare the two:

$$\text{APR} = \frac{A_0\left(1 + \frac{r}{100}\right)^m - A_0}{A_0} \times 100\%$$

$$= \left[\left(1 + \frac{r}{100}\right)^m - 1\right] \times 100\%$$

EXAMPLE

If the monthly rate of interest is 1.5%, what is (a) the nominal rate per year and (b) what is the APR?

(a) The nominal rate = 1.5% × 12 = 18%
(b) APR = [(1 + 1.5/100)^{12} − 1] × 100% = 19.56%

EXAMPLE

If the nominal rate of interest is 30% per annum but interest is compounded monthly, what is the APR?

The monthly nominal rate = 30%/12 = 2.5%
APR = [(1 + 2.5/100)^{12} − 1] × 100% = 34.49%

EXAMPLE

What monthly repayment will be required to repay a mortgage of £30 000 over 25 years given an APR of 13%?

To obtain in monthly rate, $r\%$, let

$$[(1 + r/100)^{12} - 1] \times 100\% = 13\%$$
$$(1 + r/100)^{12} - 1 = 0.13$$
$$r/100 = (1 + 0.13)^{1/12} - 1$$
$$r = 1.024$$

To calculate the monthly repayment, x (see section 9.5.3), let $A_0 = -£30\,000$, $r = 1.024$ and $t = 25 \times 12 = 300$. Then

$$x = £30\,000 \times \frac{0.01024}{1 - \dfrac{1}{(1.01024)}} = £322.37$$

9.7 REFLECTIONS ON THE ASSUMPTIONS

Throughout this chapter we have made assumptions about payments and interest, but how realistic have these been? An early assumption was that interest was paid, or money received at the end of a year — this is clearly not always the case, but the assumption was made to simplify the calculations (and the algebra!). For compound interest we have shown how to incorporate more frequent payments, and the same principle could be applied to all of the other calculations in this chapter.

The second major assumption throughout the chapter has been that interest rates remain constant for several years — this has clearly *not* been true in the UK in the 1970s and 1980s. A few contracts do involve fixed interest rates, e.g. hire-purchase agreements, but the vast majority of business contracts have variable interest rates. If we try to incorporate these variable rates into our calculations of, say, net present value, then we will need to estimate or predict future interest rates. These predictions will increase the uncertainty in the figures we calculate. As we have already noted, the higher the interest or discount rate, the less likely we are to invest in projects with a long-term payoff. However, since the interest rates charged to borrowers and lenders tend to change together over time, the opportunity cost of using or borrowing money should not be much affected.

Stable conditions are also assumed in the calculations. It is taken for granted that the general political and social structure of the Western economic system will still be the same in 5 or 10 years' time. International changes may have far-reaching domestic consequences; recent examples being the floating of currency exchange rates, the oil crises of 1973 and 1978, the debt problems of certain nations in 1983 and the fall of stock market prices in 1987.

In recent years we have seen high rates of interest as the Government's major economic instrument in its fight against inflation.

EXERCISES

1. Find two more examples of socio-political changes which would affect the stability assumption in the calculations.
2. At certain times in the past, money lending for interest (or usury) has been regarded as morally wrong. How would you justify the practice in today's world?

9.8 CONCLUSIONS

Money is often thought of in **nominal** terms, the numbers of pounds available, rather than **real** terms which is concerned with the value of money or purchasing power over time (see Chapter 5 Index Numbers). In this chapter, we have developed the concept of the time value of money: the value of money to you now depends on when you are likely to receive it. Money now presents opportunities now which we can often evaluate using an appropriate interest rate or discount rate. In assessing the value of future sums of money we also need to take account of other business factors such as present liquidity, the potential of new products and changes in the competitive environment.

9.9 PROBLEMS

1. A sum of £248 has been invested at an interest rate of 12% per annum for 4 years. What is the value of this investment, if interest is paid (a) as simple interest and (b) compounded each year?
2. An investment of £10 000 has been made on your behalf for the next 5 years. How much will this investment be worth if:
   (a) the rate of interest is 10% per annum;
   (b) interest is paid at 7% per annum for the first £1 000, 9% per annum for the next £5 000 and 12% per annum for the remainder;
   (c) the rate of interest is 9% per annum but paid on a 6 monthly basis?
3. How much would an investment of £500 accumulate to in 3 years if interest were paid at 6% per annum for the first year, 8% per annum for the second year and 12% per annum for the third year?
4. A car is bought for £5 680. It loses 15% of its value immediately and 10% per annum thereafter. How much is this car worth after 3 years?
5. A company buys a machine for £7 000. If depreciation is allowed for at a rate of 16% per annum, what will be the value of the machine in 4 years' time?
6. You are offered £400 now or £520 at the end of 5 years. You know that 8% per annum is the highest rate of interest you can get. Which offer should you accept?
7. A firm is trying to decide between two projects which have the following cash flows:

	Year			
Project	1	2	3	4
I	£10 000	£5 000	£6 000	£4 000
II	£12 000	£4 000	£4 000	£4 000

   If project I is discounted at 15% and project II at 20%, which project should be chosen?
8. A company has to replace a current production process. The current process is rapidly becoming unreliable whereas demand for the product is growing. The company must choose between alternatives to replace the process. It can buy
   (a) either a large capacity process now at a cost of £4 million, or
   (b) a medium capacity process at a cost of £2.2 million and an additional medium capacity process, also at a cost of £2.2 million, to be installed after 3 years.

   The contribution to profit per year from operating the two alternatives are:

	Contribution (£m) at year end					
	1	2	3	4	5	6
Large process	2.0	2.3	2.8	2.8	2.8	2.8
2 medium processes	2.0	2.0	2.0	2.4	2.8	2.8

   Assume a discount rate of 20%. Present a discounted cash flow analysis of this problem and decide between the alternatives.

   Comment on other factors, not taken into account in your discounted cash flow analysis, which you think may be relevant to management's decision.
9. You have decided to save £200 at the end of each year for the next 5 years. How much will you have at the end of the 5 years if you are paid interest of 10% per annum?
10. You have decided to save £200 at the beginning of each year for the next 5 years. How much will you have at the end of the 5 years if you are paid interest of 10% per annum?
11. A sum of £5 000 was invested 4 years ago. At the end of each year a further £1 000 was added. If the rate of interest paid was 12% per annum, how much is the investment worth now?
12. You require £4 000 in 5 years' time. How much will you have to invest at the end of each year if interest charged is 15% per annum?

13. A company has borrowed £5 000 to be repaid in equal end of year payments over 10 years. What will the annual repayment be if the interest charged is 15% per annum?
14. A customer credit scheme charges interest at 2% a month compounded. Calculate the true annual rate of interest, i.e. the rate which would produce an equivalent result if interest were compounded annually.
15. Calculate the annual percentage rate (APR) of (a) 1.75% per month compound, (b) 5% per quarter compound and (c) 8% per half year compound.
16. Determine the monthly rate of interest compound given an APR of 26%?
17. Given that an investment of £5000 now produces a return of £2000 at the end of years 1, 2 and 3, determine the Internal Rate of Return (IRR) using the discount rates of 9% and 10% as the basis for any calculations.

9.10 APPENDIX

Proof that

$$S = A_0\left(1+\frac{r}{100}\right)^t + \frac{x\left(1+\frac{r}{100}\right)^t - x}{\frac{r}{100}}$$

where S is the sum at the end of t years, A_0 is the initial investment and x is the amount added at the end of each year.

The value of the investment at the end of the first year is

$$S_1 = A_0\left(1+\frac{r}{100}\right)^1 + x$$

The value at the end of the second year is

$$S_2 = \left[A_0\left(1+\frac{r}{100}\right)^1 + x\right]\left(1+\frac{r}{100}\right)^1 + x$$
$$= A_0\left(1+\frac{r}{100}\right)^2 + x\left(1+\frac{r}{100}\right)^1 + x$$

If we continue in this way, it can be shown that the value after t years is

$$S = A_0\left(1+\frac{r}{100}\right)^t + x\left(1+\frac{r}{100}\right)^{t-1} + x\left(1+\frac{r}{100}\right)^{t-2} + \ldots + x$$

This can be simplified using the summation formula for a geometric progression to give

$$S = A_o\left(1+\frac{r}{100}\right)^t + x\left[\frac{1-\left(1+\frac{r}{100}\right)^t}{1-\left(1+\frac{r}{100}\right)}\right]$$

$$= A_0\left(1+\frac{r}{100}\right)^t + x\left[\frac{1-\left(1+\frac{r}{100}\right)^t}{-r/100}\right]$$

$$= A_0\left(1+\frac{r}{100}\right)^t + \frac{x\left(1+\frac{r}{100}\right)^t - x}{r/100}$$

PART THREE
CONCLUDING EXERCISE

You have £1 000 to invest for 5 years. Prepare a report which includes the following sections:

(a) an investigation of how this £1 000 could be invested;
(b) where possible, a description of each method of investment mathematically stating any assumptions made;
(c) an evaluation of each method using an appropriate criterion stating any assumptions made; and
(d) a description of the factors that may be of importance in making the investment decision.

You should consider a range of possibilities including bank deposit accounts, building society accounts, unit trusts, and the purchase of shares.

You should also consider the merits of investment in a single scheme and investment in a range of schemes.

PART FOUR
MEASURES OF UNCERTAINTY

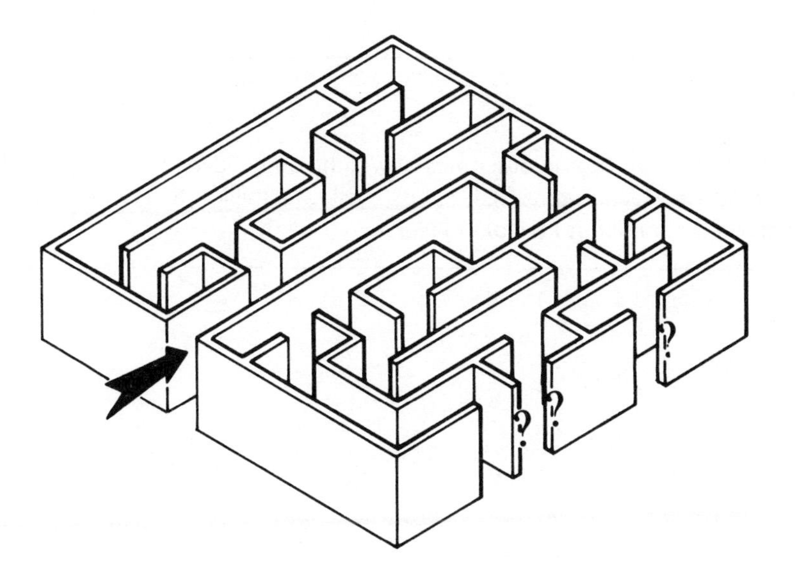

The complexity of many business problems and the uncertainty surrounding possible outcomes make some measurement of chance essential. However complete the data may seem, very few events are totally predictable. What company, for example, can be completely certain about the continuing success of its product range, or the stability of the business environment, or the adequacy of its organizational structure? If we add to this lack of predictability the incompleteness of data, the evaluation of change becomes increasingly important.

In situations where all things can be known, data collection alone can provide a basis for decision making. If we can link a set of facts to provide a certain answer (e.g. if $A = 5$ and $B = 3$ then $A + B = 8$) the problem is deterministic and can be modelled as such. If outcomes are subject to chance we still need to make sense of events (e.g. if $A = 4$ or 5 and $B = 2$ or 3 then $A + B = 6$ or 7 or 8) and attempt to understand the possible scenarios. The reality of most decision making is that we cannot be sure of the outcome, but it does make a difference knowing whether the chance of a particular outcome is 1 in 10 or 1 in 100 or 1 in 1 000.

In this part, we are concerned with introducing the basic ideas of chance and presenting some probability models. It is important to note that the determination of probability depends on the type of measurement. The distinction between discrete and continuous data, so fundamental in the presentation of data (see Chapter 2), is critical in the development of probability distributions. Discrete probability distributions are concerned with the change of 'counted' outcomes (e.g. the number of defective items) whereas continuous probability distributions are concerned with measurements where accuracy can be infinitesimally small (e.g. time to complete a task).

As you work through the following chapters, reflect on the business-related problems you have encountered and consider to what extent they were probabilistic rather than deterministic.

EXERCISES

1. List a number of events, the outcomes of which you believe are determined by chance.
2. How would you, as an individual, assess a risk of a million to 1? Give examples of risks you believe are a million to 1 or smaller.

PROBABILITY 10

Almost everyone has met, or been involved in, situations with a chance element, whether it is tossing a coin at the start of a sports match, playing cards, owning a premium bond or taking out some form of insurance. Even events which seem to have sound logical reasons for occurring may have their timing selected by chance. Other events, especially those involving large groups of people or items, often have characteristics which can be represented, or modelled, by some reference to probability.

Probability was first studied in relation to gambling, and many examples may still be drawn from the use of cards, dice, roulette wheels, etc., simply because these items are familiar to many people. However, it quickly became apparent that the ideas being investigated had a much wider application, initially in relation to rates to be charged for insurance of freight carried by sea.

Probability has found a wide range of business applications. In addition to the calculation of risk in the banking and insurance industries, probability provides the basis of many of the sampling procedures used in market research and quality control. Investment appraisal requires an assessment of risk and a measure of expected outcomes. The planning of major projects needs to take account of uncertainties, whether it is the effects of the weather on building site schedules or fashions in the market place. However, an understanding of probability will do more than just give you the right answers to certain types of problems. Probability will give you a new set of conceptual tools. Given that the outcomes of most activities are not known with certainty, i.e. are not **deterministic**, it is useful to understand them in probabilistic terms, i.e. as **stochastic** problems.

10.1 INTRODUCTION TO PROBABILITY

If we consider some results from a survey, as shown in Table 10.1, we may deduce various probabilities.

In Table 10.1 we see that 300 individuals responded to the survey, 160 of whom were male, and 140 were female. Thus if one individual were selected at random from the survey, there would be 140 chances out of 300 that it was a woman.

i.e. $$\text{Probability (female)} = \frac{140}{300}$$

$$\text{or } P\text{ (female)} = \frac{7}{15} = 0.467$$

The probability that the person selected had a salary between £6 000 and less than £10 000 would thus be 100/300: or

$$P(£6\,000 < £10\,000) = 1/3 = 0.333.$$

Looking again at the probabilities of selecting a man or a woman, we have

	Annual salary			
	less than £6000	£6000 and less than £10 000	£10 000 or more	Total
Male	30	50	80	160
Female	50	50	40	140
	80	100	180	300

Table 10.1

$$P(\text{male}) = \frac{160}{300}$$

$$P(\text{female}) = \frac{140}{300}$$

The capital P outside the brackets denotes 'that it is the probability of' whatever is described within the brackets. Thus P(male) is the probability of a male which for convenience could be abbreviated to $P(m)$. In the survey, each respondent has been classified by gender, so we are able to write

$$P(m) + P(f) = 1$$

or using the concept of 'not'

$$P(m) + P(\text{not } m) = 1$$

It follows that

$$\begin{aligned} P(m) &= 1 - P(\text{not } m) \\ &= 1 - 0.467 \\ &= 0.533 \end{aligned}$$

In this context, the manipulation above seems very obvious, but in many cases it will be easier to find the probability of something not happening than to find the probability that it does occur. For example, if items were packed in boxes of 1 000 and we wanted to find the probability that there were two or more defective items in a box, then to calculate this directly, we would need to use the following relationships:

$$\begin{aligned} &P(2 \text{ or more}) \\ &= P(2) + P(3) + P(4) + \ldots + P(999) + P(1\,000) \end{aligned}$$

However, if we notice that the only alternatives to '2 or more' are no defective items' or 'one defective item', then

$$\begin{aligned} P(2 \text{ or more}) &= 1 - P(\text{not 2 or more}) \\ &= 1 - [P(0) + P(1)] \end{aligned}$$

and this will usually be very much easier to evaluate.

Probabilities are often used to suggest what is likely to happen. If a fair coin were tossed 500 times you would expect 250 heads and 250 tails. The number of trials or samples (n) multiplied by a probability (in this case $n = 500$ and $P(\text{head}) = P(\text{tail}) = \frac{1}{2}$) gives the **expected value**. As with all probabilistic measures what is most likely to occur may not occur. You could toss a fair coin 500 times and get 251 heads or 235 heads or 300 heads; all these outcomes are possible even if they are less likely. What expectation tells you, is what will happen on average. In general:

$$\text{expectation} = (\text{probability}) \times (\text{total number of trials})$$

10.2 DEFINITIONS

In each case above we have used the classical definition of probability, because we have counted the number of ways that a person could be selected with the characteristic that we are seeking and divided this by the number of possible results that could have been obtained. For any event, E, we have:

$$P(E) = \frac{\text{no. of ways } E \text{ can occur}}{\text{total number of outcomes}}$$

This definition is widely used when trying to assess a particular situation but it is not complete. If you consider a die, there are six different faces, and this definition would suggest that $P(5) = 1/6$. However, if the die has been weighted so that, say, 6 will appear each time it is thrown, then the probability of a 5 is zero ($P(5) = 0$) despite the fact that there is one five, and there are six faces on the die. To complete the definition, we need to add that each outcome is **equally likely**; equally likely means equally probable, and thus we have a definition of probability which uses probability in that definition, i.e. a tautology. Even if we are sure that the outcomes are equally likely, the definition cannot deal with situations where there are an infinite number of outcomes. Despite these comments, most people will use this definition when considering simple probability situations.

EXERCISES

1. In a survey of 1 000 people, of which 100 were from Scotland, what is the probability when selecting one individual at random, that he or she came from Scotland?
   (Answer: $P(\text{from Scotland}) = 0.1$.)
2. If you toss two unbiased coins, what is the probability of getting two tails?
   (Answer: $P(\text{two tails}) = 0.25$.)

An alternative way to look at probability would be to carry out an experiment to determine the proportion **in the long run**, sometimes called the frequency definition of probability.

i.e.

$$P(E) = \frac{\text{no. of times } E \text{ occurs}}{\text{no. of times the experiment was conducted}}$$

This would certainly overcome the problem involving the biased die given above, since we would have $P(5) = 0$ and $P(6) = 1$. One problem with this definition is assessing how long the long run is. Experiments with a theoretical, unbiased coin do not necessarily conclude that the probability of a 'head' is $\frac{1}{2}$, even after 10 million tosses, and it can be shown that this frequency definition will not necessarily ever stabilize at some particular proportion. A second problem with this definition is that it must be possible to carry out repeated trials, whereas some situations only occur once; for example, the chance of a sales person selling more than the target set for next March.

EXERCISE

Toss a coin 50 times and count the number of heads that occur. Estimate the probability of a head for your coin.

Asking a series of questions may enable a researcher to find out people's **subjective probabilities** or their degree of confidence in certain events happening. This system will be useful if those questioned are in a position to assess the chances of the events under consideration, and will work for once-only situations. As with all sampling, the larger the number of the relevant population who take part in the survey, the better the result.

EXERCISE

Who would you question to assess the chance of a company running out of stock of one of its raw materials?

The logical difficulty of defining probability is recognized by the modern trend which bases the whole of probability theory on a number of **axioms** from which theorems are deduced. The axioms and theorems are formulated in terms of set theory and, avoiding details, some of them may be paraphrased as follows.

1. The probability of an event lies in the interval $0 \leq P(E) \leq 1$ and no other values are possible.
2. If something is certain to occur, then it has a probability of 1.
3. If two or more different outcomes of a trial or experiment cannot happen at the same time, then the probability of one or other of these outcomes occurring is the sum of the individual probabilities:

   e.g. if $P(E_1) = \frac{1}{4}$, $P(E_2) = \frac{1}{2}$

   then

   $$P(E_1 \text{ or } E_2) = \frac{1}{4} + \frac{1}{2} = \frac{3}{4}.$$

10.3 A FEW BASIC RELATIONSHIPS

Whilst the probability of a single event will be of some interest, most practical situations involve two or more events.

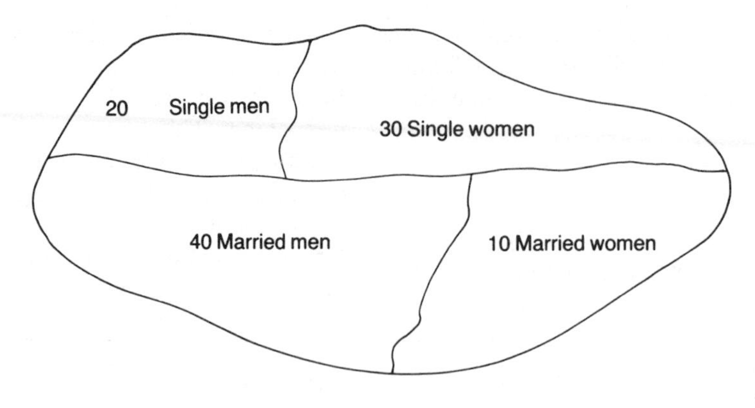

Figure 10.1

10.3.1 Mutually exclusive events

Mutually exclusive events is the situation represented in part (c) of the axiomatic definition of probability, and, as we have seen, we *add* the probabilities together to find the probability that one or other of the events will occur. For example, if a group of people consists of 20 single and 40 married men with 30 single and 10 married women, then, selecting a person at random we have:

$$P(\text{single man}) = \frac{20}{100}$$

$$P(\text{single woman}) = \frac{30}{100}$$

$$\text{and } P(\text{single person}) = \frac{20}{100} + \frac{30}{100} = \frac{50}{100} = 0.5$$

It can be seen in Figure 10.1 that 50 out of 100 are single.

10.3.2 Non-mutually exclusive events

Where one outcome has, or can have more than one characteristic, then these outcomes are said to be non-mutually exclusive. In this case it will not be possible simply to add the probabilities together as we did above to find the overall probability of one characteristic or another, since this would involve counting some outcomes twice. For example, if a group of people contains both men and women, and these people either agree or disagree with some proposition, then to find the probability of selecting a person who is either a man, or disagrees, will be

$$P(\text{man } \textit{or} \text{ disagree}) = P(\text{man}) + P(\text{disagree}) - P(\text{man } \textit{and} \text{ disagree})$$

Since the men who disagree appear in each of the first two probabilities, we need to *subtract* the probability of this group from our required probability. If, for example, there were 30 men, of whom 10 disagree with the proposition, and 70 women, of whom 40 disagree with the proposition then:

$$P(\text{man}) = \frac{30}{100}$$

$$P(\text{disagree}) = \frac{50}{100}$$

$$P(\text{man } \textit{and} \text{ disagree}) = \frac{10}{100}$$

Therefore, $P(\text{man } \textit{or} \text{ disagree})$

$$= \frac{30}{100} + \frac{50}{100} - \frac{10}{100} = \frac{70}{100} = 0.7$$

It can be seen in Figure 10.2 that there are 30 women who agree with the proposition, leaving the remaining 70 who are women or men who disagree (here we have again used the concept of *not*).

10.3.3 Independent events

When one outcome is known to have no effect on another outcome, then the events are said to be

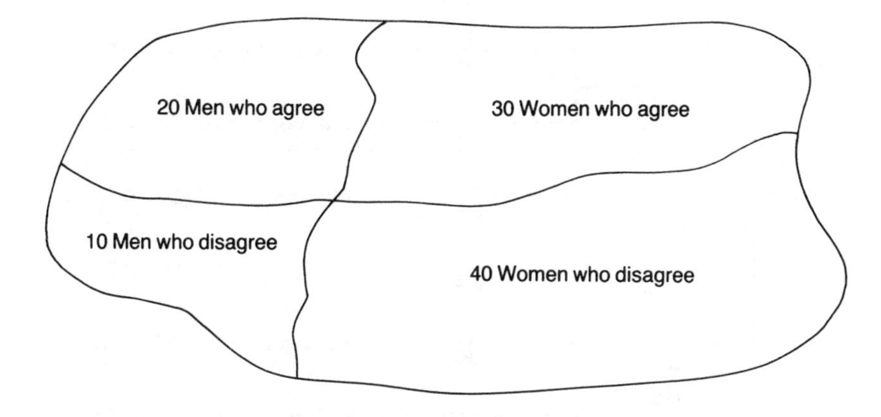

Figure 10.2

independent. For example, if the probability of a machine breaking down is 1/10 and the probability of stoppage of raw material supplies is 1/8, then we can find the probability of the two events happening together by *multiplying* the two probabilities, because the occurrence of one of these events does not affect the probability of the other. Thus,

P(breakdown *and* stoppage of supplies)

$$= \frac{1}{10} \times \frac{1}{8} = \frac{1}{80} = 0.0125.$$

10.3.4 Non-independent events

If two events are such that the outcome of one affects the probability of the outcome of the other, then the probability of the second event is said to be **dependent** on the outcome of the first. From a group of 10 people, 5 of whom are men and 5 women, the probability of selecting a man is $P(\text{man}) = 5/10 = 0.5$. If a second person is now selected from the remaining 9 people, the probabilities will depend on the outcome of the first selection.

If a man is selected first, then P (man) = 4/9 and P (woman) = 5/9.
If a woman is selected first, then P(man) = 5/9 and P(woman) = 4/9.
i.e. P(man/man) = 4/9, and P(man/woman) = 5/9

where the notation P(man/woman) is read as 'the probability of selecting a man when a woman has been selected at the first selection'. (Note that if the second selection had been from the original group of 10 people, then the two probabilities would be independent.)

These relationships can be summarized as follows:

If events A and B are mutually exclusive, then

$$P(A \text{ or } B) = P(A) + P(B)$$

If events A and B are any two events, then

$$P(A \text{ or } B) = P(A) + P(B) - P(A \text{ and } B)$$

If event A and B are independent, then

$$P(A \text{ and } B) = P(A) \times P(B)$$

If events A and B can both occur, then

$$P(A \text{ and } B) = P(A) \times P(B/A)$$

where $P(B|A)$ is the probability that B occurs given that A has already happened (often referred to as a **conditional probability**).

10.4 PROBABILITY TREES

Probability trees can be used to illustrate the combination of probabilities for a series of events which are independent or dependent. For the independent case, if the probability of outcomes A, B and C are 0.3, 0.2 and 0.5, and of outcomes X and Y are 0.6 and 0.4 then we have the situation shown in Figure 10.3. As the outcomes of each event are mutually exclusive

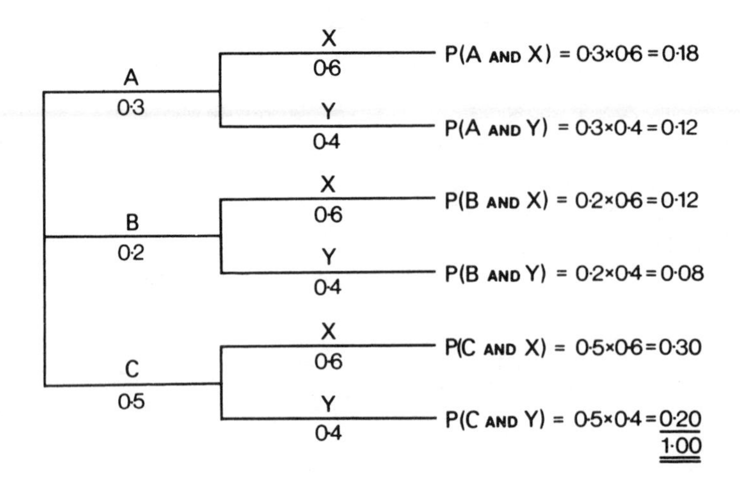

Figure 10.3

$P(A \text{ or } B \text{ or } C) = 0.3 + 0.2 + 0.5 = 1.0$
$P(X \text{ or } Y) \quad = 0.6 + 0.4 = 1.0$

We can note that the sum of all the probabilities on the right-hand side of Figure 10.3 is 1.00, and thus this represents all of the possible mutually exclusive outcomes. If we have three groups of people, a red team of 10 men and 10 women; a blue team of 7 men and 3 women and a yellow team of 4 men and 6 women, then using a two stage selection procedure firstly selecting a team, and then selecting an individual, the probability of selecting a woman will be dependent on which team is selected. In Figure 10.4 the probabilities of selecting a red, blue or yellow team are respectively 0.4, 0.4 and 0.2. The probabilities of being a man or a woman were derived from the numbers in each team (Fig. 10.4).

The probability tree can be combined with the idea of expectation in order to make a comparison between what actually happened and what we would expect to happen. Taking the information from Table 10.1 we have

$$P(\text{male}) = \frac{160}{300}$$

$$P(\text{female}) = \frac{140}{300}$$

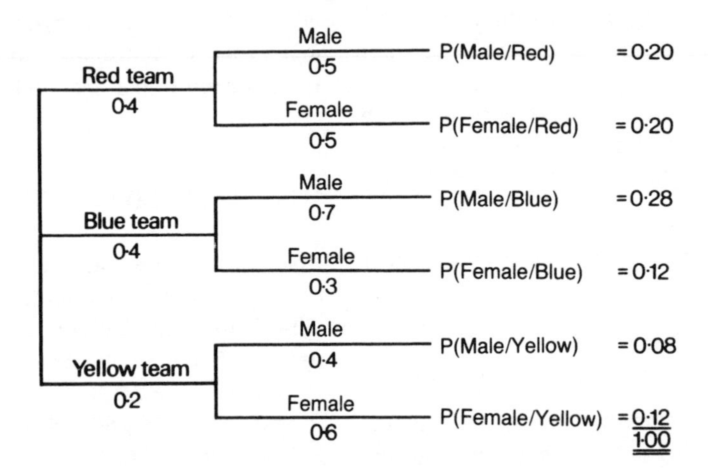

Figure 10.4

Table 10.2

	Annual salary	Probability	Expected number	Actual (from Table 10.1)
Male 160/300	less than £6 000 80/300	0.142 22	42.67	30
	£6 000 and less than £10 000 100/300	0.177 78	53.33	50
	£10 000 or more 120/300	0.213 33	64.00	80
Female 140/300	less than £6 000 80/300	0.124 44	37.33	50
	£6 000 and less than £10 000 100/300	0.155 56	46.67	50
	£10 000 or more 120/300	0.186 67	56.00	40
		1.000 00	300	300

$P(\text{less than £6 000}) = \frac{80}{300}$

$P(\text{£6 000 but less than £10 000}) = \frac{100}{300}$

$P(\text{£10 000 or more}) = \frac{120}{300}$

If we assume that these two factors are independent, we have the results in Table 10.2.

Looking down the final two columns of Table 10.2 we see that there are quite large differences between what actually happened and the expected number from the probability tree.

EXERCISE

Suggest reasons why this might happen.

(We shall return to this problem in Chapter 16).

10.5 EXPECTED VALUES AND DECISION TREES

Consider a simple game of chance. If a fair coin shows a head you win £1 and if it shows a tail you lose £2. If the game were repeated 100 times you would expect to win 50 times, that is £50 and expect to lose 50 times, that is £100. Your overall loss would be £50 or 50 pence per game **on average**. This average loss per game is referred to as **expected value** (EV) or **expected monetary value** (EMV). Given the probabilistic nature of the game, sometimes the overall loss would be more than £50, sometimes less. Rather than work with frequencies, expected value is usually determined by weighting outcomes by probabilities (see section 3.4). As we have seen, the probabilities may have been derived from relative frequency. In this simple game, the expected value of the winnings is

$$£1 \times \frac{1}{2} + -£2 \times \frac{1}{2} = -£0.50 \text{ or 50 pence}$$

where $-£2$ represents a negative win, or loss.

You will of course, never lose 50 pence in a single game, you will either win £1 or lose £2. Expected values give a long-run, average result. In general,

$$E(x) = \Sigma (x \times P(x))$$

where $E(x)$ is the expected value of x.

Suppose now, you are given two opportunities to invest your savings. The first opportunity, op-

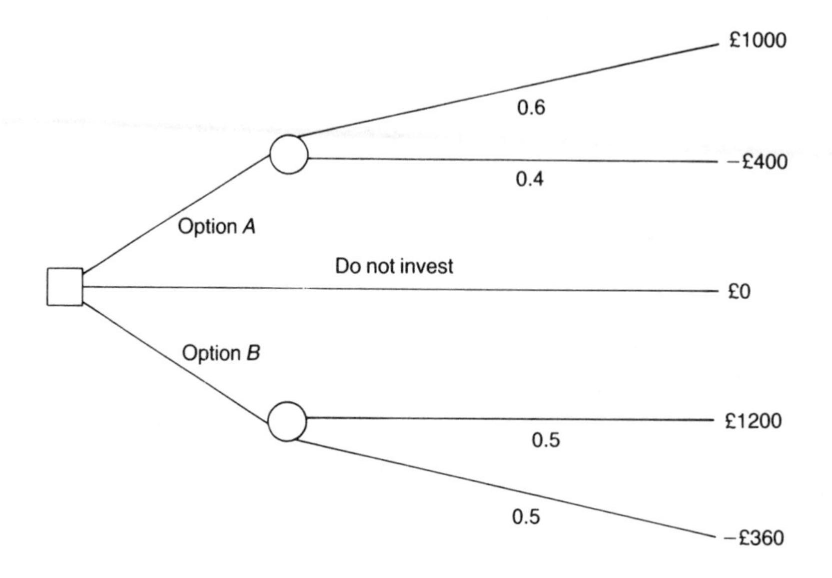

Figure 10.5

tion A, is forecast to give a profit of £1 000 with a probability of 0.6 and a loss of £400 with a probability of 0.4. The second opportunity, option B, is forecast to give a profit of £1 200 with a probability of 0.5 and a loss of £360 with a probability of 0.5. These two opportunities actually give you three choices, the third choice being not to invest. The possible decisions are represented in the **decision tree** shown in Figure 10.5.

A decision tree shows the decisions to be taken (a decision node is denoted by a □) and the possible outcomes (a chance node is denoted by a ○). The expected value associated with each chance node is:

$E(\text{option } A) = £1\,000 \times 0.6 + (-£400) \times 0.4 = £440.$
$E(\text{option } B) = £1\,200 \times 0.5 + (-£360) \times 0.5 = £420.$

On the basis of highest expected value, option A would be chosen. However, once the decision is made and if the forecasts are correct, option A will yield a profit of £1 000 or a loss of £400. It is important to recognize that £440 is an average, like 2.5 children, and may never occur as an outcome. Indeed, if you only make this type of decision once, you either win or lose. The highest expected value provides a useful decision criterion if the type of decision is being repeated many times e.g. car insurance, but may not be appropriate for a one-off decision e.g. whether to extend an existing factory. There are other decision criteria. A risk taker, for example, would be attracted to the larger possible profits of option B. A risk avoider would be deterred by possible losses and choose not to invest or choose the option with least loss i.e. option B. Different decision criteria do lead to different choices, so the selection of criteria is important.

A decision tree can include several decision nodes. Consider the following example.

EXAMPLE

A new product has been developed. The design is valued at £1 000. To launch and market the product will cost £1 500, and market research will cost a further £500. The product may be very successful, moderately successful or a failure with estimated revenues of £10 000, £4 000 and −£6 000 respectively. These revenues exclude launch and market research costs. Given that outcomes are the subject of chance, the following probabilities have been estimated:

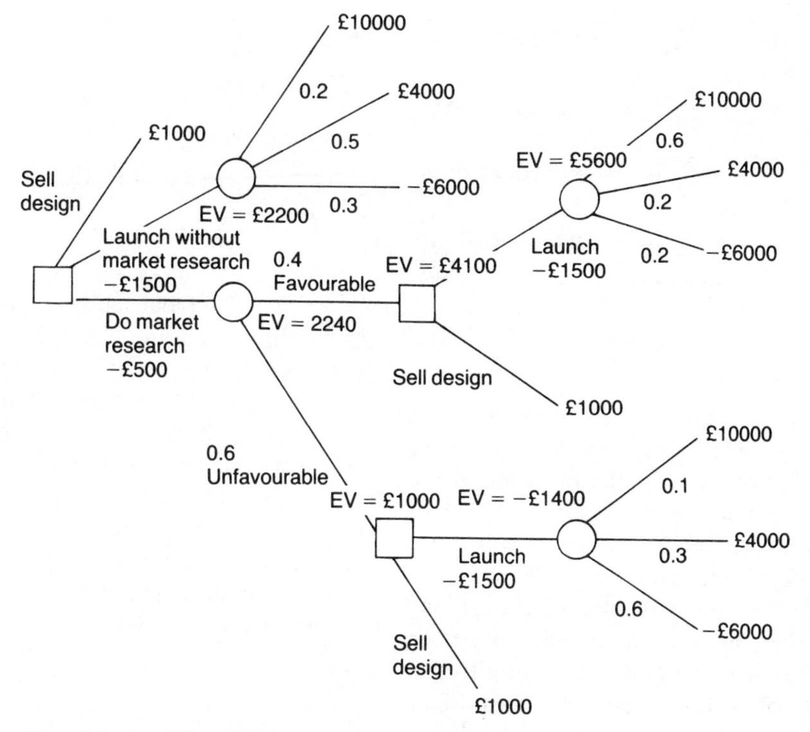

Figure 10.6

Outcome	No market research	Favourable market research	Unfavourable market research
Very successful	0.2	0.6	0.1
Mod. successful	0.5	0.2	0.3
Failure	0.3	0.2	0.6

Experience with previous products of this kind would suggest a 40% chance of a favourable market research report.

The decision tree with expected values is given as Figure 10.6.

Expected values are determined by working from the right to the left – the rollback principal. At each chance node, the expected value is calculated e.g. EV = –£1 400, and then we move back one stage. At each decision node, the highest expected value is taken, e.g. £1 000 is greater than –£29 000 (–£1400 – £1500), and used in subsequent rollback calculations. In the example given, there are two stages of decision making. The first decision is to undertake market research. The second decision depends on the outcome of the market research. If the market research is favourable, the product is launched, otherwise the design is sold.

10.6 BAYES' THEOREM

As we have seen in section 10.2, when events are dependent, the probability of a subsequent event

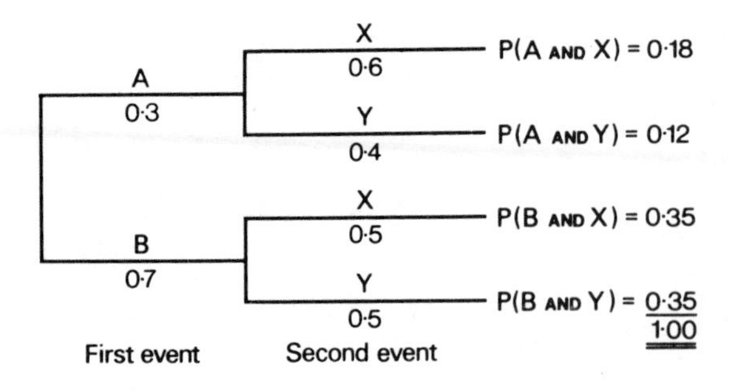

Figure 10.7

will depend on the outcome of some previous event. By using Bayes' Theorem we may move from the position of knowing the outcome of the second event, and find the probabilities of each of the outcomes of the first event, given this extra knowledge. Consider the situation in Figure 10.7.

If we know that the outcome of the second event was X, then either A or B could have been outcomes of the first event. We are trying to find the probability of each of these outcomes. Since the combinations of events (A and X) and (B and X) are mutually exclusive, these two probabilities can be added together to obtain the overall probability of X occurring.

$$\begin{aligned} P(X \mid A \text{ or } B) &= P(A \text{ and } X) + P(B \text{ and } X) \\ &= 0.18 + 0.35 \\ &= 0.53 \end{aligned}$$

If A occurred first, we have that

$P(X \mid A) = 0.6$ and $P(A) = 0.3$

which, as we have seen, gives

$P(A \text{ and } X) = P(A) \times P(X \mid A) = 0.18$

$$P(A \text{ and } X) = \frac{\text{probability obtained for outcome } X \text{ via outcome } A}{\text{probability obtained for outcome } X \text{ via any route}}$$

$$= \frac{P(A \text{ and } X)}{P(X \mid A \text{ or } B)}$$

$$= \frac{P(X|A)P(A)}{P(X|A) + P(X|B)}$$

$$= \frac{0.18}{0.53}$$

$$= 0.3396$$

and

$$P(B \mid x) = \frac{0.35}{0.53} = 0.6604$$

or more generally:

$$P(A_j \mid X) = \frac{P(X \mid A_j)\, P(A_j)}{\sum_{i=1}^{n} P(X \mid A_i) P(A_i)}$$

where A_j is the particular route for which the probability is required, and A_i ($i = 1$ to n) represents each of the routes to X (the known result).

From Figure 10.7 we have that 0.3 is the probability of A before anything happens; this is known as the **prior** probability of A. Once the outcome of the second event is known to be X, then the probability that X was reached via A is 0.3396; this is known as the **posterior** probability of A. These ideas from Bayes' Theorem are often used in decision analysis and **decision theory**.

10.7 MARKOV CHAINS

A Markov chain combines the ideas of probability with the matrix presentation shown in Chapter 7. It assumes that probabilities remain fixed over time, but that the system that is being modelled is able to change from one state to another, using these fixed values as **transition probabilities**. Consider, for example, the following transition matrix:

$$\mathbf{P} = \begin{array}{c} \\ E_1 \\ E_2 \end{array} \begin{array}{c} \begin{array}{cc} E_1 & E_2 \end{array} \\ \begin{bmatrix} 0.8 & 0.2 \\ 0.3 & 0.7 \end{bmatrix} \end{array}$$

This means that if the system is in some state labelled E_1, the probability of going to E_2 is 0.2. If

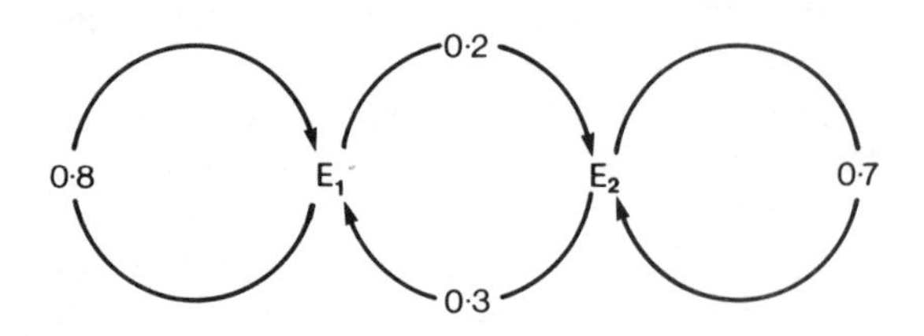

Figure 10.8

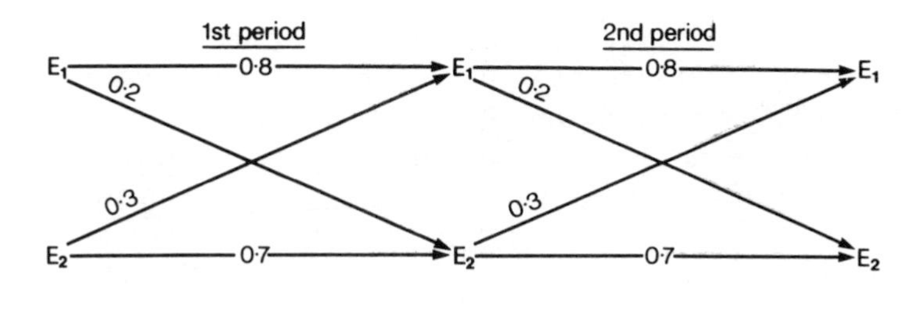

Figure 10.9

the system is at E_2, then the probability of going to E_1 is 0.3, and the probability of remaining at E_2 is 0.7. This matrix could be represented by a **directed graph** (Figure 10.8).

If we consider the movement from one state to another to happen at the end of some specific time period, and look at the passage of two of these periods we have the situation in Figure 10.9.

The probability of *ending* after two periods at E_1 if the system started at E_1 will be

$$\begin{aligned} &P(E_1 \to E_1 \to E_1) + P(E_1 \to E_2 \to E_1) \\ &= (0.8)(0.8) + (0.2)(0.3) \\ &= 0.70. \end{aligned}$$

Starting at E_1 and ending at E_2:

$$\begin{aligned} &P(E_1 \to E_1 \to E_2) + P(E_1 \to E_2 \to E_2) \\ &= (0.8)(0.2) + (0.2)(0.7) \\ &= 0.30. \end{aligned}$$

Starting at E_2 and ending at E_1:

$$\begin{aligned} &P(E_2 \to E_1 \to E_1) + P(E_2 \to E_2 \to E_1) \\ &= (0.3)(0.8) + (0.7)(0.3) \\ &= 0.45. \end{aligned}$$

Starting at E_2 and ending at E_2:

$$\begin{aligned} &P(E_2 \to E_2 \to E_2) + P(E_2 \to E_1 \to E_2) \\ &= (0.7)(0.7) + (0.3)(0.2) \\ &= 0.55. \end{aligned}$$

Thus the transition matrix for *two* periods will be:

$$\begin{array}{c} \\ E_1 \\ E_2 \end{array} \begin{array}{c} \begin{array}{cc} E_1 & E_2 \end{array} \\ \begin{bmatrix} 0.70 & 0.30 \\ 0.45 & 0.55 \end{bmatrix} \end{array}$$

but note that this is equal to $\mathbf{P}^2$, i.e. the square of the transition matrix for one period. To find the transition matrix for four periods, we would find $\mathbf{P}^4$ and so on.

The states of the system at a given instant could be an item working or not working, a company being profitable or making a loss, an individual being given a particular promotion or failing at the interview, etc. In all transition matrices, the movement over time is from the state on the left to the state above the particular column, and thus, since something must happen, the sum of any row must be equal to 1.

A state is said to be **absorbant** if it has a probability of 1 of returning to itself each time. In the matrix

$$\begin{array}{c} \\ E_1 \\ E_2 \\ E_3 \end{array} \begin{array}{c} \begin{array}{ccc} E_1 & E_2 & E_3 \end{array} \\ \begin{bmatrix} 0.4 & 0.2 & 0.4 \\ 0.2 & 0.7 & 0.1 \\ 0 & 0 & 1 \end{bmatrix} \end{array}$$

the state E_3 is an absorbant state, since each time the system reaches state E_3 it remains there. This can again be shown by a directed graph (Figure 10.10).

To use these transition matrices for predicting a future state, we need to know the initial state,

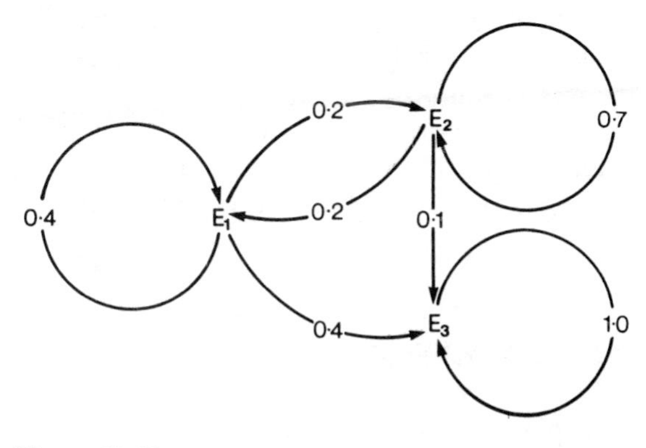

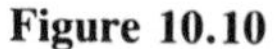
Figure 10.10

which is written in the form of a vector. For example, if state E_1 for Figure 10.8 were a company being profitable, and E_2 were a company not being profitable, then we would consider what would happen to a group of companies. If the group consists of 150 profitable companies and 50 non-profitable companies, the initial vector will be

$$\mathbf{A}_0 = \begin{matrix} E_1 & E_2 \\ [150 & 50] \end{matrix}$$

To find the situation after one time period, say 1 year, we postmultiply the initial state vector by the transition matrix:

$$[150 \quad 50]\begin{bmatrix} 0.8 & 0.2 \\ 0.3 & 0.7 \end{bmatrix} = [135 \quad 65] = \mathbf{A}_1.$$

$\mathbf{A}_1$ now represents the situation after one time period, where we would expect 135 companies to be profitable, and 65 not to be profitable. To find the situation after two time periods, we either multiply the initial state vector by P^2

$$[150 \quad 50]\begin{bmatrix} 0.7 & 0.3 \\ 0.45 & 0.55 \end{bmatrix} = [127.5 \quad 72.5] = \mathbf{A}_2.$$

or postmultiply the vector $\mathbf{A}_1$ by P:

$$[135 \quad 60]\begin{bmatrix} 0.8 & 0.2 \\ 0.3 & 0.7 \end{bmatrix} = [127.5 \quad 72.5] = \mathbf{A}_2.$$

Both calculations give a row vector labelled $\mathbf{A}_2$, which represents the expected number of companies in each state after 2 years. Note that some of the companies that are currently profitable may not have been profitable after only 1 year.

The process given above can continue with any transition matrix for any number of time periods; however, it is likely that the probabilities within the matrix will become out of date, and thus not fixed, if a prediction into the distant future is made. Markov chains are often used in manpower planning exercises and in market predictions.

10.8 CONCLUSIONS

Probability provides a measure on a scale 0 to 1 of the chance that some event will take place. At the extremes of the scale, 0 and 1, the event should not happen or will definitely happen. As with all prediction, the measures depend on the collected data and the problem perspective. You may, for example, be certain that the car journey time between Birmingham and Leeds is less than 6 hours knowing that all previous journey times were less than this. However, the journey time will depend on the type of car, the time of travel, weather conditions, the absence of mechanical failure and a host of other factors. Probability describes a very precise set of circumstances.

The rules of probability provide a way of evaluating events that may happen in sequence or in parallel. When looking at a business problem for the first time, it is useful to isolate the independent factors and the dependent factors. Probability can help you understand the consequences of the dependencies. In decision theory we need to look at problems that typically have several stages or parts.

10.9 PROBLEMS

Many of the exercises below use coins, dice and cards; this does not imply that this is the only use to which these ideas can be put, but they do provide a fairly simple mechanism for determining if you have absorbed the ideas of the previous section.

1. If you toss two coins, what is the probability of two heads?
2. If you toss two coins, what is the probability of a head followed by a tail?
3. If you toss three coins, what is the probability of a head followed by two tails?
4. If you toss three coins, what is the probability of two tails?
5. How do the probabilities in questions 1–4 change if the coins are biased so that the probability of a tail is 0.2?
6. If the experiment in question 1 were done 100 times, what is the expected number of times that two heads would occur?
7. If the experiment in question 5 were done

1 000 times, what is the expected number for each of the outcomes?

8. Construct a grid to show the various possible total scores if two dice are thrown. From this, find the following probabilities:
   (a) a score of 3;
   (b) a score of 9;
   (c) a score of 7;
   (d) a double being thrown.
9. When two dice are thrown, what is the probability of a 3 followed by a 5?
10. When a biased die (with probability of a six = 0.25) is thrown twice, what is the probability of not getting a six?
11. A die is thrown and a coin is tossed, what is the probability of a 1 and a tail?
12. In the experiment in question 11, what is the probability of a 1 or a tail or both?
13. If two coins are tossed and two dice are thrown, what is the probability of two heads and a double five?
14. From a normal pack of 52 cards, consisting of four suits each of 13 cards, on taking out one card, find the following probabilities:
   (a) an ace;
   (b) a club;
   (c) an ace or a club;
   (d) the ace of clubs;
   (e) a picture card (i.e. a jack, queen or king);
   (f) a red card;
   (g) a red king;
   (h) a red picture card.
15. What is the probability of a queen on either or both of two selections from a pack:
   (a) with replacement;
   (d) without replacement?
16. What is the probability of a queen or a heart on either or both of two cards selected from a pack without replacement?
17. A company has 100 employees, of whom 40 are men. When questioned, 60 people agreed that they were happy in their work, and of these 30 were women. Find the probabilities that:
   (a) a man is unhappy;
   (b) a woman is happy.
18. A company has three offices, *A*, *B* and *C* which have 10, 30 and 40 people in them respectively. Company policy is that three people from each office are selected for promotion interviews. This year there will be three promotions, find the probability for a person in each office of being promoted.
19. A survey of 1 000 people were asked which political party they would vote for. The respondents included 600 women. Response was as follows:

Party *A*	350
Party *B*	320
Party *C*	300
Don't know	30

   If sex and party are independent, find the expected numbers in each of the following categories:
   (a) men supporting Party *A*;
   (d) women supporting Party *C*;
   (c) women supporting either Party *A* or Party *B*.

   Would you be surprised to learn that 150 men supported Party *C*?
20. A company manufactures red and blue plastic pigs; 5% red and 10% blue are misshapen during manufacture. If the company makes equal numbers of each colour, what is the probability of selecting a misshapen pig on a random selection? How would the probability change if 60% of the pigs manufactured were blue? In a sample of three, what is the probability of getting two misshapen red pigs?
21. Two events are independent. There are three different outcomes (*A*, *B*, *C*) of the first event, with probabilities of 0.2, 0.3, and 0.5 respectively. There are two outcomes from the second event (*X* and *Y*), with probabilities of 0.1 and 0.9 respectively
   (a) Construct a probability tree to represent this situation.
   (b) What is the probability that the outcome of both events includes outcome *Y*?
22. A company produces plastic elephants in two colours for the novelty trade market.

Production in the factory is on one of three machines; 10% is on machine A, 30% on machine B, and the remainder on machine C. Machine A's production consists of 40% blue elephants and 60% pink elephants. Machine B's production consists of 30% blue elephants and 70% pink elephants. Machine C's production has 80% pink elephants with the remainder being blue.

(a) What proportion do blue elephants form of total production?

(d) If a particular elephant is pink, what is the probability it was made by machine B?

23. In a certain year the general public is divided in their support of three political parties in the ratio 35:35:30, (Con:Lab: Alli). The sex distribution of support for the parties is 60:40 for Con; 50:50 for Lab; and 45:55 for Alli, each ratio being male: female. When questioned on a current issue, male-Con were 80:20 in favour, whilst female-Con were 70:30 in favour. Male-Lab were 70:30 in favour and female-Lab were 60:40 in favour. Male-Alli were 20:80 in favour and female-Alli were 30:70 in favour.
   (a) Find the proportion of the sample who were male-Con in favour of the proposal.
   (b) If an individual is in favour of the proposition, what is the probability that they are a Con supporter.
   (c) If an individual is against the proposition, what is the probability the person is male.

24. A switch has a 0.9 probability of working effectively. If it does work, then the probability remains the same on the next occasion that it is used. If, however, it does not work effectively, then the probability it works on the next occasion is 0.1. Use a tree diagram to find:
   (a) the probability it works on three successive occasions;
   (b) if fails, but then works on the next two occasions;
   (c) on four occasions it works, fails, works, and then fails.

25. The forecasted profits from a new project have been allocated the following probabilities:

*Forecasted profit/(loss)**	*Probability*
(2 000)	0.2
(1 000)	0.2
500	0.1
1 600	0.1
3 000	0.2

*note that brackets used in this way indicate negative values

Calculate the expected profit and explain the meaning of this figure.

26. You have been given the probability distributions of possible profits from two projects, A and B:

Project A

Probability	*Profit (£)*
0.6	4 000
0.4	8 000

Project B

Probability	*Profit (£)*
0.2	2 000
0.3	2 500
0.3	4 000
0.1	8 000
0.1	12 000

Determine the expected profit from each project and state your project choice. What factors should a decision maker take into account when looking at the possible profits from these projects.

27. A small company has developed a new product for the electronics industry. The company believes that an advertising campaign costing £2 000 would give the product a 70% chance of success. It estimates that a product with this advertising support would provide a return of £11 000 if successful and a return of £2 000 if not successful. Past experience suggests that without advertising

support a new product of this kind would have a 50% chance of success giving a return of £10 000 if successful and a return of £1 500 if not successful.

Construct a decision tree and write a report advising the company on its best course of action.

28. In order to be able to meet an anticipated increase in demand for a basic industrial material a business is considering ways of developing the manufacturing process. After meeting current operating costs the business expects to make a net profit of £16 000 from its existing process when running at full capacity. All the data relates to the same period.

The Production Manager has listed the following possible courses of action.

(a) Continue to operate the existing plant and not expand to meet the new level of demand.

(b) Undertake a research programme which would cost £20 000 and has been given a 0.8 chance of success. If successful, a net profit of £60 000 is expected (before charging the research cost). If not successful a net profit of £5 000 is expected.

(c) Undertake a less expensive research programme costing £8 000 which has been given a 0.5 chance of success. If successful, a net profit of £5 000 is expected and if not successful a net profit of £4 000 is expected.

Present a decision tree. On the basis of this analysis determine the most profitable course of action. Comment on your findings.

29. A manufacturer has developed a new product and must decide whether to shelve the new product, sell the design for £60 000 or manufacture the new product. If the new product is successful then expected profits are £120 000 and if not successful £20 000.

The manufacturer also needs to decide whether or not to commission a market research survey. The cost of the survey would be £4 000.

It is accepted by the manufacturer that new products of this kind generally have a 60% chance of success. However, it is also accepted that if the results of the market research survey are favourable then the chances of success increase to 90% and if not favourable decrease to 30%. In the past, 50% of market research surveys for this type of new product have given favourable results.

Construct a tree diagram to represent the decisions the manufacturer has to make. Using the criteria of expected monetary value, advise the manufacturer as to the best course of action.

30. Draw a directed graph for each of the transition matrices given below:

(a)
$$\begin{array}{c} \\ E_1 \\ E_2 \end{array} \begin{array}{c} \begin{array}{cc} E_1 & E_2 \end{array} \\ \begin{bmatrix} 0.7 & 0.3 \\ 0.5 & 0.5 \end{bmatrix} \end{array}$$

(b)
$$\begin{array}{c} \\ E_1 \\ E_2 \\ E_3 \end{array} \begin{array}{c} \begin{array}{ccc} E_1 & E_2 & E_3 \end{array} \\ \begin{bmatrix} 1.0 & 0 & 0 \\ 0.1 & 0.8 & 0.1 \\ 0.1 & 0.8 & 0.1 \end{bmatrix} \end{array}$$

In case (b) what will eventually happen to the system?

31. For the transition matrix **P**, given below, find $\mathbf{P}^2$, $\mathbf{P}^4$, $\mathbf{P}^8$.

$$\mathbf{P} = \begin{bmatrix} 0.2 & 0.8 \\ 0 & 1.0 \end{bmatrix}$$

32. A particular market has 100 small firms, 50 medium sized firms and 10 large firms, and it has been noticed that the transition matrix from year to year is represented by the matrix given below:

Size this period	Size next period			
	Small	Medium	Large	Bankrupt
Small	0.7	0.2	0	0.1
Medium	0.3	0.5	0.1	0.1
Large	0	0.3	0.6	0.1
Bankrupt	0.1	0	0	0.9

No firms are bankrupt initially.

Find the expected number of firms in each category at the end of:

(a) one year;
(b) two years;
(c) three years;
(d) four years.

33. A firm has five levels of intake of staff, and wishes to predict the way in which staff will progress through the various grades. Data has been collected to allow the construction of a transition matrix, including those who leave the firm. This is shown below:

	1	2	3	4	5	Left
1	0.3	0.3	0.1	0.1	0.1	0.1
2	0	0.3	0.2	0.2	0.2	0.1
3	0	0	0.4	0.3	0.2	0.1
4	0	0	0	0.5	0.4	0.1
5	0	0	0	0	0.8	0.2
Left	0	0	0	0	0	1.0

Six hundred and fifty people join in a particular year, at grades of 1 to 5 as set out below:

[200 150 150 100 50]

Use this information to predict:
(a) the numbers in each grade after one year;
(b) the numbers in each grade after four years.

The wages for each grade from 1 to 5 are given below:

[50 60 80 100 120]

Find the total wage bill of the company for this cohort:
(c) as soon as they join;
(d) after four years with the company (assuming the same wage levels).

34. Within a company, an individual's probability of being promoted depends on whether or not they were promoted in the previous year. They may also be made redundant. This situation may be modelled by a Markov process, with the following transition matrix:

	This period		
Last period	*Some job*	*Promotion*	*Redundant*
Same job	0.7	0.2	0.1
Promotion	0.9	0	0.1
Redundant	0	0	1.0

If, in the last time period, 100 people retained the same job, 5 were promoted and none were made redundant, find the expected numbers in each category after:
(a) one period;
(b) two periods;
(c) three periods.

After three periods, the economy becomes more buoyant, and the threat of redundancy is lifted. The transition matrix now becomes:

	Same job	*Promotion*
Same job	0.7	0.3
Promotion	0.8	0.2

For those still remaining with the company from the original cohort, find the expected numbers in each category after:
(d) the next period, i.e. period 4;
(e) the following period, i.e. period 5;
(f) period 6.

10.10 APPENDIX

Probability occasionally gives answers which do not seem to match our own guesses. For instance, if you were asked 'What is the probability of two or more people in a group of 23 having the same birthday in terms of day and month, but not necessarily year?', what would you guess the answer to be? It is fact more than $\frac{1}{2}$!

To build up to this answer, consider a group size of two. Whenever the first person's birthday, if all birthdays are equally likely, then the probability that they have the same birthday will be 1/365 (ignoring leap years), and that they do not 364/365. For a group of 3, given the first person's birthday, the probability that the second has a different birthday is 364/365 and that the third has a different birthday still, 363/365. Thus

$$\begin{aligned} P(\text{at least 2 same}) &= 1 - P(\text{all different}) \\ &= 1 - \frac{364}{365} \times \frac{363}{365} = 0.0082 \end{aligned}$$

For a group of 4, we have:

$$P(\text{at least 2 same}) = 1 - \frac{364}{365} \times \frac{363}{365} \times \frac{362}{365} = 0.01636$$

We can continue this process, to give Table 10.3.

Table 10.3

No. of people	Probability	No. of people	Probability
2	0.002 739 7	26	0.598 24
3	0.008 024 2	27	0.626 859
4	0.016 355 9	28	0.654 46
5	0.027 135 57	29	0.680 968 5
6	0.040 462 48	30	0.706 3
7	0.056 235 7	31	0.730 45
8	0.074 335	32	0.753 347 5
9	0.094 623 8	33	0.774 97
10	0.116 948	34	0.795 3
11	0.141 141 4	35	0.814 38
12	0.167 025	36	0.832 18
13	0.194 4	37	0.848 738
14	0.223	38	0.864 06
15	0.252 9	39	0.878 2
16	0.283 6	40	0.891 23
17	0.315	41	0.903 15
18	0.346 91	42	0.914
19	0.379 1	43	0.923 9
20	0.411 4	44	0.932 885
21	0.443 688	45	0.940 97
22	0.475 695	46	0.948 25
23	0.507 297	47	0.954 77
24	0.538 34	48	0.960 597
25	0.568 699 7	49	0.965 779 6

EXERCISES

1. When will the probability reach 1?
2. If the group are all from the same country or cultural background, what factors will increase the probabilities proposed in the model given above?

DISCRETE PROBABILITY DISTRIBUTIONS 11

In the last chapter we discussed some of the basic ideas of probability, considering the possibility of an individual event, or the chance that certain events would happen at the same time, or in a particular sequence. If we wish to use these ideas in a business context, we need to look for probability models which are capable of representing a range of situations. A number of these have been recognized and are referred to generally as **probability distributions**. During a first year course, it would not be possible to discuss the whole range of distributions that might be used in particular cases, but several of those most frequently encountered are considered below.

11.1 UNIFORM DISTRIBUTION

We have already met this distribution in Chapter 1 when random numbers were used to select samples for interview or analysis. In a uniform distribution, each outcome is equally likely to occur. Thus in random number tables each of the digits 0, 1, 2, 3, 4, 5, 6, 7, 8, 9 should occur about the same number of times in a large sample. A histogram of this distribution is shown in Figure 11.1. Note that this illustrates a theoretical distribution and that in an actual sample the numbers of each digit will be only approximately the same.

To demonstrate the properties of chance on a variable following a uniform distribution, a random sample of numbers can be generated and a frequency distribution plotted. MICROSTATS allows this type of simulation (see Chapter 24) through the commands IRAN and HISTogram.

11.2 BINOMIAL DISTRIBUTION

In many cases, the variable of interest is **dichotomous**, has two parts or two outcomes. Examples include questions that only allow a 'YES' or 'NO' answer, or a classification such as male or female, or recording a component as defective or not defective. If the outcomes are also independent e.g. one respondent giving a YES answer does not influence the answer of the next respondent, then the variable is **binomial**. Consider the situation of items coming off the end of a production line, some of which are defective. If the proportion of defective items is 10% of the flow of items, we can regard the selection of a small sample as consisting of independent selections, and thus the probability of selecting a defective item will remain constant as $P(\text{defective}) = 0.1$. (Note that if there were a small, fixed number of items in total and selection were without replacement, then we would have a situation of conditional probabilities, and $P(\text{defective})$ would not remain con-

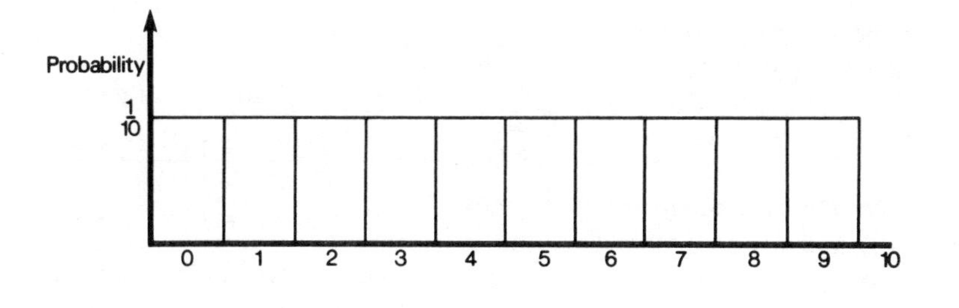

Figure 11.1

stant.) Samples will be selected from production lines to monitor quality, and thus we will be interested in the number of defective items in our sample. However, the number of defectives may be related to the size of the sample and to whether or not the process is working as expected.

EXERCISE

Which would lead to most concern, more than 1 defective in a sample of 10, or more than 10 defectives in a sample of 100?

We shall return to this exercise to compare your impressions with the theoretical results of the appropriate probability model, but first we will consider much smaller samples.

For a sample of size 1, the probability that the item selected is defective is 0.1 and the probability that it is not defective is 0.9.

For a sample of two, for each item the probabilities are as above, but we are now interested in the sample *as a whole*. There are four possibilities:

1. both items are defective;
2. the first is defective, the second is OK;
3. the first is OK, the second is defective;
4. both items are OK.

Since the two selections are independent, we can multiply the unconditional probabilities together, thus:

1. $P(\text{both defective}) = (0.1)(0.1) = 0.01$
2. $P(\text{1st defective, 2nd OK}) = (0.1)(0.9) = 0.09$
3. $P(\text{1st OK, 2nd defective}) = (0.9)(0.1) = 0.09$
4. $P(\text{both OK}) = (0.9)(0.9) = 0.81$

$$1.00$$

However, (2) and (3) both represent one defective in the sample of two. If we are not interested in the *order* in which the events occur, then:

$P(\text{2 defective}) = 0.01$
$P(\text{1 defective}) = 0.18$
$P(\text{no defectives}) = 0.81$

$$1.00$$

If p = probability of a defective, then $q = (1 - p)$ is the probability of a nondefective item, and we have:

$P(\text{2 defective}) = p^2$
$P(\text{1 defective}) = 2pq$
$P(\text{0 defective}) = q^2$

For a sample of three, the following possibilities exist (where def. is an abbreviation for defective):

1. 3 defective
2. 1st,2nd defective; 3rd OK
3. 1st,3rd defective; 2nd OK
4. 2nd,3rd defective; 1st OK
5. 1st defective; 2nd,3rd OK
6. 2nd defective; 1st, 3rd OK
7. 3rd defective; 1st,2nd OK
8. all OK

$P(\text{3 def.}) = (0.1)^3 = 0.001$
$P(\text{1,2 def.; 3 OK}) = (0.1)^2(0.9) = 0.009$
$P(\text{1,3 def.; 2 OK}) = (0.1)(0.9)(0.1) = 0.009$
$P(\text{2,3 def.; 1 OK}) = (0.9)(0.1)^2 = 0.009$
$P(\text{1 def.; 2,3 OK}) = (0.1)(0.9)^2 = 0.081$

$P(2, \text{def.}; 1,3 \text{ OK}) = (0.9)(0.1)(0.9) = 0.081$
$P(3 \text{ def.}; 1,2 \text{ OK}) = (0.9)^2(0.1) = 0.081$
$P(3 \text{ OK}) = (0.9)^3 = 0.729$

1.000

Again, we can combine events, since order is not important; (2), (3) and (4) represent two defectives; (5), (6) and (7) represent one defective. Thus:

$P(3 \text{ defectives}) = 0.001 = p^3$
$P(2 \text{ defectives}) = 0.027 = 3p^2q$
$P(1 \text{ defective}) = 0.243 = 3pq^2$
$P(0 \text{ defectives}) = 0.729 = q^3$

1.000

We could continue with the procedure, looking at sample sizes of four, five and so on, but a pattern is already emerging from the results given above. At the extreme, the probability that all of the items are defective is p^2 or p^3, and for a sample of size n, this will be p^n. With other outcomes, these consist of a series of possibilities, each with the same probability, and these are then combined. For example, in a sample of 10, the probability of 4 defectives would consist of a series of outcomes, each of which would have a probability of $(0.1)^4(0.9)^6 = p^4q^6 = 0.0000531$. (Note that four defective items means that there are also 6 items which are OK, since the sample size is 10.) The question that needs to be answered now is: 'How many of the outcomes from a sample of ten have four defective items?' To answer this question we will use the idea of **combinations**. The number of combinations of r defective items in a sample of n items is given by:

$$^nC_r = \binom{n}{r} = \frac{n!}{r!(n-r)!}$$

where ^nC_r and $\binom{n}{r}$ are the two most commonly used notations for combinations, and $n!$ is the **factorial** of n. This means, n times $(n-1)$ times $(n-2)$, etc., until 1 is reached.

For example:

$$2! = 2 \times 1 = 2$$
$$3! = 3 \times 2 \times 1 = 6$$
$$4! = 4 \times 3 \times 2 \times 1 = 24$$
$$10! = 10 \times 9 \times 8 \times 7 \times 6 \times 5 \times 4 \times 3 \times 2 \times 1 = 3\,628\,800$$

However, $0! = 1$

Note also that these factorials can be written as

$$10! = 10 \times 9 \times 8 \times 7!$$
or $$= 10 \times 9!$$
or $$= 10 \times 9 \times 8 \times 7 \times 6 \times 5 \times 4!$$

since this will help when calculating the number of combinations. (See section 11.7 for an alternative method for use with small samples.) Returning now to the sample of 10, the number of ways of getting 4 defective items in a sample will be when $n = 10$ and $r = 4$:

$$^{10}C_4 = \binom{10}{4} = \frac{10!}{4!(10-4)!} = \frac{10!}{4!\,6!}$$

If we note the highest factorial in the denominator is 6, we have

$$\binom{10}{4} = \frac{10 \times 9 \times 8 \times 7 \times 6!}{4 \times 3 \times 2 \times 1 \times 6!} = \frac{10 \times 9 \times 8 \times 7}{4 \times 3 \times 2 \times 1} = 210$$

There are 210 different ways of getting four defectives in a sample of ten, and thus

$$P(4 \text{ defective in } 10) = 210(0.1)^4(0.9)^6 = 0.011\,151$$

Looking back to the probabilities of different numbers of defectives in a sample of 3, these can now be rewritten as follows:

$P(3 \text{ defectives}) = p^3$
$P(2 \text{ defectives}) = \binom{3}{2}p^2q$
$P(1 \text{ defective}) = \binom{3}{1}pq^2$
$P(0 \text{ defectives}) = q^3$

The general formula for a binomial probability will be:

$$P(r \text{ items in a sample of } n) = \binom{n}{r}p^rq^{n-r}$$

EXAMPLE

What is the probability of more than 3 defective items in a sample of 12 items, if the probability of a defective item is 0.2?

The required probability is:

$P(4) + P(5) + P(6) + P(7) + P(8) + P(9) + P(10) + P(11) + P(12)$

but this may be written as:

$1 - [P(0) + P(1) + P(2) + P(3)]$

which will simplify the calculation. $n = 12$, $p = 0.2$ and $q = 1 - p = 0.8$ so

$$P(0) = q^{12} = (0.8)^{12} = 0.068\,719\,5$$

$$P(1) = \binom{12}{1} pq^{11} = 12(0.2)(0.8)^{11} = 0.206\,159\,4$$

$$P(2) = \binom{12}{2} p^2q^{10} = 66(0.2)^2(0.8)^{10} = 0.283\,467\,8$$

$$P(3) = \binom{12}{3} p^3q^9 = 220(0.2)^3(0.8)^9 = \frac{0.236\,223\,2}{0.794\,569\,9}$$

Therefore, the required probability is $1 - 0.794\,569\,9 = 0.220\,543\,01$ or more simply 0.22.

An alternative to this calculation would be to use tables of the cumulative Binomial distribution (see Appendix A). For example, if we require the probability of 5 or more items in a sample of 10, when $p = 0.20$, from the table we find that P(5 or more) = 0.0328.

For the same sample, if the required probability were for 5 or less, then we would look up P(6 or more) = 0.0064 and subtract this from 1:

P(5 or less) = 1 − 0.0064 = 0.9936

Returning now to the problem of whether it is a greater matter of concern to find more than 1 defective in a sample of 10, or more than 10 defectives in a sample of 100, we see that:

for a sample of 10 with $p = 0.1$, P(2 or more) = 0.2639; for a sample of 100 with $p = 0.1$, P(11 or more) = 0.4168.

EXERCISE

Determine these probabilities.

Thus, if the process is working as was proposed, giving 10% of items defective in some way, then the probability of finding 2 or more in a sample of 10 is very much lower than the probability of finding 11 or more in a sample of 100, i.e. in both cases finding more than the expected number in a sample. However, from the note on expectations in Chapter 10, we know that we are unlikely always to get the expected number in a particular sample selection. Even so, the small sample result would suggest more strongly that something was wrong with the process and would therefore be cause for more concern.

For a Binomial distribution, the mean can be shown to be np and the variance to be npq. Thus, for a sample of size 10 with a probability $p = 0.3$, the average, or expected number, of items per sample with the characteristic will be

$$np = 10 \times 0.3 = 3$$

the variance will be

$$npq = 10 \times 0.3 \times 0.7 = 2.1$$

11.3 POISSON DISTRIBUTION

While the Binomial model given above will be successful in many cases in modelling business or production situations, where the numbers involved are large and the probability of the occurrence of a characteristic is small, or where the numbers involved become infinite, then the Poisson probability model will be a better representation of the situation. The model works with the expected or average number of occurrences; if this is not given it can be found as np.

The probability model is:

$$P(x) = \frac{\lambda^x e^{-\lambda}}{x!}$$

where λ is the average number of times a characteristic occurs and x is the number of occurrences (x may be any integer from 0 to infinity). For example, if a company receives an average of three calls per 5 minute period of the working day, then we can calculate the probabilities of receiving a particular number of calls in a randomly selected 5 minute period.

The average number of calls, $\lambda = 3$ and $e^{-3} = 0.0498$, so

$$P(0 \text{ calls}) = \frac{3^0(0.0498)}{0!} = 0.0498$$

$$P(1 \text{ call}) = \frac{3^1(0.0498)}{1!} = 0.1494$$

$$P(2 \text{ calls}) = \frac{3^2(0.0498)}{2!} = 0.2241$$

$$P(3 \text{ calls}) = \frac{3^3(0.0498)}{3!} = 0.2241$$

$$P(4 \text{ calls}) = \frac{3^4(0.0498)}{4!} = 0.168075$$

As you may have noticed, there is a **recursive relationship** between any two consecutive probabilities, such that:

$$P(4 \text{ calls}) = \tfrac{3}{4} \times P(3 \text{ calls}) = \tfrac{3}{4}(0.2241) = 0.168075$$

or more generally:

$$P(N \text{ calls}) = \frac{\lambda}{N} P(N-1 \text{ calls}).$$

If the company quoted above has only four telephone lines, and calls last for at least 5 minutes, then there is a probability of

$$P(\text{no calls}) + P(1 \text{ call}) + P(2 \text{ calls}) + P(3 \text{ calls}) + P(4 \text{ calls})$$
$$= 0.0498 + 0.494 + 0.2241 + 0.2241 + 0.168075$$
$$= 0.815475$$

or approximately 0.815 of the switchboard being able to handle all incoming calls. Put another way, you would expect the switchboard to be sufficient for 81.5% of the time, but for callers to be unable to make the connection during 18.5% of the time. This raises the question of whether another line should be installed.

$$P(5 \text{ calls}) = \tfrac{3}{5} \times P(4 \text{ calls}) = 0.1008.$$

The switchboard would now be in a position to handle all calls for an *extra* 10% of the time, but whether or not this is worth while would depend upon the likely extra profits that this would create, against the cost of installation and running an extra telephone line.

Again, there is an alternative to calculating all of the probabilities each time, by using tables of cumulative Poisson probabilities (see Appendix B). For example, if the average number of faults found on a new car at its pre-delivery inspection is five, then from tables we can find that

(a) $P(3 \text{ or more}) = 0.8753$,
(b) $P(5 \text{ or more}) = 0.5595$,
(c) $P(10 \text{ or more}) = 0.0318$,

and, as before, these can be manipulated. From (a), we see that $1 - 0.8753 = 0.1247$ so that the probability of a car having less than three faults is 0.1247; or we would expect only 12.47% of cars that have pre-delivery inspections to have less than three faults.

For the Poisson distribution it can be shown that the mean and variance are both equal to λ.

11.4 POISSON APPROXIMATION TO THE BINOMIAL

Both distributions are discrete probability models, but for many values of $\lambda = np$ the Poisson model is considerably more skewed that the binomial. However, for small values of p(less than 0.1), and large values of n, it may be easier to use a Poisson distribution. (Note that if p is very small, $(1-p)$ will be close to 1 and hence $np(1-p) \simeq np = \lambda$ which is both the mean and the variance of the Poisson distribution.)

EXAMPLE

If the probability of a fault in a piece of precision equipment is 0.0001, and each completed machine has 10 000 components, what is the probability of there being two or more faults?

(a) Using Poisson distribution:
$\lambda = np = 10\,000 \times 0.0001 = 1$
$P(0) = e^{-1} = 0.3679$
$P(1) = e^{-1} = 0.3679$
$P(0) + P(1) = 0.7358$
Therefore, $P(2 \text{ or more}) = 1 - 0.7358 = 0.2642$
(b) Using Binomial distribution:

$P(0) = (0.9999)^{10\,000} = 0.3679$
$P(1) = 10\,000(0.0001)(0.9999)^{9999} = 0.3679$
$P(0) + P(1) = 0.7358$
Therefore, $P(2 \text{ or more}) = 1 - 0.7358 = 0.2642$

Comparing these two answers, it is suggested that method (a) is very much easier to work with than method (b).

11.5 CONCLUSIONS

Rather than seeing each problem as unique and requiring some time-consuming, original solution, it is useful to recognize that many problems belong to families. There are, for example, a range of problems that can be described by the Binomial distribution. When looking at a problem for the first time, try to identify the parameters (these values are always given in some form in traditional examination questions) and the assumptions that you may need to make e.g. events are independent. If your problem does match a known distribution, then clearly you have at least one well – established method of solution which may only require reference to statistical tables.

11.6 PROBLEMS

Evaluate the following expressions:

1. $\binom{3}{1}$
2. $\binom{10}{3}$ and $\binom{10}{7}$
3. $\binom{20}{6}$ and $\binom{20}{0}$
4. $\binom{10}{2}$; $\binom{10}{1}$ and $\binom{10}{0}$
5. $\binom{52}{13}$ This represents the number of different possible hands of 13 cards that could be dealt with a standard pack of playing cards.
6. Given the answer to question 5, what is the probability of getting a complete suit of cards in one of the four hands dealt from a standard pack of cards?
7. A Binomial model has $n=4$ and $p=0.6$. Find the probabilities of each of the five possible outcomes (i.e. $P(0)$—$P(4)$]. Construct a histogram of this data.
8. Attendance at a cinema has been analyzed, and shows that audiences consist of 60% men and 40% women for a particular film. If a random sample of six people were selected from the audience during a performance, find the following probabilities:
   (a) all women are selected;
   (b) three men are selected;
   (c) less than three women are selected.
9. How would the probabilities in question 8 change if the sample size were eight?
10. A quality control system selects a sample of three items from a production line. If one or more is defective, a second sample is taken (also of size three), and if one or more of these is defective then the whole production line is stopped. Given that the probability of a defective item is 0.05, what is the probability that the second sample is taken? What is the probability that the production line is stopped?
11. The probability that an invoice contains a mistake is 0.1. In an audit a sample of 12 invoices are chosen from one department; what is the probability that less than two incorrect invoices are found?
12. Find each of the Poisson probabilities from $P(0)$ to $P(5)$ for a distribution with an average of 2. Construct a histogram of this part of the distribution.
13. For a Poisson distribution with an average of 2, find the probability of $P(x>4)$ and $P(x>5)$.
14. Find the probabilities $P(0)$ to $P(4)$ for a Binomial distribution with $n=10\,000$ and $p=0.00015$.
15. The number of accidents per day on a particular stretch of motorway follows a Poisson distribution with a mean of one. Find the probabilities of 0, 1, 2, 3, 4, or more accidents on this stretch of motorway on a particular day. Find the expected number of days with 0, 1, 2, 3, 4 or more accidents in a one year period (assuming 365 days per year). If the average cost of policing an

accident is £1 000, find the expected cost of policing accidents on this stretch of motorway for a year.

16. The number of train passengers who fail to pay for their tickets in a certain region has a Poisson distribution with a mean of four. If there are 4 800 trains per month in the region, find the expected number of trains with more than three non-fare-paying passengers. Calculate the average cost to the train company during a month if the average cost of a ticket is £7.84.
17. A man has four cars for hire. The average demand on a weekday is for two cars. Assuming 312 weekdays per year, obtain the theoretical frequency distribution of the number of cars demanded during a weekday. Hence estimate to the nearest whole number, the number of days on which demand exceeds supply. (Assume demand does not surpass nine cars per day.) Would you suggest that the man buys another car?
18. (a) Items are packed into boxes of 1 000, and each item has a probability of 0.001 of having some type of fault. What is the probability that a box will contain less than three defective items?
    (b) If the company sells 100 000 boxes per year and guarantees less than three defectives per box, what is the expected number of guarantee claims?
    (c) Replacement of a box returned under the guarantee costs £150. What is the expected cost of guarantee claims?
    (d) Boxes sell at £100 but cost £60 to produce and distribute. What is the company's expected profit for sales of boxes?

11.7 APPENDIX

For a small sample it may be preferable to use **Pascal's triangle** to find the number of combinations, or the coefficients for each term in the Binomial probability model. This begins with three 1s arranged thus:

1
1 1

To find the next line, which will have three terms, the first and last will be 1s, the middle term will be the sum of the two terms just *above* it:

1
1 1
(1 + 1)
1 2 1

and this will apply to a sample size of two.

This process continues, to give the next line,

(1 + 2)(2 + 1)
1 3 3 1

which will apply for a sample of size three. The process can continue until the desired sample size is reached.

	Sample Size:
1	
1 1	
1 2 1	2
1 3 3 1	3
1 4 6 4 1	4
1 5 10 10 5 1	5
1 6 15 20 15 6 1	6
1 7 21 35 35 21 7 1	7
1 8 28 56 70 56 28 8 1	8
1 9 36 84 126 126 84 36 9 1	9
1 10 45 120 210 252 210 120 45 10 1	10
1 11 55 165 330 462 462 330 165 55 11 1	11
1 12 66 220 495 792 924 792 495 220 66 12 1	12

EXAMPLE

The number of different combinations of six defective items in samples of size 10 is the 7th term from the left in the row corresponding to sample size 10, i.e. 210.

Alternatively, the Binomial coefficient is ${}^{10}C_6$ given by

$$\binom{10}{6} = \frac{10!}{4!\,6!} = 210.$$

THE NORMAL DISTRIBUTION

12

When a variable is *continuous*, and its value is affected by a large number of chance factors, none of which predominates, then it will frequently have a **Normal distribution**. (The word 'normal' does not imply any value-judgement.) The distribution does occur frequently and is probably the most widely used statistical distribution. A Normal distribution is a *symmetrical* distribution about its mean (Figure 12.1).

Variables having a Normal distribution may be associated with production, for example the weights of jars of jam, or with human populations, for example people's height and weight, and thus clothes' sizes. These distributions are often summarized by their mean and variance (usually labelled μ and σ^2 respectively). If a variable X has a Normal distribution, this may be written as $X \sim N(\mu, \sigma^2)$. Normal distributions are characterized particularly by the areas in various sectors of the distribution. If these areas are considered as a proportion of the total area under the distribution curve, then they may also be considered as the probabilities of obtaining a value from the distribution in that sector.

Theoretically, to find the area under the distribution curve in the sector less than some value x we should need to evaluate the integral

$$\int_{-\infty}^{x} \frac{1}{\sigma\sqrt{2\pi}} \exp\left[-\frac{(x-\mu)^2}{2\sigma^2}\right] dx$$

which tends to 1 as x tends to infinity.

Fortunately there is an easier method of finding this area.

12.1 THE STANDARD NORMAL DISTRIBUTION

For any Normal distribution, if the horizontal scale can be measured in terms of the number of standard deviations away from the mean we may use published tables to find various required areas, and hence probabilities. This requires a transformation to the standard Normal distribution, which has a mean of 0 and a standard deviation of 1.

For a value of X in a Normal distribution, this transformation will be:

$$Z = \frac{X - \mu}{\sigma}$$

For example, if a variable X has a Normal distribution with a mean of 100 and a standard deviation of 10, then:

for $X = 115$, $\quad Z = \dfrac{115 - 100}{10} = 1.5;$

for $X = 75$, $\quad Z = \dfrac{75 - 100}{10} = -2.5;$

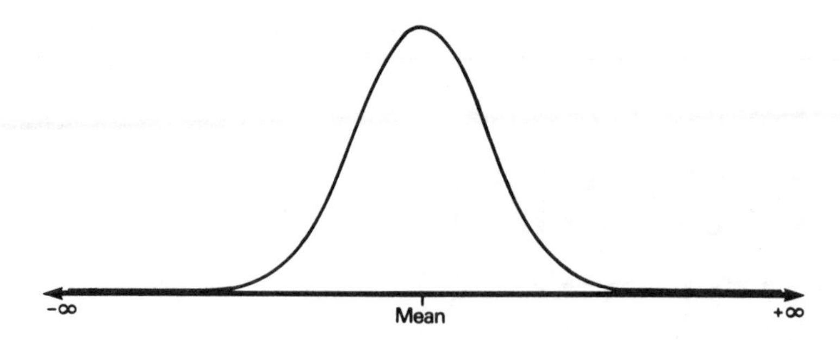

Figure 12.1 A Normal distribution

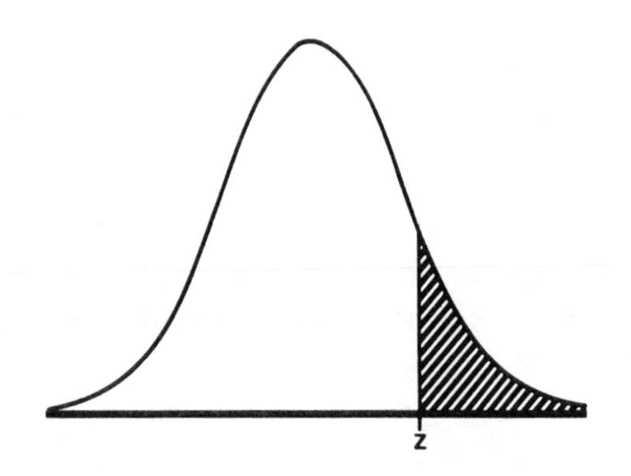

Figure 12.2

for $X = 111.3$, $Z = \dfrac{111.3 - 100}{10} = 1.13$.

The area excluded in the **right-hand tail** of the distribution is given in Appendix C and is shown in Figure 12.2.

For example, the area in the right-hand tail above $Z = 1.03$ is 0.1515. Since the total under the standard Normal curve is 1, this area is also the probability of obtaining a value from the *original* distribution more than 1.03 standard deviations above the mean. Manipulating the value from the table, we see that the probability of obtaining a value below 1.03 standard deviations above the mean is $1 - 0.1515$ (= 0.8485).

To find the probability of a value between 1 and 1.1 standard deviations above the mean, we have to subtract one from another as follows:

area above $Z = 1$ is 0.1587;
area above $Z = 1.1$ is 0.1357;
so the area between $Z = 1$ and $Z = 1.1$ is $0.1587 - 0.1357 = 0.0230$

Since the standard Normal distribution is symmetrical about its mean of 0, an area to the right of a positive value of Z will be identical to the area to the left of the corresponding negative value of Z. (Note that areas cannot be negative.) Thus to find the area between $Z = -1$ and $Z = +1$, we have:

area to left of $Z = -1$ is 0.1587
area to right of $Z = +1$ is 0.1587
area outside the range $(-1, +1)$ is $0.1587 + 0.1587 = 0.3174$
so, area between $Z = -1$ and $Z = +1$ is $1 - 0.3174 = 0.6826$.

For *any* Normal distribution, 68.26% of the values will be within one standard deviation of the mean. (Hint: it is often useful to draw a sketch of the area required by a problem and compare this with Figure 12.2.)

EXERCISE

What percentage of values will be within 1.96 standard deviations of the mean? (Answer: 95%.)

If a population is known to have a normal distribution, and its mean and variance are known, then we may use the tables to express facts about this population.

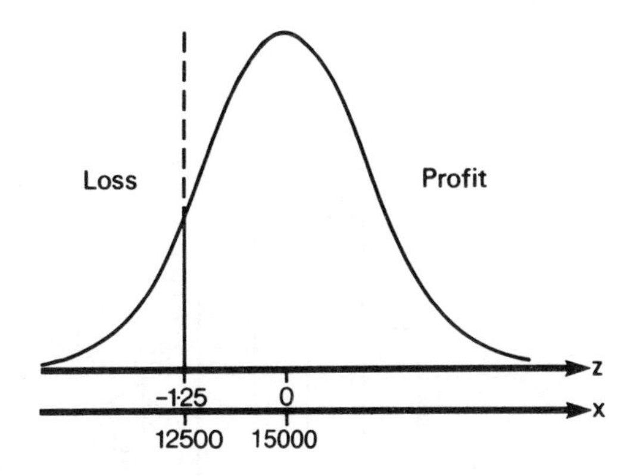

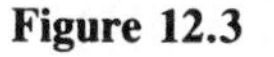
Figure 12.3

EXAMPLE

The attendance at rock concerts at a particular stadium has a normal distribution with a mean of 15 000 and a variance of 4 000 000. The promoters are able to break even at an attendance of 12 500; what proportion of concerts will make a loss?

Mean = 15 000, variance = 4 000 000, standard deviation = $\sqrt{\text{variance}}$ = 2 000.

Therefore, let $Z = \dfrac{12\,500 - 15\,000}{2000} = -1.25$

(Note that we have used two scales in Figure 12.3, one for Z and one for the original distribution; this may help in understanding some questions.)

From tables, the area of the right of $Z = +1.25$ (which is equal to the area to the left of $Z = -1.25$) is 0.1056.

Thus, 10.56% of the concerts will make a loss.

12.2 NORMAL APPROXIMATION TO THE BINOMIAL

Although the Binomial distribution is a discrete probability distribution, and the Normal distribution is continuous, it will be possible to use the Normal distribution, as an approximation to the Binomial if n is large, and $p > 0.1$. (As we saw in the last chapter if $p < 0.1$ we would use the Poisson approximation to the Binomial.) To see why this will work, consider a Binomial distribution with a probability, p, of 0.2. For various values of n, we have distributions as shown in Figure 12.4.

Looking at the various parts of Figure 12.4, we see that with $n = 2$, we have a highly skewed distribution; the mean will be $np = 0.4$. As n increases, the amount of skewness decreases: in Figure 12.4b, the mean is 2 and in Figure 12.4c, the mean is 4, and even at this stage, we are beginning to see the typical 'bell shape' of the Normal distribution curve. In Figure 12.4e, the mean is 20, and although the shape of the histogram is not exactly that of the normal curve, it is very close.

If we wish to use the Normal distribution as an approximation to the Binomial distribution, we must develop a method of moving from a discrete distribution to a continuous one. To see how to do this, look at Figure 12.5 where a curve has been super-imposed on the histogram.

Here we see that as the curve cuts through the midpoints of the blocks of the histogram, small areas such as B are *excluded*, while other areas, such as A, are *included* under the curve but not in the histogram. These areas will tend to cancel each other out. Since each block represents a whole number, often the number of successes, it can be considered as extending from 0.5 below that integer to 0.5 above. Thus in the example above, the block representing 52 successes extends from 51.5 to 52.5. In order to find the area and hence the probability for a series of outcomes, it will thus be necessary to go from 0.5 below the lowest integer to 0.5 above the highest integer. If from Figure 12.5 we wanted to find the probability of 49, 50, 51 and 52 successes, then we would need to find the area under the normal curve from $X = 48.5$ to $X = 52.5$, or if we wanted to find the probability of 52 or more successes, we should require the area to the right of $X = 51.5$.

To find areas, we must transform the X values into Z values, on the standard Normal distribution. From the previous chapter, we know that for a Binomial distribution, the mean $= np$ and the standard deviation $= \sqrt{[np(1-p)]}$; and these values can be used to calculate the Z value:

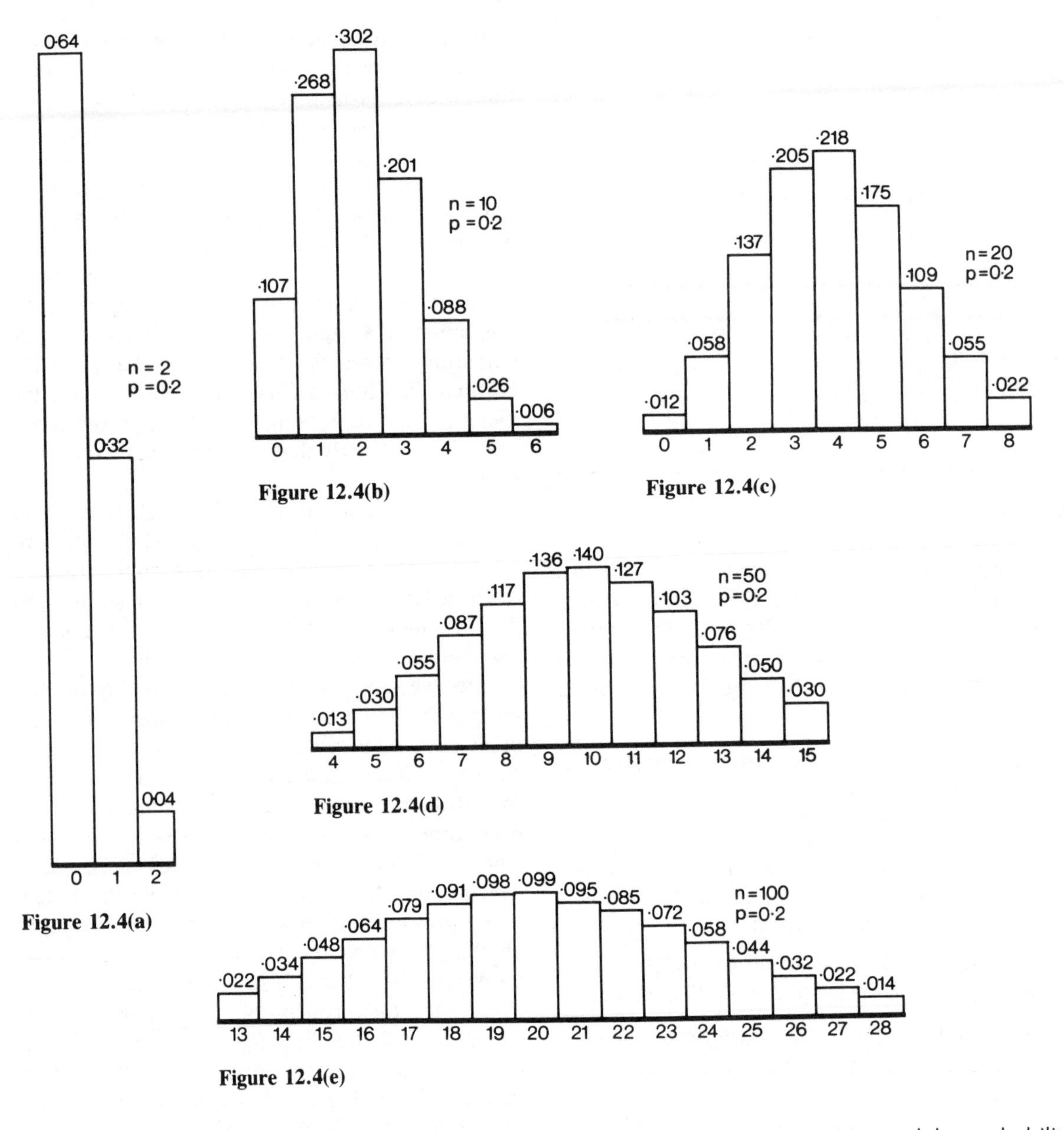

Figure 12.4(a)

Figure 12.4(b)

Figure 12.4(c)

Figure 12.4(d)

Figure 12.4(e)

$$Z = \frac{X - np}{\sqrt{[np(1-p)]}}$$

EXAMPLE

Enquiries at a travel agents lead only sometimes to a holiday booking being made. The agent needs to take 35 bookings per week to break even. If during a week there are 100 enquiries and the probability of a booking in each case is 0.4, find the probability that the agent will at least break even in this particular week.

This is a binomial situation since p is fixed (at 0.4) and each enquiry either leads to a booking or does not.

$n = 100$, $p = 0.4$, mean $= np = 40$, and

standard deviation $= \sqrt{[np(1-p)]} = \sqrt{[(100(0.4)(0.6)]} = 4.899$.

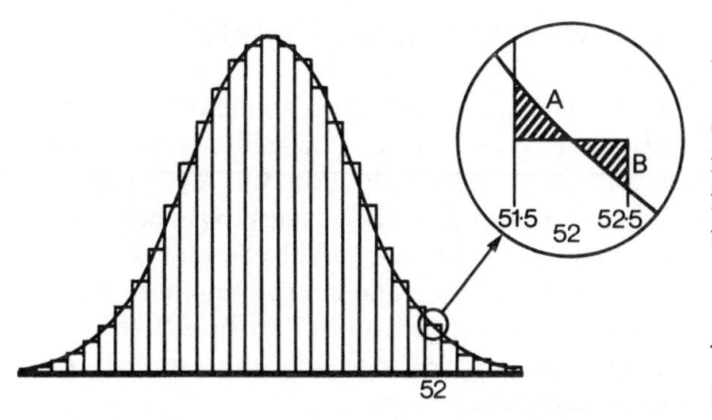

Figure 12.5

The agent will at least break even if 35 or more bookings are taken, and thus we need the area under the Normal curve to the right of $X = 34.5$. For this value of X,

$$Z = \frac{34.5 - 40}{4.899} = -1.12$$

and the area to the left of $Z = -1.12$, is 0.1314. Therefore, the area to the right of $Z = -1.12$ is $1 - 0.1314$ (= 0.8686).

The probability that the agent will at least break even is 0.8686, using the Normal approximation to the Binomial distribution.

If the Binomial probability had been computed directly we have

$$P(\text{at least break even}) = P(35) + P(36) + \dots + P(100) = 0.8697.$$

As you can see, there is very little difference in the two answers.

12.3 NORMAL APPROXIMATION TO THE POISSON

In a similar way to the Binomial distribution, as the mean, $\lambda = np$, gets larger and larger, the amount of skewness in the Poisson distribution decreases, until it is possible to use the Normal distribution. To transform values from the original distribution into Z values, we use:

$$Z = \frac{X - \lambda}{\sqrt{\lambda}}$$

(Note that it is usual to use the Normal approximation if $\lambda > 30$. Again, we should allow for the fact that we are going from a discrete to a continuous distribution.)

EXAMPLE

The average number of broken eggs per lorry load is known to be 50. What is the probability that there will be more than 70 broken eggs on a particular lorry load?

Mean = λ = 50; standard deviation = $\sqrt{\lambda} = \sqrt{50}$ = 7.071. Area required is that above X = 70.5, so the value of Z to use is as follows:

$$Z = \frac{70.5 - 50}{7.071} = 2.90$$

From tables, the area to right of $Z = 2.9$ is 0.001 87, therefore, P(more than 70 broken eggs) = 0.001 87.

12.4 COMBINATIONS OF VARIABLES

It is often useful to combine variables by either adding or subtracting values. An assembled product, for example, could be made from a number of different components, each individually described by a mean and a standard deviation. To consider the characteristics of this assembled product we will need to combine the means and standard deviations from the consituent parts. In assessing the results of a survey (see sections 14.2.1 and 14.2.2 for examples) we may need to combine results with several sources of variation. We may wish to compare the difference in annual income by region or by sex, for example.

If X and Y are two independent, Normally distributed random variables with means of μ_1 and μ_2 and variances of σ_1^2 and σ_2^2 respectively, then

for $X+Y$: mean $= \mu_1 + \mu_2$
variance $= \sigma_1^2 + \sigma_2^2$

for $X-Y$: mean $= \mu_1 - \mu_2$
variance $= \sigma_1^2 + \sigma_2^2$ (Note + sign)

If we are adding variables, the mean of the sum is the sum of the means, and the variance of the sum is the sum of the variances. We add variances, *not* standard deviations. The standard deviation is calculated by taking the square root of variance:

standard deviation $= \sqrt{(\sigma_1^2 + \sigma_2^2)}$.

EXAMPLE

An assembled product is made from two parts. The weight of each part is Normally distributed with mean and standard deviation as follows:

Part	*Mean*	*Standard deviation*
1	15	4
2	20	2

What percentage of these assembled products weighs more than 36 kilograms? Consider the assembled product as an addition of weights. We obtain

mean $= \mu = 15 + 20 = 35$
standard deviation $= \sigma = \sqrt{(4^2 + 2^2)} = 4.4721$

The corresponding Z-value is

$$Z = \frac{X-\mu}{\sigma} = \frac{36 - 35}{4.4721} = 0.22.$$

The area to the right of $Z = 0.22$ is 0.4129 so the percentage we would expect to weigh more than 36 kilograms is 41.29%.

12.5 CENTRAL LIMIT THEOREM

If the Normal distribution applied only to the situations given above, it would still be a very useful statistical distribution for modelling behaviour. However it also applies to a whole range of sampling situations which permits an even wider range of use. The interpretation of sample results will be dealt with in Part Five, but in this chapter we will indicate *why* the **sampling distributions** come about.

We will not attempt to derive this idea mathematically preferring to consider what will happen in a specific case, and then to state the general theorem. Consider a population that has a Normal distribution with a mean μ and a variance σ^2 as shown in Figure 12.6. If we took every possible sample of one from this distribution then we would just obtain the original population, and the diagram would be *exactly* as in Figure 12.6.

However, if we increase the sample size to two, and calculate the mean, there will be a change in the distribution obtained. From the use of the probability tables, we know that we are much more likely to get a sample value from somewhere near the mean, than from a point on the distribution which is a long way from the mean. Thus the probability that both values in our sample of two will be close to the mean will be considerably higher than the probability that both values will be a long way below the mean, or a long way above the mean. Again, considering every possible sample of two from the distribution, and calculating the mean, there will be more sample means close to the population mean than there were original population values, since one small value and one large value will give a mean close to the centre of the distribution.

This situation is illustrated in Figure 12.7 where we also see that the average of all of the sample

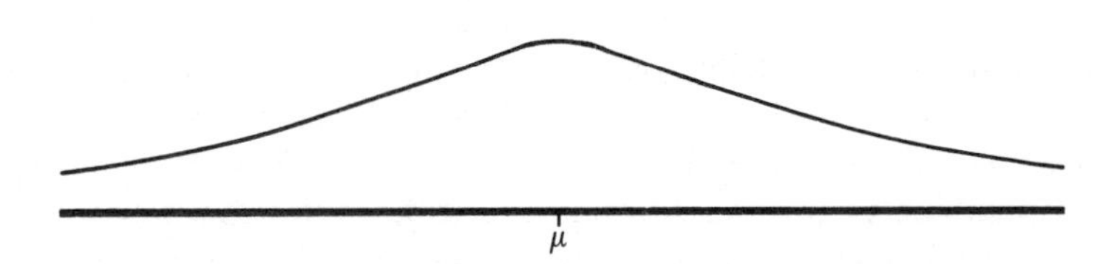

Figure 12.6

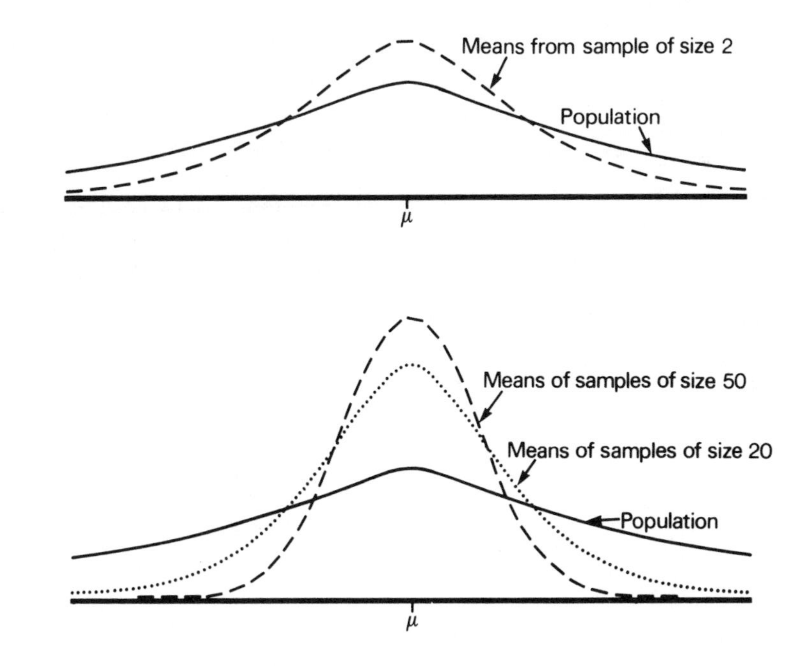

Figure 12.7

Figure 12.8

averages will be the population mean μ. As we increase the sample size, the probability of getting all of the sample values, and hence the sample average in an extreme tail of the original population distribution, becomes extremely small, while the probability of the sample mean being close to the original population mean increases.

The illustration in Figure 12.8 shows that as the sample size increases, the distribution of sample means remains a Normal distribution with μ as its mean; however, the variance of the distribution decreases as the sample size increases. It can be shown that the sample mean has the following distribution:

$$\bar{X} \sim N\left(\mu, \frac{\sigma^2}{n}\right)$$

where n is the sample size and the standard deviation of this sampling distribution or the **standard error** is given by $\sigma/\sqrt{n}$.

EXAMPLE

The time taken to travel to work by employees of a large company has a mean of 45 minutes and a standard deviation of 24 minutes. If a small sample of employees is selected at random, what is the probability that the sample average will be 50 minutes or more if:

(a) the sample size is $n = 30$?
(b) the sample size is $n = 100$?

To determine the probabilities of sample statistics, such as the mean, we need to establish the distribution concerned, in this case the Normal distribution, and the measure of spread, in this case the standard error already given.

(a) Given the sample size $n = 30$, the sample mean $\bar{x} = 50$ and the standard error $= 24/\sqrt{30} = 4.382$, we can utilize the Z transformation and consider the number of standard errors from the mean. Here,

$$Z = \frac{50 - 45}{4.382} = 1.14,$$

and using the Normal distribution tables, P(50 or more minutes) = 0.1271.

(b) When the sample size is increased to 100, the standard error becomes smaller. Standard error = $24/\sqrt{100} = 2.4$.
Using the transformation,

$$Z = \frac{50 - 45}{2.4} = 2.08,$$

and using the Normal distribution tables, P(50 or more minutes) = 0.01876.

This example illustrates the fact, that as sample size increases, the chances of getting an extreme result diminish.

The Z transformation for sample means is given by

$$Z = \frac{\bar{x} - \mu}{\sigma/\sqrt{n}};$$

we are now concerned with how many standard errors the sample mean is away from the true mean.

More generally, the Central Limit Theorem states that for any population distribution, whether it is discrete or continuous, skewed, rectangular or even multimodal, the distribution of the sample means will be approximately Normal if the sample size is sufficiently large.

In the case of a proportion of a sample, we are effectively considering a Binomial situation, and, as we saw in section 12.2, as n becomes large, the Binomial distribution can be approximately by the Normal distribution.

Thus the sampling distribution of a proportion will also be a Normal distribution. For a distribution with a population proportion π we have the distribution of the sample proportion, P, as:

$$P \sim N\left(\pi, \frac{\pi(1-\pi)}{n}\right)$$

EXAMPLE

It is known that only 40% (or 0.4 as a proportion) of a workforce favour changes in working practice. What are the chances that a random sample of the workforce will show 50% or more in favour of the changes if:

(a) a sample of 50 is selected?
(b) a sample of 100 is selected?

Again we need to adapt the Z transformation and recognize that the measure of spread of the sampling distribution, the standard error, is the square root of the variance given.

(a) Given the sample size $n = 50$, we assess the difference between the possible sample proportion of 0.5 and the known, true proportion 0.4. The standard error is given by

$$\sqrt{\frac{0.4 \times 0.6}{50}}$$

and

$$Z = \sqrt{\frac{0.5 - 0.4}{(0.4 \times 0.6)/50}} = 1.44.$$

Using tables for the Normal distribution, P(50% or more) = 0.0749 (i.e. 7%).

(b) The standard error is reduced by the increased sample size to

$$\sqrt{\frac{0.4 \times 0.6}{100}}$$

and

$$Z = \sqrt{\frac{0.5 - 0.4}{(0.4 \times 0.6)/100}} = 2.04.$$

Using tables for the Normal distribution P(50% or more) = 0.02068 (i.e. 2%).

This example illustrates again that the chances of an extreme sample result are reduced if the sample size is increased. It is assumed, of course, that the selection method is random, and the sampling frame and questionnaire valid.

The Z transformation for a sample proportion p is given by

$$Z \frac{p - \pi}{\sqrt{\frac{\pi(1-\pi)}{n}}}$$

where p is the sample proportion. In this case we are concerned with how many standard errors the sample proportion is away from the true proportion. It is worth noting that many problems of this kind are specified in terms of percentages and can be managed in exactly the same way.

It is not being suggested that in any situation all possible samples would be selected, but if we know the **theoretical distribution** of sample means or sample proportions, then we can compare this with the one, particular sample that we have taken.

12.6 CONCLUSIONS

In terms of the range of applications and the development of theory, the Normal distribution is the most important continuous distribution by far. It not only describes many observed distributions e.g. people's height and weight, but also describes the distribution of many sample statistics provided the sample size is reasonably large. Reasonably large in this context is 30 or more.

It is worth recalling that data collected on a business problem is likely, in many cases, to be skewed rather than Normal. Typically, the distribution of income or wealth is skewed (see section 4.3). However, the distribution of the sample statistics e.g. the mean, are likely to follow the Normal distribution regardless of the shape of the parent (or population) distribution.

12.7 PROBLEMS

1. A Normal distribution has a mean of 30 and a standard deviation of 5; find the Z values equivalent to the X values given below:
   (a) 35; (b) 27; (c) 22.3; (d) 40.7; (e) 30.
2. Use the tables of areas under the standard Normal distribution (given in Appendix C) to find the following areas:
   (a) to the right of $Z = 1$;
   (b) to the right of $Z = 2.85$;
   (c) to the left of $Z = 2$;
   (d) to the left of $Z = 0.1$;
   (e) to the left of $Z = -1.7$;
   (f) to the left of $Z = -0.3$;
   (g) to the right of $Z = -0.85$;
   (h) to the right of $Z = -2.58$;
   (i) between $Z = 1.55$ and $Z = 2.15$;
   (j) between $Z = 0.25$ and $Z = 0.75$;
   (k) between $Z = -1$ and $Z = -1.96$;
   (l) between $Z = -1.64$ and $Z = -2.58$;
   (m) between $Z = -2.33$ and $Z = 1.52$;
   (n) between $Z = -1.96$ and $Z = +1.96$.
3. Find the Z value, such that the standard Normal curve area:
   (a) to the right of Z is 0.0968;
   (b) to the left of Z is 0.3015;
   (c) to the right of Z is 0.4920;
   (d) to the right of Z is 0.99266;
   (e) to the left of Z is 0.9616;
   (f) between $-Z$ and $+Z$ is 0.95;
   (g) between $-Z$ and $+Z$ is 0.9.
4. Invoices at a particular depot have amounts which follow a normal distribution with a mean of £103.60 and a standard deviation of £8.75.
   (a) What percentage of invoices will be over £120.05?
   (b) What percentage of invoices will be below £92.75?
   (c) What percentage of invoices will be between £83.65 and £117.60?
   (d) What will be the invoice amount such that approximately 25% of invoices are for greater amounts?
   (e) Above what amount will 90% of invoices lie?
5. Items coming from the end of a production line are measured for their diameter. The measurements have a mean of 1 450 mm and a variance of 0.25 mm^2.
   (a) To meet quality control checks, the items must be between 1 448.5 mm and 1 451.5 mm. What proportion will meet these standards?
   (b) At a later stage in the production process, this item has to fit into a hole with a diameter which has a Normal distribution with a mean of 1 451 mm and a standard deviation of 1 mm. What proportion of these holes will have diameters over 1448.5?
   (c) What proportion of holes will have diameters above 1 451.5 mm?
6. Thirty percent of the general public have

bought a certain item in the last month. If a sample of 1 000 people is selected at random find the following probabilities:
   (a) more than 310 have bought the product;
   (b) less than 295 have bought the product;
   (c) more than 285 have bought the product.
7. On average, one in 10 000 people have the skills and talents to become major successes as singers. If a sample of 20 000 people were selected at random from a particular region, calculate the probability that there would be:
   (a) no suitable people;
   (b) more than three suitable people.
8. (a) In a very large population there are 40% of the people who support a change in government policy on regional aid. If a sample of 1 000 people is chosen at random, what is the probability that this sample will contain over 425 people who support a change in policy?
   (b) What is the probability that less than 390 people support a change in policy?
9. The average number of customers in a shop per week is 256. Calculate the probability of there being
   (a) more than 240 customers in a week;
   (b) less then 280 customers in a week;
   (c) 234 to 290 customers in a week.
10. A process yields 15% defective items. If 180 items are randomly selected from the process, what is the probability that the number of defectives is 30 or more?
11. If the probability of a smoker is 0.3, determine the probability of more than 20 smokers:
   (a) in a randomly selected sample of 80;
   (b) at a conference of 70 doctors.
   State any assumptions made.
12. In a certain manufacturing process, 25% of the items produced are classified as seconds. If 5 000 items have been produced, what is the probability that at least 1 300 items will be available as seconds for a sale?
13. A switchboard receives 42 calls per minute on average. Estimate the probability that there will be at least:
   (a) 40 calls in the next minute;
   (b) 50 calls in the next minute.
14. The average demand for a particular item from stock is found to be 45 per week. Estimate the probability of a demand for 50 or more:
   (a) in the next week;
   (b) in the next 2 weeks.
15. The times taken, in minutes, to complete three tasks, machining, assembly and packaging, have been recorded as follows:

	Mean	*Standard deviation*
Machining	25	7.5
Assembly	15	5.5
Packaging	15	4.5

   If a particular job requires all three tasks, what is the probability that the job will take no more than an hour?
16. It has been estimated that the average weekly wage in a particular industry is £172 and that the standard deviation is £9.
   (a) What is the probability that a random sample of 10 employees will have an average weekly wage of £180 or more?
   (b) What is the probability that an individual will have an average weekly wage of £180 or more if it can be assumed that wages follow a normal distribution?
17. An industrial chemical, that comes in granular form, is packaged and sold to the trade in 20 kg plastic sacks. To minimize the number of underweight sacks, the filling process has been adjusted so that the mean is 21 kg. The weight of the sacks varies because of the nature of the product and this has been measured by a standard deviation of 2.5 kg.
   (a) Determine the probability of an underweight sack.
   (b) If the sacks can also be bought in batch quantities of 30, determine the probability that the average weight per sack in the batch is less than 20 kg.

(c) Comment on your results.

18. It has been claimed that only 45% of customers find changes to an invoicing system an improvement. Assuming this is the case, determine the probability that a market research survey of 100 customers, will show that 50% or more report an improvement. What are the implications for the design of the survey?

PART FOUR
CONCLUDING EXERCISES

1. Select a situation where a queue forms awaiting service, and arrange to observe the behaviour of the queue over ten 10-minute periods. Note the arrival times of customers.

   When you have collected this data, draw up a table of the amount of time between arrivals in each of the 10-minute periods, constructing histograms to illustrate the data.

   Pool all of the data which you have collected, and calculate the probabilities of new customers arriving in under 1 minute, under 2 minutes, and so on. Does this distribution match any of the distributions discussed in this section?

   From your probability distribution, construct a table of expected percentages of customers arriving in each of the various periods.

   Observe the same queue on one further occasion, calculating the percentages of customers arriving for various time periods and compare your new results with the expected percentages.

   What may you conclude from this exercise?
2. Using MICROSTATS or a similar computer package, generate a sample of 10 random numbers between 1 and 100. Determine the mean and median for this sample. Repeat the exercise 30 times. Construct a histogram for the 30 sample means and the 30 sample medians.

   Repeat the above procedure using samples of 50 random numbers and again using samples of 100 random numbers. In each case construct a histogram for the sample means and the sample medians.

   What conclusions can you draw from a comparison of the constructed histograms?

PART FIVE STATISTICAL INFERENCE

The majority of the book so far has been concerned with describing sets of data, whether they are derived from the whole population or from a small sample of that population. This description has a useful part to play in business decision making, but the rôle of statistical inference is considerably more significant.

Here we are attempting to draw conclusions about the values of certain measures or characteristics relating to the whole population from the values obtained from a sample of that population. In Chapter 1 we considered the practical aspects of taking samples in terms of random and quota designs, and different methods of eliciting information from respondents. Here we will consider how we might use this information, once it has been summarized by the methods outlined in Part Two. The derivation of the relationships used in this part depend on the probability models developed in Part Four.

Samples provide *estimates* only of the true population values or population parameters. How close these estimates are to the population parameters will depend upon the size of the sample selected and the amount of variability in the original population. It will also be necessary to decide how certain we want to be about the results we obtain, since it is not feasible to be 100% certain about our results without conducting a census of the population (and even then we would have to assume that everyone took part, and that they all told the truth!).

This relationship between sample size, variability of the population and degree of confidence in the results is the key to understanding the chapters in this part of the book.

EXERCISE

Give examples of where estimated values are used to provide business, political and social information.

CONFIDENCE INTERVALS 13

Sampling, as we have seen in Chapter 1, is concerned with the collection of data from a (usually small) group selected from a defined, relevant population. Various methods are used to select the sample from this population, the main distinction being between those methods based on random sampling and those which are not. In the development of statistical sampling theory it is assumed that the *samples used are selected by simple random sampling*: although the methods developed in this, and subsequent chapters are often applied to other sampling designs.

Sampling theory applies whether the data is collected by interview, postal questionnaire or observation. However, as you will be aware, there are ample opportunities for bias to arise in the methods of extracting data from a sample, including the percentage of non-respondents. These aspects must be considered in interpreting the results together with the statistics derived from sampling theory.

Sampling theory (see section 12.5) provides a basis for understanding how the results from a sample may be interpreted in relation to the parent population; in other words, what conclusions can be drawn about the population on the basis of the sample results obtained. For example, if a company conducted a market research survey in Bury and found that 30% of their customers would like to try a new flavour of their ice cream, what useful conclusions could be drawn about all existing customers in Bury, or about existing and potential customers elsewhere? (Whether useful conclusions can be drawn will depend upon the sample design and the original problem definition.) If a manufacturer of ladies hats knows that the average head size is 6.75 inches, what sizes and variations should be produced?

13.1 STATISTICAL INFERENCE

Only a census which had 100% response and where everyone told the truth would allow us to be absolutely certain of our results; and even then, only at that point in time. This situation is unlikely to arise (!) and thus it is difficult to justify the cost, both in terms of money and time, of a census. Often results are required quickly, for example, in aiming to predict the result of an election, or the number of defectives in a production process and, in these circumstances, there is no time to conduct a census. Fortunately a census is rarely needed since a body of theory has grown up which will allow us to draw conclusions about a population from the results of a sample survey. This is **statistical inference**.

Statistical inference draws upon the probability results which we have seen in Part Four, especial-

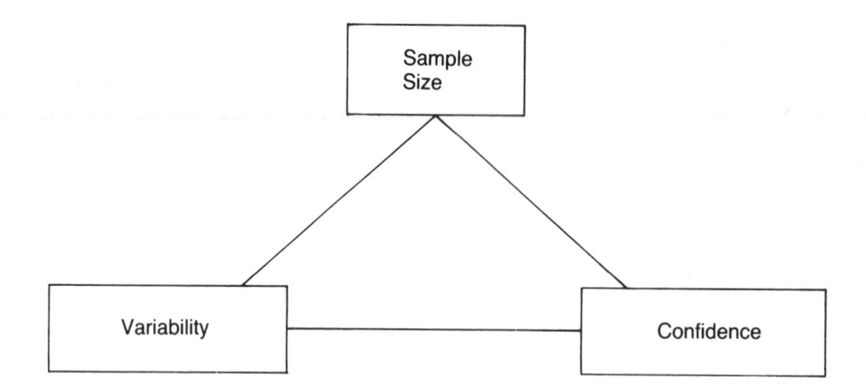

ly the Normal distribution. It can be shown that given a few basic conditions, the statistics derived from a sample will follow a Normal distribution. To understand statistical inference it is necessary to recognize that three basic factors will affect our results; these are:

1. the size of the sample;
2. the variability in the relevant population;
3. the level of confidence we wish to have in the results.

Increases in sample size will generally make the results more accurate (i.e. closer to the results which would be obtained from a census), but this is not a simple linear relationship so that doubling the sample size does not double the level of accuracy. Very small samples, for example under 30, tend to behave in a slightly different way from larger samples and we will consider these in Chapter 14. In practice, sample sizes can range from about 300 to 3 000. Many national samples for market research or political opinion polling, require a sample size of about 1 000 and a policy on non-response.

If there was no variation in the original population, then it would only be necessary to take a sample of one, for example if everyone in the country had the same opinion about a certain government policy, then knowing the opinion of one individual would be enough. However, we do not live in such a homogeneous (boring) world, and there are likely to be a wide range of opinions on such issues as government policy; the design of the sample will need to ensure that the full range of opinions is represented. Even items which are supposed to be exactly alike turn out not to be so. For example, items coming off the end of a production line should be identical but there will be slight variations due to machine wear, temperature variation, quality of raw materials, skill of the operators, etc.

Since we cannot be 100% certain of our results, there will always be a risk that we will be wrong, we therefore need to specify how big this risk will be. Do you want to be 99% certain you have the right answer, or would 95% certain be sufficient? How about 90% certain? As we will see in this chapter, the lower the risk you are willing to accept of being wrong, the less exact the answer is going to be.

13.2 INFERENCE ON THE POPULATION MEAN

As we are now dealing with statistics from samples and making inferences to populations we need a notational system to distinguish between the two. Greek letters will be used to refer to population parameters, μ(mu) for the mean and σ(sigma) for the standard deviation, and N for the population size; whilst ordinary (arabic) letters will be used for sample statistics, $\bar{x}$ for the mean, s for the standard deviation, and n for the sample size.

13.2.1 Confidence intervals

When a sample is selected from a population the arithmetic mean may be calculated in the usual

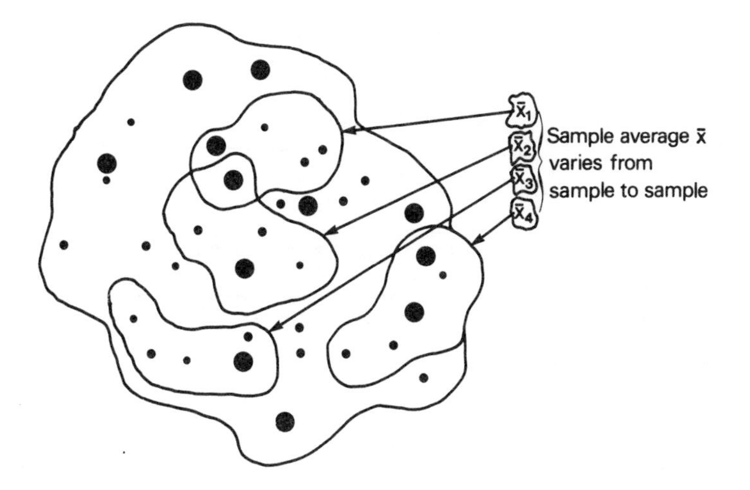

Figure 13.1 A population of different sized 'dots'

way by dividing the sum of the values by the number in the sample. If a second sample is selected, and its mean calculated, it is very likely that a different value for the sample mean will be obtained. Further samples will yield more values for the sample mean. Note that the population mean is always the same throughout this process, it is only the different samples which give different answers. This is illustrated in Figure 13.1.

Since we are obtaining different answers from each of the samples, it would not be reasonable to just assume that the population mean was equal to any of the sample means. In fact each sample mean is said to provide a **point estimate** for the population mean, but it has virtually no probability of being exactly right; if it were, this would be purely by chance. We may estimate that the population mean lies within a small interval around the sample mean; this interval representing the **sampling error**. Thus the population mean is estimated to lie in the region:

$\bar{x} \pm$ sampling error

Thus we are attempting to create an **interval estimate** for the population mean. You should recall from Chapter 12 that the area under a distribution curve can be used to represent the probability of a value being within an interval. We are therefore in a position to talk about the population mean being within the interval with a calculated probability.

As we have seen in section 12.5, the distribution of sample means will follow a normal distribution, at least for large samples, and have a mean equal to the population mean and a standard deviation equal to $\sigma/\sqrt{n}$.

The **Central Limit Theorem** (for means) states that if a simple random sample of size n ($n > 30$) is taken from a population with mean μ and a standard deviation σ, the sampling distribution of the sample mean is approximately Normal with mean μ and standard deviation $\sigma\sqrt{n}$. (This standard deviation is usually referred to as the **standard error** when we are talking about the sampling distribution of the mean.) Note that this is a more general result than that shown in Chapter 12, since it does not assume anything about the shape of the population distribution; it could be any shape.

From our knowledge of the Normal distribution we know that 95% of the distribution lies within 1.96 standard deviations of the mean. Thus for the distribution of sample means, 95% of these will lie in the interval $\mu \pm 1.96\sigma/\sqrt{n}$ as shown in Figure 13.2. This may also be written as a probability statement:

$$P\left(\mu - 1.96\frac{\sigma}{\sqrt{n}} \leq \bar{x} \leq \mu + 1.96\frac{\sigma}{\sqrt{n}}\right) = 0.95.$$

This is a fairly obvious and uncontentious statement which follows directly from the Central Limit Theorem. As you can see, a larger sample

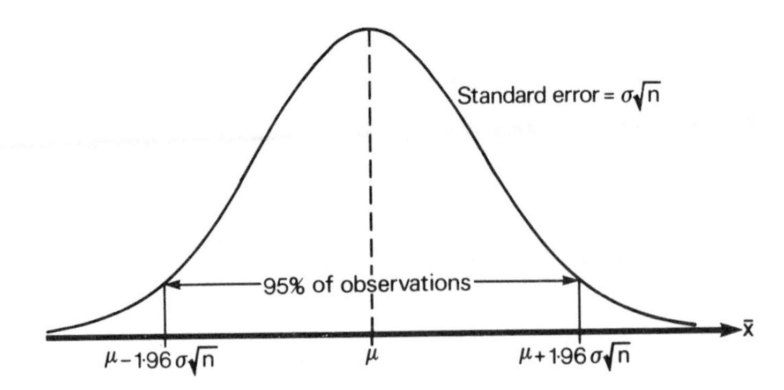

Figure 13.2 The distribution of sample means

size would narrow the width of the interval (since we are dividing by root n), whilst if we were to increase the percentage of the distribution included, by increasing the 0.95, we would need to increase the 1.96 values, and thus the interval would get wider.

By re-arranging the probability statement we can produce a 95% confidence interval for the population mean:

$$\mu = \bar{x} \pm 1.96 \frac{\sigma}{\sqrt{n}}.$$

This is the form of the confidence interval which we will use, but is worth stating what it says in words. It claims that the true population mean (which we do not know) will lie within 1.96 standard errors of the sample mean with a 95% level of confidence. In practice you would only take a single sample, but this result utilizes the Central Limit Theorem to allow you to make the statement about the population mean. There is also a 5% chance that the true population mean lies outside this confidence interval, for example, the data from sample 3 in Figure 13.3.

EXAMPLE

A random sample of 100 accounts at a bank branch were checked and the average amount held was found to be £253. If it is known that the standard deviation for all accounts held at the bank branch is £70, calculate a 95% confidence interval for the mean.

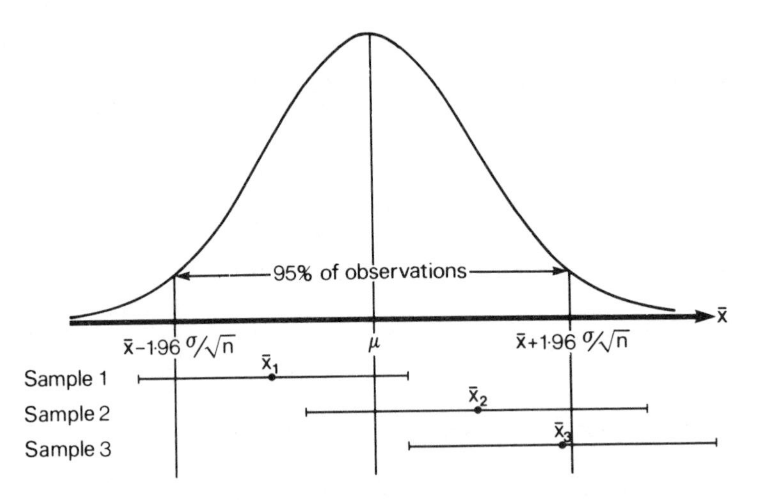

Figure 13.3 Confidence interval from different samples

The sample size is n = 100, the sample mean, $\bar{x} = 253$ and the population standard deviation, $\sigma = 70$. By substituting into the formula given above, we have:

$$\mu = £254 \pm 1.96 \times \frac{£70}{\sqrt{100}}$$
$$= £253 \pm £13.72$$

or we could write this as

$£239.28 < \mu < £266.72$.

Thus we estimate that the true population mean, that is the average balance on accounts at the bank branch, is between £239.28 and £266.72 with 95% confidence. There is a 5% chance that the true mean lies outside of this interval.

So far our calculations have attempted to estimate the unknown population mean from the known sample mean using a result found directly from the Central Limit Theorem. However, looking again at our formula, we see that it uses the value of the population standard deviation, σ and if the population mean is unknown it is highly unlikely that we would know this value. To overcome this problem we may substitute the sample estimate of the standard deviation, s, but unlike the examples in Chapter 3, here we need to divide by $(n-1)$ rather than n. This follows from a separate result of sampling theory which states that the sample standard deviation calculated in this way is a better estimator of the population standard deviation than that using a divisor of n. (Note that we do not intend to prove this result which is well documented in a number of mathematical statistics books.)

The structure of the confidence interval is still valid provided that the sample size is fairly large. Thus the 95% confidence interval which we shall use, will be:

$$\mu = \bar{x} \pm 1.96 \frac{s}{\sqrt{n}}$$

For a 99% confidence interval, the formula would be:

$$\mu = \bar{x} \pm 2.576 \frac{s}{\sqrt{n}}$$

EXAMPLES

1. A sample of 80 households was selected from a large population. If the average amount spent per week on a certain product was found to be £5.90 and the sample standard deviation was £1.20, calculate the 95% confidence interval for the mean of the population.

   The sample statistics are $n = 80$, $\bar{x} = 5.90$, and $s = 1.20$. By substitution into the formula, the 95% confidence interval is given by:

   $$\mu = £5.90 \pm 1.96 \times \frac{£1.20}{\sqrt{80}}$$
   $$= £5.90 \pm 0.263$$

   or we could write this as:

   $£5.637 \leqslant \mu \leqslant £6.163$.

2. A random sample of 200 components was selected from a production process to estimate their expected length of life. The sample mean was calculated to be 300 hours and the sample standard deviation found to be 8 hours. Calculate the 95% and 99% confidence intervals for the population mean.

   The sample statistics are $n = 200$, $\bar{x} = 300$ and $s = 8$ hours. By substitution into the formula the 95% confidence interval is

   $$\mu = 300 \pm 1.96 \times \frac{8}{\sqrt{200}}$$
   $$= 300 \pm 1.1 \text{ hours}$$

   or $298.9 \leqslant \mu \leqslant 301.1$.

   Similarly, the 99% confidence interval given

   $$\mu = 300 \pm 2.576 \times \frac{8}{\sqrt{200}}$$
   $$= 300 \pm 1.5 \text{ hours}$$

   or $298.5 \leqslant \mu \leqslant 301.5$.

As this last example illustrates, the more certain we are of the result (i.e. the higher the level of confidence), the wider the interval becomes. That is, the sampling error becomes larger. Sampling error depends on the probability excluded in the extreme tails of the Normal distribution, and so, as the confidence level increases, the amount excluded in the tails becomes smaller. This example also illustrates a further justification for sampling, since the measurement itself is destructive, and thus if all items were tested, there would be none left to sell.

13.2.2 Confidence intervals using survey data

Extending our examples of calculating a confidence interval for the population mean, we will now show how you would move from the basic, tabulated survey data to a confidence interval.

Suppose the results of an incomes survey were given as in Table 13.1. We can calculate the mean and sample standard deviation using

$$\bar{x} = \frac{\Sigma fx}{n}$$

$$s = \sqrt{\left[\frac{\Sigma f(x-\bar{x})^2}{n-1}\right]} \text{ or } s = \sqrt{\left[\frac{\Sigma fx^2}{n-1} - \frac{(\Sigma fx)^2}{n(n-1)}\right]}$$

The sample standard deviation, s, sometimes denoted by $\hat{\sigma}$, is being used as an **estimator** of the population standard deviation σ. The sample standard deviation will vary from sample to sample in the same way that the sample mean, $\bar{x}$, varies from sample to sample. The sample mean will sometimes be too high or too low, but on average will equal the population mean μ. You will notice that the distribution of sample means $\bar{x}$ in Figure 13.2 is symmetrical about the population mean μ. In contrast, if we use the divisor n, the sample standard deviation will on average be less than σ. To ensure that the sample standard deviation is large enough to estimate the population standard deviation σ reasonably, we use the divisor $(n-1)$. The calculations are shown in Tables 13.2 and 13.3.

Confidence intervals are obtained by the substitution of sample statistics, e.g. using the results from Tables 13.2 or 13.3, the 95% confidence interval is:

$$\mu = \bar{x} \pm 1.96 \frac{s}{\sqrt{n}}$$

Table 13.1

Weekly income	*Frequency*
£100 but less than £200	10
£200 but less than £300	28
£300 but less than £400	42
£400 but less than £600	50
£600 but less than £1 000	20
	150

Table 13.2

Weekly income	*Frequency*	*x*	*fx*	$f(x-\bar{x})^2$
£100 but less than £200	10	150	1 500	772 840
£200 but less than £300	28	250	7 000	887 152
£300 but less than £400	42	350	14 700	255 528
£400 but less than £600	50	500	25 000	259 200
£600 but less than £1 000	20	800	16 500	2 767 680
	150		64 200	4 942 400

$$\bar{x} = \frac{\Sigma fx}{n} = \frac{£64\,200}{150} = £428$$

$$s = \sqrt{\left[\frac{\Sigma f(x-\bar{x})^2}{n-1}\right]} = \sqrt{\frac{£4\,942\,400}{149}} = £182.13$$

Table 13.3

Weekly income	*Frequency*	*fx*	*fx*^{2}
£100 but less than £200	10	1 500	225 000
£200 but less than £300	28	7 000	1 750 000
£300 but less than £400	42	14 700	5 145 000
£400 but less than £600	50	25 000	12 500 000
£600 but less than £1 000	20	16 000	12 800 000
	150	64 200	32 420 000

$$\bar{x} = \frac{\Sigma fx}{n} = \frac{£64\,200}{150} = £428$$

$$s = \sqrt{\left[\frac{\Sigma fx^2}{n-1} - \frac{(\Sigma fx)^2}{n(n-1)}\right]} = \sqrt{\left[\frac{£32\,420\,000}{149} - \frac{(£64\,200)^2}{150 \times 149}\right]} = £182.13.$$

$$= £428 \pm 1.96 \times \left(\frac{£182.13}{\sqrt{150}}\right)$$
$$= £428 \pm £29.147$$

13.2.3 Sample size

As we have seen, the size of the sample selected has a significant bearing upon the actual width of the confidence interval that we are able to calculate from the sample results. If this interval is too wide, it may be of little use, for example, for a confectionery company to know that the weekly expenditure on a particular type of chocolate was between £0.60 and £3.20. Users of sample statistics require a level of accuracy in their results. From our calculations above, the confidence interval is given by

$$\mu = \bar{x} \pm z \times \frac{s}{\sqrt{n}}$$

where z is the value from the Normal distribution tables (for a 95% interval this is 1.96). We could re-write this as

$$\mu = \bar{x} \pm e$$

and now

$$e = z \times \frac{s}{\sqrt{n}}$$

From this we can see that the error (e) is determined by the z value, the standard deviation and the sample size. As the sample size increases, so the error decreases, but to halve the error we would need to quadruple the sample size (since we are dividing by the square root of n).

Re-arranging this formula gives:

$$n = \left(\frac{zs}{e}\right)^2$$

and we thus have a method of determining the sample size needed for a specific error level, at a given level of confidence. Note that we would have to estimate the value of the sample standard deviation, either from a previous survey, or from a pilot study.

EXAMPLE

What sample size would be required to estimate the population mean for a large file of invoices to within £0.50 with 95% confidence, given that the estimated standard deviation of the values of the invoices is £6?

To determine the sample size for a 95% confidence interval, let $z = 1.96$ and, in this case, $e = 0.50$ and $s = 6$. By substitution, we have:

$$n = \left(\frac{1.96 \times 6}{0.50}\right)^2$$
$$= 553.19$$

and we would round up to 554 invoices to be checked, using a random sample.

13.3 INFERENCE ON THE POPULATION PERCENTAGE

13.3.1 Confidence intervals

In the same way that we have used the sample mean ($\bar{x}$) to estimate a confidence interval for the population mean (μ), we can now use the percentage with a certain characteristic in a sample (p) to estimate the percentage with that characteristic in the whole population (π). Sample percentages will vary from sample to sample from a given population (in the same way that sample means varied), and for large samples, this will be in accordance with the Central Limit Theorem.

The **Central Limit Theorem** for percentages states that, if a simple random sample of size n ($n > 30$) is taken from a population with a percentage π having a particular characteristic, then the sampling distribution of the sample percentage, p, is approximated by a Normal distribution with a mean of π and a standard error of

$$\sqrt{\frac{\pi(100 - \pi)}{n}}$$

The 95% confidence interval for a percentage will be given by:

$$\pi = p \pm 1.96 \times \sqrt{\frac{\pi(100 - \pi)}{n}}$$

as shown in Figure 13.4.

The probability statement would now be:

$$P\left(\pi - 1.96\sqrt{\frac{\pi(100 - \pi)}{n}} < P < \pi + 1.96\sqrt{\frac{\pi(100 - \pi)}{n}}\right) = 0.95$$

but a more usable format is:

$$\pi = p \pm 1.96\sqrt{\frac{\pi(100 - \pi)}{n}}$$

Unfortunately, this contains the value of the population percentage, π, on the right-hand side of the equation, and this is precisely what we are trying to estimate. Therefore we substitute the value of the sample percentage, p. Thus the 95% confidence interval that we will use, will be given by

$$\pi = p \pm 1.96\sqrt{\frac{p(100 - p)}{n}}$$

A 99% confidence interval for a percentage would be given by

$$\pi = p \pm 2.58\sqrt{\frac{p(100 - p)}{n}}$$

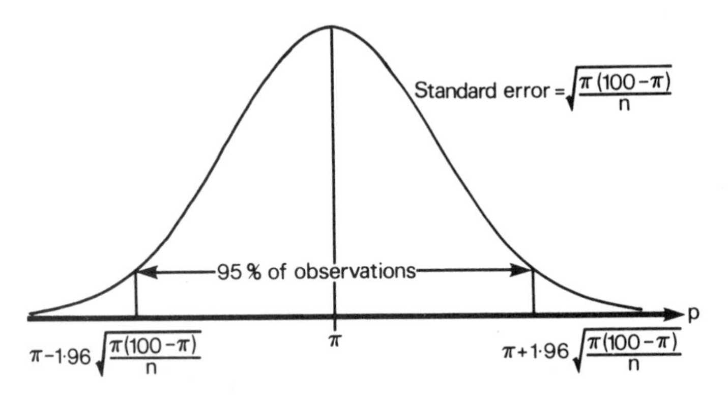

Figure 13.4 The distribution of sample percentages

Interpretation of these confidence intervals is exactly the same as the interpretation of confidence intervals for the mean.

EXAMPLE

A random sample of 100 invoices has been selected from a large file of company records. If 12 were found to be incorrect, calculate a 95% confidence interval for the true percentage of invoices that are incorrect.

The sample percentage of incorrect invoices was $p = 12\%$. This sample statistic is used to estimate the percentage incorrect, π, for the whole set of records. By substituting into the formula for a 95% confidence interval, we have:

$$\pi = 12\% \pm 1.96\sqrt{\frac{12 \times 88}{100}}$$

$$\pi = 12\% \pm 6.4\%$$

or we could write

$$5.6\% \leqslant \pi \leqslant 18.4\%$$

13.3.2 Sample size

As with the confidence interval for the mean, when we are considering percentages, we will often wish to specify the amount of acceptable error in the final result. If we look at the form of the error, we will be able to determine the appropriate sample size. The error is given by

$$e = z\sqrt{\frac{p(100-p)}{n}}$$

and re-arranging this gives

$$n = \left(\frac{z}{e}\right)^2 \times p \times (100-p)$$

The value of p used will either be a reasonable approximation a value from a previous survey, or from a pilot study.

EXAMPLE

In a pilot survey of 100 invoices, randomly selected from a large file, 12 were found to be incorrect. What sample size would it be necessary to take if we wish to produce an estimate of the percentage of incorrect invoices in the whole file to within $\pm$ 3% with a 95% level of confidence?

Here we may use the result of the pilot study as our estimate of p to use in the formula; now $p = 12\%$ and the value of z is 1.96. Substituting, we have

$$n = \left(\frac{1.96}{3}\right)^2 \times 12 \times 88$$

$$= 450.75$$

Thus to achieve the required level of accuracy we would need to take a random sample of 451 from the whole file of invoices.

Where no information is available about the appropriate value of p to use in the calculations, we would use a value of 50%. Looking at the information given in Table 13.4, we can see that at a value of $p = 50\%$ we have the largest possible standard error, and thus the largest sample size. This will be the safest approach where we have no prior knowledge.

EXAMPLE

What sample size would be required to produce an estimate for a population percentage to within

Table 13.4 The size of standard error

p	$(100-p)$	$\sqrt{\left(\frac{p(100-p)}{n}\right)}$
10	90	$\sqrt{(900/n)}$
20	80	$\sqrt{(1\,600/n)}$
30	70	$\sqrt{(2\,100/n)}$
40	60	$\sqrt{(2\,400/n)}$
50	50	$\sqrt{(2\,500/n)}$

$\pm 3\%$ with 95% confidence, if no prior information were available?

In this case, we would let $p = 50\%$ and assume a 'worst possible case'. By substituting into the formula, we have:

$$n = \left(\frac{1.96}{3}\right)^2 \times 50 \times 50$$
$$= 1067.11$$

So, to achieve the required accuracy, we would need to take a random sample of size 1 068.

Comparing the last two examples, we see that in both cases the level of confidence specified is 95%, and that the level of acceptable error to be allowed is plus or minus 3%. However, because of the different assumption that we were able to make about the value of p in the formula, we arrive at very different values for the required sample size. This shows the value of having some prior information, since, for the cost of a small pilot survey we are able to reduce the main sample size to approximately 45% of the size it would have been without that information. An added bonus, when we are questioning human populations, is that the pilot survey also allows us to test the questionnaire to be used.

An alternative to the usual procedure of a pilot survey followed by the main survey is to use a **sequential sampling procedure**. This involves a relatively small sample being taken first, and then further numbers are added as better and better estimates of the parameters become available. In practice, sequential sampling requires the continuation of interviews until results of sufficient accuracy have been obtained.

13.4 THE FINITE POPULATION CORRECTION FACTOR

In all of the previous sections of this chapter we have assumed that we are dealing with a relatively small sample of the whole population, and we have seen that samples of about 1 000 people are sufficient to create confidence intervals with an error of $\pm 3\%$ no matter what the size of the whole population. Some populations are small in themselves, maybe only having twenty or thirty members, and in these cases it may be possible to conduct a census. However, even for populations as small as this, we may still decide to collect information on a sample of the population; in these cases the sample will represent a substantial proportion of the population.

As the proportion of the population included in the sample increases, so the size of the error will decrease, a result you should expect, since if we take a 100% sample (i.e. a census) then there will be no error due to sampling. To take this into account in our calculations we need to correct the estimate of the standard error by multiplying it by the **finite population correction factor**. This is given by the following formula:

$$\sqrt{\left(1 - \frac{n}{N}\right)}$$

where n is the sample size, and N is the population size. As you can see, as the value of n approaches the value of N, the value of the bracket gets closer and closer to zero, thus making the size of the standard error smaller and smaller. As the value of n becomes smaller and smaller in relation to N, the value of the bracket gets nearer and nearer to one, and the standard error gets closer and closer to the value we have used previously.

Where the finite population correction factor is used, the formula for a 95% confidence interval becomes:

$$\mu = \bar{x} \pm 1.96\sqrt{\left(1 - \frac{n}{N}\right)} \times \frac{s}{\sqrt{n}}$$

EXAMPLE

Suppose a random sample of 30 wholesalers used by a toy producer order on average 10 000 cartons of crackers each year. The sample showed a standard deviation of 1 500 cartons. In total, the manufacturer uses 40 wholesalers. Find a 95% confidence interval for the average size of annual order to this manufacturer.

Here, $n = 30$, $N = 40$, $\bar{x} = 10\,000$, and $s = 1\,500$. Substituting these values into the formula, we have:

$$\mu = 10\,000 \pm 1.96\sqrt{\left(1 - \frac{30}{40}\right)} \times \frac{1\,500}{\sqrt{30}}$$
$$= 10\,000 \pm 268.384$$

which we could write as:

$$9\,731.616 \leqslant \mu \leqslant 10\,268.384$$

If no allowance had been made for the high proportion of the population selected as the sample, the 95% confidence interval would have been:

$$\mu = 10\,000 \pm 1.96\sqrt{\frac{1\,500}{N30}}$$
$$= 10\,000 \pm 536.768$$

or

$$9\,463.232 \leqslant \mu \leqslant 10\,536.768$$

which is considerably wider.

13.5 CONFIDENCE INTERVAL FOR THE MEDIAN — LARGE SAMPLE APPROXIMATION

As we have seen in Chapter 3, the arithmetic mean is not always an appropriate measure of average. Where this is the case, we will often want to use the median. (To remind you, the median is the value of the middle item of a group, when the items are arranged in either ascending or descending order.) Reasons for using a median may be that the data is particularly skewed, for example, income or wealth data, or it may lack calibration, for example, the ranking of consumer preferences. Having taken a sample, we still need to estimate the errors or variation due to sampling, and to express this in terms of a confidence interval, as we did with the arithmetic mean.

Since the median is determined by ranking all of the observations and then counting to locate the middle item, the probability distribution is discrete (the confidence interval for a median can thus be determined directly using the Binomial distribution). If the sample is reasonably large ($n > 30$), however, a large sample approximation will give adequate results (see Chapter 12 for the Normal approximation to the Binomial distribution).

Consider the ordering of observations by value, as shown below:

$X_1, X_2, X_3, \ldots, X_n$
where $X_i \leqslant X_{i+1}$.

The median is the middle value of this ordered list, corresponding to the $(n+1)/2$ observation. The confidence interval is defined by an upper ordered value (u) and a lower ordered value (l). for a 95% confidence interval, these values are located using:

$$u = \frac{n}{2} + 1.96\frac{\sqrt{n}}{2}$$
$$l = \frac{n}{2} - 1.96\frac{\sqrt{n}}{2} + 1$$

where n is the sample size.

EXAMPLE

Suppose a random sample of 30 people has been selected to determine the median amount spent on groceries in the last seven days. Results are listed in the table below:

2.50	2.70	3.45	5.72	6.10	6.18
7.58	8.42	8.90	9.14	9.40	10.31
11.40	11.55	11.90	12.14	12.30	12.60
14.37	15.42	17.51	19.20	22.30	30.41
31.43	42.44	54.20	59.37	60.21	65.27

The median will now correspond to the $(30+1)/2 = 15$th observation. Its value being found by averaging the 15½th and 16th observations:

$$\text{median} = \frac{11.90 + 12.14}{2} = 12.02.$$

The sample median is a **point estimate** of the population median. A 95% confidence interval is determined by locating the upper and lower boundaries.

$$u = \frac{30}{2} + 1.96\frac{\sqrt{30}}{2} = 20.368$$

$$l = \frac{30}{2} - 1.96\frac{\sqrt{30}}{2} + 1 = 10.632$$

thus the upper bound is defined by the 21st value (rounding up) and the lower bound by the 10th value (rounding down). By counting through the set of sample results we can find the 95% confidence interval for the median to be:

$$9.14 \leqslant \text{median} \leqslant 17.51.$$

This is now an **interval estimate** for the median.

13.6 CONCLUSIONS

Sample data give us some idea about the values of parameters in the parent population, but, as we have seen, different sample results can be consistent with the same population parameters. Rather than use just a single point, for example, the sample mean to estimate the population value, a confidence interval suggests that the population value lies between certain limits derived from that sample value. In making a point estimate (i.e. using the sample mean as our estimate of the population mean) we will only be correct by chance; in fact, there is a very good chance that we will get the answer wrong! An interval increases the probability of including the correct answer within our interval estimate; and more than this, we can specify what that probability is by the way in which we construct the confidence interval (i.e. 95% or 99% or any other value we choose).

13.7 PROBLEMS

1. A sample of 50 second-hand cars was selected to estimate the average selling price. The sample mean was £2 400 and sample standard deviation £800. Calculate the 95% confidence interval of the mean. Describe how you could reduce the sampling error in this case.
2. The mileages recorded for a sample of company vehicles during a given week yielded the following data:

138	164	150	132	144	125	149	157
146	158	140	147	136	148	152	144
168	126	138	176	163	119	154	165
146	173	142	147	135	153	140	135
161	145	135	142	150	156	145	128

   (a) Calculate the mean and standard deviation, and construct a 95% confidence interval.
   (b) What sample size would be required to estimate the average mileage to within ± 3 miles with 95% confidence? State any assumptions made.
3. The number of breakdowns each day on a section of road were recorded for a sample of 250 days as follows:

Number of breakdowns	*Number of days*
0	100
1	70
2	45
3	20
4	10
5	5
	250

   Calculate the 95% and the 99% confidence intervals. Explain your results.
4. The average weekly overtime earnings from a sample of workers from a particular service industry were recorded as follows:

Average weekly overtime earnings (£)	*Number of workers*
under 1	10
1 but under 2	29
2 but under 5	17

5 but under 10	12
10 or more	3
	80

(a) Calculate the mean, standard deviation and the 95% confidence interval for the mean.
(b) What sample size would be required to estimate the average overtime earnings to within ± £0.50 with a 95% confidence interval?

5. In a survey of 1000 electors, 20% were found to favour party X.
   (a) Construct a 95% confidence interval for the percentage in favour of party X.
   (b) What sample size would be required to estimate the percentage in favour of party X to within ± 1% within a 95% confidence interval?
6. What sample size would be required if you wanted to estimate the percentage of homes with gas central heating to within ± 5% with a 95% confidence interval if:
   (a) a previous survey had shown the percentage to be approximately 42%;
   (b) no prior information were available?
7. Describe how you would design a survey to estimate the percentage of homes in the UK in need of major repairs.
8. A small club has 50 members and the committee wishes to find their views on the introduction of a life membership fee. They are able to randomly select twenty members and find that only three of these support the idea. Construct a 95% confidence interval for the percentage of members who are in favour of a life membership fee.
9. Two hundred people work for a small engineering company and 49 out of a random sample of 50 workers are in favour of a new bonus scheme. Construct a 95% confidence interval for the percentage of the whole workforce who are in favour of the scheme. Would it surprise you to learn that 8 people were against the scheme?
10. A sample of 35 workers was randomly selected from a workforce of 110 to estimate the average amount spent weekly at the canteen. The sample mean was £5.40 and the sample standard deviation £2.24.
    (a) Calculate a 95% confidence interval for the mean.
    (b) What sample size would be required to estimate the mean to within ± £0.50 with a 95% confidence interval?
11. Using the car mileage data from question 2, determine the median and a 95% confidence interval for the median. Compare your answer with that obtained in question 2.

CONFIDENCE INTERVALS: FURTHER DEVELOPMENTS 14

In the previous chapter we have seen how to construct a simple confidence interval for a mean and for a percentage from a single set of sample results. Generalizing these intervals, we see that they have the form:

population parameter = sample statistic
± a sampling error

and that we are estimating the value of the population parameter, the mean or percentage, with a certain degree of confidence. Sampling error is described by the variation of the sampling distribution, and is measured by the standard error. The larger the degree of confidence which we require in our answer, the larger the value of the sampling error, and thus, the wider the confidence interval. We have also seen that the sampling distribution of both the sample mean and of the sample percentage is described by the Normal distribution; a two-sided 95% confidence interval being given by:

sample statistic ± 1.96 standard errors

In this chapter we will extend this analysis to consider other forms of confindence interval. However you will notice that the basic structures set out above can still be applied here.

14.1 ONE-SIDED CONFIDENCE INTERVALS

In some cases samples are taken to assess whether the population parameter is either above or below a certain limit. This may be where small items are being packaged by weight and we wish to check that the machinery is not putting too little weight into each packet. Alternatively, it may be that, whilst we accept that some defective items will be produced by a production process, we wish to institute a quality control process to assure ourselves that the percentage defective is not becoming too high; in other words, is not greater than some pre-set limit. If we are to focus attention on only one end of the distribution, then we need to place all of the area outside of the confidence interval at that end of the sampling distribution. For a 95% confidence interval this will mean that we wish to cut off 5% of the area at one end (or in one tail) of the distribution. From Chapter 12 and Appendix C we know that the z value which cuts off the top 5% of a Normal distribution is 1.645. Thus there will be two versions of the one-sided confidence interval:

1. To find the value which the population parameter is *above*, with 95% confidence, we use:

sample statistic − 1.645 standard errors

2. To find the value which the population parameter is *below* with 95% confidence, we use:

sample statistic + 1.645 standard errors

14.1.1 One-sided confidence intervals for the mean

Using the rules expressed above we can create one-sided confidence intervals for the arithmetic mean, given sample statistics for the mean ($\bar{x}$), standard deviation (s) and the sample size (n). The lower boundary is given by:

$$\mu > \bar{x} - 1.645 \frac{s}{\sqrt{n}}$$

and so we are 95% confident that the true population mean lies above this value. If we wish to be 99% confident that the true population mean lies above the calculated value, then we would need to use a value of 2.33 rather than 1.645 (since 2.33 cuts off the top 1% of a Normal distribution).

Similarly, the upper boundary is given by:

$$\mu < \bar{x} + 1.645 \frac{s}{\sqrt{n}}$$

EXAMPLE

A machine lubrication oil is supplied to an engineering company in 5-litre tins. In an attempt to explain recent shortages of this oil, the company randomly selected 20 cans from the stores. The mean was found to be 5.1 litres and the standard deviation 0.25 litres. Construct an appropriate 95% confidence interval.

The lower boundary, using a one-sided 95% confidence interval can be calculated thus:

$$\mu > 5.1 - 1.645 \times \frac{0.25}{\sqrt{20}} \text{ litres}$$
$$\mu > 5.1 - 0.09 \text{ litres}$$
$$\mu > 5.01 \text{ litres}$$

In contrast the result from a two-sided 95% confidence would have been:

$$\mu = 5.1 \pm 0.11 \text{ litres}$$
or $$4.99 \leq \mu \leq 5.21 \text{ litres}$$

The one-sided confidence interval, in this case, has allowed a more precise statement to be made about the average contents of the tins.

14.1.2 One-sided confidence interval for percentages

As with the previous section, the construction of these confidence intervals is simple once the criteria have been established. Here all we need to know is the percentage in the sample (p) and the sample size (n) before the intervals can be constructed. Interpretation of the upper and lower boundaries is basically the same as for the mean.

The lower boundary is given by:

$$\pi > p - 1.645 \sqrt{\left[\frac{p(100-p)}{n}\right]}$$

and hence we are 95% confident that the true population percentage lies above this calculated figure. This might be useful in a marketing context where calculations have shown that a product is only viable if it can achieve a certain percentage of market share. If this statistic is above the minimum market share, then there is a 95% chance that the product will be viable (at least on the criterion of market share).

Similarly, the upper bound is given by:

$$\pi < p + 1.645 \sqrt{\left[\frac{p(100-p)}{n}\right]}$$

This could be the maximum acceptable percentage of defective items from a production process.

Again, different levels of confidence can be achieved by changing the z-value taken from the Normal distribution tables (Appendix C).

EXAMPLE

A sample of 40 components is randomly selected from those available for production, and 5 are found to be defective. In response to comments that the percentage of defectives is too high, construct an appropriate 95% confidence interval.

The sample size $n = 40$ and the percentage $p = (5/40) \times 100 = 12.5\%$. We can establish an upper boundary by substitution, 95% confidence interval:

$$\pi < 12.5\% + 1.645 \times \sqrt{\left(\frac{12.5 \times 87.5}{40}\right)}$$
$$< 21.10\%$$

The two-sided, 95% confidence interval would be:

$$\pi = 12.5 \pm 10.25$$
$$\text{or} \quad 2.25\% \leqslant \pi \leqslant 22.75\%$$

Again the one-sided confidence interval has allowed a more precise statement about how high the true percentage defective is likely to be.

EXERCISES

1. An operator of fleet vehicles wishes to estimate the largest number of days he can expect to lose through maintenance and repair each year. A sample of 40 service records is randomly selected. If the sample mean is 8 days, and the standard deviation 5 days, construct an appropriate 95% confidence interval.
   (Answer 95% confidence interval $\mu < 9.3$ days.)
2. In a survey of 600 electors, 315 claimed they would vote for party X. Construct a 95% confidence interval to show the minimum support party X can expect.
   (Answer: 95% confidence interval $\pi >$ 49.15%.)

14.2 THE DIFFERENCES BETWEEN INDEPENDENT SAMPLES

In the whole of Chapter 13 and in the preceeding sections of this chapter we have been concerned only with a single sample, trying to estimate the limits between which the true population statistic will lie. In many cases of survey research, as well as these estimates, we also wish to make comparisons between groups in the population, or between seemingly different populations. In other words, we want to make comparisons between two sets of sample results. This could be to test a new machining process in comparison to an existing one by taking as sample of output from each. Similarly we may want to compare consumers in the North with those in the South.

In this section we will make these comparisons by calculating the difference between the sample statistics derived from each sample. We will also assume that the two samples are *independent* and that we are dealing with large samples. (For information on dealing with small samples see section 14.3) Although we will not derive the statistical theory behind the results we use, it is important to note that the theory relies on the samples being independent and that the results do not hold if this is not the case. For example, if you took a single sample of people and asked them a series of questions, and then two weeks later asked the same people another series of questions, the samples would not be independent and we could not use the confidence intervals shown in this section. We will consider related samples in Chapter 16.

One result from statistical sampling theory states that although we are taking the **difference** between the two sample parameters (the means or percentages), we *add* the variances. This is because the two parameters are themselves variable (see section 13.1) and thus the measure of variability needs to take into account the variability of both samples.

14.2.1 The difference of means

The format of a confidence interval remains the same as before:

population parameter = sample statistic ± sampling error

but now the population parameter is the difference between the population means $(\mu_1 - \mu_2)$, the sample statistic is the difference between the sample means ($\bar{x}_1 - \bar{x}_2$) and the sampling error consists of the z-value from the Normal distribution tables multiplied by the root of the sum of the sample variances. This sounds like quite a mouthful (!) but is fairly straightforward to use with a little practice.

The 95% confidence interval for the difference of means is given by the following formula:

$$\mu_1 - \mu_2 = \bar{x}_1 - \bar{x}_2 \pm 1.96 \times \sqrt{\left(\frac{s_1^2}{n_1} + \frac{s_2^2}{n_2}\right)}$$

where the subscripts denote sample 1 and sample 2. (Note the relatively obvious point that we must keep a close check on which sample we are dealing with at any particular time.)

EXAMPLE

A sample of 75 packets of cereals was randomly selected from the production process and found to have a mean of 500 g and standard deviation of 20 g. A week later a second sample of 50 packets of cereal was selected, using the same procedure and found to have a mean of 505 g and standard deviation of 16 g. Construct a 95% confidence interval for the change in the average weight of cereal packets.

The summary statistics are as follows:

Sample	*Sample 2*
$n_1 = 75$	$n_2 = 50$
$x_1 = 500$	$x_2 = 505$
$s_1 = 20$	$s_2 = 16$

By substitution, the 95% confidence interval is:

$$\mu_1 - \mu_2 = (500\,\text{g} - 505\,\text{g}) \pm 1.96 \times \sqrt{\left[\frac{(20)^2}{75} + \frac{(16)^2}{50}\right]}\,\text{g}$$

$$= -5\,\text{g} \pm 6.34\,\text{g}$$

or

$$-11.34 \leq \mu_1 - \mu_2 \leq 1.34\,\text{g}$$

As this range includes zero, we cannot be 95% confident that there has been a decrease or increase. The observed change in the average weight could be explained by inherent variation in the sample results.

In this example, a confidence interval has been used to describe the measurement or measurements achieved. A related approach is significance testing (see Chapter 15) which provides a method of deciding on whether there has been an increase, decrease or no change.

14.2.2 The difference of percentages

Here we need to know only the two sample sizes and the two sample percentages in order to estimate the difference in the population percentages. Using the structure from the introduction to this section, we can find a formula for this confidence interval. This will be:

$$\pi_1 - \pi_2 = p_1 - p_2 \pm 1.96\sqrt{\left[\frac{p_1(100-p_1)}{n_1} + \frac{p_2(100-p_2)}{n_2}\right]}$$

where the subscripts denote sample 1 and sample 2.

EXAMPLE

A sample of 120 housewives was randomly selected from those reading a particular magazine, and 18 were found to have purchased a new household product. Another sample of 150 housewives was randomly selected from those not reading the particular magazine, and only 6 were found to have purchased the product. Construct a 95% confidence interval for the difference in the purchasing behaviour.

The summary statistics are as follows:

Sample 1	*Sample 2*
$n_1 = 120$	$n_2 = 150$
$p_1 = \frac{18}{120} \times 100 = 15\%$	$p_2 = \frac{6}{150} \times 100 = 4\%$

By substitution, the 95% confidence interval is:

$$\pi_1 - \pi_2 = (15\% - 4\%) \pm 1.96 \sqrt{\left(\frac{15\% \times 85\%}{120} + \frac{4\% \times 96\%}{150}\right)}$$

$$= 11\% \pm 7.1\%$$

or $3.9\% \leq \pi_1 - \pi_2 \leq 18.1\%$.

The range does not include any negative value or zero, suggesting that the readership of this particular magazine has made a *statistically significant* impact on purchasing behaviour.

The significance of results will reflect the survey design. The width of the confidence interval (or the chance of including zero) will decrease as:

1. the size of sample or samples is increased; and
2. the percentage becomes larger or smaller than 50% (see Table 13.4).

You should also note that all variations used with confidence intervals from a single sample can also be applied to the difference between two samples. Thus we can construct confidence intervals with different levels of confidence, for example, 90%, 99% or even 99.9%. Similarly, we could construct one-sided confidence intervals for the difference between two sample statistics.

EXERCISES

1. An operator of fleet vehicles wishes to compare the service costs at two different garages. Records from one garage show that for the 70 vehicles serviced the mean cost was £55 and the standard deviation £9. Records from the other garage show that for the 50 vehicles serviced the mean cost was £52 and the standard deviation £12. Construct a 95% confidence interval for the difference in servicing costs.
(Answer: 95% confidence interval, $\mu_1 - \mu_2 =$ £3 ± £3.94.)
2. In a survey of 600 electors, 315 claimed they would vote for party X. A month later, in another survey of 500 electors, 290 claimed they would vote for party X. Construct a 95% confidence interval for the difference in voting.
(Answer: 95% confidence interval, $\pi_1 - \pi_2 = -5.5\% \pm 5.89\%$.)

14.3 THE *t*-DISTRIBUTION

In all of the previous sections we have assumed that either the population standard deviation (σ) was known (an unlikely event), or that the sample size was sufficiently large so that the sample standard deviation, s, provided a good estimate of the population value (see section 13.1.2). Where these criteria are not met, we are not able to assume that the sampling distribution is a Normal distribution, and thus the formulae developed so far will not apply. As we have seen in section 13.1.3, we are able to calculate the standard deviation from sample data, but where we have a *small* sample, the amount of variability will be *higher*, and as a result, the confidence interval will need to be *wider*.

If you consider the case where there is a given amount of variability in the population, when a large sample is taken, it is likely to pick up examples of both high and low values, and thus the variability of the sample will reflect the variability of the population. When a small sample is taken from the same population, the fewer values available make it less likely that all of the variation is reflected in the sample. Thus a given standard deviation in the small sample would imply more variability in the population than the same standard deviation in a large sample.

Even with a small sample, if the population standard deviation is known, then the confidence intervals can be constructed using the Normal distribution as:

$$\mu = \bar{x} \pm z \frac{\sigma}{\sqrt{n}}$$

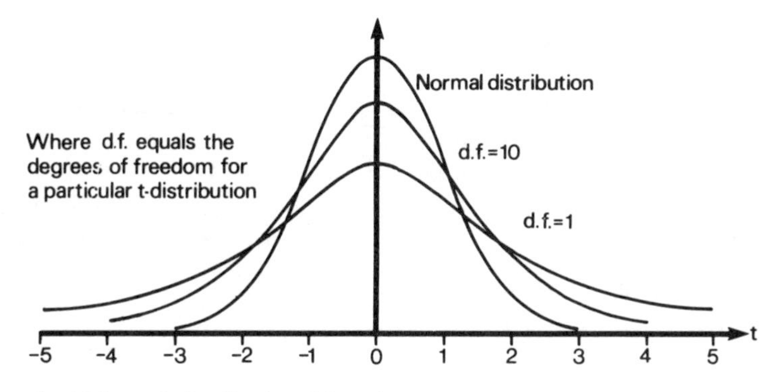

Figure 14.1 The standard Normal distribution (z) and the t-distribution

where z is the critical value taken from the Normal distribution tables.

Where the value of the population standard deviation is *not* known we will use the t-distribution to calculate a confidence interval. Thus:

$$\mu = \bar{x} \pm t \frac{s}{\sqrt{n}}$$

where t is a critical value from the t-distribution. (The derivation of why the t-distribution applies to small samples is beyond the scope of this book, and of most first year courses, but the shape of this distribution as described below, gives an intuitive clue as to its applicability.)

The shape of the t-distribution is shown in Figure 14.1. You can see that it is still a symmetrical distribution about a mean (like the Normal distribution), but that it is wider. In fact it is a misnomer to talk about *the* t-distribution, since the width and height of a particular t-distribution varies with the number of **degrees of freedom**. This new term is related to the size of the sample, being represented by the letter ν (pronounced nu), and being equal to $n-1$ (where n is the sample size). As you can see from the diagram, with a small number of degrees of freedom, the t-distribution is wide and flat; but as the number of degrees of freedom increases, the t-distribution becomes taller and narrower. As the number of degrees of freedom increases, the t-distribution tends to the Normal distribution.

Values of the t-distribution are tabulated by degrees of freedom and are shown in Appendix D but to illustrate the point about the relationship to the Normal distribution, consider Table 14.1. We know that to exclude 2.5% of the area of a Normal distribution in the right-hand tail we would use a z-value of 1.96. Table 14.1 shows the comparative values of t for various degrees of freedom.

Table 14.1 Percentage points of the t-distribution

ν	*Percentage excluded in right-hand tail area* $2\frac{1}{2}\%$
1	12.706
2	4.303
5	2.571
10	2.228
30	2.042
∞	1.960

Before using the t-distribution, let us consider an intuitive explanation of degrees of freedom. If a sample were to consist of one observation we could estimate the mean (take the average to be that value) but could make no estimate of the variation. If the sample were to consist of two observations, we would have only one measure of difference or one degree of freedom. Degrees of freedom can be described as the number of independent pieces of information. In estimating variation around a single mean the degrees of freedom will be $n-1$. If we were estimating the variation around a line on a graph (see section

17.5) the degrees of freedom would be $n-2$ since two parameters have been estimated to position the line.

The 95% confidence interval for the mean from sample data when σ is unknown takes the form

$$\mu = \bar{x} \pm t_{0.025} \frac{s}{\sqrt{n}}$$

where $t_{0.025}$ excludes $2\frac{1}{2}$% of observations in the extreme right-hand tail area.

EXAMPLE

A sample of six company vehicles was selected from a large fleet to estimate the average annual maintenance cost. The sample mean was £340 and the standard deviation £60. Calculate the 95% confidence interval for the population (fleet) mean.

The summary statistics are: $n = 6$, $\bar{x} = £340$ and $s = £60$.

In this case, the degrees of freedom are $\nu = n - 1 = 5$, and the critical value from the t-distribution 2.571 (see Table 14.2 or Appendix D). By substitution, the 95% confidence interval is given by

$$\mu = £340 \pm 2.571 \times \frac{£60}{\sqrt{6}}$$

$$= £340 \pm £62.98.$$

or $£277.02 \leqslant \mu \leqslant £402.98$.

If the sampling error is unacceptably large we would need to increase the size of the sample.

EXERCISES

1. A sample of 15 employees was randomly selected from a workforce to estimate the average travel time to work. If the sample mean was 55 minutes and the standard deviation 13 minutes calculate the 95% and the 99% confidence intervals.
   (Answer: 95% confidence interval, $\mu = 55 \pm 7.20$;
   99% confidence interval, $\mu = 55 \pm 9.99$.)
2. The sales of a particular product in one week were recorded at five randomly selected shops as 16, 82, 29, 31 and 54. Calculate the sample mean, standard deviation (using a divisor of $n - 1$) and 95% confidence interval.
   (Answer: $\bar{x} = 42.4$, $s = 26.02$;
   95% confidence interval, $\mu = 42.4 \pm 32.30$.)

We have illustrated the use of the t-distribution for estimating the 95% confidence interval for a population mean from a small sample. Similar reasoning will allow calculation of a 95% confidence interval for a population percentage from a single sample, or variation of the level of confidence by changing the value of t used in the calculation.

Where two small independent samples are involved, and we wish to estimate the difference in either the means or the percentages, we can still use the t-distribution, but now the number of degrees of freedom will be related to both sample sizes.

$$\nu = n_1 + n_2 - 2$$

and it will also be necessary to allow for the sample sizes in calculating a pooled standard error for the two samples.

In the case of estimating a confidence interval for the difference between two means the pooled standard error is given by:

$$s_p = \sqrt{\frac{(n_1 - 1)s_1^2 + (n_2 - 1)s_2^2}{n_1 + n_2 - 2}}$$

and the confidence interval is

$$(\mu_1 - \mu_2) = (\bar{x}_1 - \bar{x}_2) \pm ts_p\sqrt{\frac{1}{n_1} + \frac{1}{n_2}}$$

the t value being found from the tables, having

$$\nu = n_1 + n_2 - 2$$

degrees of freedom. A theoretical requirement of this approach is that both samples have variability of the same order of magnitude.

EXAMPLE

Two processes are being considered by a manufacturer who has been able to obtain the following figures relating to production per hour. Process A produced 110.2 units per hour as the average from a sample of 10 hourly runs. The standard deviation was 4. Process B had 15 hourly runs and gave an average of 105.4 units per hour, with a standard deviation of 3. The summary statistics are as follows:

Process A	*Process B*
$n_1 = 10$	$n_2 = 15$
$x_1 = 110.2$	$x_2 = 105.4$
$s_1 = 4$	$s_2 = 3$

Thus the pooled standard error for the two samples is given by

$$s_p = \sqrt{\frac{(10-1)4^2 + (15-1)3^2}{10+15-2}}$$
$$= 3.42624$$

There are $\nu = 10 + 15 - 2 = 23$ degrees of freedom, and for a 95% confidence interval, this gives a *t*-value of 2.069 (see Appendix D). Thus the 95% confidence interval for the difference between the means of the two processes is

$$(\mu_1 - \mu_2) = (110.2 - 105.4) \pm 2.069\,(3.42624)\sqrt{\frac{1}{10} + \frac{1}{15}}$$
$$= 4.8 \pm 2.894$$

or $1.906 \leqslant (\mu_1 - \mu_2) \leqslant 7.694$.

A summary of the notation and standard errors used in calculating confidence intervals is given in Table 14.2.

Table 14.2 Notation and standard errors

Sample statistic	*Population parameter*	*Sample estimate of standard error*
$\bar{x}$	μ	$\frac{s}{\sqrt{n}}$
p	π	$\sqrt{\frac{p(100-p)}{n}}$
Large Samples:		
$\bar{x}_1 - \bar{x}_2$	$\mu_1 - \mu_2$	$\sqrt{\frac{s_1^2}{n_1} + \frac{s_2^2}{n_2}}$
$p_1 - p_2$	$\pi_1 - \pi_2$	$\sqrt{\frac{p_1(100-p_1)}{n_1} + \frac{p_2(100-p_2)}{n_2}}$
Small Samples:		
$\bar{x}$	μ	$\frac{s}{\sqrt{n}}$
p	π	$\sqrt{\frac{p(100-p)}{n}}$
$\bar{x}_1 - \bar{x}_2$	$\mu_1 - \mu_2$	$s_p = \sqrt{\frac{(n_1-1)s_1^2 + (n_2-1)s_2^2}{n_1+n_2-2}}$ and $\sigma = s_p\sqrt{\frac{1}{n_1} + \frac{1}{n_2}}$

14.4 CONCLUSIONS

We have now taken the concept of confidence intervals a step further by considering more than one sample and also, situations where the sample is small. These confidence intervals allow us to set limits on the likely values of the parent population values, based upon information obtained from the sample or samples. Confidence intervals add description to sample results. However, we also use sample results to test our ideas about a particular population and this is considered in the next chapter.

14.5 PROBLEMS

1. A contractor supplies sand to the building industry by the cubic yard. It is accepted that the amount delivered in each lorry load is going to vary for a number of reasons, but to check that sufficient quantities are forthcoming, the contractor has agreed that the weight should be recorded for a sample of 50 lorry loads. It was found that the sample mean was 20 cubic yards and the sample standard deviation was 1.5 cubic yards. Construct an appropriate 95% confidence interval.
2. A store manager is concerned about the amount owing by customers in the store's credit scheme. In a sample of 60 customers it was found that the average amount owing was £12.43 and the standard deviation was £6.50. Construct an appropriate 95% confidence interval.
3. Following the promotion of a new product, a brand manager would like to know how much a potential customer is now spending each week on a competing product. In a survey of 450 potential customers the mean was calculated to be £0.80 and the standard deviation to be £0.28. Construct a 95% confidence interval to show 'at least' how much is being spent on average.
4. Following the promotion of a new product, a brand manager would like to know what percentage of potential customers are now aware of this product. In a survey of 450 potential customers, 432 were found to be aware of this new product. Construct a 95% confidence interval to show the minimum level of awareness.
5. It has been claimed that party Y can expect no more than 10% of the vote. If 13 out of a sample of 150 electors say they will vote for party Y is there evidence to support the claim?
6. The average weekly overtime earnings from a sample of workers this year and last year were recorded as follows:

	Average weekly overtime earnings (£)	
Sample statistics	*last year*	*this year*
number	80	90
mean	3.00	3.50
standard deviation	2.98	3.06

   Construct a 95% confidence interval for the increase in average weekly overtime earnings.
7. The number of breakdowns each day on two sections of road, section A and section B were recorded independently for a sample of 250 days as follows:

	Number of days	
Number of breakdowns	*Section A*	*Section B*
0	100	80
1	70	65
2	45	57
3	20	31
4	10	11
5	5	6
	250	250

   Construct a 95% confidence interval for the difference in the average number of breakdowns on the two sections of road and comment on the structure of the test.
8. In a survey of 1000 electors, 600 in the North and 400 in the South, 22% were found to

favour party X in the North and 18% to favour party X in the South. Construct a 95% confidence interval to show the regional difference.

9. In a sample of 200 cars produced by company A, 42 were found to have faults whilst in a sample of 230 cars produced by company B, 46 were found to have faults. Is there any evidence to suggest a significant difference in the percentage of cars with faults produced by company A and company B?
10. To estimate the average cost of window replacement, 11 quotes were obtained for a typical semi-detached house. The mean was £1 259 and the standard deviation £153. Construct a 95% confidence interval for the mean.
11. The time taken to complete the same task was recorded for seven participants in a training exercise as follows:

Participant	1	2	3	4	5	6	7
Time taken (in minutes)	8	7	8	9	7	7	9

Construct a 95% confidence interval for the average time taken to complete the task.

12. A survey of expenditure on a manufacturer's product has found that the average amount spent in the South is £28 per month with a variance of £5.30. In the North the average was £24 per month with a variance of £3.40. The sample sizes were 10 and 14 respectively. Construct a 95% confidence interval for the difference in the average amount spent in the two regions.
13. A company claims that women take more time off through illness than men. The union wish to test this claim and are able to select two random samples from the workforce; a sample of 17 women and a sample of 11 men. From the personnel records it is found that for the sample of women, the mean amount of days illness is 6.3 with a variance of 5.8. For the men the mean is 5.1 days with a variance of 2.6. To help the union in arguing its case, construct a 99% confidence interval for the difference in time lost through illness between men and women.

SIGNIFICANCE TESTING 15

Significance testing aims to make statements about a population parameter, or parameters, on the basis of sample evidence. This description could also apply to confidence intervals, but the emphasis is slightly different when we are dealing with significance tests. As you would expect, since the two ideas are so closely linked, all of the assumptions that we have made in Chapters 13 and 14 about the underlying sampling distribution remain the same for significance tests.

The emphasis here is on testing whether a set of sample results support, or are consistent with, some fact or supposition about the population. We are thus concerned to get a 'Yes' or 'No' answer from a significance test; either the sample results *do* support the supposition, or they *do not*. Since we are dealing with samples, and not a census, we can never be 100% sure of our results (since some vital evidence may not have been collected in the sample); we will therefore conduct these significance tests at a particular level, for example 5% or 1%. These percentages represent the chance of drawing the wrong conclusion.

The idea we are testing may have been obtained in the past from a census or survey, or it may be an assertion that we put forward, or has been put forward by someone else. Similarly, it could be a critical value in terms of product planning, for example, if a new machine is to be purchased but the process is only viable with a certain level of through-put, we would want to be sure that the company can sell, or use, that quantity before commitment to the purchase of that particular type of machine. Advertising claims could also be tested using this method by taking a sample and testing if the advertised characteristics, for example strength, length of life, or percentage of people preferring this brand, are supported by the evidence of the sample. In developing statistical measures we may want to test if a certain parameter is significantly different from zero; this has useful applications in the development of regression and correlation models (see Chapters 17, 18 and 19).

Taking the concept a stage further, we may have a set of sample results from two regions giving the percentage of people purchasing a product, and want to test whether there is a significant difference between the two percentages. Similarly, we may have a survey which is carried out on an annual basis and wish to test if there has been a significant shift in the figures since last year. In the same way that we created a confidence interval for the differences between two samples, we can also test for such differences.

Finally we will also look at the situation where only small samples are available and consider the application of the *t*-distribution to significance testing, another concept to which we will return in Chapters 17 and 18.

15.1 SIGNIFICANCE TESTING USING CONFIDENCE INTERVALS

Significance testing is concerned with accepting or rejecting ideas. These ideas are known as **hypoth-**

eses. If we wish to test one in particular, we refer to it as the **null hypothesis**. The term 'null' can be thought of as meaning no change or no difference.

As a procedure, we would first state a null hypothesis; something we wish to judge as true or false on the basis of statistical evidence. We would then check whether or not the null hypothesis was consistent with the confidence interval. If the null hypothesis was contained within the confidence interval it would be accepted, otherwise, it would be rejected. A confidence interval can be regarded as a set of acceptable hypotheses.

To illustrate significance testing consider an example from Chapter 13.

EXAMPLE

A sample of 80 housewives was randomly selected from a large population to estimate the average amount spent weekly on a particular product. The sample mean was found to be £1.40 and the standard deviation £0.15.

Assuming the sample size is large enough to justify an assumption of Normality, the 95% confidence interval can be constructed as follows:

$$\mu = £1.40 \pm 1.96 \times \frac{£0.15}{\sqrt{80}} = £1.40 \pm £0.033$$

or $£1.367 \leqslant \mu \leqslant £1.433$

Now suppose that the purpose of the survey was to test a store manager's view that the average amount spent on this product was £1.50. If the confidence interval can be regarded as the set of acceptable hypotheses, then the null hypothesis (denoted by H_0) that the average was £1.50 is expressed as

H_0: μ = £1.50

and must be rejected.

The values of the null hypothesis that we can accept or reject are shown in Figure 15.1. In rejecting the store manager's view that the average could be £1.50 we must also accept that our decision could be wrong. There is a 5% chance that the average is greater than £1.433 or less than £1.367. As we shall see (section 15.3) there is a probability of making a wrong decision and that the acceptance or rejection of a null hypothesis is a matter of balancing risks.

15.2 HYPOTHESIS TESTING FOR SINGLE SAMPLES

Hypothesis testing is merely an alternative name for significance tests, the two being used interchangeably. This name does stress that we are testing some supposition about the population,

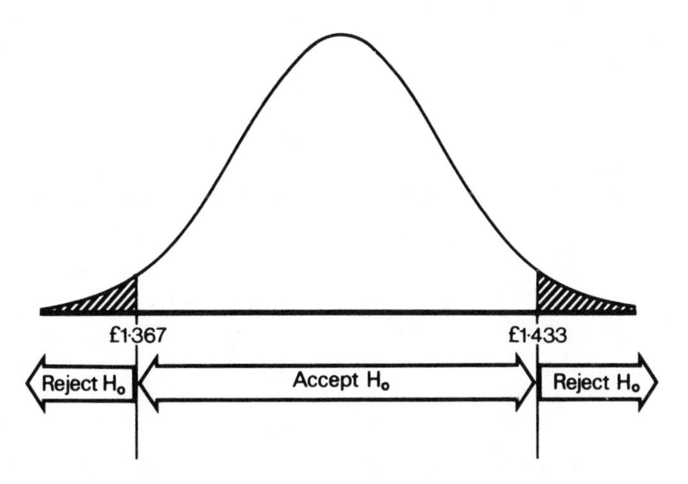

Figure 15.1 Acceptance and rejection regions

which we can write down as a hypothesis. In fact, we will have two hypotheses whenever we conduct a significance test; one relating to the supposition that we are testing, and one which describes the alternative situation.

The first hypothesis relates to the claim, supposition or previous situation and is usually called the **null hypothesis** and labelled as H_0. It implies that there has been no change in the value of the parameter that we are testing from that which previously existed; for example, if the average spending per week on beer last year amongst 18- to 25-year olds was £10.53, then the null hypothesis would be that it is still £10.53. If it has been claimed that 75% of consumers prefer a certain flavour, then the null hypothesis would be that the population percentage is equal to 75%.

The null hypothesis could be written out as a sentence, but it is more usual to abbreviate it to

H_0: $\mu = \mu_0$

for a mean where μ_0 is the claimed or previous population mean. For a percentage we write:

H_0: $\pi = \pi_0$

where π_0 is the claimed or previous population percentage.

The second hypothesis summarizes what will be the case if the null hypothesis is not true. It is usually called the **alternative hypothesis** (fairly obviously!), and is labelled as H_A or as H_1 depending on which text you follow: we will use the H_1 notation here. This alternative hypothesis is usually not specific, in that it does not usually specify the exact alternative value for the population parameter, but rather, it just says that some other value is appropriate on the basis of the sample evidence; for example, the mean amount spent is not equal to £10.53, or the percentage preferring this flavour is not 75%.

As before, this hypothesis could be written out as a sentence, but a shorter notation is usually preferred. We write

H_1: $\mu \neq \mu_0$

for a mean where μ_0 is the claimed or previous population mean. For a percentage we write

H_1: $\pi \neq \pi_0$

where π_0 is the claimed or previous population percentage.

Whenever we conduct an hypothesis test, we assume that the null hypothesis is true whilst we are doing the test, and then come to a conclusion on the basis of the figures that we calculate during the test.

Since the sampling distribution of a mean or a percentage (for large samples) is given by the Normal distribution, we can use tables (in Appendix C) to determine the probability that the sample result that we have obtained could have been found if the null hypothesis were true. To do this we calculate the z-value for the sample result, look it up in the tables, and extract the probability. It is more usual however to divide the Normal distribution diagram into sections or areas, and to see whether the z-value falls into a particular section; then we may either accept or reject the null hypothesis.

Most tests are conducted at the 5% level of significance, and you should recall that in the normal distribution, the z-values of +1.96 and −1.96 cut off a total of 5% of the distribution, 2.5% in each tail. If a calculated z-value is between −1.96 and +1.96, then we accept the null hypothesis; if the calculated z-value is below −1.96 or above +1.96, we reject the null hypothesis in favour of the alternative hypothesis. This situation is illustrated in Figure 15.2.

If we were to conduct the test at the 1% level of significance, then the two values used to cut off the tails of the distribution would be +2.58 and −2.58.

For each test that we wish to conduct, the basic layout and procedure will remain the same, although some of the details will change, depending upon what exactly we are testing. A proposed layout is given below, and we suggest that by following this you will present clear and understandable significance tests (and not leave out any steps).

Step	*Example*
1. State hypotheses.	H_0: $\mu = \pi_0$
	H_1: $\mu \neq \pi_0$

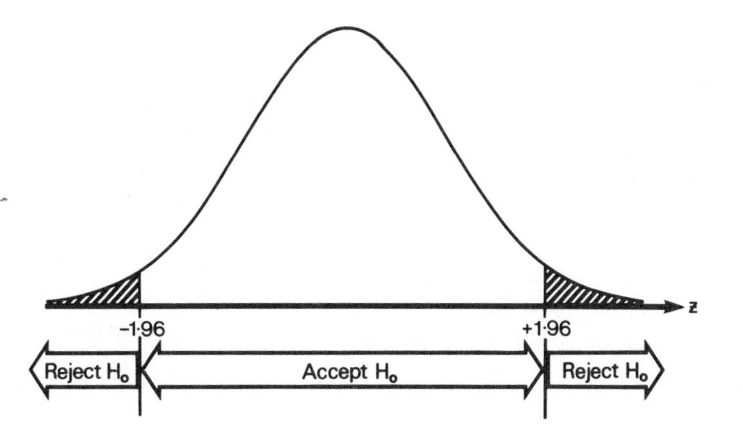

Figure 15.2 Acceptance and rejection regions

2.	State significance level.	5%
3.	State critical (cut-off) values.	−1.96 +1.96
4.	Calculate the test statistic (z).	Varies for each test, but say 2.5 for example
5.	Compare the z value to the critical values.	In this case it is above + 1.96.
6.	Come to a conclusion.	Here we would reject H_0.
7.	Put your conclusion into English.	The sample evidence does not support the original claim that the mean was the specified value.

15.2.1 A test statistic for a population mean

At step 4 of the significance testing procedure above, we need to calculate a z-value on the basis of the sample information, for which we need a formula for z. In the case of testing for a population mean, this formula is:

$$z = \frac{\bar{x} - \mu}{\sigma/\sqrt{n}}$$

As we saw in Chapter 13, the population standard deviation (σ) is rarely known, and so we use the sample standard deviation (s) in its place. We are assuming that the null hypothesis is true, and so $\mu = \mu_0$, and the formula will become:

$$z = \frac{\bar{x} - \mu_0}{s/\sqrt{n}}$$

We are now in a position to carry out a test.

EXAMPLE

A store manager believes that the average amount spent per week on jam and marmalade is £1.50. A random sample of 80 shoppers was selected from a large population and asked about the amount spent per week on these commodities. Results from the survey showed that the average amount spent per week on jam and marmalade was £1.40, with a standard deviation of £0.15. Does this lend support to the store manager's opinion at the 5% significance level?

Step 1: The null hypothesis is based upon the store manager's belief, that:

H_0: $\mu = £1.50$.

The alternative hypothesis is any other answer, and can be stated as:

H_1: $\mu \neq £1.50$

Step 2: As stated in the question, this level is 5%.
Step 3: The critical values are −1.96 and +1.96
Step 4: Using the formula given above, the z-value can be calculated as:

$$z = \frac{1.40 - 1.50}{0.15/\sqrt{80}}$$
$$= -5.96.$$

Step 5: The calculated value (−5.96) is below the lower critical value (−1.96).

Step 6: We may therefore *reject* the null hypothesis at the 5% level of significance.

Step 7: The sample evidence does not support the store manager's belief that shoppers spend an average of £1.50 per week on jam and marmalade. The amount spent is *significantly different* from £1.50.

Although there is only a 10 pence difference between the claim and the sample mean in this example, the test shows that the sample result is significantly different from the claimed value. In all hypothesis testing, it is not the *absolute* difference between the values which is important, but the *number of standard errors* that the sample value is away from the claimed value. Statistical significance should not be confused with the notion of business significance or importance.

15.2.2 A test for a population percentage

Here the process will be identical to the one followed above, except that the formula used to calculate the z-value will need to be changed. (You should recall from Chapter 13 that the standard error for a percentage is different from the standard error for a mean.)
The formula will be:

$$z = \frac{p - \pi_0}{\sqrt{\left[\frac{\pi_0(100 - \pi_0)}{n}\right]}}$$

where p is the sample percentage and π_0 is the claimed population percentage (remember that we are assuming that the null hypothesis is true).

EXAMPLE

An auditor claims that 10% of invoices for a company are incorrect. To test this claim a random sample of 100 invoices are checked, and 12 are found to be incorrect. Test, at the 5% significance level, if the auditor's claim is supported by the sample evidence.

Step 1: The hypotheses can be stated as:

$$H_0: \pi_0 = 10\%$$
$$H_1: \pi_0 \neq 10\%$$

Step 2: The significance level is 5%.

Step 3: The critical values are −1.96 and +1.96.

Step 4: The sample percentage is $\frac{12}{100} \times 100 =$ 12%

$$z = \frac{12 - 10}{\sqrt{\frac{10\ (100 - 10)}{100}}}$$
$$= 0.67$$

Step 5: The calculated value falls between the two critical values.

Step 6: We therefore *cannot reject* the null hypothesis.

Step 7: The evidence from the sample is consistent with the auditor's claim that 10% of the invoices are incorrect.

Notice that in this example, when the calculated value falls between the two critical values, the answer is to *not reject* the claim, rather than to *accept* the claim. This is because we only have the sample evidence to work from, and we are aware that it is subject to sampling error. The only way to firmly accept the claim would be to check *every* invoice (i.e. carry out a census) and work out the actual population percentage which are incorrect.

When we calculated a confidence interval for the population percentage (in section 13.3.1) we used a slightly different formulation of the sampling error. There we used

$$\sqrt{\left[\frac{p(100 - p)}{n}\right]}$$

because we did not know the value of the population percentage; we were using the sample percentage as our best estimate, and also as the only available value. When we come to significance tests, we have already made the assumption that the null hypothesis is true (whilst we are conducting the test), and so we can use the hypothesized value of the population percentage in the calculations. The formula for the sampling error will thus be

$$\sqrt{\left[\frac{\pi_0(100-\pi_0)}{n}\right]}$$

In many cases there would be very little difference in the answers obtained from the two different formulae, but it is good practice (and shows that you know what you are doing) to use the correct formulation.

15.3 ONE-SIDED SIGNIFICANCE TESTS

So far we have merely considered hypothesis tests where the true population parameter *is* equal to some value, or it *is not*. This may be a useful test if the only consideration is whether or not the claimed value is likely to be true, but, generally, we would want to specify whether the real value is above or below the claimed value in those cases where we are able to reject the null hypothesis.

One-sided tests will allow us to do exactly this. The method employed, and the appropriate test statistic which we calculate, will remain exactly the same; it is the hypotheses and the interpretation of the answer which will change. Suppose that we are investigating the purchase of cigarettes, and know that the percentage of the adult population who regularly purchased last year was 34%. If a sample is selected, we do not want to know only whether the percentage purchasing has changed, but rather, whether it has decreased (or increased). Before carrying out the test it is necessary to decide which of these two propositions you wish to test.

If we wish to test whether or not the percentage has decreased, then our hypotheses would be:

null hypothesis H_0: $\pi = \pi_0$
alternative hypothesis H_1: $\pi < \pi_0$.

where π_0 is the actual percentage in the population last year. This may be an appropriate hypothesis test if you were working for a health lobby.

If we wanted to test if the percentage had increased, then our hypotheses would be:

null hypothesis H_0: $\pi = \pi_0$
alternative hypothesis H_1: $\pi > \pi_0$.

This could be an appropriate test if you were working for a manufacturer in the tobacco industry and were concerned with forecasting future demand patterns.

To carry out the test, we will want to concentrate the chance of rejecting the null hypothesis at one end of the Normal distribution; where the significance level is 5%, then the critical value will be −1.645 (i.e. the cut off value taken from the Normal distribution tables) for the hypotheses H_0: $\pi = \pi_0$, H_1: $\pi < \pi_0$; and +1.645 for the hypotheses H_0: $\pi = \pi_0$, H_1: $\pi > \pi_0$. (Check these figures from Appendix C.) Where the significance level is set at 1%, then the critical value becomes either −2.33 or +2.33. In terms of answering examination questions, it is important to read the wording very carefully to determine which type of test you are required to perform.

The interpretation of the calculated z-value is now merely a question of deciding into which of two sections of the Normal distribution it falls.

15.3.1 A one-sided test for a population mean

Here the hypotheses will be in terms of the population mean, or the claimed value. Consider the following examples.

EXAMPLE

A manufacturer of batteries has assumed that the average expected life is 299 hours. As a result of recent changes to the filling of the batteries, the manufacturer now wishes to test if the average life has increased.

A sample of 200 batteries was taken at random from the production line and tested. Their average life was found to be 300 hours with a standard deviation of 8 hours. You have been asked to carry out the appropriate hypothesis test at the 5% significance level.

Step 1: H_0: $\mu = 299$
H_1: $\mu > 299$
Step 2: The significance level is 5%.
Step 3: The critical value will be +1.645.
Step 4:

$$z = \frac{300 - 299}{8/\sqrt{200}} = 1.77.$$

(Note that we are still assuming the null hypothesis to be true while the test is conducted.)
Step 5: The calculated value is larger than the critical value.
Step 6: We may therefore reject the null hypothesis.
Note that had we been conducting a two–sided hypothesis test, then we would have been unable to reject the null hypothesis, and so the conclusion would have been that the average life of the batteries had not changed.)
Step 7: The sample evidence supports the supposition that the average life of the batteries has increased by a significant amount.

A further idea can be drawn from this example. Although the significance test has shown that there has been a significant increase in the average length of life of the batteries, this may not be an important conclusion for the manufacturer. For instance, it would be quite misleading to use it to back up an advertising campaign which claimed that 'our batteries now last even longer!'. Whenever hypothesis tests are used it is important to distinguish between statistical significance and importance.

EXAMPLE

The management of a company claim that the average weekly earning of their employees is £450. A shop steward disputes this figure, believing that the average earnings are somewhat lower. In order to test this claim, a random sample of 150 employees is selected from the pay-roll, and their gross pay noted. The sample results show an average of £428, with a standard deviation of £183.13. Use an appropriate significance test, at the 5% level, to determine which side of this argument you agree with.

1. H_0: μ = £450
   H_1: μ < £450
2. Significance level is 5%.
3. Critical value is −1.645.
4. $z = \frac{428 - 450}{182.13/\sqrt{150}} = -1.48.$
5. $-1.645 < -1.48$
6. Cannot reject H_0.
7. The sample does not support the claim that average weekly earnings in this company are below £450.

15.3.2 A one-sided test for a population percentage

Here the methodology is exactly the same as that employed above, and so we just provide two examples of its use.

EXAMPLE

A small political party expects to gain 20% of the vote in by-elections. A particular candidate from the party is about to stand in a by-election in Derbyshire South East, and has commissioned a survey of 200 randomly selected voters in the constituency. If 44 of those interviewed said that they would vote for this candidate in the forthcoming by-election, test whether this would be significantly above the national party's claim. Use a test at the 5% significance level.

1. H_0: $\pi = 20\%$
   H_1: $\pi > 20\%$
2. Significance level = 5%.
3. Critical value = +1.645.
4. Sample percentage $= \frac{44}{200} \times 100 = 22\%$

$$z = \frac{22 - 20}{\sqrt{\frac{20(100 - 20)}{200}}} = 0.7071$$

5. $0.7071 < 1.645$.
6. Cannot reject H_0.
7. There is no evidence that the candidate will do better than the national party's claim.

Here there may not be evidence of a statistically significant vote above the national party's claim (a sample size of 1083 would have made this 2% difference statistically significant), but if the candidate does manage to achieve 22% of the vote, and it is a close contest over who wins the seat between two other candidates, then the extra 2% could be very important.

EXAMPLE

A company claims that 4% of the components supplied by Walkways Ltd. are defective, but this is disputed by Walkways themselves. In order to test whether the claim is valid, a random sample of 500 components is selected, and each one carefully checked. From this sample it is found that 12 are defective. Use an appropriate test, at the 5% level, to find out if the company's claim is valid.

1. H_0: $\pi = 4\%$
   H_1: $\pi < 4\%$
2. Significance level = 5%.
3. Critical value = −1.645.
4. Sample percentage $= \frac{12}{500} \times 100 = 2.4\%$

$$z = \frac{2.4 - 4}{\sqrt{\frac{4(100 - 4)}{500}}} = -1.826.$$

5. $-1.826 < -1.645$.
6. Reject H_0.
7. The sample evidence suggests that the company is wrong in claiming that 4% of Walkways' supplies are defective.

15.3.3 Producers' risk and consumers' risk

One-sided tests are sometimes referred to as testing **producers' risks** or **consumers' risks**. If we are looking at the amount of a product in a given packet or lot size, then the reaction to variations in the amount may be different from the two groups. Say, for instance, that the packet is sold as containing 100 items. If there are more than 100 items per packet, then the producer is effectively giving away some items, since only 100 are being charged for. The producer, therefore, is concerned not to overfill the packets but still meet legal requirements. In this situation, we might presume that the consumer is quite happy, since some items come free of charge. In the opposite situation of less than 100 items per packet, the producer is supplying less than 100 but being paid for 100 (this, of course, can have damaging consequences for the producer in terms of lost future sales), while the consumers receive less than expected. Given this scenario, one would expect the producer to conduct a one-sided test using H_1: $\mu > \mu_0$, and a consumer group to conduct a one-sided test using H_0: $\mu < \mu_0$. There may, of course, be legal implications for the producer in selling packets of 100 which actually contain less than 100. (In practice most producers will, in fact, play it safe and aim to meet any minimum requirements.)

15.4 TYPES OF ERROR

We have already seen, throughout this section of the book, that samples can only give us a partial view of a population; there will always be some chance that the true population value really does lie outside of the confidence interval, or that we will come to the wrong decision when conducting a significance test. In fact these probabilities are specified in the names that we have already used — a 95% confidence interval and a test at the 5% level of significance. Both imply a 5% chance of being wrong.

If you consider significance tests a little further, however, you will see that there are, in fact, two

Table 15.1 Possible results of a hypothesis test

	Accept H_0	*Reject H_0*
If H_0 is correct	Correct decision	Type I error
If H_0 is not correct	Type II error	Correct decision

different ways of getting the wrong answer. You could throw out the claim when it is, in fact, true; or you could fail to reject it when it is, in fact, false. It becomes important to distinguish between these two types of error.

As you can see from Table 15.1, the two different types of error are referred to as Type I and Type II.

A **Type I error** is the rejection of a null hypothesis when it is true. This probability is known as the **significance level** of the test. This is usually set at either 5% or 1% for most business applications, and is decided upon *before* the test is conducted.

A **Type II error** is the failure to reject a null hypothesis which is false. The probability of this error cannot be determined before the test is conducted.

Consider a more everyday situation. You are standing at the kerb, waiting to cross a busy main road. You have four possible outcomes:

1. If you decide not to cross at this moment, and there is something coming, then you have made a correct decision.
2. If you decide not to cross, and there is nothing coming, you have wasted a little time, and made a Type II error.
3. If you decide to cross and the road is clear, you have made a correct decision.
4. If you decide to cross and the road isn't clear, as well as the likelihood of injury, you have made a Type I error.

The error which can be controlled is the Type I error; and this is the one which is set before we conduct the test.

Consider the sampling distribution of z, consistent with the null hypothesis, H_0: $\mu = \mu_0$, illustrated as Figure 15.3.

The two values for the test-statistic, A and B, are both possible, with B being less likely than A. The construction of this test, with a 5% significance level, would mean the acceptance of the null hypothesis when A was obtained and its rejection when B was obtained. Both values could be attributed to an alternative hypothesis, H_1: $\mu = \mu_1$, which we accept in the case of B and reject for A. It is worth noting that if the test statistic follows a sampling distribution we can never be certain about the correctness of our decision. What we can do, having fixed a significance level (Type I error) is construct the rejection region to minimize Type II error. Suppose the alternative hypothesis was that the mean for the population, μ, was not μ_0 but a larger value μ_1. This we would state as:

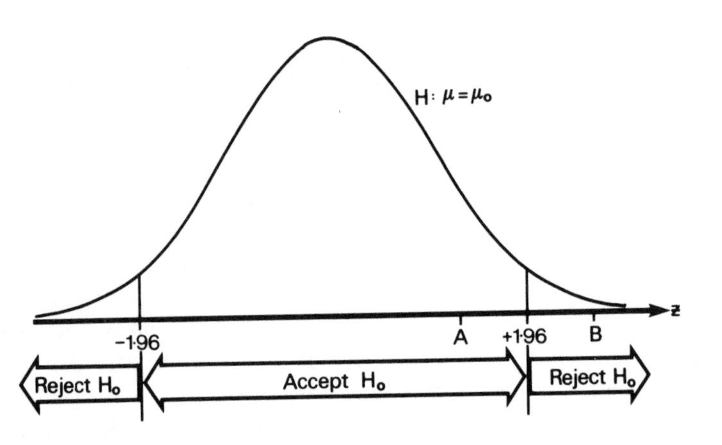

Figure 15.3 The sampling distribution assuming the null hypothesis to be correct

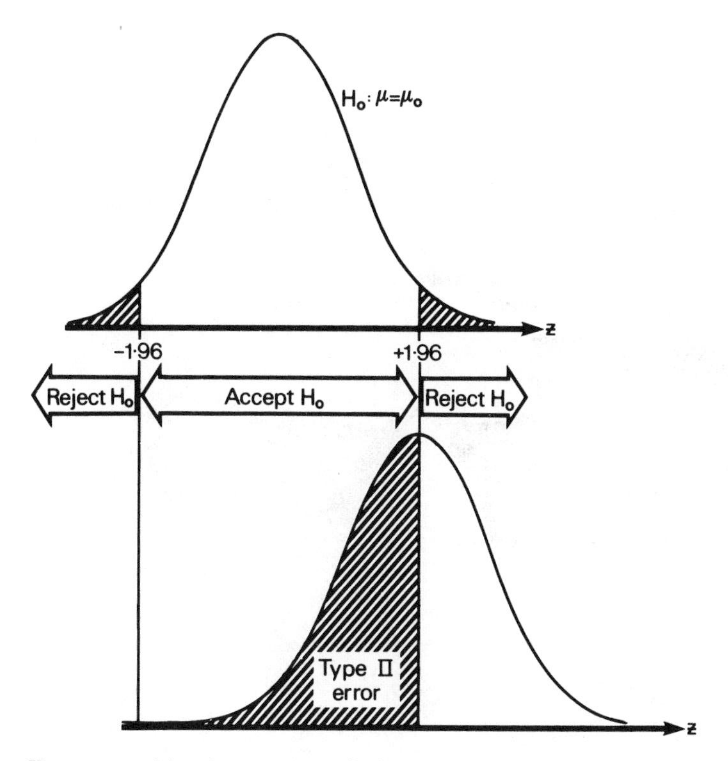

Figure 15.4 The Type II error resulting from a two-tailed test

H_1: $\mu = \mu_1$ where $\mu_1 > \mu_0$

or, in the more general form:

H_1: $\mu > \mu_0$

If we keep the same acceptance and rejection regions as before (Figure 15.3), the Type II error is as shown in Figure 15.4.

As we can see, the probability of accepting H_0 when H_1 is correct can be relatively large. If we test the null hypothesis H_0: $\mu = \mu_0$ against an alternative hypothesis of the form H_1: $\mu < \mu_0$ or H_1: $\mu > \mu_0$, we can reduce the size of the Type II error by careful definition of the rejection region. If the alternative hypothesis is of the form H_1: $\mu < \mu_0$, a critical value of $z = -1.645$ will define a 5% rejection region in the left-hand tail, and if the alternative hypothesis is of the form H_1: $\mu > \mu_0$, a critical value of $z = 1.645$ will define a 5% rejection region in the right-hand tail. The reduction in Type II error is illustrated in Figure 15.5.

It can be seen from Figure 15.5 that the test statistic now rejects the null hypothesis in the range 1.645 to 1.96 as well as values greater than 1.96. If we construct one-sided tests, there is more chance that we will reject the null hypothesis in favour of a 'more radical' alternative that the parameter has a larger value (or has a smaller value) than specified when that alternative is true. Note that if the alternative hypothesis was of the form H_1: $\mu < \mu_0$, the rejection region would be defined by the critical value $z = -1.645$ and we would reject the null hypothesis if the test statistic took this value or less.

15.5 HYPOTHESIS TESTING WITH TWO SAMPLES

All of the tests of hypothesis which we have used so far have made comparisons between some known, or claimed, population value and a set of

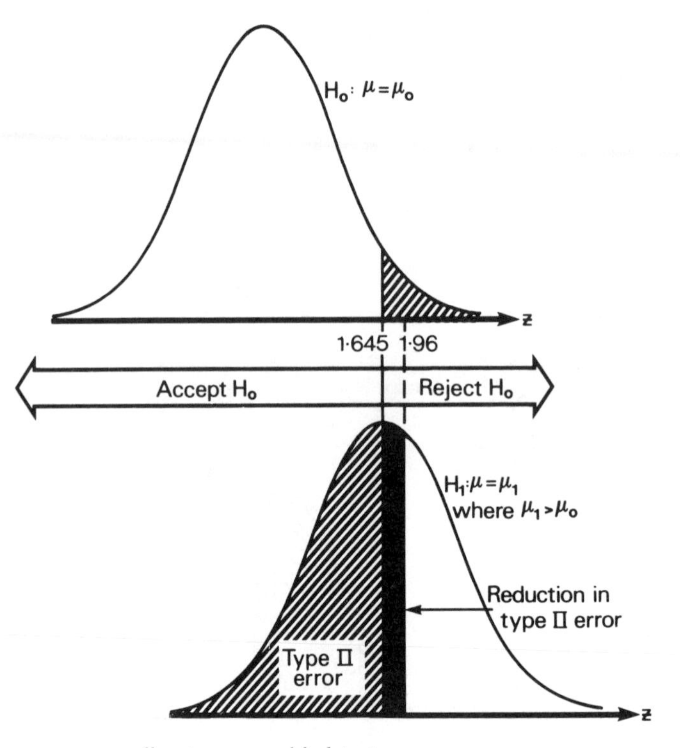

Figure 15.5 Type II error corresponding to a one-sided test

sample results. In many cases we may know little about the actual population value, but will have available the results of another survey. This could be a situation where we have two surveys conducted in different parts of the country, or at different points in time. Our concern now, is to determine if the two sets of results are consistent with each other (come from the same parent population) or whether there is a difference between them. Such comparisons may be important in a marketing context, looking at whether or not there are different perceptions of the product in different regions of the country. They could be used in performance appraisal of employees to answer questions on whether the average performance in different regions is, in fact, different. They could be used in assessing a forecasting model, to see if the values have changed significantly from one year to the next.

Again we can use the concepts and methodology developed in previous sections of this chapter. Tests may be two–sided to look for a difference, or one–sided, to look for a significant increase (or decrease). Similarly, tests may be carried out at the 5% or the 1% level, and the interpretation of the results will depend upon the calculated value of the test statistic and its comparison to a critical value. It will be necessary to rewrite the hypotheses to take account of the two samples and to find a new formulation for the test statistic.

15.5.1 Tests for a difference of means

In order to be clear about which sample we are talking about, we will use the suffixes 1 and 2 to refer to sample 1 and sample 2. Our assumption is that the two samples do, in fact, come from the same population, and thus the population means associated with each sample result will be the same. This assumption will give us our null hypothesis:

H_0: $\mu_1 = \mu_2$ or $\mu_1 - \mu_2 = 0$.

The alternative hypothesis will depend upon whether we are conducting a two-sided test or a one-sided test. If it is a two-sided test, the alternative hypothesis will be

H_1: $\mu_1 \neq \mu_2$ or $\mu_1 - \mu_2 \neq 0$,

and if it is a one-sided test, it would be either

H_1: $\mu_1 > \mu_2$ or $\mu_1 - \mu_2 > 0$

or

H_1: $\mu_1 < \mu_2$ or $\mu_1 - \mu_2 < 0$.

We could also test whether the difference between the means could be assigned to a specific value, for instance, that the difference in take home pay between two groups of workers was £25; here we would use the hypotheses:

H_0: $\mu_1 - \mu_2 = £25$
H_1: $\mu_1 - \mu_2 \neq £25$.

The test statistic used is closely related to the formula developed in section 14.2.1 for the confidence interval for the difference between two means. It will be

$$z = \frac{(x_1 - x_2) - (\mu_1 - \mu_2)}{\sqrt{\frac{s_1^2}{n_1} + \frac{s_2^2}{n_2}}}$$

but if we are using a null hypothesis which says that the two population means are equal, then the second bracket on the top line will be equal to zero, and the formula for z will be simplified to

$$z = \frac{(\bar{x}_1 - \bar{x}_2)}{\sqrt{\frac{s_1^2}{n_1} + \frac{s_2^2}{n_2}}}.$$

The example below illustrates the use of this test.

EXAMPLE

A random sample of 75 packets of cereal was selected from the production process of a company and found to have a mean of 500 grammes and a standard deviation of 20 grammes. After a technical change to the production process, a second random sample of 50 packets was selected, and this was found to have a mean of 505 grammes and a standard deviation of 16 grammes. Has there been a significant change in the weight of the packets at the 1% level of significance?

1. We assume no change, so

   H_0: $(\mu_1 - \mu_2) = 0$,

   and we are testing for a difference, so

   H_1: $(\mu_1 - \mu_2) \neq 0$.
2. Significance level is 1%.
3. Critical values are -2.58 and $+2.58$.
4. $z = \frac{500 - 505}{\sqrt{\frac{20^2}{75} + \frac{16^2}{50}}} = -1.546.$
5. $-2.58 \leqslant -1.546 \leqslant +2.58$
6. Therefore we cannot reject H_0.
7. There is no significant difference in weight.

15.5.2 Tests for a difference in population percentage

Again, adopting the approach used for testing the difference of means, we will modify the hypotheses, using π instead of p for the population values, and redefine the test statistic. Both one- and two-sided tests may be carried out.

The test statistic will be

$$z = \frac{(p_1 - p_2) - (\pi_1 - \pi_2)}{\sqrt{\frac{\pi_1(100 - \pi_1)}{n_1} + \frac{\pi_2(100 - \pi_2)}{n_2}}}$$

but if the null hypothesis is $\pi_1 - \pi_2 = 0$, then this is simplified to

$$z = \frac{(p_1 - p_2)}{\sqrt{\frac{\pi_1(100 - \pi_1)}{n_1} + \frac{\pi_2(100 - \pi_2)}{n_2}}}$$

EXAMPLE

A manufacturer believes that a new promotional campaign will increase favourable perceptions of the product by 10%. Before the campaign, a random sample of 500 consumers showed that 20% had favourable attitudes towards the product. After the campaign, a second random sample of 400 consumers found favourable reactions amongst 28%. Use an appropriate test at the 5% level of significance to find if there has been a 10% improvement in favourable perceptions.

1. Since we are testing for a specific increase, the hypotheses will be:

   H_0: $(\pi_2 - \pi_1) = 10\%$
   H_1: $(\pi_2 - \pi_1) < 10\%$.

2. The significance level is 5%.
3. The critical value is -1.645
4. $z =$

$$\frac{(28 - 20) - (10)}{\sqrt{\dfrac{20(100 - 20)}{500} + \dfrac{28(100 - 28)}{400}}} = -0.697$$

5. $-1.645 < -0.697$
6. We cannot reject H_0.
7. The sample supports a 10% increase in favourable views.

15.6 HYPOTHESIS TESTING WITH SMALL SAMPLES

15.6.1 Single samples

As we saw in Chapter 14, the basic assumption that the sampling distribution for the sample parameters is a Normal distribution only holds when the samples are large. Once we turn to small samples, we need to use a different sampling distribution: the t-distribution (see section 14.3). Apart from this difference, which implies a change to the formula for the test statistic and also to the table of critical values, all of the tests developed so far in this chapter may be applied to small samples.

For a single sample, the test statistic for a population mean is calculated by using:

$$t = \frac{\bar{x} - \mu}{s/\sqrt{n}}$$

with $(n - 1)$ degrees of freedom.

For a single sample, the test statistic for a population percentage will be calculated using:

$$t = \frac{p - \pi_0}{\sqrt{\dfrac{\pi_0(100 - \pi_0)}{n}}}$$

and, again there will be $(n - 1)$ degrees of freedom. (Note that, as with the large sample test, we use the null hypothesis value of the population percentage in the formula.)

Below are a series of examples to illustrate the use of the t-distribution in hypothesis testing.

EXAMPLE

A lorry manufacturer claims that the average annual maintenance cost for its vehicles is £500. The maintenance department of one of their customers believes it to be higher, and to test this randomly selects a sample of six lorries from their large fleet. From this sample, the mean annual maintenance cost was found to be £555, with a standard deviation of £75. Use an appropriate hypothesis test, at the 5% level, to find if the manufacturer's claim is valid.

1. H_0: $\mu = £500$
   H_1: $\mu > £500$
2. Significance level is 5%.
3. Degrees of freedom $= 6 - 1 = 5$.
   Critical value $= 2.015$.
4. $t = \dfrac{555 - 500}{75/\sqrt{6}} = 1.796$
5. $1.796 < 2.015$.
6. Therefore we cannot reject H_0.
7. The sample evidence does not suggest that maintenance costs are more than £500 per annum.

EXAMPLE

A company had a 50% market share for a newly developed product last year. It believes that as more entrants start to produce this type of product, its market share will decline, and in order to monitor this situation, decides to select a random sample of 15 customers. Of these, six have bought the company's product. Carry out a test at the 5% level to find if there is sufficient evidence to suggest that their market share is now below 50%.

1. H_0: $\mu = 50\%$
   H_1: $\mu < 50\%$
2. Significance level = 5%.
3. Degrees of freedom = 15 − 1 = 14.
   Critical value = −1.761.
4. The sample percentage is $\frac{6}{15} \times 100 = 40\%$

$$t = \frac{40 - 50}{\sqrt{\frac{50(100 - 50)}{15}}} = -0.775$$

5. −1.761 < −0.775.
6. We cannot reject H_0.
7. It appears that the company still has a 50% market share.

15.6.2 Two samples

Where we have two samples, we will again use similar tests to section 15.5, but with formulae developed using the sampling errors developed in Chapter 14.

In the case of estimating a confidence interval for the difference between two means the pooled standard error (assuming that both samples have similar variability), was given by:

$$s_p = \sqrt{\frac{(n_1 - 1)\, s_1^2 + (n_2 - 1)\, s_2^2}{n_1 + n_2 - 2}}.$$

Using this, we have the test statistic:

$$t = \frac{(x_1 - x_2) - (\mu_1 - \mu_2)}{s_p\sqrt{\frac{1}{n_1} + \frac{1}{n_2}}}$$

with $(n_1 + n_2 - 2)$ degrees of freedom.

EXAMPLE

A company has two factories, one in the UK and one in Germany. The head office staff feel that the German factory is more efficient the the British one, and to test this select two random samples. The British sample consists of 20 workers who take an average of 25 minutes to complete a standard task. Their standard deviation is 5 minutes. The German sample has 10 workers who take an average of 20 minutes to complete the same task, and the sample has a standard deviation of 4 minutes. Use an appropriate hypothesis test, at the 1% level, to find if the German workers are more efficient.

1. H_0: $(\mu_1 - \mu_2) = 0$
   H_1: $(\mu_1 - \mu_2) > 0$
2. Significance level = 1%.
3. Degrees of freedom = 20 + 10 − 2 = 28.
   Critical value = 2.467.
4. $s_p = \sqrt{\left[\frac{(19 \times 5^2) + (9 \times 4^2)}{20 + 10 - 2}\right]} = 4.70182.$

$$t = \frac{25 - 20}{4.70182\sqrt{\left(\frac{1}{20} + \frac{1}{10}\right)}} = 2.746.$$

5. 2.746 > 2.467.
6. We therefore reject H_0.
7. It appears that the German workers are more efficient at this particular task.

A special case arises when we are considering tests of the difference of means if the two samples are related in such a way that we may pair the observations. This may arise when two different people assess the same series of situations, for example, different interviewers assessing the same group of candidates for a job. The example below illustrates the situation of **matched pairs**.

EXAMPLE

Seven applicants for a job were interviewed by two personnel officers who were asked to give marks on a scale of 1 to 10 to one aspect of the candidates' performance. A summary of the marks given is shown below.

	Marks given to candidate:						
	I	*II*	*III*	*IV*	*V*	*VI*	*VII*
Interviewer A	8	7	6	9	7	5	8
Interviewer B	7	4	6	8	5	6	7

Text if there is a significant difference between the standards being applied by the two interviewers at the 5% level.

1. Since we are working with matched pairs, we need only look at the differences between the two scores. The hypotheses will be:

   H_1: $\mu = 0$
   H_1: $\mu \neq 0$

2. The significance level is 5%.
3. The number of degrees of freedom is $(7 - 1) = 6$ and the critical value is 2.447.

Table 15.2 The calculation of summary statistics for recorded differences of marks

Interviewer A	*Interviewer B*	*Difference (x)*	$(x - \bar{x})^2$
8	7	1	0
7	4	3	4
6	6	0	1
9	8	1	0
7	5	2	1
5	6	−1	4
8	7	1	0
		7	10

4. We now need to calculate the summary statistics from the paired samples.

$$\bar{x} = \frac{\Sigma x}{n} = \frac{7}{7} = 1$$

$$s = \sqrt{\left[\frac{\Sigma(x - \bar{x})^2}{n - 1}\right]} = \sqrt{\frac{10}{6}} = 1.2910$$

$$t = \frac{1 - 0}{1.2910/\sqrt{7}} = 2.0494$$

5. Now $2.0494 < 2.447$.
6. We cannot reject the null hypothesis.
7. There is no evidence that the two interviewers are using different standards in their assessment of this aspect of the candidates.

The use of such matched pairs is common in market research and psychology. If, for reasons of time and cost, we need to work with small samples, results can be improved using paired samples since it reduces the between-sample variation.

15.7 CONCLUSIONS

In this chapter we have introduced the topic of hypothesis testing by looking at the more commonly used tests. As you will have seen, the actual mechanics (number-crunching) involved in these tests is fairly straight-forward, and usually easily performed on a calculator. What is a little more involved is setting up the correct hypotheses and deciding what the numerical answer means.

Since we are inevitably dealing with sample data when we conduct such tests of hypothesis, we can never be 100% sure that we have come to the *right* conclusion; but by using an appropriate significance level, we can, at least, be sure of the chance of having made a mistake. Conversely, even where the test statistic is considerably above the critical value, there is still a small chance that the null hypothesis is correct.

Finally we should mention again that the results of a significance test need to be interpreted in the light of the situation you are considering. Showing that there is a statistically significant difference between two values may be very important to your organisation, but there are situations where it will not matter. Statistical tests only have meaning within a context. It could be argued that such tests should *not* be conducted if the presence or absence of a difference is immaterial to the company.

15.8 PROBLEMS

1. A machine is supposed to be adjusted to produce components to a dimension of 2.000 inches. In a sample of 50 components the mean was found to be 2.001 inches and the standard deviation to be 0.003 inches. Is there evidence to suggest that the machine is set too high?
2. A production manager believes that 100 hours are required to complete a particular task. The accountant, however, disagrees and believes that the task takes less time. To test the view of the accountant, the time taken to complete the task was recorded on 60 occasions by an independent observer. The mean was found to be 96 hours and the standard deviation to be 2.5 hours. Formulate and perform an appropriate test at the 5% significance level.
3. A manufacturer sells their product by weight and states on the packaging that the weight is 375 grammes. The quality control inspector suspects that the average weight of packets is, in fact, higher than this and institutes a survey to check these suspicions. In a random sample of 100 packets, the average weight is found to be 379 grammes with a standard deviation of 14.2 grammes. Test at the 5% level of significance whether this result confirms the suspicions of the quality control inspector.
4. A supermarket sets standards for its checkout operators of serving 50 customers per shift. The evening shift claim that they are not able to meet this standard since customers tend to buy more on evening shopping expeditions. A survey of 40 evening operator shifts finds an average of 47.2 customers served with a standard deviation of 1.3. Test at the 5% level of significance if the operators are significantly below the standard set by the company.
5. A manufacturer claims that only 2% of the items produced are defective. If 7 defective items were found in a sample of 200 would you accept or reject the manufacturer's claim?
6. A photocopying machine produces copies, 18% of which are faulty. The supplier of the machine claims that by using a different type of paper the percentage of faulty copies will be reduced. If 45 are found to be faulty from a sample of 300 using the new paper, would you accept the claim of the supplier?
7. Directors of a company claim that 90% of the workforce support a new shift pattern which they have suggested. A random survey of 100 people in the workforce finds 85 in favour of the new scheme. Test at the 5% level if there is a significant difference between the survey results and the claim made by the directors. If there is a statistical difference, does it really matter to the company?
8. When introducing new products to the market place a particular company has a policy that a minimum of 40% of those trying the product at the test market stage should express their approval of it. Testing of a new product has just been completed with a sample of 200 people, of whom 78 expressed their approval of the product. Does this result suggest that significantly less than 40% of people approve of the product? (Conduct your test at the 5% level of significance.)
9. A dispute exists between workers on two production lines. The workers on production line A claim that they are paid less than those on production line B. The company investigates the claim by examining the pay of 70 workers from each production line. The results were as follows:

Sample statistics	*Production line* A	B
Mean	£193	£194.50
Standard deviation	£6	£7.50

Formulate and perform an appropriate test.

10. A manufacturer of soft drinks is currently operating in France and wishes to expand production to enter the Italian market. It is suspected that the typical Italian consumer in the target segment of the market purchases more soft drinks per week than a similar French person. In order to test this hypothesis, two surveys are conducted. In France a sample of 50 people finds an average of 5 purchases per week with a variance of 2.3. The Italian survey finds an average of 5.6 purchases per week with a variance of 1.4 from a sample of 100 people. What conclusions may the manufacturer draw from these survey results?
11. Market awareness of a new chocolate bar has been tested by two surveys, one in the Midlands and one in the South East. In the Midlands of 150 people questioned, 23 were aware of the product, whilst in the South East 20 out of 100 people were aware of the chocolate bar. Test at the 5% level of significance if the level of awareness is higher in the South East.
12. Two models of washing machine, the Washit and the Supervat, were tested for a specified range of faults. In a sample of 200 Washit washing machines, 60 were found to have such faults and in a sample of 250 Supervat washing machines, 52 were found to have such faults. Do you consider the two models of washing machine to be equally prone to such faults?
13. A sample of 10 job applicants were asked to complete a mathematics test. The average time taken was 28 minutes and the standard deviation was 7 minutes. If the test had previously taken job applicants on average 30 minutes, is there any evidence to suggest that the job applicants are now able to complete the test more quickly?
14. Records were kept for 7 cars to test whether they could achieve a fuel consumption of 56 miles to a gallon of petrol. The results obtained were 52, 49, 57, 53, 54, 55 and 53 miles to the gallon. Formulate and perform an appropriate test at the 5% significance level.
15. A claim has been made that 20% of people are capable of benefiting from a new language learning method developed by a certain company. A consumer group selected a random sample of 14 people who try the new method and, of these, only two are seen to benefit. Test the company's claim on the basis of these survey results.
16. A slot machine manufacturer claims that there is a pay-out on half of the occasions that the machine is used. After incurring substantial losses it is decided to check the machine installed in the 'Man of War' public house. The machine is played 18 times and a pay-out given on eleven of these. Test if the machine is set to give more than the claimed percentage of pay-outs.
17. The times taken to complete a task of a particular type were recorded for a sample of eight employees before and after a period of training as follows:

Employee	*Time to complete task (minutes)* *Before training*	*After training*
1	15	13
2	14	15
3	18	15
4	14	13
5	15	13
6	17	16
7	13	14
8	12	12

Test whether the training is effective.

18. Several universities and polytechnics conduct surveys of the careers of their graduates. One polytechnic has found that in a survey of 21 former students of the Business Studies degree, the average salary one year after graduation was £11 372 with a variance of 137. From a survey of 17 former students

on the Accountancy degree course, the comparable figures were £11 519 and 221. Test at the 5% level of significance if the former Accountancy degree students earn more than the former Business Studies degree students.

19. The hypothesis H_0: $\mu = 20$ is to be tested against the hypothesis H_1: $\mu = 21$. The test is based on a sample of 100 at the 5% significance level and the sample standard deviation is 4. What is the Type II error if the rejection region is (a) both tails of the Normal distribution, and (b) only the right tail of the Normal distribution?

NON-PARAMETRIC TESTS 16

In Chapter 15 we have considered a series of hypothesis tests which are designed to look at an individual parameter from a sample and then compare it with a known, or supposed, value from the population. Such tests are extremely important in the development of statistical theory and for testing of many sample results, but they do not cover all types of data, particularly when parameters cannot be calculated in a meaningful way. We therefore need to develop other tests which will be able to deal with such situations; a small range of these are covered in this chapter.

Parametric tests require the following conditions to be satisfied.

1. A null hypothesis can be stated in terms of parameters.
2. A level of measurement has been achieved that gives validity to differences.
3. The test statistic follows a known distribution.

It is not always possible to define a meaningful parameter. For instance, what is an average eye-colour? Equally it is not always possible to give meaning to differences in values, for instance, if brands of soft drink are ranked in terms of value for money or taste.

Where the conditions listed above cannot be met, **non-parametric** tests may be appropriate, but note that for some circumstances, there may be no suitable test. As with all tests of hypothesis, it must be remembered that even when a test result is significant in statistical terms, there may be situations where it has no importance in practice. A non-parametric test is still a hypothesis test, but rather than considering just a single parameter of the sample data, it looks at the overall distribution and compares this to some known or expected value, usually based upon the null hypothesis.

16.1 CHI-SQUARED TESTS

This non-parametric test probably bears the closest resemblance to the tests we have used so far. As before, we will define hypotheses, calculate a test statistic, and compare this to a value from tables in order to decide whether or not to reject the null hypothesis. As the name may suggest, the statistic calculated involves squaring values, and thus the result can be positive only.

We shall look at two particular applications of the chi-squared test. The first considers survey data, usually from questionnaires, and tries to find if there is an association between the answers given to a pair of questions. Secondly, we will look at using a chi-squared test to check whether a particular set of data follow a known statistical distribution.

16.1.1 Tests of association

When analyzing the results of a questionnaire, the first step is usually to find out how many responses there are to each alternative answer to each question. (This could be done by using HIST in the MICROSTATS package.) Such information will allow us to calculate the percentages of people who do a certain thing or hold a particular view. The next step will be to produce cross-tabulations of results between two questions, as shown below.

Suppose that two questions had given the following sets of answers:

Qu. 5 What type of property do you live in?

House	150
Flat	100
Bedsit	45
Other	5
	300

Qu. 13 How often do you do this?

Once a month	40
Once a week	200
Twice a week	50
More often	10
	300

This tells us that 150 people (or 50%) live in a house and also that 40 people (or 13.3%) perform this activity once a month. It does not tell us how many both live in a house and perform the activity once a month! To find this information we need to divide up the 150 people who live in houses on the basis of their answers to question 13; thus filling in the table given below. (Each box in the table is referred to as a cell.)

This would be an extremely boring and time consuming job to do manually, but any statistical package will perform the task very quickly. (In MICROSTATS you use the CONT command.)

The result may look like the second table below.

We are now in a stronger position to relate the two answers, but, because different numbers of people live in each of the types of accommodation, it is not immediately obvious if different behaviours are associated with their type of resi-

			Type of property		
How often	*House*	*Flat*	*Bedsit*	*Other*	*Total*
Once a month					40
Once a week					200
Twice a week					50
More often					10
TOTAL	150	100	45	5	300

			Type of property		
How often	*House*	*Flat*	*Bedsit*	*Other*	*Total*
Once a month	30	5	4	1	40
Once a week	110	80	8	2	200
Twice a week	5	10	33	2	50
More often	5	5	0	0	10
TOTAL	150	100	45	5	300

dence. A chi-squared test will allow us to find if there is a statistical association between the two sets of answers; and this, together with other information, may allow the development of a proposition that there is a causal link between the two.

To carry out the test we will follow the seven steps used in Chapter 15.

Step 1

State the hypotheses:

H_0: There is *no* association between the two sets of answers.
H_1: There *is* an association between the two sets of answers.

Step 2

State the significance level. As with a parametric test, the significance level can be set at various values, but for most business data it is usually 5%.

Step 3

State the critical value. The chi-squared distribution varies in shape with the number of degrees of freedom (in a similar way to the *t*-distribution), and thus we need to find this value before we can look up the appropriate critical value.

Consider the empty table on the previous page. There are four rows and four columns, giving a total of sixteen cells. Each of the row and column totals is fixed (i.e. these are the actual numbers given by the frequency count for each question), and thus the individual cell values must add up to the appropriate totals. In the first row, we have freedom to put any numbers into three of the cells, but the fourth is then fixed because all four must add to the (fixed) total (i.e. 3 degrees of freedom). The same will apply to the second row (i.e. 3 more degrees of freedom). And again to the third row (3 more degrees of freedom). Now all of the values on the fourth row are fixed because of the totals (0 degrees of freedom). Totalling these, we have $3 + 3 + 3 + 0 = 9$ degrees of freedom for this table.

There is a short cut! If you take the number of rows minus one and multiply by the number of columns minus one you get the number of degrees of freedom.

$$v = (r - 1) \times (c - 1)$$

Using the tables in Appendix E, we can now state the critical value as 16.9.

Step 4

Calculate the test statistic. The chi-squared statistic is given by the following formula:

$$\chi^2 = \sum\left[\frac{(O - E)^2}{E}\right]$$

where

O = the observed cell frequencies (the actual answers);
E = the expected cell frequencies (if the null hypothesis is true);

Finding the expected cell frequencies takes us back to some simple probability rules, since the null hypothesis makes the assumption that the two sets of answers are *independent* of each other. If this is true, then the cell frequencies will depend only upon the totals of each row and column.

Consider the first cell of the table (i.e. the first row and the first column). The number of people living in houses is 150 out of a total of 300, and thus the probability of someone living in a house is

$$\frac{150}{300} = 0.5.$$

The probability of 'once a month' is

$$\frac{40}{300} = 0.13333.$$

Thus the probability of living in a house, *and* 'once a month' is

$$0.5 \times 0.13333 = 0.066665.$$

Since there are three hundred people in the sample, one would expect there to be

$0.066665 \times 300 = 19.9995$

people who fit the category of the first cell. (Note that the observed value was 30.)

Again there is a short cut! Look at the way in which we have found the expected value.

$$19.9995 = \frac{150}{300} \times \frac{40}{300} \times 30$$

$$= \frac{\text{row total}}{\text{grand total}} \times \frac{(\text{column total})}{(\text{grand total})} \times (\text{grand total})$$

$$= (\text{row total} \times \text{column total})/(\text{grand total})$$

We need to complete this process for the other cells in the table, but remember that, because of the degrees of freedom, you only need to calculate nine of them, the rest being found by subtraction.

Statistical packages will, of course, find these expected values very quickly. In MICROSTATS you use the CHIS command and the expected values are displayed. In some versions of SPSS you need to include a specific option command to have the expected frequencies displayed.

Once this has been done, we could construct a table of expected frequencies, as shown at the foot of the page.

If we were to continue this test by hand calculation, we would now calculate the chi-squared value using the following table.

O	E	$(O-E)$	$(O-E)^2$	$\frac{(O-E)^2}{E}$
30	20	10.0	100.00	5.00
5	13.3	−8.3	68.89	5.18
4	6	−2.0	4.00	0.67
1	0.7	0.3	0.09	0.13
110	100	10.0	100.00	1.00
80	66.7	13.3	176.89	2.65
8	30	−22.0	484.00	16.13
2	3.3	−1.3	1.69	0.51
5	25	−20.0	400.00	16.00
10	16.7	−6.7	44.89	2.69
33	7.5	25.5	650.25	86.70
2	0.8	1.2	1.44	1.80
5	5	0.0	0.00	0.00
5	3.3	1.7	2.89	0.88
0	1.5	−1.5	2.25	1.50
0	0.2	−0.2	0.04	0.20
			chi-squared =	141.04

(Note that the rounding has caused a slight error here. If the original table is analyzed using MIC-ROSTATS, the chi-squared value is 140.875.)

Step 5

Compare the calculated value and the critical value: 140.875 > 16.9

Step 6

Come to a conclusion. We already know that chi-squared cannot be below zero. If all of the expected cell frequencies were exactly equal to the observed cell frequencies, then the value of chi-squared would be zero. Any differences between the observed and expected cell frequencies may be due to either sampling error or to an association between the answers; the larger the

			Type of property		
How often	*House*	*Flat*	*Bedsit*	*Other*	*Total*
Once a month	20	13.3	6	0.7	40
Once a week	100	66.7	30	3.3	200
Twice a week	25	16.7	7.5	0.8	50
More often	5	3.3	1.5	0.2	10
TOTAL	150	100	45	5	300

(Note that all numbers have been rounded for ease of presentation.)

differences, the more likely it is that there is an association. Thus, if the calculated value is *below* the critical value, we will be *unable to reject* the null hypothesis, but if it is *above* the critical value, we *reject* the null hypothesis.

In this example, the calculated value is above the critical value, and thus we reject the null hypothesis.

Step 7

Put the conclusion into English:

There appears to be an association between the type of property in which people live and how often they perform this activity.

An adjustment

In fact, although the basic methodology of the test is correct, there is a problem. One of the basic conditions for the chi-squared test is that all of the expected frequencies must be *above five*. This is not true for our example! In order to make this condition true, we need to combine adjacent categories until their expected frequencies are equal to five or more.

To do this, we will combine the two categories 'Bedsit' and 'Other' to represent all non house or flat dwellers; it will also be necessary to combine 'Twice a week' with 'More often' to represent anything above once a week. Our new cross-tabulation will now be as follows:

How often	*Type of property* House	Flat	Other
Observed frequencies:			
Once a month	30	5	5
Once a week	110	80	10
More often	10	15	35
Expected frequencies:			
Once a month	20	13.3	6.7
Once a week	100	66.7	33.3
More often	30	20	10

The number of degrees of freedom now becomes $(3 - 1) \times (3 - 1) = 4$, and the value of chi-squared (from tables) is 9.49.

Re-computing the value of chi-squared from Step 4, we have a value of approximately 107.6, which is still substantially above the critical value of chi-squared. However, in other examples, the amalgamation of categories may affect the decision. In practice, one of the problems of meeting this condition is deciding which categories to combine, and deciding what, if anything, the new category represents.

This has been a particularly long example since we have been explaining each step as we have gone along. Performing the tests is much quicker in practice, even if a computer package is not used.

EXAMPLE

Purchases of different strengths of lager are thought to be associated with the gender of the drinker and a brewery has commissioned a survey to find if this is true. Summary results are shown below.

	Strength High	Medium	Low
Male	20	50	30
Female	10	55	35

1. H_0: No association between gender and strength bought.
   H_1: An association between the two.
2. Significance level is 5%.
3. Degrees of freedom $= (2 - 1) \times (3 - 1) = 2$.
   Critical value $= 5.99$.
4. Find totals:

	Strength High	Medium	Low	Total
Male	20	50	30	100
Female	10	55	35	100
Total	30	105	65	200

Expected frequency for Male and High Strength is

$$\frac{100 \times 30}{200} = 15.$$

Expected frequency for Male and Medium Strength is

$$\frac{100 \times 105}{200} = 52.5.$$

Continuing in this way the expected frequency table can be completed as follows:

	Strength			
	High	*Medium*	*Low*	*Total*
Male	15	52.5	32.5	100
Female	15	52.5	32.5	100
Total	30	105	65	200

Calculating chi-squared:

O	*E*	*(O − E)*	$(O-E)^2$	$\frac{(O-E)^2}{E}$
20	15	5	25	1.667
10	15	−5	25	1.667
50	52.5	−2.5	6.25	0.119
55	52.5	2.5	6.25	0.119
30	32.5	−2.5	6.25	0.119
35	32.5	2.5	6.25	0.119
				Chi-squared = 3.810

5. 3.81 < 5.99.
6. Therefore we *cannot reject* the null hypothesis.
7. There appears to be no association between the gender of the drinker and the strength of lager purchased at the 5% level of significance.

16.1.2 Tests of goodness-of-fit

If data has been collected and seems to follow some pattern, it would be useful to identify that pattern and to determine whether it follows some (already) known statistical distribution. If this is the case, then many more conclusions can be drawn about the data. (We have seen a selection of statistical distributions in Chapters 11 and 12) The chi-squared test provides a suitable method for deciding if the data follows a particular distribution, since we have the observed values and the expected values can be calculated from tables (or by simple arithmetic). For example, do the sales of whiskey follow a Poisson distribution? If the answer is 'yes', then sales forecasting becomes a much easier process.

Again, we will work our way through examples to clarify the various steps taken in conducting goodness-of-fit tests. The statistic used will remain as:

$$\chi^2 = \sum\left[\frac{(O-E)^2}{E}\right]$$

where

O = the observed frequencies;
E = the expected (theoretical) frequencies.

16.1.2.1 Test for a uniform distribution

You may recall the uniform distribution from Chapter 11; it implies that each item or value occurs the same number of times. Such a test would be useful where we want to find if several individuals are all working at the same rate, or if sales of various 'reps' are the same. Suppose that we are checking on the number of tasks completed in a set time by five machine operators and the following data has been made available.

Machine operator	*Number of tasks completed*
Alf	27
Bernard	31
Chris	29
Dawn	27
Eric	26
Total	140

1. State the hypotheses:

   H_0: All operators complete the same number of tasks.

H_1: All operators do not complete the same number of tasks.

(Note that the null hypothesis is just another way of saying that the data follows a uniform distribution.)

2. The significance level will be taken as 5%.
3. The degrees of freedom will be the number of cells *minus* the number of parameters required to calculate the expected frequencies *minus* one. Here $\nu = 5 - 0 - 1 = 4$. Therefore (from tables) the critical value is 9.49.
4. Since the null hypothesis proposes a uniform distribution, we would expect all of the operators to complete the same number of tasks in the allotted time. This number is

$$\frac{\text{Total tasks completed}}{\text{Number of operators}} = \frac{140}{5} = 28.$$

We can then complete the table:

O	E	$(O - E)$	$(O - E)^2$	$\frac{(O - E)^2}{E}$
27	28	−1	1	0.0357
31	28	3	9	0.3214
29	28	1	1	0.0357
27	28	−1	1	0.0357
26	28	−2	4	0.1429
			chi-squared =	0.5714

5. $0.5714 < 9.49$.
6. Therefore we do not reject H_0.
7. There is no evidence that the operators work at different rates in the completion of these tasks.

16.1.2.2 *Test for Binomial or Poisson distribution*

A similar procedure may be used in this case, but in order to find the theoretical values we will need to know one parameter of the distribution. In the case of a Binomial distribution this will be the probability of success, p (see Chapter 11). For a Poisson distribution it will be the mean (λ).

EXAMPLE

The components produced by a certain manufacturing process have been monitored to find the number of defective items produced over a period of 96 days. A summary of the results is contained in the following table:

Number of defective items:	0	1	2	3	4	5
Number of days:	15	20	20	18	13	10

You have been asked to establish whether or not the rate of production of defective items follows a Binomial distribution, testing at the 1% level of significance.

1. H_0: the distribution of defectives is Binomial.
   H_1: the distribution is not Binomial.
2. The significance level is 1%
3. The number of degrees of freedom is

   No. of cells − No. of parameters − 1
   $= 6 - 1 - 1$
   $= 4$

   From tables (Appendix E) the critical value is 13.3 (but see below for modification of this value).
4. In order to find the expected frequencies, we need to know the probability of a defective item. This may be found by first calculating the average of the sample results.

$$\bar{x} = \frac{(0 \times 15) + (1 \times 20) + (2 \times 20) + (3 \times 18) + (4 \times 13) + (5 \times 10)}{96}$$

$$= \frac{216}{96}$$

$$= 2.25.$$

From Chapter 10, we know that the mean of a binomial distribution is equal to np, where n is maximum number of defectives, and so

$$p = \frac{\bar{x}}{n} = \frac{2.25}{5} = 0.45.$$

Using this value and the Binomial formula we can now work out the theoretical values, and hence the expected frequencies. The Binomial formula states

$$P(r) = \binom{n}{r} p^n(1 - p)^{n-r}.$$

r	P(r)	Expected frequency = 96 × P(r)	
0	0.0503	4.83	24.6
1	0.2059	19.77	
2	0.3369	32.34	
3	0.2757	26.47	
4	0.1127	10.82	12.59
5	0.0185	1.77	
	1.0000	96.00	

Note that because two of the expected frequencies are less than five, it has been necessary to combine adjacent groups. This means that we must modify the number of degrees of freedom, and hence the critical value.

Degrees of freedom = 4 − 1 − 1 = 2
Critical value = 9.21

We are now in a position to calculate chi-squared.

r	O	E	(O − E)	$(O - E)^2$	$\frac{(O - E)^2}{E}$
0, 1	35	24.60	10.40	108.16	4.3967
2	20	32.34	−12.34	152.28	4.7086
3	18	26.47	−8.47	71.74	2.7103
4, 5	23	12.59	10.41	108.37	8.6075
				chi-squared =	20.4231

5. 20.4231 > 9.21.
6. We therefore reject the null hypothesis.
7. There is no evidence to suggest that the production of defective items follows a Binomial distribution at the 1% level of significance.

EXAMPLE

Items are taken from the stockroom of a jewellery business for use in producing goods for sale. The owner wishes to model the number of items taken per day, and has recorded the actual numbers for a hundred day period. The results are shown below.

Number of items	*Number of days*
0	7
1	17
2	26
3	22
4	17
5	9
6	2

If you were to build a model for the owner, would it be reasonable to assume that withdrawals from stock follow a Poisson distribution? (Use a significance level of 5%)

1. H_0: the distribution of withdrawals is Poisson.
   H_1: the distribution is not Poisson.
2. Significance level is 5%.
3. Degrees of freedom = 7 − 1 − 1 = 5.
   Critical value = 11.1 (but see below).
4. The parameter required for a Poisson distribution is the mean, and this can be found from the sample data.

$$\bar{x} = \frac{(0 \times 7) + (1 \times 17) + (2 \times 26) + (3 \times 22) + (4 \times 17) + (5 \times 9) + (6 \times 2)}{100}$$

$$= \frac{260}{100} = 2.6 = \lambda \text{ (lambda)}$$

We can now use the Poisson formula to find the various probabilities and hence the expected frequencies:

$$P(x) = \frac{\lambda^x e^{-\lambda}}{x!}$$

x	P(x)	Expected frequency P(x) × 100	
0	0.0743	7.43	
1	0.1931	19.31	
2	0.2510	25.10	
3	0.2176	21.76	
4	0.1414	14.14	
5	0.0736	7.36	12.26
*6 or more	0.0490	4.90	
	1.0000	100.00	

(*Note that, since the Poisson distribution goes off to infinity, in theory, we need to account for all of the possibilities. This probability is found by summing the probabilities from 0 to 5, and subtracting the result from one.)

Since one expected frequency is below 5, it has been combined with the adjacent one. The degrees of freedom therefore are now $6 - 1 - 1 = 4$, and the critical value is 9.49. Calculating chi-squared, we have:

(x)	O	E	(O − E)	$(O - E)^2$	$\frac{(O-E)^2}{E}$
0	7	7.43	−0.43	0.1849	0.0249
1	17	19.31	−2.31	5.3361	0.2763
2	26	25.10	0.90	0.8100	0.0323
3	22	21.76	0.24	0.0576	0.0027
4	17	14.14	2.86	8.1796	0.5785
5+	11	12.26	−1.26	1.5876	0.1295
				chi-squared =	1.0442

5. $1.0442 < 9.49$.
6. We therefore cannot reject H_0.
7. The evidence from the monitoring suggests that withdrawals from stock follow a Poisson distribution.

16.1.2.1 Test for the Normal distribution

This test can involve more data manipulations since it will require grouped (tabulated) data, and the calculation of *two* parameters (the mean and the standard deviation) before expected frequencies can be determined. (In some cases, the data may already be arranged into groups.)

EXAMPLE

The results of a survey of 150 people contain details of their income levels and are summarized here. Test at the 5% significance level if this data follows a normal distribution.

Weekly income	Number of people
under £100	30
£100 but under £200	40
£200 but under £300	45
£300 but under £500	20
£500 but under £900	10
over £900	5
	150

1. H_0: The distribution is Normal.
   H_1: The distribution is not Normal.
2. Significance level is 5%.
3. Degrees of freedom $= 6 - 2 - 1 = 3$.
   Critical value (from tables) = 7.81.
4. The mean of the sample = £266 and the standard deviation = £239.02. (N.B. see section 13.2.2 for formulae.)

   To find the expected values we need to:
   (a) convert the original group limits into z-scores (by subtracting the mean and then dividing by the standard deviation);
   (b) find the probabilities by using the Normal distribution tables (Appendix C), and
   (c) find our expected frequencies.

   This is shown in the table below.

Weekly income	z	Probability	Expected frequency = prob × 150	
under £100	−0.70	0.2420	36.3	
£100 but under £200	−0.28	0.1471	22.065	
£200 but under £300	0.14	0.1666	24.99	
£300 but under £500	0.98	0.2708	40.62	
£500 but under £900	2.65	0.15948	23.922	} 24.525
over £900		0.00402	0.603	
		1.00000	150	

(Note that the last two groups are combined since the expected frequency of the last group is less than five. This changes the degrees of freedom to $5 - 2 - 1 = 2$, and the critical value to 5.99.)

Thus we can now calculate the chi-squared value:

O	E	(O − E)	$(O - E)^2$	$\frac{(O-E)^2}{E}$
30	36.3	−6.300	39.6900	1.0934
40	22.065	17.935	321.6642	14.5780

45	24.99	20.01	400.4001	16.0224
20	40.62	−20.62	425.1844	10.4674
15	24.525	−9.525	90.7256	3.6993
			chi-squared =	45.8605

5. $45.8605 > 5.99$
6. We therefore reject the null hypothesis.
7. There is no evidence that the income distribution is Normal.

16.1.3 Summary note

It is worth noting the following characteristics of chi-squared.

(a) χ^2 is only a symbol; the square root of χ^2 has no meaning.

(b) χ^2 can never be less than zero. The squared term ensures positive values.

(c) χ^2 is concerned with comparisons of observed and expected frequencies (or counts). We therefore only need a classification of data to allow such counts and not the more stringent requirements of measurement.

(d) If there is a close correspondence between the observed and expected frequencies, χ^2 will tend to a low value attributable to sampling error, suggesting the correctness of the null hypothesis.

(e) If the observed and expected frequencies are very different we would expect a large positive value (not explicable by sampling errors alone), which would suggest that we reject the null hypothesis in favour of the alternative hypothesis.

16.2 MANN–WHITNEY U TEST

This non-parametric test deals with two samples which are independent and may be of different sizes. It is the equivalent of the *t*-test which we considered in Chapter 15. Where the samples are small (<30) we need to use tables of critical values (Appendix G) to find whether or not to reject the null hypothesis; but where the sample size is large, we can use a test based on the Normal distribution.

The basic premise of the test is that once all of the values in the two samples are put into a single ordered list, if they come from the same parent population, then the rank at which values from Sample 1 and Sample 2 appear will be by chance (at random). If the two samples come from different populations, then the rank at which sample values appear will not be random and there will be a tendency for values from one of the samples to have lower ranks than values from the other sample. We are thus testing for different *locations* of the two samples.

Whilst we will show how to conduct this test by hand calculation, packages such as SPSS-X, SPSS-PC and MINITAB will perform the test by using the appropriate commands.

16.2.1 Small sample test

Consider the situation where samples have been taken from two branches of a chain of stores. The samples relate to the daily takings and both branches are situated in city centres. We wish to find if there is any difference in turnover between the two branches.

Branch 1: £235, £255, £355, £195, £244,
£240, £236, £259, £260
Branch 2: £240, £198, £220, £215, £245

1. H_0: the two samples come from the same population.
   H_1: the two samples come from different populations.
2. We will take a significance level of 5%.
3. To find the critical level for the test statistic, we look in the tables (Appendix G) and locate the value from the two sample sizes. Here the sizes are 9 and 5, and so the critical value of U is 8.
4. To calculate the value of the test statistic, we need to rank all of the sample values, keeping a note of which sample each value came from.

Rank	Value	Sample
1	195	1
2	198	2
3	215	2
4	220	2
5	235	1
6	236	1
7.5	240	1
7.5	240	2
9	244	1
10	245	2
11	255	1
12	259	1
13	260	1
14	355	1

(Note that in the event of a tie in ranks, an average is used.)

We now sum the ranks for each sample:

Sum of ranks for Sample 1 = 78.5.
Sum of ranks for Sample 2 = 26.5.

We select the smallest of these two, i.e. 26.5 and put it into the following formula:

$$T = S - \frac{n_1(n_1 + 1)}{2}$$

where S is the smallest sum of ranks, and n_1 is the number in the sample whose ranks we have summed.

$$T = 26.5 - \frac{5\,(5 + 1)}{2} = 11.5.$$

5. $11.5 > 8$.
6. Therefore we reject H_0.
7. Thus we conclude that the two samples come from different populations.

16.2.2 Large sample test

Consider the situation where the awareness of a company's product has been measured amongst groups of people in two different countries. Measurement is on a scale of 0–100, and each group has given a score in this range; the scores are shown below.

Country A: 21, 34, 56, 45, 45, 58, 80, 32, 46, 50, 21, 11, 18, 89, 46, 39, 29, 67, 75, 31, 48
Country B: 68, 77, 51, 51, 64, 43, 41, 20, 44, 57, 60

Test to discover whether the level of awareness is the same in both countries.

1. H_0: the levels of awareness are the same. H_1: the levels of awareness are different.
2. We will take a significance level of 5%.
3. Since we are using an approximation based on the Normal distribution, the critical values will be ± 1.96.
4. Ranking the values, we have:

Rank	Value	Sample	Rank	Value	Sample
1	11	A	16.5	46	A
2	18	A	18	48	A
3	20	B	19	50	A
4.5	21	A	20.5	51	B
4.5	21	A	20.5	51	B
6	29	A	22	56	A
7	31	A	23	57	B
8	32	A	24	58	A
9	34	A	25	60	B
10	39	A	26	64	B
11	41	B	27	67	A
12	43	B	28	68	B
13	44	B	29	75	A
14.5	45	A	30	77	B
14.5	45	A	31	80	A
16.5	46	A	32	89	A

Sum of ranks of A = 316.
Sum of ranks of B = 212. (minimum)
Therefore

$$T = 212 - \frac{11\,(11 + 1)}{2} = 146.$$

$$\text{Mean} = \frac{n_1 n_2}{2} = \frac{21 \times 11}{2} = 115.5.$$

$$\text{Standard error} = \frac{n_1 \times n_2\,(n_1 + n_2 + 1)}{12} = \frac{21 \times 11 \times (21 + 11 + 1)}{12} = 25.204.$$

Therefore

$$z = \frac{146 - 115.5}{25.204}$$
$$= 1.21$$

5. $1.21 < 1.96$.
6. Therefore we cannot reject the null hypothesis.
7. There appears to be no difference in the awareness of the product between the two countries.

16.3 WILCOXON TEST

This test is the non-parametric equivalent of the t-test for matched pairs and is used to identify if there has been a change in behaviour. This could be used when analyzing a set of panel results, where information is collected both before and after some event (for example, an advertising campaign) from the same people.

Here the basic premise is that whilst there will be changes in behaviour, or opinions, the ranking of these changes will be random if there has been no overall change (since the positive and negative changes will cancel each other out). Where there has been an overall change, then the ranking of those who have moved in a positive direction will be different from the ranking of those who have moved in a negative direction.

As with the Mann–Whitney test, where the sample size is small we shall need to consult tables to find the critical value (Appendix H); but where the sample size is large we can use a test based on the Normal distribution.

Whilst we will show how to conduct this test by hand calculation, again packages such as SPSS-X, SPSS-PC and MINITAB will perform the test by using the appropriate commands.

16.3.1 Small sample test

Consider the situation where a small panel of 8 members have been asked about their perception of a product before and after they have had an opportunity to try it. Their perceptions have been measured on a scale, and the results are given below.

Panel member	*Before*	*After*
A	8	9
B	3	4
C	6	4
D	4	1
E	5	6
F	7	7
G	6	9
H	7	2

You have been asked to test if the favourable perception (shown by a high score) has changed after trying the product.

1. H_0: There is no difference in the perceptions.
   H_1: There is a difference in the perceptions.
2. We will take a significance level of 5%, although others could be used.
3. The critical value is found from tables (Appendix H), here it will be 4.
4. To calculate the test statistic we find the differences between the two scores and rank them by absolute size (i.e. ignoring the sign). Any ties are ignored.

Before	*After*	*Difference*	*Rank*
8	9	+1	2
3	4	+1	2
6	4	−2	4
4	1	−3	5.5
5	6	+1	2
7	7	0	ignore
6	9	+3	5.5
7	2	−5	7

Sum of positive ranks = 11.5.
Sum of negative ranks = 16.5.

We select the minimum of these (i.e. 11.5) as our test statistic.

5. $11.5 > 4$.
6. Therefore we can reject the null hypothesis.
7. There has been a change in perception after trying the product.

16.3.2 Large sample test

Consider the following example. A group of workers have been packing items into boxes for some time and their productivity has been noted. A training scheme is initiated and the workers' productivity is noted again one month after the completion of the training. The results are shown below.

Person	*Before*	*After*	*Person*	*Before*	*After*
A	10	21	N	40	41
B	20	19	O	21	25
C	30	30	P	11	16
D	25	26	Q	19	17
E	27	21	R	27	25
F	19	22	S	32	33
G	8	20	T	41	40
H	17	16	U	33	39
I	14	25	V	18	22
J	18	16	W	25	24
K	21	24	X	24	30
L	23	24	Y	16	12
M	32	31	Z	25	24

1. H_0: there has been no change in productivity. H_1: there has been a change in productivity.
2. We will use a significance level of 5%.
3. The critical value will be ±1.96, since the large sample test is based on a Normal approximation.
4. To find the test statistic, we must rank the differences, as shown below:

Person	*Before*	*After*	*Difference*	*Rank*
A	10	21	+11	23.5
B	20	19	−1	5.5
C	30	30	0	ignore
D	25	26	+1	5.5
E	27	21	−6	21
F	19	22	+3	14.5
G	8	20	+12	25
H	17	16	−1	5.5
I	14	25	+11	23.5
J	18	16	−2	12
K	21	24	+3	14.5
L	23	24	+1	5.5
M	32	31	−1	5.5
N	40	41	+1	5.5
O	21	25	+4	17
P	11	16	+5	19
Q	19	17	−2	12
R	27	25	−2	12
S	32	33	+1	5.5
T	41	40	−1	5.5
U	33	39	+6	21
V	18	22	+4	17
W	25	24	−1	5.5
X	24	30	+6	21
Y	16	12	−4	17
Z	25	24	−1	5.5

(Note the treatment of ties in absolute values when ranking. Also note that n is now equal to 25.)

Sum of positive ranks = 218.
Sum of negative ranks = 107 (minimum)

The mean is given by:

$$\frac{n(n+1)}{4}$$

$$= \frac{25\,(25+1)}{4}$$

$$= 162.5.$$

The standard error is given by:

$$\frac{n(n+1)(2n+1)}{24}$$

$$= \sqrt{\frac{25\,(25+1)(50+1)}{24}}$$

$$= 37.165$$

Therefore the value of z is given by

$$z = \frac{107 - 162.5}{37.165}$$

$$= -1.493.$$

5. $-1.96 < -1.493$
6. Therefore we cannot reject the null hypothesis.
7. A month after the training there has been no change in the productivity of the workers.

16.4 RUNS TEST

This is a test for randomness in a dichotomized variable, for example gender is either male or female. The basic assumption is that if gender is unimportant then the *sequence* of occurrence will be random and there will be no long runs of either male or female in the data. Care needs to be taken over the order of the data when using this test, since if this is changed it will affect the result. The sequence could be chronological, for example the date on which someone was appointed if you were checking on a claimed equal opportunity policy. The runs test is also used in the development of statistical theory, for example, looking at residuals in time series analysis.

Whilst we will show how to conduct this test by hand calculation, packages such as SPSS-X, SPSS-PC and MINITAB will perform the test by using the appropriate commands.

EXAMPLE

With equal numbers of men and women employed in a department there have been claims that there is discrimination in choosing who should attend conferences. During April, May and June of last year the gender of those going to conferences was noted and is shown below.

Date	*Person attending*
April 10	Male
April 12	Female
April 16	Female
April 25	Male
May 10	Male
May 14	Male
May 16	Male
June 2	Female
June 10	Female
June 14	Male
June 28	Female

1. H_0: there is no pattern in the data (i.e. random order)
   H_1: there is a pattern in the data.
2. We will take a significance level of 5%.
3. Critical values are found from tables (Appendix I). Here, since there are 6 men and 5 women and we are conducting a two-tailed test, we can find the values of 3 and 10.
4. We now find the number of runs in the data:

Date	*Person attending*	*Run no.*
April 10	Male	1
April 12	Female	
April 16	Female	2
April 25	Male	
May 10	Male	
May 14	Male	
May 16	Male	3
June 2	Female	
June 10	Female	4
June 14	Male	5
June 28	Female	6

   Note that a run may constitute just a single occurrence (as in run 5) or may be a series of values (as in run 3). The total number of runs is 6.
5. $3 < 6 < 10$
6. Therefore we cannot reject the null hypothesis.
7. There is no evidence that there is discrimination in the order in which people are chosen to attend conferences.

16.5 CONCLUSIONS

Many of the tests which we have considered use the ranking of values as a basis for deciding whether or not to reject the null hypothesis, and this means that we can use non-parametric tests where only ordinal data is available. Such tests do not suggest that the parameters, such as the mean and variance, are unimportant, but that we do not need to know the underlying distribution in order to carry out the test. They are also called **distribution free tests**.

In general, non-parametric tests are less efficient than parametric tests, since larger samples are needed to reach the same level of confidence, but in many cases they are the only type of test that can be applied.

This chapter does not purport to have dealt with all of the non-parametric tests, but has given an overview of how such tests differ from those shown in Chapter 15 and illustrated this by using four such tests.

16.6 PROBLEMS

1. The respondents of a survey were classified by magazine read and income as follows:

Magazine read	*under 10 000*	*Annual income (£) 10 000 and under 15 000*	*15 000 and over*
A	28	60	57
B	12	40	53

   Test the hypothesis that magazine choice is independent of the level of income using a 5% level of significance.

2. In a survey concerned with changes in working procedures the following table was produced:

	Opinon on changes in working procedures		
	in favour	*opposed*	*undecided*
Skilled workers	21	36	30
Unskilled workers	48	26	19

   Test the hypothesis that the opinion on working procedures is independent of whether workers are classified as skilled or unskilled.

3. The table below gives the number of claims made in the last year by the 9650 motorists insured with a particular insurance company:

Number of claims	*Insurance groups*			
	I	*II*	*III*	*IV*
0	900	2 800	2 100	800
1	200	950	750	450
2 or more	50	300	200	150

   Is there an association between the number of claims and the insurance group?

4. A production process uses four machines in its three-shift operation. A random sample of breakdowns was classified according to machine and the shift in which the breakdown occurred:

Shift	*Machine*			
	A	*B*	*C*	*D*
1	10	11	8	9
2	16	9	13	11
3	12	9	14	9

   Is there reason to doubt the independence of shift and machine breakdown?

5. A random sample of 500 units is taken from each day's production and inspected for defective units. The number of defectives recorded in the last working week were as follows:

Day	*Number of defectives*
Monday	15
Tuesday	8
Wednesday	5
Thursday	5
Friday	12

   Test the hypothesis that the difference between the days is due to chance.

6. The number of breakdowns each day on a section of road were recorded for a sample of 250 days as follows:

Number of breakdowns	*Number of days*
0	100
1	70
2	45
3	20
4	10
5	5
	250

   Test whether a Poisson distribution describes the data.

7. The number of car repairs completed each month in the last year were recorded as follows:

Month	*Number of repairs*	*Month*	*Number of repairs*
January	95	July	108
February	98	August	95
March	92	September	94
April	90	October	92
May	83	November	97
June	102	December	94

Is there reason to believe that the number of car repairs does not follow a uniform distribution?

8. The demand for hire cars from a specialist company has been tabulated for the last 100 working weeks.

Demand for hire cars	*Number of weeks*
0	39
1	32
2	19
3	10

Does demand follow a known distribution?

9. The average weekly overtime earnings from a sample of workers from a particular service industry were recorded as follows:

Average weekly overtime earnings (£)	*Number of workers*
under 1	19
1 but under 2	29
2 but under 5	17
5 but under 10	12
10 or more	3
	80

Do average weekly overtime earnings follow a Normal distribution?

10. Eggs are packed into cartons of six. A sample of 90 cartons is randomly selected and the number of damaged eggs in each carton counted.

Number of damaged eggs	*Number of cartons*
0	52
1	15
2	8
3	5
4	4
5	3
6	3

Does the number of damaged eggs in a carton follow a Binomial distribution?

11. Applications for posts in a major company are received from both men and women. The company has an equal opportunities policy but suspects that women are inhibited from applying in some cases. It has therefore monitored the gender of applicants for a series of posts and the results are shown below.

	Number of applicants
Female	5, 10, 11, 21, 15, 17, 21, 10, 14,
Male	30, 8, 8, 12, 20, 22, 32, 25, 8, 6, 12, 6

(Note that some posts had no women applicants.)

Use a Mann–Whitney test at the 5% level of significance to find if there is a significant difference in the number of applications from men and women.

12. Appointments depend on qualifications, experience and personal qualities. In an effort to distinguish the rôle of experience, a placement agency have noted the number of years of management experience held by people sent to interviews for senior posts. It has tabulated this against whether or not they obtained the job. The results are shown in the following table.

	Number of years of management experience
Appointed	14, 16, 19, 40, 21, 10, 6, 11, 30, 35
Not appointed	17, 23, 3, 2, 1, 7, 5, 24, 15, 7, 5, 20, 1, 3, 9, 12, 8, 1, 4, 2, 13

Use a Mann–Whitney test to find if the experience of those appointed is greater than that of those not appointed.

13. Various positions on a committee became vacant and elections were held. Votes could be split into 'Right' and 'Left'; and those given to the successful candidates are shown below.

Position	*A*	*B*	*C*	*D*	*E*	*F*	*G*	*H*	*I*
'Right'	10	30	27	42	39	16	14	15	28
'Left'	17	12	31	29	50	11	18	33	9

Use a Mann–Whitney test to find if the new committee is mainly supported by the 'Right'.

14. Attitudes to 'Green' issues can be characterized into conservative and radical. During the building of a new airport, the percentages who held radical views on the possible harmful effects on the environment were canvassed in local towns and villages, and are recorded in the following table.

	Percentages holding radical views
Villages	31, 31, 50, 10, 12, 17, 19, 17, 22, 22, 23, 27, 42, 17, 5, 6, 8, 24, 31, 15
Towns	30, 18, 25, 41, 37, 30, 29, 43, 51

Test at the 1% level of significance if there is a difference in the level of support for a radical view between the villages and the towns.

15. A national charity has recently held a recruitment drive and to test if this has been successful has monitored the membership of ten area groups before the campaign and three months after the campaign.

Local area group:	*A*	*B*	*C*	*D*	*E*	*F*	*G*	*H*	*I*	*J*
Before	25	34	78	49	39	17	102	87	65	48
After	30	38	100	48	39	16	120	90	60	45

Use a Wilcoxon test, at the 5% level of significance, to find if there has been an increase in local group membership.

16. Local public support for a company's environmental policy was tested in the areas close to its various factories before and after a national advertising campaign aimed at increasing the company's environmentally friendly image. The results from the twenty factories are shown in the accompanying table.

Factory	*Before*	*After*	*Factory*	*Before*	*After*
1	50	53	11	63	66
2	48	53	12	62	63
3	30	28	13	70	70
4	27	25	14	61	60
5	49	50	15	57	60
6	52	56	16	51	50
7	48	47	17	44	40
8	54	59	18	42	41
9	58	59	19	47	50
10	60	62	20	30	24

Use a Wilcoxon test at the 5% significance level to find if there has been a change in local support for the company's policy. (NB. Use the Normal approximation.)

17. Employees from a company are tested before and after a training course. The results for ten employees are shown below.

Employee	*Before*	*After*
A	10	14
B	12	13
C	13	14
D	15	14
E	17	18
F	17	19
G	18	16
H	9	15
I	5	4
J	3	1

Test at the 2.5% level of significance if there has been a general increase in the employees' abilities following the training course.

18. A panel survey asked participants to rate the appearance of a company's product before a change in packaging and again after the change had been introduced. Each participant rated the product on a scale of 0–100.

Participant	*Before*	*After*	*Participant*	*Before*	*After*
A	80	85	K	37	40
B	75	82	L	55	68
C	90	91	M	80	88
D	65	68	N	85	95
E	40	34	O	17	5
F	72	79	P	12	5
G	41	30	Q	15	14
H	10	0	R	23	25
I	16	12	S	34	45
J	22	16	T	61	80

Carry out a Wilcoxon test at the 5% level of significance to find if the new packaging has a more favourable reaction amongst consumers.

19. To test the claimed equal opportunities policy of a company the ethnic origin of successful candidates has been monitored for one year. (Candidates are classified merely as 'English' and 'Non-English'.)

Date of Appointment	*Ethnic origin*
1st February	E
7th February	Non-E
4th March	Non-E
5th March	Non-E
30th March	E
5th May	Non-E
10th June	E
18th June	E
22nd August	Non-E
29th August	E
1st September	Non-E
14th November	Non-E
18th December	E

Use a runs test at the 5% level of significance to test if there is any bias in selection of candidates. What criticisms would you have about the way in which the monitoring was undertaken?

20. Items produced by a certain process can be classified as acceptable (A) or defective (D). Twenty three such items from a particular machine are checked and the sequence below was obtained.

D, D, D, D, A, A, A, A, A, A, A, D, D, D, D, A, A, A, A, D, D, D, D.

Test if the sequence is random.

PART FIVE CONCLUDING EXERCISE

A shopping survey has been conducted in a particular town and the results are given in the file DATA2 on the accompanying disk. The questionnaire used is reproduced here.

Shopping Survey

Questionnaire Number: . . . .   Note Sex: Male/Female

1. How often do you visit this shopping centre?
   Once a day   Once a week
   Twice a week   Less often than this
2. Have you ever visited Hamblug's shop? Yes/No
   IF NO, go to Qu. 8
3. Have you been there today? Yes/No
   IF NO, go to Qu. 8
4. Did you buy anything there? Yes/No
   IF NO, go to Qu. 8
5. Did you buy anything in the food department? Yes/No
   IF NO, go to Qu. 8
6. Which of the following items did you buy?
   Fresh vegetables Yes/No
   Fresh fruit Yes/No
   Fresh meat Yes/No
   Preprepared meals Yes/No
   Frozen food Yes/No
7. Approximately how much did you spend on food at Hamblug's today? (TICK ONE)
   under £1 ☐
   £1 but under £3 ☐
   £3 but under £5 ☐
   £5 but under £10 ☐
   £10 but under £20 ☐
   £20 or more
8. How many people do you buy food for?
   (WRITE IN NUMBER) . . . .
9. Which area do you live in? (TICK ONE)
   Astrag ☐   Duffold ☐
   Baldon ☐   East Heckham ☐
   Cleardon ☐   Other ☐

The coding shown in the table opposite was used.

1. Use the data to find out about the shopping habits of these respondents, especially in relation to Hamblug's store. It would be useful to find out how many people gave each answer to each question (use the HIST command in Microstats).
2. From this survey, estimate the 95% confidence interval for the percentage of shoppers who have ever visited Hamblug's.
3. Estimate the 95% confidence interval for the average amount spent on food in Hamblug's by all shoppers.
4. Conduct a hypothesis test at the 5% level of significance to find if shoppers who buy food at Hamblug's spend above £17 on average.
5. Conduct a hypothesis test at the 5% level of significance to find if a higher percentage of shoppers buy fresh vegetables than buy frozen food.

Column Number	*Description*	*Coding used*
1	Respondent number	
2	Sex	1 = male; 2 = female
3	Qu.1	1 = once a day 2 = twice a week 3 = once a week 4 = less often
4	Qu.2	1 = Yes; 2 = No
5	Qu.3	1 = Yes; 2 = No; 9 = Not asked
6	Qu.4	1 = Yes; 2 = No; 9 = Not asked
7	Qu.5	1 = Yes; 2 = No; 9 = Not asked
8	Qu.6 (a)	1 = Yes; 2 = No; 9 = Not asked
9	Qu.6 (b)	1 = Yes; 2 = No; 9 = Not asked
10	Qu.6 (c)	1 = Yes; 2 = No; 9 = Not asked
11	Qu.6 (d)	1 = Yes; 2 = No; 9 = Not asked
12	Qu.6 (e)	1 = Yes; 2 = No; 9 = Not asked
13	Qu.7	0.55 under £1 1.5 £1 but under £3 4.0 £3 but under £5 7.5 £5 but under £10 15.0 £10 but under £20 35.0 £20 and more 0 Not asked (Note use of mid-points here)
14	Qu.8	number given
15	Qu.9	1 = Astrag 2 = Baldon 3 = Cleardon 4 = Duffold 5 = East Heckham 6 = Other

6. Find the cross-tabulation of amount spent (Qu.7) by number of people shopped for (Qu.8), and thus conduct a chi-squared test for association at the 5% level of significance. (NB. You will need to exclude those who did not buy food by copying the relevant parts of column 13 into a new column. When you produce the cross-tabulation, put it into some empty columns, so that you can conduct the test.)

7. Find a cross-tabulation of amount spent (Qu.7) by area lived in (Qu.9), and thus conduct a chi-squared test for association at the 5% level of significance.
(NB. You will need to exclude those who did not buy food by copying the relevant parts of column 13 into a new column. When you produce the cross-tabulation, put it into some empty columns, so that you can conduct the test.)

PART SIX
RELATING TWO OR MORE VARIABLES

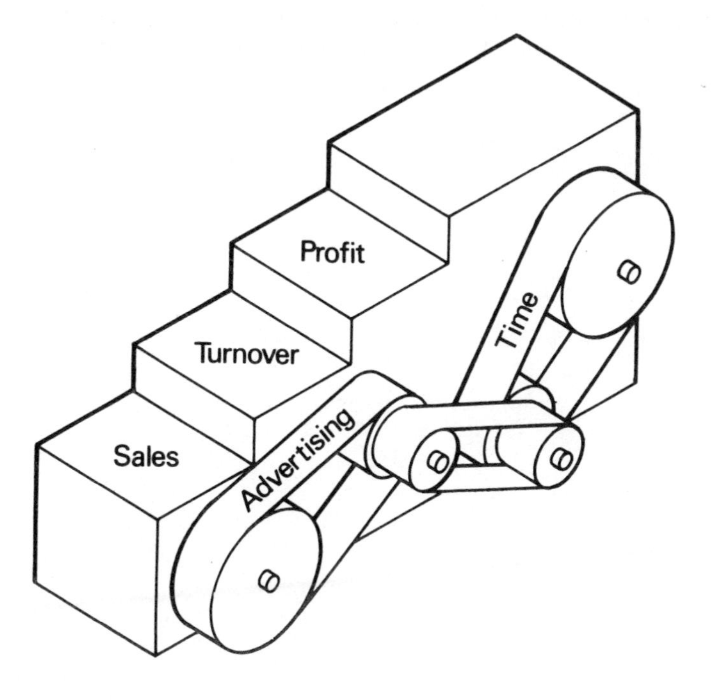

Most of the techniques and applications that have been discussed so far have been concerned with describing or analyzing the behaviour of a single variable. In this part of the book we will consider ways of relating variables together, looking for movements in one which are associated with movements in another, or maybe, just related to the passage of time. In considering various aspects of business, for example economics, accountancy or marketing, you will have seen that most outcomes are the result of a whole host of inter-related factors. This section looks at ways of identifying such relationships and goes on to attempt to predict the outcome of one variable if the values of other variables are known. Such prediction is essential in business to allow planning for production, stock control or the development of marketing strategies.

EXERCISES

1. For each of the following variables state the most important single factor which you think determines its value:
   (a) the balance of payments figure;
   (b) new car registrations;
   (c) deaths from heart disease;
   (d) your income in 10 years' time.
   List other factors which will affect these variables and in each case state whether or not it might be possible to collect quantitative data on these factors.
2. Find examples of two series from government publications (e.g. *Economic Trends* or the *Monthly Digest of Statistics*) which are presented in a seasonally adjusted format.
3. Suggest two things that you do which vary depending on the time of year.

CORRELATION 17

Numerical information on almost every aspect of life seems to be collected by someone; the major source of collected data in the UK is the government statistical service. Most of this data is published, in one form or another, and thus is available to business to help in their planning or analysis. Businesses themselves, whether they are profit making or non-profit making, also produce vast amounts of data. Such data may be collected at a single point in time, as with a survey, or at regular time intervals, as with the export figures.

It seems obvious that there will be relationships between various sets of data, and our aim in this chapter is to discover if a relationship exists between two sets of data, and if so, how strong it is, whether it applies to the data at any point in time or only for certain periods, and whether or not it is better than some other relationship.

17.1 SCATTER DIAGRAMS

Visual presentation of information, as we have seen in Chapter 2, allows an immediate impression to be gained about some data set. Where we are dealing with two variables, the appropriate method of illustration is the scatter diagram. A scatter diagram allows us to show two variables together, one on each axis, each pair being represented by a cross on a graph. Where there is only a very limited number of pairs of data, little can be inferred from such a diagram, but in most business situations there will be a large number of observations and we may distinguish some pattern in the picture obtained. The lack of a pattern however, may be just as significant.

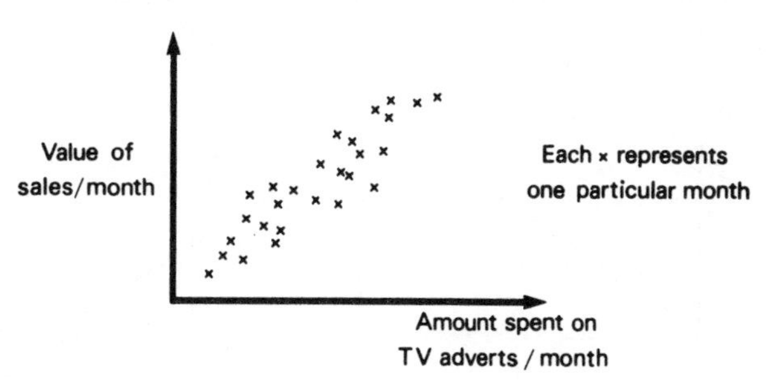

Figure 17.1

Looking at Figure 17.1 we can see that, in general, the larger the amount spent on TV advertising the larger the sales value. This is by no means a deterministic relationship and does not imply a necessary cause and effect: it just means that in months with high sales value, there is usually a large amount spent on TV advertising.

Figure 17.2 shows the relationship between the number of employees employed with a fixed capital investment and the total output of the plant: here we see that there is somewhat more of a relationship than in the previous example, since the points lie within a much narrower band. This type of result is expected if the law of diminishing returns is true. The law says that as more and

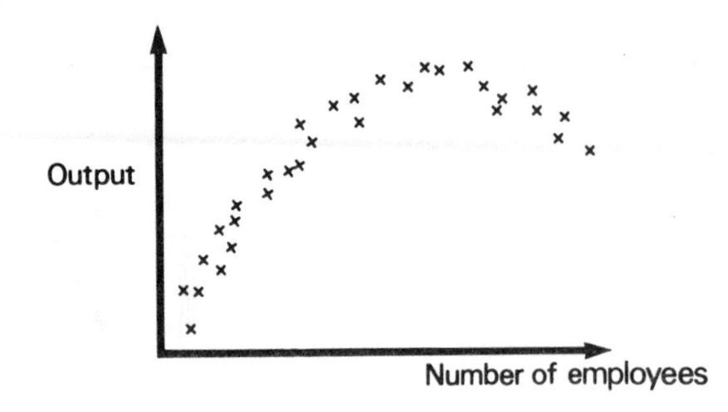

Figure 17.2

more of one factor of production is used, with fixed amounts of the other factors, there will come a point when total output will fall.

Having established that some relationships can be seen from scatter diagrams, we now need to find some 'extreme' relationships so that we may compare future diagrams with given standards.

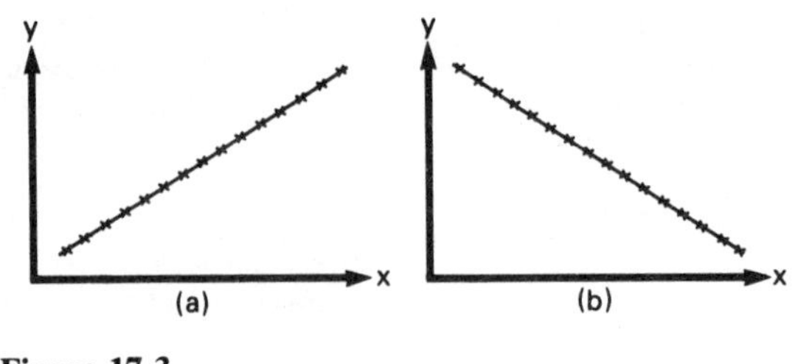

Figure 17.3

Looking at the two diagrams in Figure 17.3, we see situations where all of the data points fall on a straight line. Such a relationship is known as a *perfect* or *determinstic* linear relationship. Part (a) of the diagram shows a positive linear relationship, since as the value of x increases, so does the value of y; in fact, for a one unit rise in x, there is always a given increase in y. A negative linear relationship is illustrated in part (b) of Figure 17.3, where an increase in the value of x is matched by a decrease in the value of y.

Although the signs of the relationships shown in Figure 17.3 are different, they both represent a perfect relationship, and thus can be thought of a one extreme. The opposite of a perfect relationship between x and y would be one where there is no relationship between the two variables, and this is illustrated in Figure 17.4 below.

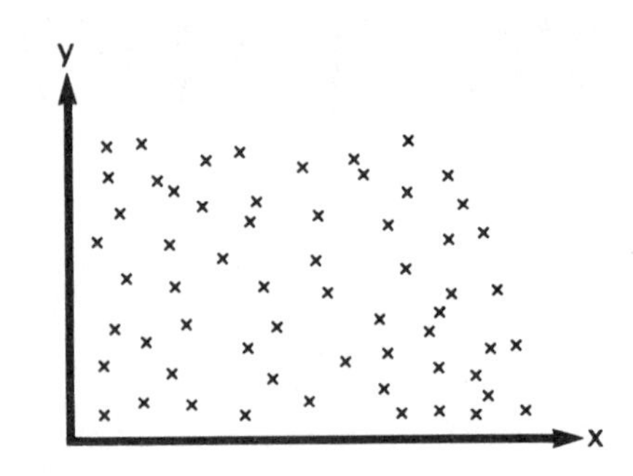

Figure 17.4

These three cases form a basis for comparison with actual data which we may collect, or which may be generated by a business. It is unlikely that any real situation will give an exact match to any one of these scatter diagrams, but they will allow us to begin our interpretation of the relationship before going into detailed calculations. (They may also help to avoid silly mistakes such as having a scatter diagram which shows a positive relationship and, through a small error, producing a numerical calculation which gives a negative answer.)

Scatter diagrams are often dismissed in favour of numerical calculations, but they do have a significant rôle to play in the analysis of bivariate relationships. Where the scatter diagram is similar to Figure 17.4 there may be little point in going on to work out a numerical relationship since no amount of information on likely values of x would allow you to make predictions of the behaviour of y. Similarly, if a particular scatter diagram is similar to Figure 17.2, there may be no point in calculating a linear relationship since the diagram shows a non-linear situation. Thus scatter diagrams can save considerably in the time needed to analyze a situation.

Scatter diagrams may be obtained from MICROSTATS by using the PLOT command. If we have the x values in Column 1 and the y values in column 2, then we would type: PLOT C2 C1.

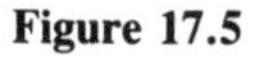
Figure 17.5

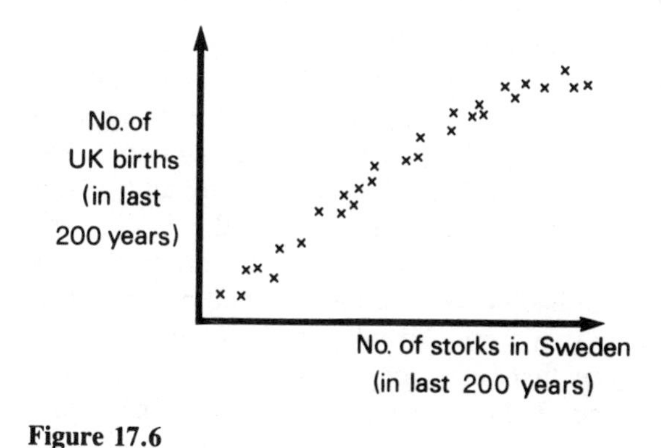

Figure 17.6

17.2 CAUSE AND EFFECT RELATIONSHIPS

It is very tempting, having plotted a scatter diagram such as Figure 17.5 which suggests a strong linear relationship, then to go on to say that wage rises cause rises in prices, i.e. inflation. Others, however would argue from this diagram that it is price rises that cause wage rises.

EXERCISE

Is either assertion correct? Are they both correct?

Whilst it is unlikely that there will ever be an answer to these questions that everyone can agree with, some consideration of cause and effect will help in understanding the relationship.

Only one of the variables can be a cause at a particular point in time, but they could both be effects of a common cause! If you increase the standard of maternity care in a particular region, you will, in general, increase the survival rate of babies: it is easy to identify the direction of the relationship as it is extremely unlikely that an increasing survival rate would encourage health authorities to increase spending and improve standards! But would you agree on the direction of causation in Figure 17.6?

Here there appears to be a correlation, but it is spurious as *both* variables are changing over *time*. Time itself cannot be a cause, except perhaps in the ageing process, but where a situation of this type occurs, or where we have two effects of a **common cause**, there will be no way of controlling, or predicting, the behaviour of one variable by taking action on the other variable. An example of two effects of a common cause would be ice-cream sales and the grain harvest, both being affected by the weather.

Whilst it cannot be the intention of this chapter to look deeply into the philosophy of cause and effect it will help to pick out a few conclusions from that subject. David Hume, in the eighteenth century, suggested that

> 'all reasonings concerning matters of fact seem to be founded on the relation of cause and effect'

He then went on to prove that this is not rationally justified. Which matters are justified in using cause and effect has interested many philosophers.

Conditions, or variables, can be divided into two types, **stable** and **differential**, and it is the second type which is most likely to be identified as a cause. New roadworks on the M1 at Watford Gap will be cited as a cause of traffic jams tailing back to Coventry, since the roadworks are seen as a changed situation from the normally open three lane road. Drug abuse will be cited as a cause of death or illness, since only taking the recommended dosage is seen as normal (in some cases this will be a nil intake e.g. heroin), and thus the alteration of normal behaviour is seen as causing the illness or death. There is also the temporal ordering of events; an 'effect should not precede a cause' and usually the events should be close together in time. A long time interval leads us to think of chains of causality. It is important to note the timing of events, since their recording must take place at the same time; for example, the studies by Doll and Hill of smoking and deaths from lung cancer conducted in the UK in the 1950s, which asked how many cigarettes per day people had smoked in the past and not just how many they were currently smoking. Some events may at first sight appear to be the 'wrong' way around. If most newspapers and commentators are predicting a sharp rise in VAT at a forthcoming budget in, say, 6 weeks' time, then many people will go out to buy fridges, washing machines and video recorders before budget day. If asked, they will say that the reason for their purchase is the rise in VAT in the budget in 6 weeks' time; however, if we look a little more closely, we see that it is their expectation of a VAT rise, and not the rise itself, which has caused them to go out and buy consumer durables. This however, leads us to a further problem: it is extremely difficult to measure expectations!

A final, but fundamental question that must be answered is 'does the relationship make sense?'

Returning now to the problem in Figure 17.5, we could perhaps get nearer to cause and effect if we were to lag one of the variables, so that we relate wage rises in one month to price rises in the previous month: this would give some evidence of prices as a cause. The exercise could be repeated with wage rises lagged one month behind price increases — giving some evidence of wages as a cause. In a British context of the 1970s and 1980s both lagged relationships will give evidence of a strong association. This is because the wage-price-wage spiral is quite mature and likely to continue, partly through cause and effect, partly through expectations and partly through institutionalized factors.

EXERCISES

1. Give two other examples of variables between which there is a spurious correlation apart from those mentioned in the text.
2. What are the stable and differential conditions in relation to a forest fire?

17.3 MEASURING LINEAR ASSOCIATION

Pictorial representation of the relationship between two variables is used as a first stage in the process of developing a model of the behaviour of the variables. To illustrate a verbal presentation which aims to show an overall picture of the situation, scatter diagrams may be sufficient, but for a more detailed analysis we must take the next step and find a way of measuring the strength of the relationship. Such measurement will enable us to make comparisons between different models or proposed explanations of the behaviour of the variables. The measure which we will calculate is called the **coefficient of correlation**.

As we have already seen in Figure 17.3, an extreme type of relationship is one where all of the points of the scatter diagram lie on a straight line. Data which gave the scatter diagram shown in Figure 17.3(a) would have a correlation coefficient of $+1$, since y increases as x increases; whilst the correlation coefficient from Figure 17.3(b) would be -1. Where there is no relationship between the variables, the coefficient of correlation would be equal to zero (for example in Figure 17.4). Each of these situations is very unlikely to exist in practice. Even if there were to be a perfect, linear relationship between two variables, there would probably be some measurement errors and the data which we collected would give an answer between -1 and $+1$. Data generated by businesses and the government is often affected by the same underlying movements and disturbances and this will mean that even where there is no real relationship between two variables (i.e. no cause and effect), there may still be a correlation coefficient that differs from zero.

Finally, before we begin the process of calculation, we should make a distinction between cardinal and ordinal data. Although the method of calculation is basically the same, we must remember that the type of data will affect the way in which we interpret the answers calculated. Note that when we are using a statistical package, such as MICROSTATS, we may be able to use exactly the same commands whichever type of data we have; if, however, you have to perform the calculation by hand, there is a simplified formula which applies only to the case of ordinal data. (Note also that certain corrections may be applied to the calculation of the correlation coefficient from ordinal data by some statistical packages.)

17.3.1 Rank correlation

Ordinal data consists of values defined by the position of the data in an ordered list (a rank) and may be applied to situations where no numerical measure can be made, but where best and worst, or most favoured and least favoured can be identified. They may also be used for international comparisons where the use of exchange rates to bring all of the data into a common currency would not be justified. Rankings are often applied where people or companies are asked to express preferences or to put a series of items into an order. In these situations it is no longer possible to say how much better first is than second; it may be just fractionally better, or it may be outstandingly better. We therefore need to exercise caution when interpreting a rank correlation coefficient (or Spearman's correlation coefficient) since similar ranking, and hence fairly high correlation may represent different views of the situation.

Table 17.1 Results of a survey which asked American and European companies to rank the following areas in relation to their importance for company success

	America	*Europe*
Research and development	4	2
Age of plant	2	4
Advertising spend	3	1
Promotional strategy	7	6
Product packaging	8	7
Salaries	5	8
Infrastructure of country	6	5
Performance of the economy	1	3

Suppose the rankings in Table 17.1 show how the companies on the two continents view the various areas in relation to their success. We are trying to answer the question 'Are there similarities in the rankings given by the two groups of companies?', or 'Is there an association between the rankings?'. To do this we will use Spearman's formula:

$$r_S = 1 - \frac{6\Sigma d^2}{n(n^2-1)}.$$

The first step is to find the difference in the rank given to each item (d = American − European), which will give us a series of positive and negative numbers which add up to zero. To overcome this, we will square each number and then sum these squared values before putting the answer into the formula. This calculation is shown in Table 17.2. Such calculations are easily performed by hand, or could be done on a spreadsheet.

Table 17.2

	America	*Europe*	*d*	d^2
Research and development	4	2	2	4
Age of plant	2	4	−2	4
Advertising spend	3	1	2	4
Promotional strategy	7	6	1	1
Product packaging	8	7	1	1
Salaries	5	8	−3	9
Infrastructure of country	6	5	1	1
Performance of the economy	1	3	−2	4
			0	28

Using the formula given above, we now have

$$\begin{aligned} r_s &= 1 - \frac{6 \times 28}{8(64-1)} \\ &= 1 - \frac{168}{504} \\ &= 1 - 0.3333 \\ &= 0.6667. \end{aligned}$$

This answer suggests that there is a fairly high degree of agreement between the two groups of companies on the most important factors for success; however the limitations mentioned above must be borne in mind.

Rank correlation would not, normally, be used with cardinal data since information would be lost by moving from actual measured values to simple ranks. However, if one variable was measured on the cardinal scale and the other was measured on the ordinal scale, ranking both and then calculating the rank correlation coefficient may prove the only practical way of dealing with these mixed measurements. Rank correlation also finds applications where the data is cardinal but the relationship is non-linear.

In MICROSTATS we would place one set of ranks in column 1 and the other in column 2; we can then use the command, CORR C1 C2.

17.3.2 Correlation for cardinal data

Where we do have quantitative measures of the variables which do not rely on subjective judgements we can calculate a correlation coefficient developed by Pearson from which we may draw much firmer conclusions. Such a coefficient will define the amount of association between the two sets of values, a value close to either +1 or −1 indicating a high degree of correlation. Remember however that a high degree of correlation does not necessarily mean that a cause and effect relationship exists between the variables under consideration.

Pearson's coefficient of correlation is given by

$$r = \frac{n\Sigma xy - \Sigma x \Sigma y}{\sqrt{\{[n\Sigma x^2 - (\Sigma x)^2][n\Sigma y^2 - (\Sigma y)^2]\}}}$$

Although this formula looks very different from the Spearman's formula, they can be shown to be the same (see section 17.10), since in the Spearman's case, the sum of x is equal to the sum of y and the sum of x^2 is equal to the sum of y^2. For those who are more technically minded, the formula is the covariance of x and y (how much they vary together), divided by the root of the product of the variance of x and the variance of y (how much they each, individually, vary). The derivation is shown in section 17.9.

Taking an example of the turnover and the profit levels of ten companies from a particular industry given in Table 17.3 we can calculate the correlation coefficient between them.

Table 17.3 Turnover and profit figures

Turnover (£m)	*Profit (£m)*
30.0	3.0
25.5	2.8
6.7	1.1
45.2	5.3
10.5	0.6
16.7	2.1
20.5	2.1
21.4	2.4
8.3	0.9
70.5	7.1

If these two sets of figures are put into two columns of MICROSTATS we may use the command CORRelation C1 C2 to obtain the answer

Table 17.4 Calculation of the correlation coefficient

Turnover (x)	*Profit (y)*	x^2	y^2	xy
30.0	3.0	900.00	9.00	90.00
25.5	2.8	650.25	7.84	71.40
6.7	1.1	44.89	1.21	7.37
45.2	5.3	2 043.04	28.09	239.56
10.5	0.6	110.25	0.36	6.30
16.7	2.1	278.89	4.41	35.07
20.5	2.1	420.25	4.41	43.05
21.4	2.4	457.96	5.76	51.36
8.3	0.9	68.89	0.81	7.47
70.5	7.1	4 970.25	50.41	500.55
255.3	27.4	9 944.67	112.30	1 052.13

$$r = \frac{10 \times 1052.13 - (255.3)(27.4)}{\sqrt{\{[10 \times 9944.67 - (255.3)^2][10 \times 112.3 - (27.4)^2]\}}}$$

$$= \frac{10521.3 - 6995.22}{\sqrt{\{[99446.7 - 65178.09][1123 - 750.76]\}}}$$

$$= \frac{3526.08}{\sqrt{\{34268.61 \times 372.24\}}} = \frac{3526.8}{3571.575}$$

$$= 0.98726$$

that their correlation coefficient is 0.9873. Alternatively, we could place the figures into a spreadsheet and then create columns to calculate the various summations that we require for the formula. You may, of course, be expected to be able to find the answer by using hand calculation, and this is shown here. (Note that the columns shown in Table 17.4 are those that you would obtain when using a spreadsheet for the calculation.)

Thus there is a high degree of positive correlation between turnover and profit levels in the companies in this industry since the coefficient is close to plus one.

EXERCISE

Obtain a scatter diagram for this data.

17.4 MEASURING NON-LINEAR ASSOCIATION

So far we have been limited to simple linear relationships, but many interconnections in business and economics (and elsewhere) are not linear. In order to measure association, we will still use Pearson's correlation coefficient, but will adjust it to allow for the non-linearity of the data. We must return to the scatter diagram in order to decide upon the type of non-linearity present. (It is beyond the scope of this book to look at every type of non-linearity, but a few of the most commonly met types are given below.)

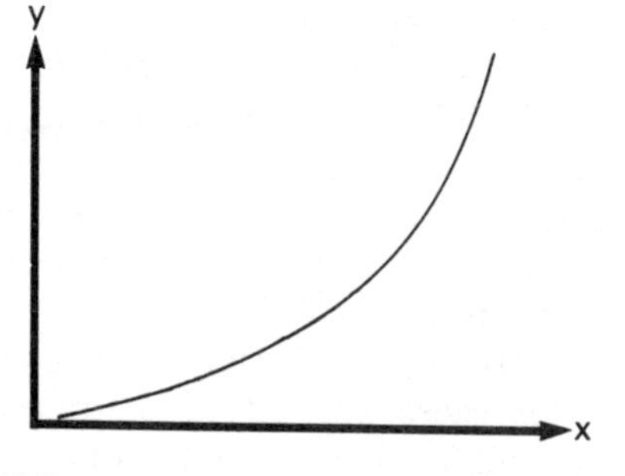

Figure 17.7

If the scatter diagram shows the relationship in Figure 17.7, then we will find that by taking the log of y, we get a linear function as in Figure 17.8. When performing the calculations, we will take the log of y and then use this value exclusively, as in Table 17.5. (Note that we could use either natural logs or logs to base 10. Here logs to base 10 are used, to two decimal places.)

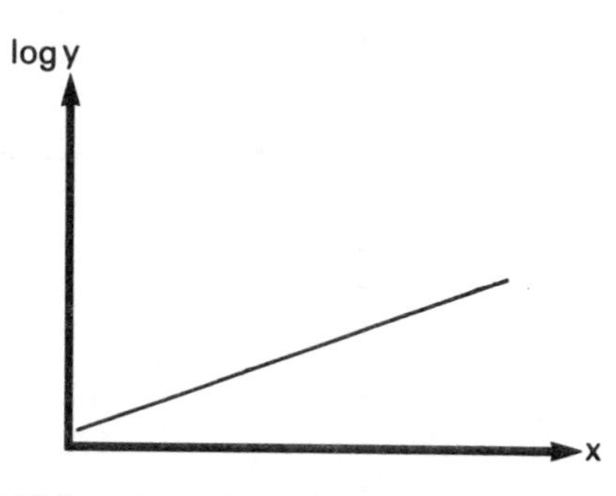

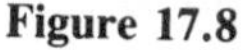

Figure 17.8

Table 17.5

x	y	$\log_{10}y$	x^2	$(\log_{10}y)^2$	$x(\log_{10}y)$
1	20	1.30	1	1.6900	1.30
2	30	1.48	4	2.1904	2.96
3	50	1.70	9	2.8900	5.10
7	100	2.00	49	4.0000	14.00
10	150	2.18	100	4.7524	21.80
11	200	2.30	121	5.2900	25.30
14	260	2.42	196	5.8564	33.88
14	400	2.60	196	6.7600	36.40
16	400	2.60	256	6.7600	41.60
17	700	2.85	289	8.1225	48.45
95		21.43	1221	48.3117	230.79

$$r = \frac{(10)(230.79) - (95)(21.43)}{\sqrt{[\{(10)(1221) - 95^2][(10)(48.3117) - 21.43^2]\}}}$$

$$= \frac{272.05}{\sqrt{\{3185 \times 23.8721\}}}$$

$$= 0.9866 \qquad \text{(with } x \text{ and } y,\ r = 0.884)$$

We have carried out a **transformation to linearity**.

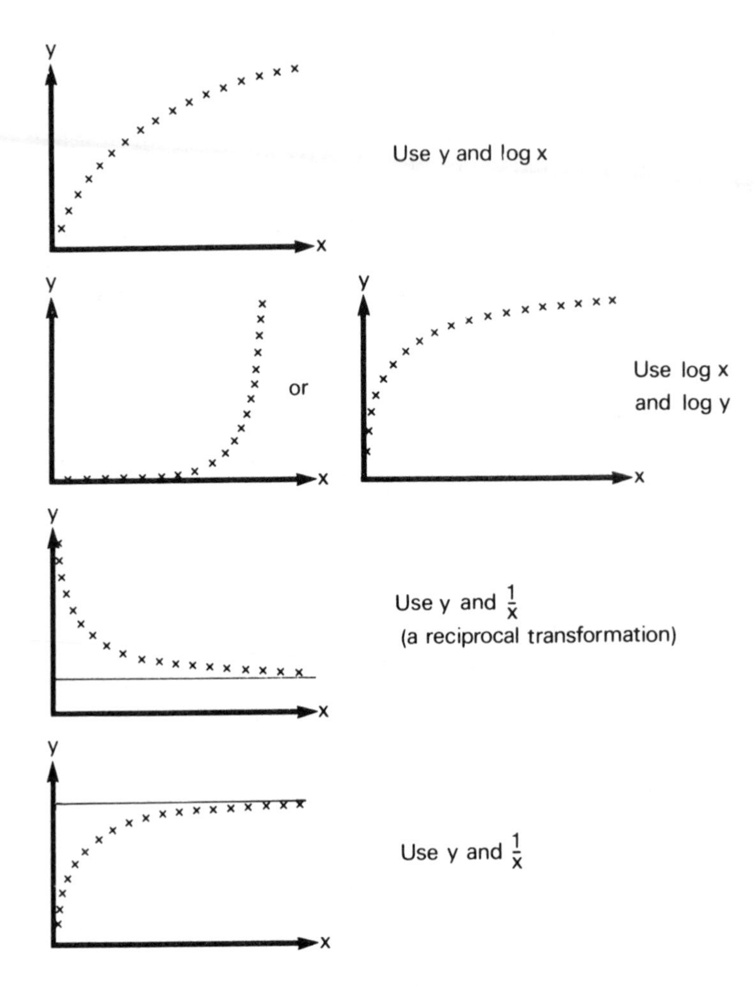

Figure 17.9

Figure 17.9 illustrates some other common transformations which yield linear correlation when the untransformed data have the forms shown in these scatter diagrams.

These scatter diagrams are only given as guidance in selecting a transformation for a particular problem on which you are working. If there is high correlation, it will usually be easy to select the most appropriate transformation, but when there is a low correlation, signified by a high scatter of points, it may be necessary to use trial and error to find the 'best' transformation. This is considerably less difficult than it appears at first, since most computer systems have a regression and correlation package available, which will also transform the data in many different ways.

17.5 TESTING THE SIGNIFICANCE OF THE CORRELATION

We have talked about 'high' and 'low' correlations, but the statistical significance of a particular numerical value will vary with the number of pairs of observations available to us. The smaller the number of observations, the higher must be the value of the correlation coefficient to establish association between the two sets of data, and thus, more generally, between the two variables under consideration. Showing that there is a significant correlation will still not be conclusive evidence for cause and effect but it will be a necessary condition before we propound a theory of this type.

As we have seen in Chapter 15, a series of values can be thought of as a sample from a wider population, and thus we can use a test of hypothesis to evaluate the results we have obtained. In the same way that we were able to test a sample mean, we may also test a correlation coefficient; here the test being to find if the value obtained is significantly greater than zero. The correlation coefficient is treated as the sample value, r, and the population (or true) value of correlation is represented by the letter ρ (rhō). Statistical theory shows that whilst the distribution of the test statistic for r (see step 4 below) is symmetrical, it is narrower than a Normal distribution, and in fact follows a t-distribution.

Using the methodology proposed in Chapter 15, we may conduct the test. Suppose that we have obtained a linear correlation coefficient of 0.65 from ten pairs of data.

1. State hypotheses. $H_0: \rho = 0$
   $H_1: \rho > 0$
   (Note that we are using a one-tailed test.)
2. State the significance level: we will use 5%.
3. State the critical value: 1.86.
   (Note that there are $(n-2) = 8$ degrees of freedom for this test, and that the critical value is obtained from Appendix D.)
4. Calculate the test statistic:

$$t = \frac{r\sqrt{(n-2)}}{\sqrt{(1-r^2)}}$$
$$= \frac{0.65\ \sqrt{(10-2)}}{(1-0.65^2)}$$
$$= \frac{1.8385}{0.5775}$$
$$= 3.184$$

5. Compare the values: $3.184 > 1.86$.
6. Conclusion: we may reject H_0.
7. Put the conclusion into English. Since the value of the calculated statistic is above the critical value we may conclude that there is a correlation between the two sets of data which is significantly above zero.

It would be possible to use a two-tailed test to try to show that there was a correlation between the two sets of data, but in most circumstances it is preferable to use a one-tailed test.

With the increasing use of computer packages it has become very easy to construct models of data and to obtain correlation coefficients between pairs of data. A situation frequently faced is where we are trying to find a variable associated with some variable we are trying to predict. Data is obtained and the correlation calculated but the question arises as to whether a model with a slightly higher correlation is significantly better than one with a slightly lower correlation.

Similarly, we may have data from different regions or countries and want to know if there is more correlation between two specific variables in one of these.

Suppose that we have obtained the following information:

Sample 1: From a sample of 10 companies in England, the correlation between turnover and profit is 0.76.

Sample 2: From a sample of 15 companies in Scotland, the correlation between turnover and profit is 0.81.

Is there more association between these two variables in Scottish companies than in English companies?

Using our established testing procedure:

1. Hypotheses: H_0: $\rho_1 = \rho_2$
   H_1: $\rho_2 < \rho_2$
2. Significance level is 5%.
3. Critical value is -1.645.
   (Note that here we revert to using values from the Normal distribution.)
4. Calculate the test statistic: (*this is fairly lengthy*)
   (a) Calculate the values of z_r.

$$z_{r1} = 0.5 \ln \left[\frac{(1+r_1)}{(1-r_1)}\right]$$

$$z_{r2} = 0.5 \ln \left[\frac{(1+r_2)}{(1-r_2)}\right]$$

where ln is the natural log.
For our data this gives:

$$z_{r1} = 0.5 \ln \left[\frac{(1+0.76)}{(1-0.76)}\right] = 0.9962.$$

$$z_{r2} = 0.5 \ln \left[\frac{(1+0.81)}{(1-0.81)}\right] = 1.1270.$$

(b) Calculate the values of the standard deviations:

$$\sigma_{z1} = \sqrt{\frac{1}{(n_1 - 3)}}$$

$$\sigma_{z2} = \sqrt{\frac{1}{(n_2 - 3)}}$$

For our data this gives:

$$\rho_{z1} = \sqrt{\frac{1}{(10-3)}} = 0.37796$$

$$\rho_{z2} = \sqrt{\frac{1}{(15-3)}} = 0.28868$$

(c) Use the formula for testing:

$$z = \frac{z_{r1} - z_{r2}}{\sigma_{z1}^2 + \sigma_{z2}^2}$$

Here,

$$z = \frac{0.9962 - 1.1270}{0.37796^2 + 0.28868^2} = \frac{-0.1308}{0.47559} = -0.275.$$

5. Compare the values: $-0.275 > -1.645$.
6. Conclusion: we cannot reject H_0.
7. Put conclusion into English. Since the calculated value is above the critical value we are unable to reject the null hypothesis, and must therefore conclude that the difference between the two correlation coefficients is due to sampling error.

17.6 THE COEFFICIENT OF DETERMINATION

Whilst the correlation coefficient is used as a measure of the association between two sets of data, probably a more common statistic is the **coefficient of determination**. This statistic will allow us to take an extra step in analyzing the relationship. It is defined to be

$$\frac{\text{variation in the variable explained by the association}}{\text{total variation in the variable}}$$

Fortunately this is not difficult to find since it can be shown to be equal to the value of the coefficient of correlation squared. (Note that this only makes sense for cardinal data.) The value obtained is usually multiplied by 100 so that the answer may be quoted as a percentage. Many statistical packages will automatically give the value of r^2 when told to find correlation and regression for a set of data.

The coefficient of determination will always be positive, and can only take values between 0 and 1 (or 0 and 100 if quoted as a percentage), and will have a lower numerical value than the coefficient of correlation. For example, if the correlation were $r = 0.93$, then the coefficient of determination would be, $r^2 = 0.8649$. This would be interpreted as 86.49% of the variation in one of the variables being explained by the association between them, whilst the remaining 13.51% is explained by other factors. (Merely stating this is useful when answering examination questions where you are asked to comment on your answer.)

The value of the coefficient of determination is often used in deciding whether or not to continue the analysis of a particular set of data. Where it has been shown that one variable only explains 10% or 20% of the behaviour of another variable, there seems little point in trying to predict the future behaviour of the second from the behaviour of the first. This point can easily be missed with the simplicity of making predictions using computer packages. However, if x explains 95% of the behaviour of y, then predictions are likely to be fairly accurate.

Interpretation of the value of r^2 should be made with two provisos in mind. Firstly, the value has been obtained from a specific set of data over a given range of values. This means that it does not necessarily apply to other ranges of values. In general, it may well apply to data close to the range of values used, but is much less likely to apply to data well away from this range. Inspection of the scatter diagram may help in determining where it does apply. Secondly, the value obtained for r^2 does not give evidence of cause and effect, despite the fact that we talk about *explained* variations. Explained here refers to *explained by the analysis* and is purely an arithmetic answer.

EXERCISES

1. Why is it important to know the number of observations when interpreting a correlation coefficient?
2. If a firm has studied the relationship between its sales and the price which it charges for its product over a number of years and finds that $r^2 = 0.75$, how would you interpret this result? Is there enough evidence to suggest a cause and effect relationship?

17.7 CONCLUSIONS

Correlation analysis is a useful first step when linking two variables together. Where it can be shown that there is an association between two sets of data it will be reasonable to take further steps in the analysis, particularly for cardinal data. Whilst a high correlation cannot be said to provide evidence of a causal relationship between the two variables, some would argue that it is a necessary, rather than a sufficient, condition for proposing such a relationship to exist. In other words, having a high correlation does not mean that you have a cause and effect relationship, but if you think that there should be such a relationship (maybe from some suggested theory), then you should be able to find a high correlation between the variables.

17.8 PROBLEMS

1. Calculate Spearman's coefficient of correlation for the following data:

x	1	2	3	4	5
y	2	1	4	5	3

2. Find a correlation coefficient using the following ordinal data:

x	1	2	3	4	5	6	7	8
y	3	8	7	5	6	1	4	2

3. A group of people have been ranked in terms of both their mechanical/technical ability and their tact in handling people and their rankings are given below. Find if there is a correlation between the two rankings.

Person	*Mechanical/technical*	*Tact*
A	1	7
B	4	8
C	7	1
D	8	5
E	5	6
F	2	4
G	3	2
H	6	3

4. A group of athletes were ranked before a recent race and their positions in the race noted. Details are given below. Find the correlations between the rankings and the positions.

Ranking	1	2	3	4	5	6	7	8	9	10
Finishing position	3	5	2	1	10	4	9	7	8	6

5. Ranking of individuals or events can be a very personal activity. Suggest two characteristics, or events, which individuals would rank (a) similarly, (b) differently.
6. Using the cardinal data given below, find Pearson's coefficient of correlation.

x	1	2	3	4	5
y	3	6	10	12	14

7. Take the data given below and construct a scatter diagram. Find the correlation coefficient and the coefficieint of determination for this data.

x	10	12	14	16	18	20	22	24	26	28
y	25	24	22	20	19	17	13	12	11	10

8. A farmer has recorded the number of fertilizer applications to each of the fields in one section of the farm, and, at harvest time, records the weight of crop per acre. The results are given in the accompanying table.

x	1	2	4	5	6	8	10
y	2	3	4	7	12	10	7

Find the correlation between fertilizer applications and weight of crop per acre. Draw a scatter diagram from the data. Using your two results, what advice, if any, could you give to the farmer?

9. Determine the correlation coefficient for the following data by inspection.

x	11	12	13	14	15	16	17	18	19	20
y	4	4	4	4	4	4	4	4	4	4

If you are unable to see the answer, use MICROSTATS to find the correlation coefficient.

10. During the making of certain electrical components each item goes through a series of heat processes. The length of time spent in this heat treatment is related to the useful life of the component. To find the nature of this relationship a sample of twenty components are selected from the process and tested to destruction and the results are presented below.

Time in process (minutes)	*Length of life (hours)*
25	2005
27	2157
25	2347
26	2239
31	2889
30	2942
32	3048
29	3002
30	2943
44	3844
41	3759
42	3810
41	3814
44	3927
31	3110
30	2999
55	4005
52	3992
49	4107
50	3987

(a) Construct a scatter diagram from this data.
(b) Use MICROSTATS (or another suitable package) to find the correlation between the time spent in the heat processes and the length of useful life. Hence find the coefficient of determination.
(c) Compare your scatter diagram with those in Figure 16.9 and thus decide upon a suitable transformation for the data. Calculate the new correlation coefficient and the new coefficient of determination.
(d) Comment on the results that you have found.

11. Construct a scatter diagram for the following data:

x	1	2	3	4	5	6	7	8	9	10
y	10	10	11	12	12	13	15	18	21	25
x	11	12	13	14	15	16	17	18	19	20
y	26	29	33	39	46	60	79	88	100	130

Find the coefficient of correlation and the coefficient of determination. Now find the log of y and re-calculate the two statistics. How would you interpret your results?

12. Costs of production have been monitored for some time within a company and the following data found:

Production level (000)	*Average total cost (£000)*
1	70
2	65
3	50
4	40
5	30
6	25
7	20
8	21
9	20
10	19
11	17
12	18
13	18
14	19
15	20

(a) Construct a scatter diagram for the data.
(b) Calculate the coefficient of determination and explain its significance for the company.
(c) Is there a better model than the simple linear relationship which would increase the value of the coefficient of determination? If your answer is 'yes', calculate the new coefficient of determination.
(d) What factors would affect the average total cost other than the production level?

17.9 DERIVATION OF THE CORRELATION COEFFICIENT

Pearson's correlation coefficient is

$$\frac{\text{covariance of } x \text{ and } y}{\sqrt{[(\text{variance of } x)(\text{variance of } y)]}}$$

$$= \frac{\frac{1}{n}\Sigma(x-\bar{x})(y-\bar{y})}{\sqrt{\left[\frac{1}{n}\Sigma(x-\bar{x})^2 \frac{1}{n}\Sigma(y-\bar{y})^2\right]}}$$

$$= \frac{\frac{1}{n}[\Sigma(xy - x\bar{y} - \bar{x}y + \bar{x}\bar{y})]}{\frac{1}{n}\sqrt{[\Sigma(x^2 - 2x\bar{x} + \bar{x}^2)\Sigma(y^2 - 2y\bar{y} + \bar{y}^2)}}$$

$$= \frac{\Sigma xy - \frac{1}{n}\Sigma x\Sigma y - \frac{1}{n}\Sigma x\Sigma y + \frac{n}{n^2}\Sigma x\Sigma y}{\sqrt{\left\{\left[\Sigma x^2 - \frac{2}{n}(\Sigma x)^2 + \frac{n}{n^2}(\Sigma x)^2\right]\left[\Sigma y^2 - \frac{2}{n}(\Sigma y)^2 + \frac{n}{n^2}(\Sigma y)^2\right]\right\}}}$$

$$= \frac{\Sigma xy - \frac{1}{n}(\Sigma x)(\Sigma y)}{\sqrt{\left\{\left[\Sigma x^2 - \frac{1}{n}(\Sigma x)^2\right]\left[\Sigma y^2 - \frac{1}{n}(\Sigma y)^2\right]\right\}}}$$

or if we multiply top and bottom by n, we have:

$$r = \frac{n\Sigma xy - \Sigma x\Sigma y}{\sqrt{\{[n\Sigma x^2 - (\Sigma x)^2][n\Sigma y^2 - (\Sigma y)^2]\}}}$$

17.10 ALGEBRAIC LINK BETWEEN SPEARMAN'S AND PEARSON'S COEFFICIENTS

Pearson's coefficient is:

$$r = \frac{n\Sigma xy - \Sigma x\Sigma y}{\sqrt{\{[n\Sigma x^2 - (\Sigma x)^2][n\Sigma y^2 - (\Sigma y)^2]\}}}$$

Now ranked data (both x and y) will only take on integer values from 1 to n and there are formulae for the sum of the first n natural numbers and the sum of their squares. These are:

$$\Sigma x = \Sigma y = \frac{n(n+1)}{2}$$

and

$$\Sigma x^2 = \Sigma y^2 = \frac{n(n+1)(2n+1)}{6}$$

and therefore

$$r = \frac{n\Sigma xy - (\Sigma x)^2}{n\Sigma x^2 - (\Sigma x)^2}.$$

Also

$$\Sigma d = \Sigma(x-y)$$

so

$$\Sigma d^2 = \Sigma(x-y)^2 = \Sigma x^2 + \Sigma y^2 - 2\Sigma xy$$
$$= 2\Sigma x^2 - 2\Sigma xy.$$

Rearranging gives:

$$\Sigma xy = \Sigma x^2 - 0.5\Sigma d^2$$
$$= \frac{n(n+1)(2n+1)}{6} - \frac{\Sigma d^2}{2}$$

Putting all of these together gives:

$$r = \frac{n\frac{(n(n+1)(2n+1)}{6} - \frac{\Sigma d^2}{2} - \frac{n^2(n+1)^2}{4}}{\frac{n(n(n+1)(2n+1))}{6} - \frac{n^2(n+1)^2}{4}}$$

Multiplying the top and bottom by 12 gives:

$$r = \frac{2n^2(n+1)(2n+1) - 6n\Sigma d^2 - 3n^2(n+1)^2}{2n^2(n+1)(2n+1) - 3n^2(n+1)^2}$$

Collecting terms gives:

$$r = \frac{2n^2(n+1)(2n+1) - 3n^2(n+1)^2 - 6n\Sigma d^2}{2n^2(n+1)(2n+1) - 3n^2(n+1)^2}$$

$$= 1 - \frac{6n\Sigma d^2}{2n^2(n+1)(2n+1) - 3n^2(n+1)^2}$$

The bottom line of the fraction can be simplified to $n^2(n^2-1)$ so

$$r = 1 - \frac{6n\Sigma d^2}{n^2(n^2-1)}$$

$$1 - \frac{6\Sigma d^2}{n(n^2-1)}$$

REGRESSION 18

In the previous chapter we have shown a way of deciding if a relationship exists between two sets of data by calculating the coefficients of correlation and determination. Where such a relationship does exist, businesses will want to know what that relationship is. If we are able to specify the relationship in the form of an equation, then we will be able to make predictions about the behaviour of one variable when we know, or suspect, values of the other variable. The equation itself may also enable a business to understand how the relationship works, and thus to take steps to change a variable it can control, for example, the price charged for its product, in order to influence a variable it cannot control, for example, the product sales that are achieved. Such an equation may also indicate by how much the controllable variables need to be changed in order to achieve the desired effects on the uncontrollable variables. Without help from such analysis, a business may decide to try to boost sales by reducing price and find that it can no longer cope with the demand for its product at the new price; it then raises the price again to reduce demand. Alternatively, the reduction in price, in some markets, may lead to very little change in sales if there is strong brand loyalty, and it may have been better to have increased the promotional spending of the company.

Computer packages, such as MICROSTATS make it very easy to find equations which link two variables together, but remember to take note of the correlation coefficient, since if there is little correlation between the data sets, the equation will be meaningless. The methods described in this chapter are only applicable to cardinal data.

18.1 LINEAR REGRESSION

Initially we will look at the situation of trying to fit a straight line to the data which we have available. Such straight lines are surprisingly powerful tools in the analysis of both business and macroeconomic relationships and will prove to be sufficient in many situations. Even when the whole of the relationship cannot be represented by a single straight line, it is quite likely that sections (i.e. between certain values) will be approximately linear.

The first step in such a procedure will be to look at the scatter diagram which we have constructed to find if the relationship is approximately linear. Where it is, then it may be tempting to just draw on a line with a ruler but this will lead to many different lines being drawn as different people attempt to guess at the best line through the data. What we need is some method of identifying the **best line** through the data so that any disagreement is over the interpretation of the results rather than the positioning of the line. To do this it will be necessary to define *best*.

To define best we may reconsider the reason

for trying to estimate the regression line. Our objective is to be able to predict the behaviour of one of the variables, say y, from the behaviour of the other, say x. So the best line will be the one which gives us the best prediction of y. Best will now refer to how close the predictions of y are to the actual values of y. Since we only have the data pairs which have given us the scatter diagram we will have to use those values of x to predict the corresponding values of y, and then make an assumption that if the relationship holds for the current data, that it will hold for other values of x.

If you consider Figure 18.1, which shows part of a scatter diagram, you will see that a straight line has been superimposed on to the scatter of points and the **vertical** distance, d, has been marked between one of the points and the line. This shows the difference, at one particular value of x, between what actually happened to the other variable, y, and what would be predicted to happen if we use the linear function, $\hat{y}$ (pronounced y hat). This is the error that would be made in using the line for prediction.

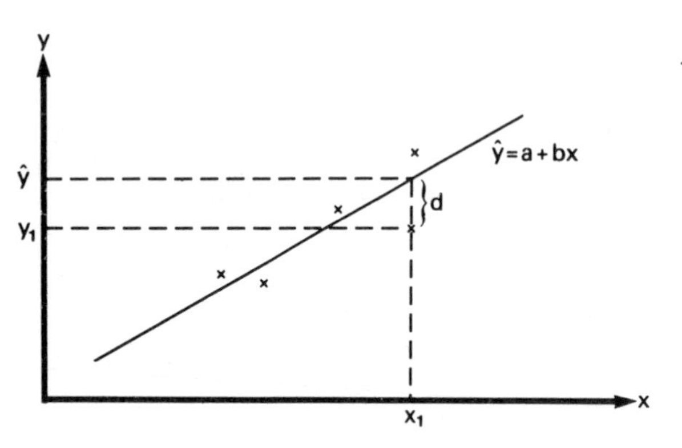

Figure 18.1

To get the best line for predicting y we want to make all of these errors as small as possible and, as we have seen in Chapter 8, to minimize something we can use calculus (see section 18.9 for complete derivation of formulae). Working through this process will give values for a and b and thus we will have identified a specific linear function through the data. The line we have identified may be called by a number of names, the **line of best fit**, the **least squares line** (since it minimizes the sum of squared differences of observed values from the regression line; see section 18.9), or the **regression line of y on x** (since it is the line which was derived from a desire to predict y values from x values).

The formulae for estimating a and b are:

$$b = \frac{n\Sigma xy - \Sigma x \Sigma y}{n\Sigma x^2 - (\Sigma x)^2}$$
$$a = \bar{y} - b\bar{x}$$

giving the regression line: $\hat{y} = a + bx$.

If we compare the formula for b with the correlation formula for cardinal data given in Chapter 17:

$$r = \frac{n\Sigma xy - \Sigma x \Sigma y}{\sqrt{\{[n\Sigma x^2 - (\Sigma x)^2][n\Sigma y^2 - (\Sigma y)^2]\}}}$$

then we see that their numerators are identical, and the denominator of b is equal to the first bracket of the denominator for r. Thus the calculation of b, after having calculated the correlation coefficient, will be a simple matter.

EXAMPLE

In Chapter 17 we considered pairs of data relating turnover to profits (17.3.2) where we treated turnover as x and profit as y. The various summations were as follows:

$\Sigma x = 255.30$, $\Sigma y = 27.40$,
$\Sigma x^2 = 9944.67$, $\Sigma xy = 1052.13$, and $n = 10$

Correlation was $r = 0.9881$.

Using this data, we may now find the equation of the regression line of profits on turnover (y on x) by using the formulae given above.

$$b = \frac{(10)(1052.13) - (255.30)(27.40)}{(10)(9944.67) - (255.30)^2}$$
$$= \frac{3526.08}{34\,268.61}$$
$$= 0.102895$$
$$a = \frac{27.40}{10} - 0.102895\frac{(255.30)}{10}$$
$$= 0.113091$$

Thus the equation is

$y = 0.113 + 0.103x$

(You should check this result by using MICROSTATS or some other suitable package, or even by hand!)

Using MICROSTATS we would have the x values in column 1 and the y values in column 2 and could then type in REGRESSION C2 C1 or REGR C2 C1, to obtain the results. In MINITAB we would need to type REGR C2 1 C1 since it is necessary to tell the package how many x variables there are (here there is only one!)

18.2 GRAPH OF THE REGRESSION LINE

Having identified the specific regression line that applies to a particular set of points, it is usual to place this on the scatter diagram. Since we are dealing with a linear function, we only need two points on the line, and these may then be joined using a ruler. The two points could be found by substituting values of x into the regression equation and working out the values of $\hat{y}$, but there is an easier way. For a straight line, $y = a + bx$, the value of a is the intercept on the y-axis, so this value may be plotted. From the formula for calculating a (or from the proof in section 18.9) we see that the line goes through the point $(\bar{x}, \bar{y})$, which has already been calculated, so that this may be plotted. The two points are then joined together (Fig. 18.2).

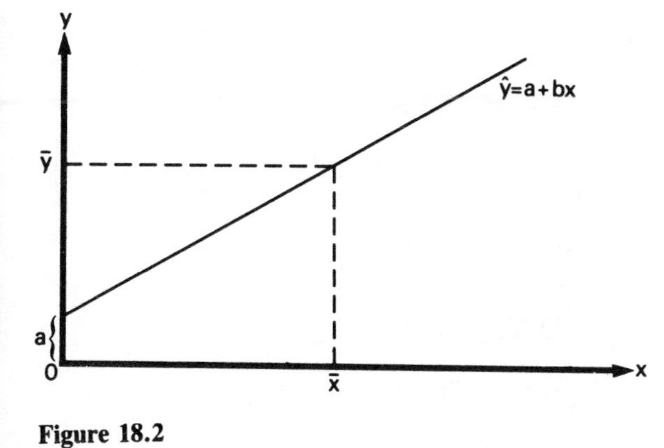

Figure 18.2

EXERCISE

Plot the regression line on the scatter diagram that you drew for the exercise in section 17.3.2.

18.3 PREDICTION FROM THE REGRESSION LINE

Prediction is one of the major objectives of regression analysis since we are attempting to find how one of the variables will behave given certain values for the other variable. This may be in the future, or may relate to values of the x variable for which we do not, currently, have any y variable values. In deriving the regression line given in section 17.1 we emphasized that it was for predicting y values, so to make prediction we can simply substitute x values into this equation.

For example, if we know a company in the industry has a turnover of £35 million (i.e. $x = 35$), then we can predict that its profits would be given by:

$$\begin{aligned} y &= 0.113 + 0.103x \\ &= 0.113 + 0.103\,(35) \\ &= 0.113 + 3.605 \\ &= 3.718 \end{aligned}$$

So the profits will be £3.718 million.

How good this prediction will be depends upon two factors, the value of the correlation coefficient and the value of x used. If the correlation coefficient is close to 1 (or -1), then all of the points on the scatter diagram are close to the regression line and thus we would expect the prediction to be fairly accurate. However, with a low correlation coefficient, the points will be widely scattered away from the regression line, and thus we will have less faith in the prediction. For this reason it may not be worth calculating a regression line if the correlation coefficient is small.

The value of x used to make the prediction will also affect the accuracy of that prediction. If the value is close to the average value of x, then the prediction is likely to be more accurate than if the value used is remote from $\bar{x}$. In Chapter 13 we looked at confidence intervals for certain statistics, and the same idea can be applied to the regression line when we consider prediction. Note that we will have to calculate the confidence interval for *each* value of x that we use in prediction since the width of the interval depends upon a standard deviation for the prediction which itself varies. At first sight these formulae look extremely daunting, but they use numbers that have already been calculated!

The 95% confidence interval will be:

$$\hat{y}_0 \pm t_{\sigma,2\frac{1}{2}\%}\,\hat{\sigma}_p$$

where $\hat{y}_0$ is the value of y obtained by putting x_0 into the regression equation, $\hat{\sigma}_p$ is the standard deviation of the prediction (see below) and $t_{\delta,2\frac{1}{2}\%}$ is a value from the t-distribution, obtained from tables and using both tails of the distribution with $\delta = n - 2$, the number of degrees of freedom.

To find the standard deviation of the prediction, $\hat{\sigma}_p$, we will use the following formulae:

$$\hat{\sigma}^2 = \frac{1}{n(n-2)}\left\{\frac{[n\Sigma x^2 - \Sigma x)^2][n\Sigma y^2 - (\Sigma y)^2] - (n\Sigma xy - \Sigma x\Sigma y)^2}{n\Sigma x^2 - (\Sigma x)^2}\right.$$

and

$$\hat{\sigma}_p = \hat{\sigma} \times \sqrt{\left[\frac{1}{n} + \frac{n(x_0 - \bar{x})^2}{n\Sigma x^2 - (\Sigma x)^2}\right]}$$

These formulae can be simplified if we look back to the formula for correlation;

$$r = \frac{n\Sigma xy - \Sigma x\Sigma y}{\sqrt{\{[n\Sigma x^2 - (\Sigma x)^2][n\Sigma y^2 - (\Sigma y)^2]\}}}$$

let A equal the numerator, B equal the first bracket of the denominator and C equal the second bracket of the denominator. Then

$$r = \frac{A}{\sqrt{(B \times C)}}$$

Using the same notation, we have:

$$\hat{\sigma}^2 = \frac{1}{n(n-2)}\left(\frac{B \times C - A^2}{B}\right)$$

and

$$\hat{\sigma}_p = \hat{\sigma} \times \sqrt{\left[\frac{1}{n} + \frac{n(x_0 - \bar{x})^2}{B}\right]}$$

Before going on to look at an example which uses these formulae, consider the bracket $(x_0 - \bar{x})$ which is used in the second formula. x_0 is the value of x that we are using to make the prediction, and thus the value of this bracket gets larger the further we are away from the mean value of x; this in turn will change the value of $\hat{\sigma}_p$ which explains why the confidence intervals will vary in width, with the value of x used.

EXAMPLE

From the previous example we have established that the regression line of y on x is

$$y = 0.113 + 0.103x$$

and that the correlation coefficient is

$$r = \frac{3526.08}{34268.61 \times 372.24} = 0.98726$$

so that:

$A = 3526.08$; $B = 34268.61$; $C = 372.24$; $n = 10$; $\Sigma x = 255.3$.

If we now wish to predict the y (profit) value for an x (turnover) of £35 million, we know that the point prediction is £3.718 million (see above), and we can establish a confidence interval by using the appropriate formulae:

$$\begin{aligned}
\hat{\sigma}^2 &= \frac{1}{10(10-2)}\left[\frac{(34\,268.61)(372.24) - (3\,526.08)^2}{34\,268.61}\right] \\
&= \frac{1}{80}\left[\frac{12\,756\,147.39 - 12\,433\,240.17}{34\,268.61}\right] \\
&= \frac{1}{80}\left[\frac{322\,907.22}{34\,268.61}\right] \\
&= \frac{1}{80}(9.422828) \\
&= 0.11778535
\end{aligned}$$

$\hat{\sigma} = 0.3431987$

$$\hat{\sigma}_p = 0.3431987 \times \sqrt{\left[\frac{1}{10} + \frac{10\,(35 - 25.53)^2}{34\,268.61}\right]}$$
$$= 0.3431987 \times \sqrt{[0.1 + 0.026169985]}$$
$$= 0.3431987 \times \sqrt{0.126169985}$$
$$= 0.3431987 \times 0.35520415$$
$$= 0.1219$$

This is the standard deviation of the prediction from $x_0 = 35$. To find the confidence interval, we need to know the *t*-value from the tables (Appendix D) for $n - 2$ degrees of freedom. For a 95% confidence interval this is 2.306.

Thus the 95% confidence interval for the prediction of profit from a turnover of £35 million is:

£(3.718 ± 2.306 × 0.1219) million
£(3.718 ± 0.2811) million
£3.4369 million to £3.9991 million

We are 95% confident that the profit for a company in this industry with a turnover of £35 million will be between £3.4369 million and £3.9991 million.

Using a spreadsheet, we can calculate the lower and upper 95% confidence limits for various values of x_0, as shown below:

x_0	$\hat{y}$	t	σ_p	*lower*	*upper*
5	0.63	2.306	0.1621	0.25	1.00
10	1.14	2.306	0.1417	0.82	1.47
15	1.66	2.306	0.1249	1.37	1.95
20	2.17	2.306	0.1133	1.91	2.43
25	2.69	2.306	0.1086	2.44	2.94
30	3.20	2.306	0.1116	2.95	3.46
35	3.72	2.306	0.1219	3.44	4.00
40	4.23	2.306	0.1378	3.92	4.55
45	4.75	2.306	0.1575	4.38	5.11
50	5.26	2.306	0.1799	4.85	5.68
55	5.78	2.306	0.2040	5.31	6.25
60	6.29	2.306	0.2294	5.76	6.82
65	6.81	2.306	0.2556	6.22	7.40
70	7.32	2.306	0.2824	6.67	7.97
75	7.84	2.306	0.3097	7.12	8.55
80	8.35	2.306	0.3373	7.58	9.13
85	8.87	2.306	0.3652	8.03	9.71
90	9.38	2.306	0.3932	8.48	10.29
95	9.90	2.306	0.4215	8.93	10.87

These values can now be graphed, together with the values of the point predictions, i.e. the regression line itself. This is shown in Figure 18.3.

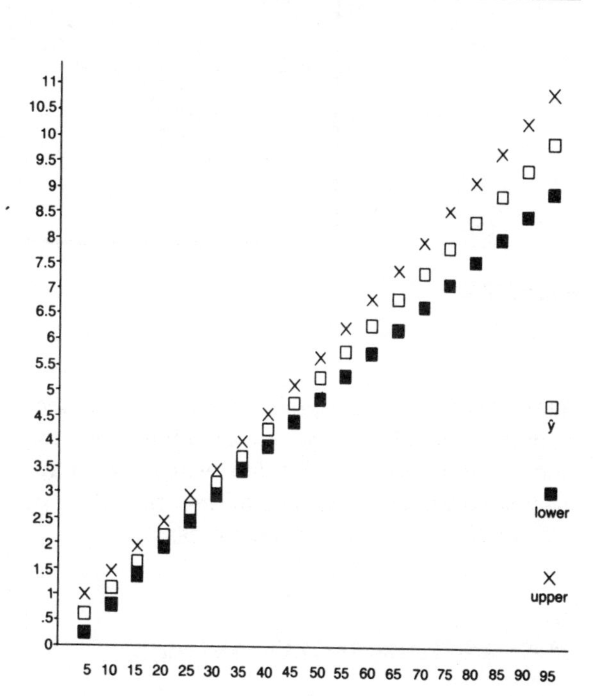

Figure 18.3

Prediction is often broken down into two sections:

1. predictions from x values that are within the original range of x values are called **interpolation**;
2. predictions from outside this range are called **extrapolation**.

As you can just see in Figure 18.3, the further away x_0 is from the average value of x, the wider the confidence interval. The reason that it is difficult to see in Figure 18.3 is that there is such a high correlation between x and y. A stylized representation is shown in Figure 18.4 to emphasize the point.

Interpolation is seen as being relatively accurate (depending upon the value of the correlation coefficient), whilst extrapolation is seen as less accurate. Even where the width of the confidence

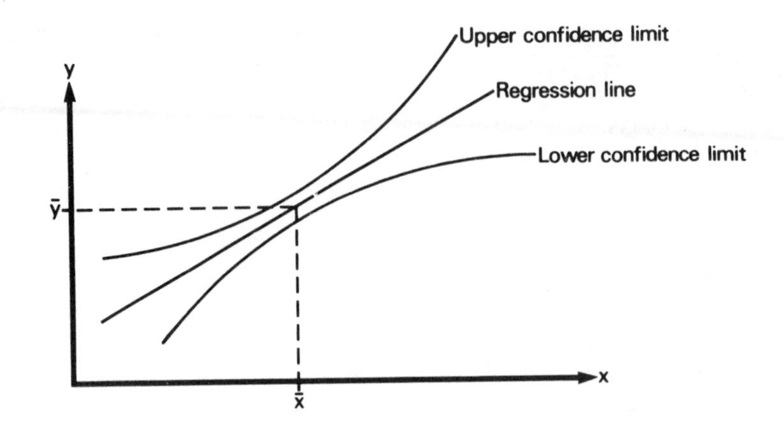

Figure 18.4

interval is deemed acceptable, we must remember that in extrapolation we are assuming that the linear regression relationship which we have calculated will apply outside of the original range of x values. This may not be the case!

18.4 ANOTHER REGRESSION LINE

When we were trying to decide how to define the best line through the scatter diagram we used the vertical distances between the points and the linear function. What would happen if we had used *horizontal* distances? Doing this would mean that we were trying to predict x values from a series of y values, but again we would want to minimize the errors in prediction. Following through a similar line of logic using horizontal distances gives:

if

$$\hat{x} = c + my$$

then

$$m = \frac{n\Sigma xy - \Sigma x\Sigma y}{n\Sigma y^2 - (\Sigma y)^2}$$

and

$$c = \bar{x} - m\bar{y}$$

The result of this calculation is called the regression line of x on y. Note that this also goes through the point $(\bar{x}, \bar{y})$, but in most circumstances, the line will be different from the regression line of y on x. The only exception is when the correlation coefficient is equal to $+1$ or -1, since there we have a completely deterministic relationship and all of the points on the scatter diagram are on a single straight line. Figure 18.5 shows two regression lines superimposed on to a scatter diagram.

EXAMPLE

Using the data on turnover and profit, we had:

$\Sigma x = 255.3$; $\Sigma y = 27.4$; $\Sigma xy = 1052.13$; $\Sigma y^2 = 112.3$; $n = 10$.

Thus:

$$m = \frac{10\,(1052.13) - (255.3)(27.4)}{10\,(112.3) - (27.4)^2}$$
$$= \frac{3526.08}{372.24}$$
$$= 9.472598$$

and

$$c = 25.53 - 9.472598\,(2.74)$$
$$= -0.424919.$$

so

$$x = -0.425 + 9.473y$$

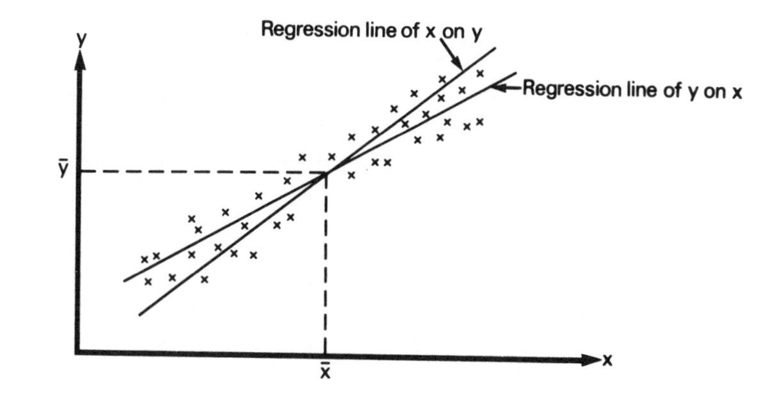

Figure 18.5

For many sets of data it is only necessary to calculate one of the regression lines, since it will only make sense to predict one of the two variables. If we are using data on advertising expenditure and sales of a product, then we wish to be able to predict sales for a particular level of advertising expenditure.

18.5 INTERPRETATION

As stated above, the major reason for wishing to identify a regression relationship is to make predictions about what will happen to one of the variables (the **endogenous** variable) at some value of the other variable (often called the **exogenous** variable). However, it is also possible to interpret the parts of the regression equation. According to the simple Keynesian model of the economy, consumption expenditure depends upon the level of disposable income. If data are collected on each of these variables over a period of time, then a regression relationship can be calculated. For example:

consumption expenditure = 2000 + 0.85(disposable income)

From this we see that even if there is no disposable income, then consumption expenditure is 2000 to meet basic needs, this being financed from savings or from government transfer payments. Also, we can see that consumption expenditure increases by 0.85 for every increase in disposable income of 1. If the units used were pounds, then 85 pence of each pound of disposable income is spent on consumption and 15 pence is saved. Thus 0.85 is the average propensity to consume (APC). This type of relationship will work well for fairly narrow ranges of disposable income, but the correlation coefficient is likely to be small if all households within an economy are considered together, since those on low disposable incomes will tend to spend a higher proportion of that income on consumption than those on very high disposable incomes.

18.6 NON-LINEAR RELATIONSHIPS

As we saw in section 17.4, some relationships are non-linear, and thus we will need to be able to calculate a regression relationship that will allow prediction from a non-linear function. If the correlation coefficient has already been calculated, then little additional calculation is needed to obtain the regression equation.

Extending the example in section 17.4, we continue to use the log of y instead of y:

$$r = \frac{272.05}{\sqrt{(3185 \times 23.8721)}}$$
$$= 0.9866.$$

For log y on x

$$b = \frac{272.05}{3185}$$
$$= 0.08542$$

and

$$a = \frac{\Sigma(\log_{10}y)}{n} - b\bar{x}$$
$$= \frac{21.43}{10} - 0.08542(9.5)$$
$$= 2.143 - 0.81149$$
$$= 1.33151$$

So the log-linear regression relationship is

$$\log_{10}y = 1.33151 + 0.08542x.$$

NB. although this is a linear relationship, if you wish to plot it on a scatter diagram, you will have to work out several individual values of $\hat{y}$ and join these together to form a curve.

To predict from this relationship, if $x_0 = 9$ then

$$\log_{10}\hat{y} = 1.33151 + 0.08542 \times 9$$
$$= 1.33151 + 0.76878$$
$$= 2.10029.$$

and by taking the antilog

$$\hat{y} = 125.98.$$

Finding the regression line for other non-linear transformations will follow a similar pattern.

18.7 CONCLUSIONS

Regression analysis is one of the most widely used techniques in business since it is concerned with prediction. As we have seen in the chapter, where there is a strong correlation between two data sets, then regression allows such prediction to take place. For most purposes, linear regression will be sufficient; however the methods can be extended to include non-linear relationships. Computer packages have made the calculation of regression relationships relatively easy, but they should not be used blindly as applying a linear function to a non-linear situation will result in useless analysis and predictions.

18.8 PROBLEMS

1. Use the data given below to create a scatter diagram.

x	1	2	3	4	5
y	3	6	10	12	14

Find the regression line of y on x and place this onto the scatter diagram.

2. Find the regression line of y on x for the data given below.

x	10	12	14	16	18	20	22	24	26	28
y	25	24	22	20	19	17	13	12	11	10

Construct a scatter diagram and place your regression line onto the graph.

3. A farmer wishes to predict the number of tons per acre of crop which will result from a given number of applications of fertilizer. Data has been collected and is shown below.

Fertilizer applications	1	2	4	5	6	8	10
Tons per acre	2	3	4	7	12	10	7

Find a suitable regression relationship to help the farmer in making the required prediction, and from your result predict the number of tons per acre from 7 fertilizer applications.

4. Determine the regression line of y on x from the following data without calculation.

x	11	12	13	14	15	16	17	18	19	20
y	4	4	4	4	4	4	4	4	4	4

5. Use the data given below to find the regression line of y on x and the regression line of Log y on x.

x	1	2	3	4	5	6	7	8	9	10
y	10	10	11	12	12	13	15	18	21	25
x	11	12	13	14	15	16	17	18	19	20
y	26	29	33	39	46	60	79	88	100	130

Construct a scatter diagram and place both of your regression lines onto it.

6. During the making of certain electrical components each item goes through a series

of heat processes. The length of time spent in this heat treatment is related to the useful life of the component. To find the nature of this relationship a sample of twenty components are selected from the process and tested to destruction and the results are presented below.

Length of life (hours)		*Time in process (minutes)*	
25	2005	41	3759
27	2157	42	3810
25	2347	41	3814
26	2239	44	3927
31	2889	31	3110
30	2942	30	2999
32	3048	55	4005
29	3002	52	3992
30	2943	49	4107
44	3844	50	3987

(a) Find the regression line of useful life on time spent in process.
(b) Predict the useful life of a component which spends 33 minutes in process.
(c) Predict the useful life of a component which spends 60 minutes in process.
(d) Using a suitable transformation, find a regression relationship which has a higher coefficient of determination. (Hint, look back to question 10 at the end of Chapter 17.)
(e) From this new relationship, predict the useful life of a component which spends 33 minutes in process.
(f) From this new relationship, predict the useful life of a component which spends 60 minutes in process.

7. Costs of production have been monitored for some time within a company and the following data found:

Production level (000s)	*Average total cost (£000s)*
1	70
2	65
3	50
4	40
5	30
6	25
7	20
8	21
9	20
10	19
11	17
12	18
13	18
14	19
15	20

(a) Find the regression line of average total cost on production level.
(b) Place this onto a scatter diagram.
(c) Predict the average total cost if the production level were:
(i) 8 500 units;
(ii) 16 000 units;
(iii) 20 000 units.
(d) Find an alternative model to relate average total cost to production level and make the same predictions again using the new model.

18.9 APPENDIX

Equation of the straight line to predict y values is:

$$\hat{y} = a + bx$$

Thus for a value x_1, we have

$$\hat{y}_1 = a + bx_1$$

and

$$\begin{aligned} d_1 &= y_1 - \hat{y}_1 \\ &= y_1 - (a + bx) \\ &= y_1 - a - bx_1 \end{aligned}$$

where d is the difference between an observed y and its value predicted by the regression equation.

This vertical distance may be defined, and calculated as above, for each point on the scatter diagram. To minimize the prediction error, we may sum these vertical distances; however, if the observed y values deviate from the regression line in a random manner, then:

$$\Sigma d = 0$$

To overcome this problem, we may square each value of d, and then minimize the sum of these squares:

$$d_1 = y_1 - a - bx_1$$
$$d_1^2 = (y_1 - a - bx_1)^2$$

$$S = \sum_{i=1}^{n} d_i^2$$
$$= \sum_{i=1}^{n} (y_i - a - bx_i)^2$$

To minimize S we need to differentiate the function with respect to a and b (since the values of x and y are fixed by the problem we are trying to solve). Thus:

$$\frac{\partial S}{\partial a} = -2 \sum_{i=1}^{n} (y - a - bx_i) = 0$$

$$\frac{\partial S}{\partial b} = -2 \sum_{i=1}^{n} x_i(y_i - a - bx_i) = 0$$

These are the **first order conditions**, and if they are rearranged, we have the normal equations:

$$\Sigma y \; = na + b\Sigma x \qquad 18.1$$
$$\Sigma xy = a\Sigma x + b\Sigma x^2. \qquad 18.2$$

These equations could be used as a pair of simultaneous equations each time that we wanted to identify the values of a and b, but it is usually more convenient to find the general solution for a and b to give the following formulae.

Multiply equation 18.1 by $(\Sigma x)/n$:

$$\frac{\Sigma x \Sigma y}{n} = a\Sigma x + \frac{b}{n}(\Sigma x)^2 \qquad 18.3$$

Subtracting equation 18.3 from equation 18.2:

$$\Sigma xy - \frac{\Sigma x \Sigma y}{n} = b\Sigma x^2 - \frac{b}{n}(\Sigma x)^2$$
$$= b\left[\Sigma x^2 - \frac{(\Sigma x)^2}{n}\right]$$

So

$$n\Sigma xy - \Sigma x \Sigma y = b[n\Sigma x^2 - (\Sigma x)^2]$$

or

$$b = \frac{n\Sigma xy - \Sigma x \Sigma y}{n\Sigma x^2 - (\Sigma x)^2}$$

Rearranging equation 17.1 gives:

$$a = \frac{\Sigma y - b\Sigma x}{n}$$

or

$$a = \bar{y} - b\bar{x}$$

MULTIPLE REGRESSION AND CORRELATION 19

In the last two chapters we have developed the ideas of regression and correlation. Correlation allowed us to measure the relationship between two variables, and to put a value on the strength of that relationship. Regression determines the parameters of an equation which relates the variables, and in that way provides a method to predict of the behaviour of one variable from the behaviour of the other.

In this chapter we consider the effects of adding one or more variables to the equation, so that the dependent variable, (usually y) is now predicted by the behaviour of two or more variables. In most cases this will increase the amount of correlation in the model and will tend to give 'better' predictions. An obvious question to ask now is that if two independent variables give us better results than one, then why not try three, or four, or five, or even more? For most relationships in business and economics our answer would be 'Yes!'. Since we are dealing with complex relationships which are not fully explained by theory, using extra, relevant variables may well give us better results.

Once we move to several variables on the right-hand side of the regression equation, difficulties can arise which may mean that the answer we obtain does not necessarily give us the best results. We will consider some of these issues within this chapter.

Even where there is some theoretical background which suggests that several variables should be used to explain the behaviour of the dependent variable, it may not be that easy to get the necessary data. Take, for instance, the sales of a product. Economic theory suggests that the price of that product will explain some of the variations in sales; and this should be easy to find. However, the simple economic model is based on *ceteris paribus*, and once we allow for variations in other variables we will be looking for data on the prices of other products, the tastes and preferences of consumers, the income levels of consumers, advertising and promotion of the product, and advertising and promotion of competitive products. There may also be a rôle for the level of activity in the economy including the growth rate, levels of unemployment, rate of inflation, and expectations of future levels and rates of these variables. Some of this data such as that on tastes and preferences, is likely to be very difficult to obtain.

It is not feasible to develop multiple regression relationships by hand for more than very simple data, and so we will assume that you have at least some access to a computer system running a suitable package. The MICROSTATS package which comes with this book will provide you with a suitable computer program and we will use some of the output in this chapter to illustrate the examples and explain how one might interpret such printout. Similar printouts are obtainable from both MINITAB and SPSS. Whilst, in the past, this topic was excluded from introductory courses, the wide use of computer packages now makes it easily accessible to most students.

19.1 A BASIC MODEL (2 VARIABLES)

As a first step we will re-work a two-variable regression model, adding some extra comments on the type of printout obtained from typical statistical packages. (The following examples are taken from an SPSS-X printout.)

Table 19.1 Simple data

x:	60	85	110	95	140	160	80	40	55	90	115	120	180	95
y:	25	20	35	40	60	55	45	15	20	30	40	50	70	45

Taking some very simple data (14 observations on 2 variables, as shown in Table 19.1) we can use a standard package. We could obtain a result as in Figure 19.1 from typical packages.

We use these figures to check that we are using the data we expect; had we obtained a mean for x of 35 we would suspect that the wrong data was being read by the computer.

The matrix in Figure 19.2 just tells us that there is a correlation of 0.903 between x and y; as we move on to larger models the rôle of this matrix will change.

The first part of Figure 19.3 is recognizable from Chapter 17 as the correlation between x and y and the coefficient of determination. It can be shown that the coefficient of determination is biased upwards, and the 'Adjusted R square' figure represents an unbiased value. The standard error is a measure of the overall variation in the model.

The equation of the model can be read from the first two columns at the bottom of Figure 19.3

$$\hat{y} = 1.428634 + 0.371929x$$

where the column headed 'B' gives the estimated values of the equation parameters.

The two coefficients may be subjected to hypothesis tests to determine if they are non-zero, the usual hypotheses being:

$$H_0: \beta_1 = 0$$
$$H_1: \beta_1 > 0,$$

a one-tailed test. This t-test is performed by the package, with the t-values shown for each coefficient and a significance level given. Assuming

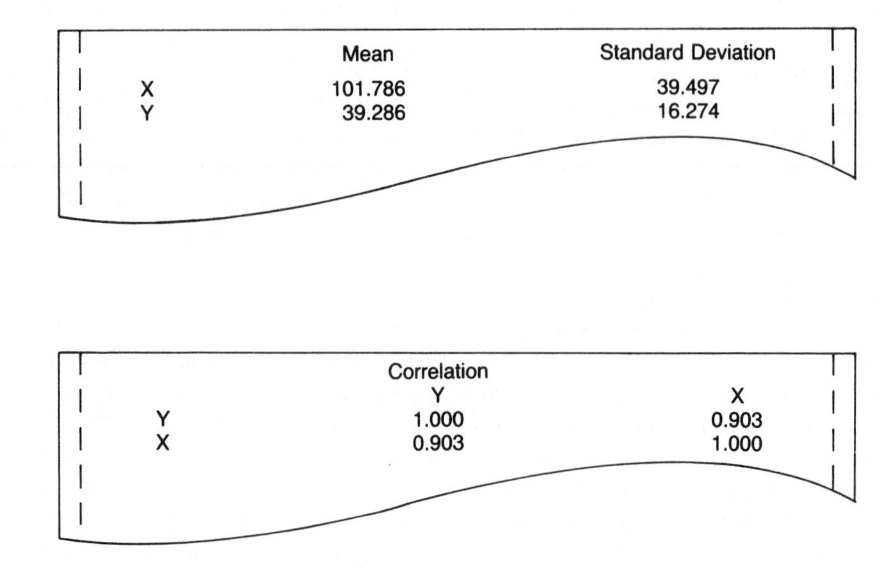

	Mean	Standard Deviation
X	101.786	39.497
Y	39.286	16.274

Figure 19.1

Correlation	Y	X
Y	1.000	0.903
X	0.903	1.000

Figure 19.2

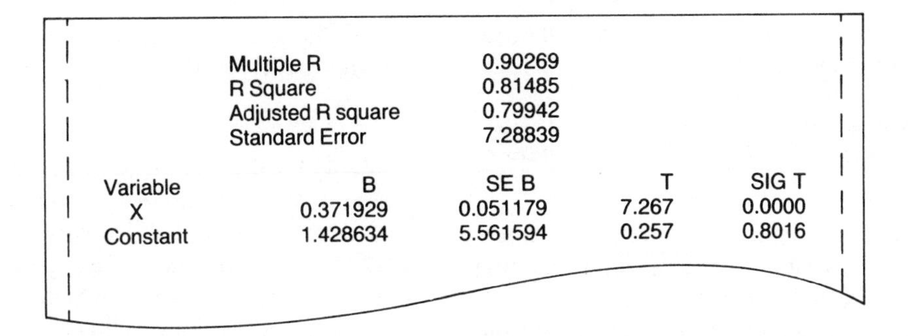

Multiple R	0.90269			
R Square	0.81485			
Adjusted R square	0.79942			
Standard Error	7.28839			

Variable	B	SE B	T	SIG T
X	0.371929	0.051179	7.267	0.0000
Constant	1.428634	5.561594	0.257	0.8016

Figure 19.3

that we are testing at the 5% level of significance, then if the value of 'SIG T' is below 0.05, we may reject the null hypothesis and conclude that the value is non-zero. (In other words, we have identified a significant relationship.) When the value of 'SIG T' is above 0.05 we cannot reject H_0.

In the example given, we reject H_0 for the coefficient of x, but we cannot reject H_0 for the constant in the equation.

19.2 ADDING AN EXTRA VARIABLE

There are five basic assumptions that are normally made when calculating a multiple regression relationship:

1. that we are dealing with a linear function of the independent variables plus a disturbance term, $\hat{y} = \beta_1 + \beta_2 X_2 + \beta_3 X_3 + \ldots\ldots \beta_n X_n + u$
2. that this disturbance term has a mean of zero;
3. that there is a constant variance in the model;
4. that the independent variables are fixed, i.e. non-stochastic;
5. that there is no significant linear relationship between the independent variables.

We will assume that you are using a computer package to find such multiple regression relationships and will only look at results, rather than the formulae used in hand calculation. All of these assumptions are incorporated in the calculations used by computer packages, but by studying the printouts obtained we will be able to tell if they have been met in a particular case (see section 19.3).

Table 19.2 Company data

Time	*Sales (Y)*	*Price (X2)*	*Marketing spend (X3)*	*Index of economic activity (X4)*	*Index of unit cost (X5)*
1	986	1.8	0.4	100	100.0
2	1025	1.9	0.4	103	101.1
3	1057	2.1	0.5	104	101.1
4	1248	2.2	0.7	106	106.2
5	1142	2.2	0.6	102	106.3
6	1150	2.3	0.7	103	108.4
7	1247	2.4	0.9	107	108.7
8	1684	2.2	1.1	110	114.2
9	1472	2.3	0.9	108	113.8
10	1385	2.5	1.1	107	114.1
11	1421	2.5	1.2	104	115.3
12	1210	2.6	1.4	99	120.4
13	987	2.7	1.1	97	121.7
14	940	2.8	0.8	98	119.8
15	1001	2.9	0.7	101	118.7
16	1025	2.4	0.9	104	121.4
17	1042	2.4	0.7	102	121.3
18	1210	2.5	0.9	104	120.7
19	1472	2.7	1.2	107	122.1
20	1643	2.8	1.3	111	124.7

Using the data given in Table 19.2 (DATA3 on disk) we will build a series of multiple regression models to show the effects of adding extra variables. Our aim will be to make predictions of sales.

A first step may be to look at each variable in relation to sales. Using MICROSTATS we may quickly establish the following 2-variable relationships:

sales = 965.4 + 104.54 price $r^2 = 0.01866$
sales = 786.3 + 492.66 marketing spend $r^2 = 0.40426$
sales = −3988.8 + 50.13 index of economic activity $r^2 = 0.7157$
sales = 448.1 + 6.75 index of unit costs $r^2 = 0.05623$

On the criteria of highest coefficient of determination, the best single variable for predicting sales is the index of economic activity.

To improve our model we could now add the variable 'marketing spend'. Assuming that we have 'time' in column C1, 'sales' in C2, 'price' in C3, 'marketing spend' in C4, 'index of economic activity' in C5, and 'index of unit cost' in C6, we can now type the following into MICROSTATS at the command line:

MREGRESSION C2 2 C4 C5 (or MREG may be used)

(The dependent variable is given first followed by a number which is the number of independent variables. The other two column numbers tell the package which variables to use as independent variables.)

This gives the relationship shown in Figure 19.4.

Looking at this result we can note several differences from the output from simple regression.

1. Each coefficient in the equation has three figures below it:
   (a) the standard error of the coefficient;
   (b) the value obtained by a t-test; and
   (c) a probability value.

   Provided that the probability value is below 0.05 the coefficient passes a one-tailed *t*-test using the hypotheses:

   H_0: $\beta_j = 0$ $\quad H_1$: $\beta_j > 0$
2. We now use a capital R^2 to denote the coefficient of determination: this is to show that we have a multiple regression relationship.
3. There is a second value given ($\bar{R}^2$ pronounced R bar squared) which is the adjusted value of R^2. It has been shown that R^2 is biased upwards, thus giving a falsely good impression of the relationship, and this adjusted value is an unbiased estimate of the explanatory power of the model.
4. A new statistic is given, the Durbin–Watson statistic (DW); we will deal with its interpretation in the next section.
5. A correlation matrix is given which shows the inter-relationships between all of the variables; again we will deal with this in section 19.3.

The index of economic activity explained 71.57% of the behaviour of sales; by adding marketing spend, we can now explain 87.186% ($\bar{R}^2$) of this behaviour: an increase of 15.616%. This is considerably less than the 40.426% which marketing spend explains on its own and shows that we cannot simply add the individual values of r^2 together.

If two variables explain more of the behaviour of sales than one variable, three should explain even more, and four more still. Adding variables to the model cannot decrease the value of R^2 but

```
              Sales = -3526.9 +  42.88 C4 +  332.94 C5
                      (514.03)   (5.076)     (66.383)
                  t    -6.861     8.446       5.015
               Prob     .0000     .0000       .0001

     R² = 0.88535       R̄² = 0.87186         DW = 1.83666

                     C2           C4            C3
              C2    1.000        0.846         0.636
              C4    0.846        1.000         0.285
              C3    0.636        0.285         1.000
```

Figure 19.4

```
C2 = -2638.8 - 68.31 C3 + 478.50 C4 + 39.46 C5 - 4.35 C6
     (693.48)  (119.05)  (99.87)     (5.21)     (4.85)
t =   -3.805    -0.574    4.791      7.571     -0.897
Prob   .0017     .5746    .0002      .0000      .3836

R² = 0.95266          R̄² = 0.90755          DW = 2.46098
```

	C2	C3	C4	C5	C6
C2	1.000	0.137	0.636	0.846	0.237
C3	0.137	1.000	0.656	−0.073	0.855
C4	0.636	0.656	1.000	0.285	0.738
C5	0.846	−0.073	0.285	1.000	0.015
C6	0.237	0.855	0.738	0.015	1.000

Figure 19.5

do not necessarily raise it. Since the adjusted R^2 figure takes into account the number of variables in the model, its value may decrease as more variables are added.

EXERCISE

Use the data given in Table 19.2 together with the MICROSTATS package to find the following relationships:

1. Sales on price and marketing spend;
2. Sales on marketing spend and index of unit cost;
3. Sales on price, marketing spend and index of economic activity.

(Answers:

$C2 = 1476.7 - 377.59\ C3 + 743.61\ C4 \qquad \bar{R}^2 = 0.48900$

$C2 = 2187.9 + 784.96\ C4 - 14.54\ C6 \qquad \bar{R}^2 = 0.46685$

$C2 = -2950.8 - 142.78\ C3 + 439.5\ C4 + 39.75\ C5 \qquad \bar{R}^2 = 0.88432$

Putting all of the variables into the model (MREG C2 4 C3–C6) we obtain the results shown in Figure 19.5.

Now we are explaing 90.755% of the variation in sales by using all of the available variables, but only two of the independent variables pass the *t*-test (C4 and C5).

19.3 ECONOMETRIC PROBLEMS

Once we move from dealing with two variable models, to those with three or more variables, not only do we increase the complexity of the calculations in order to identify the equation which relates the variables together, but we also find that other issues arise about the relationships between some or all of the variables. These problems can be summarized under the mnemonic 'MALTHUS'.

M – multicollinearity
A – autocorrelation
L – lack of data
T – time and cost constraints
H – heteroskedasticity
U – under-identification
S – specification

We will look briefly at each of these problems, but emphasise that a full treatment of such topics is beyond the scope of this book, and suggest that you consult a text in econometrics.

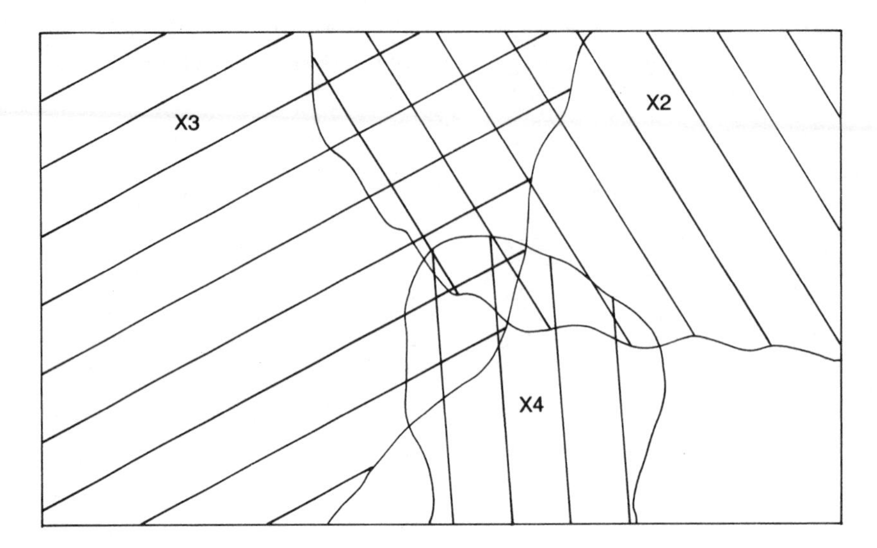

Figure 19.6 Multicollinearity

19.3.1 Multicollinearity

This term refers to the inter-relatedness of the variables on the right-hand side of the equation; the independent variables. In an ideal world, there would be no inter-relationship between the independent variables, but since we are not in an ideal world, but are dealing with business and economic problems, some inter-relationships will exist. Consider the situation shown in Figure 19.6.

The outer box represents the *total* variation in the dependent variable and the shaded areas represent the variation 'explained' by the individual independent variables X2, X3 and X4. As we can see, there is considerable overlap between X2 and X3, and this overlap represents multicollinearity. There is little overlap between X4 and X2, showing that there is little multicollinearity. Computer packages do not produce such a diagram, but they do produce a correlation matrix which relates all of the variables together. Multicollinearity is where there is a high correlation between two or more independent variables. Table 19.3 shows such a matrix.

From this we see that we have perfect correlation on the diagonal from top right to bottom left (1.00) since this is a variable correlated with itself.

Table 19.3 Correlation matrix

	Y	X2	X3	X4
Y	1.00	0.75	0.65	0.45
X2	0.75	1.00	0.86	0.23
X3	0.65	0.86	1.00	0.33
X4	0.45	0.23	0.33	1.00

We need high correlations in the first column since this represents the correlation of each variable, individually, with Y. The other values should, ideally, be low, but the 0.86 shows that there is multicollinearity between variables X2 and X3. There is no specific value at which we would say multicollinearity exists; it is a matter of judgement.

If multicollinearity exists in a model then the coefficients of the independent variables may be unstable, especially if we try to use the equation to forecast after there has been a policy change. Such a change is likely to change the multicollinearity between the independent variables, and thus make the model invalid.

Where multicollinearity exists we could delete some of the independent variables from the model, in order to remove the effects of the correlations, or we could try adding more data in an attempt to find the underlying relationships. Providing that there is no policy change, multicollinearity will not seriously affect the predictions from the model in the short run.

19.3.2 Autocorrelation

When building a multiple regression model we assume that the disturbance terms are all independent of each other. If this is not the case, then the model is said to suffer from autocorrelation. We will use the error terms generated by the computer program to look for autocorrelation with regression models.

Autocorrelation may arise when we use quarterly or monthly data, since there will be a seasonal effect which is similar in successive years, and this will mean that there is some correlation between the error terms. The basic test for autocorrelation is the **Durbin–Watson test**, which is automatically calculated by most computer programs. This statistic can only take values between 0 and 4, with an ideal value being 2; indicating the absence of autocorrelation. Since we are unlikely to get a value exactly equal to 2, we need to consult tables (see Appendix J) as we analyze a multiple regression model.

Suppose that we have an equation calculated from 25 observations which has 5 independent variables. Using the steps set out in Chapter 15 for conducting hypothesis tests we have:

1. State hypotheses: H_0: $\rho = 0$; no autocorrelation.
   H_1: $\rho > 0$; positive autocorrelation.

   Note that the test is *always* a one-tailed test and that we therefore can test for either positive autocorrelation or negative autocorrelation, but not both at once.
2. State the significance level: 5%, say.
3. State the critical value(s):
   These will vary with the number of observations ($n = 25$) and the number of independent variables (usually labelled k, here equal to 5). From the tables we find two values:

   $d_l = 0.95$ and $d_u = 1.89$.
4. Calculate the test statistic:
   This is done automatically by the computer program.
5. Compare the test statistic to the table values:
   It is easiest here to use a line to represent the distribution, as in the diagram below:

   ⇐Reject H_0⇒⇐Inconclusive⇒⇐Cannot reject H_0

   0 ——— 0.95 ——— 1.89 ——— 2
6. Come to a conclusion:
   If the calculated value is below the lower limit (0.95) then we reject the null hypothesis and if it is above the upper limit we cannot reject the null hypothesis. Any value between the limits leads to an inconclusive result, and we are unable to say if autocorrelation exists.
7. Put the conclusion into English:
   Where we reject the null hypothesis, then we say that autocorrelation exists in the model and that this may lead to errors in prediction.

It is possible to take a series of extra steps which will help to remove autocorrelation from the model, but we refer you to more advanced texts for details of these methods.

To see the evidence for autocorrelation we may add an extra item to the command for multiple regression in MICROSTATS by putting a column number in brackets after the command, for example:

```
MREG  C2  4  C3  C4  C6  C7  (C9)
```

This will put the residuals into column 9 and we can then create a graph of these against time by using the PLOT command.

19.3.3 Lack of data

Many data sets are incomplete or subject to review and alteration. Lack of data may represent a situation where data has only been collected for a short period, for example, a company which has only been trading for a year. When new statistics are calculated by government these series will not be usable in modelling for some years.

A related problem is that of unsuitable data.

Data may be available on the variable which you are trying to model, but the definition may be different from the one you wish to use. An example of this situation would be the official statistics on income and expenditure of the personal sector of the economy where the definition includes unincorporated businesses, charities and trade unions.

19.3.4 Time and cost constraints

In an educational context, there may be little cost pressure to complete the modelling process, at least in money terms. In a business context there is considerably more pressure, since predictions are needed to plan production and distribution decisions give a time constraint whilst the overall budget and potential savings give a cost constraint.

19.3.5 Heteroskedasticity

We have assumed that the variation in the data remains of the same order throughout the model, i.e. a constant variance. If this is not the case, then the model is said to suffer from heteroskedasticity. It is most prevelant in time series models which deal with long periods of data and usually has an increasing variance over time. This is illustrated by a scatter diagram in Figure 19.7 below.

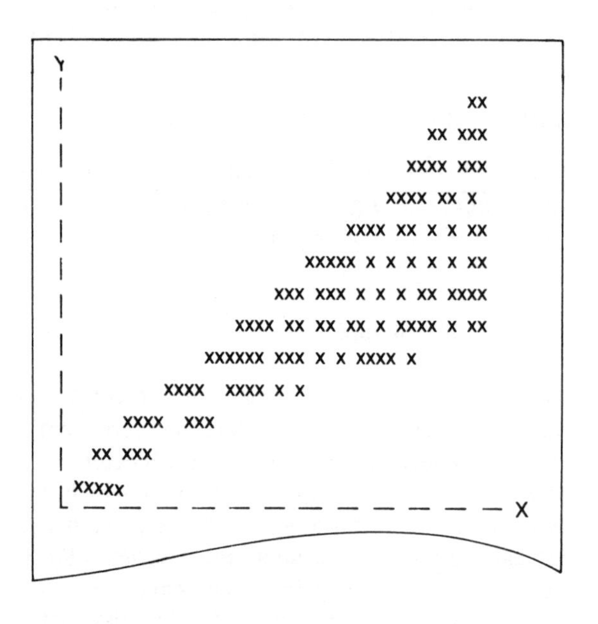

Figure 19.7 Heteroskedasticity

Heteroskedasticity in a model will lead to problems of prediction since the equation predicts the mean value, and as we can see from the diagram, the variation about that mean is increasing.

19.3.6 Under-identification

This is a problem which affects multiple equation models, and as such is outside the scope of this book. It represents a situation where it is not possible to identify which equation has been estimated by the regression analysis, for example in market analysis both the demand function and the supply function have the quantity as a function of price, tastes and preferences, etc.

19.3.7 Specification

This problem relates to the choice of variables in the equation and to the 'shape' of the equation. When constructing business models, we rarely know for certain which variables are relevant. We may be sure of some, and unsure about others. If we build a model and leave out relevant variables, then the equation will give biased results. If we build a model which includes variables which are not relevant in this case, then the variance will be higher than it would otherwise be. Neither case is ideal.

Similar problems arise over the 'shape' of the equation; should we use a linear function, or should we try a non-linear transformation?

In practice we never know whether we have the exact specification of the model! It is usually prudent to build a model with too many variables, rather than too few, since the problem of increased variance may be easier to deal with than the problem of biased predictions.

19.4 ANALYSIS OF A MULTIPLE REGRESSION MODEL

As an example we will take a macro-economic model of inflation which uses 54 quarterly data points taken from the 1970s and 1980s in the UK. Various lags have been built into the model to allow for the delayed effects of the changes in one variable on another.

The final model gave the information shown in Figure 19.8 where

X2 = change in the money supply, lagged by 9 quarters
X3 = change in basic wages, lagged by 4 quarters
X4 = growth in the economy, lagged by 10 quarters
X5 = investment, lagged by 7 quarters.

Looking through the results we can see that the model explains only 67.824% of the variations in inflation, there is thus 32.176% which must be explained by factors which we have not taken into account. Each of the coefficients of the independent variables passes a one-tailed *t*-test at the 5% level and thus each has a significant effect on the inflation figures.

Looking at the correlation matrix we can see that whilst there is a relatively high level of correlation between the independent variables and inflation (except for X4), that there is little interrelationship between the independents. The model, therefore, does not suffer from multicollinearity.

Performing a Durbin–Watson test at the 5% level, we have:

$d_{\alpha} = 1.41$ and $d_{u} = 1.72$

and a statistic from the model of 0.74194. We must therefore reject the null hypothesis (H_0: $\rho = 0$) in favour of the alternative hypothesis that the model suffers from *positive* autocorrelation. Obtaining a graph of the residuals gives Figure 19.9 in which you can see the pattern created. If no autocorrelation existed, then this graph would be a random series of points.

At this level it is not really possible to test for heteroskedasticity.

Finally we may question the specification of the model. Several variables identified by economic theory are missing from the model, for example, the level of unemployment. Adding extra variables may increase the explanatory power of the model but at the expense of breaking some more of the assumptions. For example, adding a variable to represent unemployment to the model leads to a small amount of multicollinearity whilst the coefficient fails the *t*-test. There is still autocorrelation present in the model.

19.5 USE OF MULTIPLE REGRESSION

Having established a multiple regression relationship from historical data, we usually wish to

Inflation =	−23.1	+ 0.299 X2	+ 0.328 X3	+ 0.0026 X4	+ 0.416 X5
	(7.794)	(0.075)	(0.072)	(0.00081)	(0.156)
t =	−2.964	3.977	4.565	3.242	2.662
Prob	.0047	.0002	.0000	.0021	.0105

n = 54 $R^2 = 0.70252$ $\bar{R}^2 = 0.67824$ DW = 0.74194

	Inflation	X2	X3	X4	X5
Inflation	1.000	0.599	0.699	0.394	0.511
X2	0.599	1.000	0.397	0.136	0.132
X3	0.669	0.397	1.000	0.199	0.387
X4	0.394	0.136	0.199	1.000	0.197
X5	0.511	0.132	0.387	0.197	1.000

Figure 19.8 Inflation model

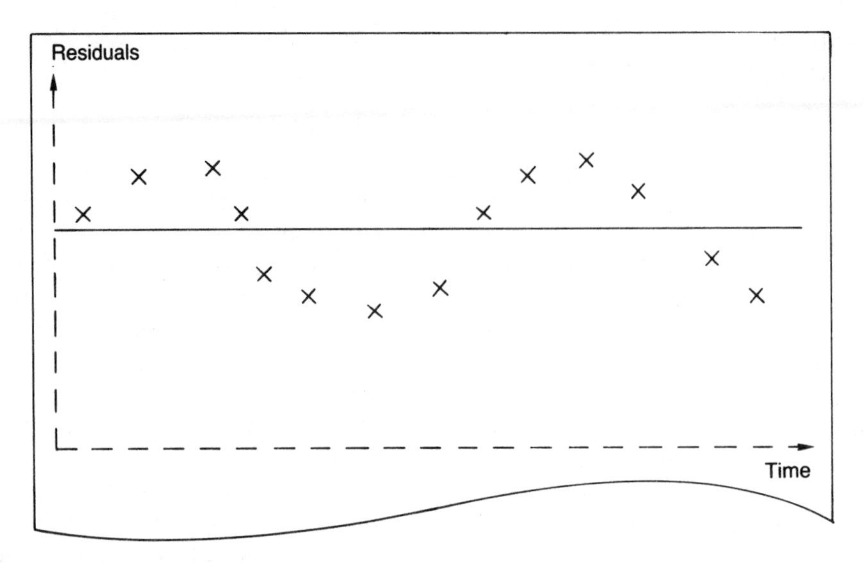

Figure 19.9 Residuals of the inflation model

use this to make predictions about the future. As in any model of this type, if the value of R^2 is low, or the specification of the model is very poor, then the predictions will be worthless.

Where we have used lagged relationships, we will be able to use historic data relating to the independent variables to make predictions of the independent variable by an amount of time equal to the shortest lag. We do this by substituting into the equation that we have found. An example is shown below in Table 19.4.

In Table 19.4 we are at time period 24 with all data available up until that point. The data values marked with a star (*) are those which are used to make the prediction of Y for period 24. If we now substitute the next value of X2 (5.5), X3 (102.2) and X4 (2.8) into the equation, then we find a prediction for period 25 (527.43). A similar process will give the prediction for period 26. No further predictions can be made from this model until more data becomes available (i.e. in period 25).

Table 19.4
$Y_t = 20 + 6.3\ X2_{t-2} + 4.7\ X3_{t-3} - 2.7\ X4_{t-4}$

Time	*Y*	*Prediction*	*X2*	*X3*	*X4*
20	520		4.7	103.7	2.7*
21	527		4.9	102.6*	2.8
22	531		5.3*	102.2	2.9
23	535		5.5	101.4	2.9
24	530	528.32	5.7	103.1	2.6
25		527.43			
26		524.66			

To predict where we have not used lagged relationships or where the lags were very short, we would need to make predictions of the independent variables, and then feed these predictions into the equation. By putting the data, the multiple regression equation and the predictions of the independent variables into a spreadsheet we can build a management model which will allow us to see the effects of changing variables which are under our control, and the effects of different forecasts of those variables which are not under our control.

Suppose that a company has built a multiple regression model to predict sales in value terms, and that the equation is

Company sales value
= 200
− 0.04 × (Price–price of company B's product)
+ 0.03 × Advertising expenditure
+ 0.01 × Total industry advertising expenditure

Table 19.5

Time	*Sales*	*Price*	*B Price*	*Advertising expenditure*	*Total advertising expenditure*	*Quality rating*	*Quality rating B*	*Growth*	*U/E*
1	3 059.62	10	11	20 000	200 000	5	4	3.20	12.30
2	3 173.22	10	11	21 000	210 000	5	4	3.00	12.30
3	3 293.28	11	11	22 050	220 500	5	4	2.80	12.30
4	3 412.01	12	11	23 152.50	231 525	5	4	2.50	12.20
5	3 529.78	12	12	24 310.13	243 101.25	5	4	2.10	12.10
6	3 524.38	12	12	25 525.63	255 256.31	5	4.50	0.10	10.50
7	3 656.86	13	13	26 801.91	268 019.13	5	4.50	−0.30	12.10
8	3 839.11	13	14	28 142.01	281 420.08	5	4.50	−0.20	12.60
9	4 079.14	14	15	29 549.11	295 491.09	5	5	0.50	13.20
10	4 361.21	15	15	31 026.56	310 265.64	5	5	1.60	13.60

+ 0.5 × Quality rating of company's product
− 0.1 × Quality rating of B's product
+82 × Growth rate of the economy
− 0.4 × Unemployment rate in the economy

and that this achieved a very high R^2 value. The data from the last ten time periods (quarters) is shown in Table 19.5.

The company has, in the past, increased its advertising budget at 5% per time period and this is in line with the increase in total advertising spending by the industry. We can therefore predict the next value of 'company advertising' by multiplying the current value by 1.05. The same will be true for total advertising expenditure in the industry.

If we assume that 5 is the highest quality rating it is possible to achieve, and that neither this company nor company B is likely to reduce quality, then we can continue the rating of 5 into the foreseeable future.

The government has forecast that growth will be 2% for the next four time periods, and then will be 3%.

Unemployment is predicted to be related to growth such that:

unemployment = 12.5 + (growth − growth last period)

An internal report suggests that the company price will be 15 in period 11, 16 in periods 12, 13, 14, and 15; 17 in and periods 16, 17 and 18; and 18 in periods 19 and 20. A similar report acquired from the competitor, without their knowledge, suggests that company B's prices will be 15 in period 11; 16 in periods 12, 13, 14, and 15; and 17 in periods 16, 17, 18, 19 and 20.

Placing all of this information into a spreadsheet will give the predictions shown in Table 19.6

EXERCISE

Build a spreadsheet using the multiple regression relationship shown above and confirm the results in Table 19.6. Now vary the price charged for the product and note the effect on sales. You could also try varying the advertising expenditure of the company and the predictions of the growth rate of the economy.

19.6 CONCLUSIONS

Multiple regression analysis is a very powerful statistical technique which will enable us to make better predictions about the behaviour of the dependent variable than simple two variable models. The increased complexity, however, raises problems of interpretation of the results

Table 19.6

Time	*Sales*	*Price*	*B Price*	*Advertising expenditure*	*Total advertising expenditure*	*Quality rating*	*Quality Rating B*	*Growth*	*U/E*
11	4 595.97	15	15	32 577.89	325 778.93	5	5	2	12.90
12	4 807.88	16	16	34 206.79	342 067.87	5	5	2	12.50
13	5 030.23	16	16	35 917.13	359 171.27	5	5	2	12.50
14	5 263.69	16	16	37 712.98	377 129.83	5	5	2	12.50
15	5 590.42	16	16	39 598.63	395 986.32	5	5	3	13.50
16	5 848.17	17	16	41 578.56	415 785.64	5	5	3	12.50
17	6 118.47	17	17	43 657.49	436 574.92	5	5	3	12.50
18	6 402.25	17	17	45 840.37	458 403.66	5	5	3	12.50
19	6 700.17	18	17	48 132.38	481 323.85	5	5	3	12.50
20	7 013.03	18	17	50 539.00	505 390.04	5	5	3	12.50

since in business situations, not all of the underlying assumptions are met. Multiple regression should not be used blindly; it is necessary to perform a series of tests on the model before it can be used to make predictions. As with any technique, it is necessary to ask questions about the reliability and accuracy of the data before accepting the results of the model.

19.7 PROBLEMS

1.

Y	10	12	15	17	19	22	24	27	29	30
X_2	1	1	2	2	3	4	4	5	5	6
X_3	10	9	8	7	6	5	4	3	2	1

(a) Use the data given above to find the regression relationship between Y and X_2

(b) Use the data given above to find the regression relationship between Y and X_3.

(c) Use the data given above to find the regression relationship between Y and X_2 and X_3.

(d) From your answer to part (c) test each of the coefficients to find if they are non-zero.

(e) Find the coefficient of multiple determination.

2. A brewer is interested in the amount spent per week on alcohol and the factors which influence this. Data has been collected on the amount spent, the income of the head of household (in thousands per year) and the household size for twenty families and the results are presented below.

Amount spent per week	*Income of head of household*	*Household size*
20	6	1
17	5	2
5	10	1
0	14	4
3	25	2
8	10	5
14	21	1
19	17	1
32	29	2
17	14	3
9	7	1
8	9	3
4	14	2
20	19	1
10	13	1
9	10	2
7	9	3
14	11	3
59	34	6
7	10	2

(a) Find the regression relationship of amount spent on income and family size.

(b) Carry out a statistical analysis of this result.
(c) What would be the problems for the brewer in interpreting this answer?

3. A personnel and recruitment company wishes to build a model of likely income level and identifies three factors which it feels are important. Data collected on twenty clients gives the following results.

Income level	*Years of post 16 education*	*Years in post*	*No. of previous jobs*
15	2	5	0
20	5	3	1
17	5	7	2
9	2	2	0
18	5	8	2
24	7	4	3
37	10	11	2
24	5	7	1
19	6	4	0
21	2	8	4
39	7	12	2
24	8	8	1
22	5	6	2
27	6	9	1
19	4	4	1
20	4	5	2
24	5	2	3
23	5	6	1
17	4	3	4
21	7	4	1

(a) Find the regression relationship for predicting the income level from the other variables.
(b) Assess the statistical quality of this model.
(c) Would you expect the company to find the model useful?
(d) What other factors would you wish to build into such a model?

4. A manufacturer of diesel engines for small boats wished to predict the overall demand for the product. This demand was a derived

Variable	Mean	Standard deviation
UKBOAT	−3.3387	21.5417
TIME	16.5000	9.3808
SWF4	−8.9716	7.6931
MLR4	18.4537	40.4900
GDP4	1.9006	3.3965

UKBOAT = −12.88 − 1.244SWF4 − 0.232MLR4 + 0.82GDP4 + 0.067TIME

		SWF4	MLR4	GDP4	TIME
st.err		(.509)	(.0915)	(1.0865)	(.4058)
t		2.444	2.546	0.755	0.165

Correlation matrix:

	UKBOAT	TIME	SWF4	MLR4	GDP4
UKBOAT	1.000	−0.165	−0.354	−0.293	0.121
TIME	−0.165	1.000	0.366	0.014	−0.196
SWF4	−0.354	0.366	1.000	−0.251	−0.234
MLR4	−0.293	0.014	−0.251	1.000	0.244
GDP4	0.121	−0.196	−0.234	0.244	1.000

Durbin–Watson statistic = 1.73816
$R^2 = 0.544$ $\bar{R}^2 = 0.296$
Number of observations = 32.

Figure 19.10 Model of boat sales

demand from the sales of new boats and sales of replacement engines. Data was collected on a series of macro-economic variables and built into an econometric model; the results of which are presented in Figure 19.10.

UKBOAT = UK boat sales;
TIME = dummy variable;
SWF4 = exchange rate to Swiss franc with lag of one year;
MLR4 = minimum lending rate with lag of one year;
GDP4 = percentage change in gross domestic product of UK with lag of one year.

Assess this model in both statistical terms and in terms of its likely usefulness to the company.

5. A company has developed a model of its sales and asks you to build a spreadsheet model to allow predictions to be made. The specification of the model is as follows:

Sales value = Sales volume × Price
Sales volume
= 10 000
−20 000 (Price-average price in the industry)
+30 × Advertising expenditure
+20 × Promotional expenditure
+ a seasonal factor
+25 000 × Growth rate in the economy
−250 000 × No. of companies in the industry
−100 000 × Unemployment rate in the economy
−2 × Advertising expenditure of other companies
−1.5 × Promotional spending of other companies

The following data is available for the first ten time periods:

Time	*Price*	*Average price*	*Advertising expenditure*	*Promotional advertising*
1	1.35	1.25	145 000.00	80 000.00
2	1.35	1.25	146 450.00	81 600.00
3	1.45	1.25	147 914.50	83 232.00
4	1.45	1.30	149 393.65	84 896.64
5	1.45	1.30	150 887.58	86 594.57
6	1.45	1.30	152 396.46	88 326.46
7	1.45	1.30	153 920.42	90 092.99
8	1.45	1.30	155 459.63	91 894.85
9	1.45	1.35	157 014.22	93 732.75
10	1.45	1.35	158 584.36	95 607.41

The seasonal factors are

−100 000
−200 000
−150 000
+450 000

repeated through time. The number of companies is:

8 in periods 1 to 9
9 in periods 10 to 15
10 in periods 16 to 25

Data on external factors for the first ten periods:

Time	*Growth*	*U/E*	*Others' Advertising*	*Others' Promotion*
1	2.90	14	600 000.00	250 000.00
2	2.80	14.10	612 000.00	260 000.00
3	2.70	14.10	624 240.00	270 400.00
4	2.40	14.30	636 724.80	281 216.00
5	2.10	14.30	649 459.30	292 464.64
6	1.40	14.70	662 448.48	304 163.23
7	0.30	15.10	675 697.45	316 329.75
8	0.30	14	689 211.40	328 982.94
9	−0.10	14.40	702 995.63	342 142.26
10	−0.70	14.60	717 055.54	355 827.95

(a) Use this information to build a spreadsheet model and obtain a graph of sales value for the ten periods.
(b) Use the information provided below to extend the spreadsheet to predict sales for 25 time periods, obtaining a graph of sales value for the whole period.
(c) What are the implications of your predictions?
Write a brief report (400 words maximum)

(d) Vary price, advertising spend and promotional spend (within reasonable limits) to smooth out the sales value figures. Write a brief report on your suggested policy for the company (400 words maximum)

Extra information:

Time	*Price*	*Average price*	*Growth*
11	1.50	1.35	−1.30
12	1.50	1.35	−1.40
13	1.50	1.45	−0.09
14	1.50	1.45	−0.50
15	1.60	1.45	−0.10
16	1.60	1.45	0.10
17	1.60	1.55	0.30
18	1.60	1.55	0.70
19	1.65	1.55	1.30
20	1.65	1.55	1.50
21	1.70	1.55	1.70
22	1.70	1.65	2.20
23	1.70	1.65	2.30
24	1.70	1.65	2.20
25	1.70	1.65	2.00

Note also the following:

(a) Advertising expenditure continues to grow at 1% per time period.

(b) Promotional spending continues to grow at 2% per time period.

(c) Unemployment = 14 + (Growth_t − Growth_{t-1})

(d) Others advertising spend increases by 2% per time period.

(e) Other promotional spend increases by 4% per time period.

TIME SERIES 20

Most things that interest businesspeople and economists can be seen to vary over time. This variability may take the form of a gradual movement continually in the same direction, or, more usually, as a series of apparently haphazard oscillations. A few of the haphazard movements seem to move and change direction purely in relation to things that are happening at the present time, for example the *Financial Times* 100 share index. Others seem to be proceeding in some general direction when viewed over fairly lengthy time periods, even though short-term fluctuations are frequent. Still other variables tend to behave in a particular way at certain points in time, perhaps every spring, or once every 9 years. These observations, while interesting in themselves for a particular variable that affects a company, are purely descriptive of what has happened in the past.

Can we use these observations for more general situations? For those series which react immediately to current events only, it is unlikely that much can be done but record the history of the variation. However, many business and economic variables exhibit a combination of the other factors noted above and we shall aim to draw these ideas together to try to *explain* the behaviour of these series over time. Whilst this explanation may be very useful in helping to show what has happened in the past, for business, a more fundamental aim of analyzing a time series is to try to *predict* what will happen in the future by projecting the patterns identified in the past. This involves a fundamental assumption that these patterns will still be relevant in the future. For the near future, this assumption is likely to be true, but for the distant future there are so many things that could, and will, change to affect the particular variable in which we are interested that the assumption can only be viewed as a rough guide to likely events.

An example of this problem may be drawn from population studies: we need to project population figures into the future in order to plan the provision of housing, schools, hospitals, roads and other public utilities. How far ahead do we need to project the figures? For schools we need to plan 5–6 years ahead, to allow time to design the buildings, acquire the land and train teachers in the case of expansion of provision. For contraction, the planning horizon is somewhat shorter. In the case of other social service needs, for example the increasing number of elderly in the UK over the next 30–40 years, advance warning will allow consideration of how these needs are to be met, and who is to finance the provision.

If we are willing to accept the assumption of a continuing pattern, plus the restriction of only limited prediction, then we may build models of the behaviour of a variable over time and use them to project into the future.

20.1 TIME SERIES MODELS

Despite the apparently random nature of a graph of a variable over time there is often an underlying pattern. This pattern can be shown to consist of various elements, or factors, which are combined together in some way to reproduce the original time series. In this section we will discuss the four factors which make up a time series and the two methods of combining them into a model. There is no one model which will be perfect in every situation.

20.1.1 The factors

1. Trend (T) — this is a broad, underlying movement in the data which represents the general direction in which the figures are moving. It can be identified in a number of different ways, as shown in section 20.2 below.
2. Seasonal factors (S) — these are regular fluctuations which take place within one complete period. If the data are quarterly, then they are fluctuations specifically associated with each quarter; if the data are daily, then fluctuations associated with each day. An example would be the demand for electronic games. If we have quarterly data for sales, then there will usually be the highest level of sales in the fourth quarter each year (because of Christmas), with perhaps the lowest levels in the third quarter each year (because families are on holiday). These are discussed in section 20.3.
3. Cyclical factors (C) — this is a longer term regular fluctuation which may take several years to complete. The most famous example of a cycle is the trade cycle in economic activity in the UK that was observed in the late nineteenth century. Whilst this cycle, which lasted approximately 9 years, does not exist at the end of the twentieth century, there are other cycles which do affect businesses and the economy in general. These are discussed in section 20.4.
4. Random factor (R) — many other factors affect a time series, and their overall effect is usually small. However, from time to time, they do have a significant, but unpredictable, effect on the data. For example, if we are interested in new house starts, then occasionally there will be a particularly low figure due to a more than usually severe winter. Despite advances in weather forecasting, these are not yet predictable. The effects of these non-predictable factors will be gathered together in this random, or residual, factor. There is further discussion of these in section 20.5.

20.1.2 The additive model

In the additive model all of the elements are added together to give the original or actual data (A).

$A = T + S + C + R$

For many models there will not be sufficient data to identify the cyclical element, and thus the model will be reduced to

$A = T + S + R$

Since the random element is unpredictable, we shall make a *working assumption* that its overall value, or average value, is 0.

The additive model will be most appropriate where the variations about the trend are of similar magnitude in the same period of each year or week, as in Figure 20.1.

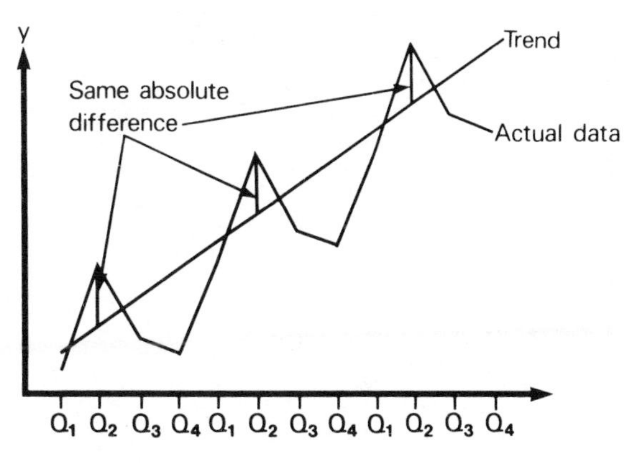

Figure 20.1

20.1.3 The multiplicative model

In the multiplicative model the main, or predictable elements of the model are multiplied together; the random element may be either multiplied together with this product:

$$A = T \times S \times C \times R$$

(here A and T are actual quantities while S, C and R are ratios) or may be added to the product:

$$A = T \times S \times C + R$$

(here A, T and R are actual quantities while S and C are ratios).

In the second case, the random element is still assumed to have an average value of 0, but in the former case the assumption is that this average value is 1. Again, the lack of data will often mean that the cyclical element cannot be identified, and thus the models will become:

$A = T \times S \times R$ and $A = T \times S + R$

The multiplicative model will be most appropriate for situations where the variations about the trend are the same proportionate size (or percentage) of the trend in the same period of each year or week, as in Figure 20.2.

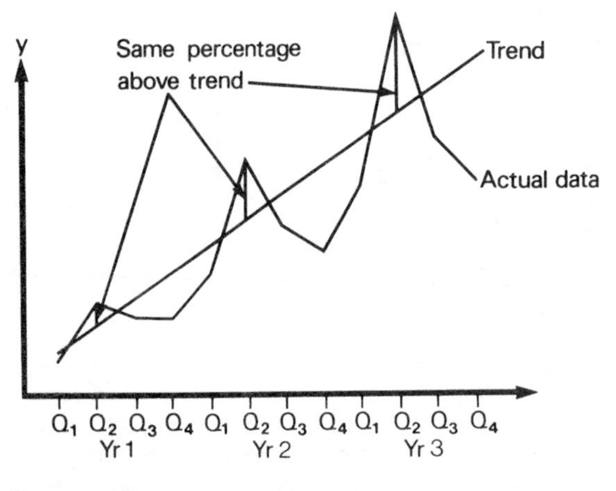

Figure 20.2

Note that the illustrations to this section have used linear trends for clarity, but the arguments apply equally to non-linear trends. In the case where the trend is a horizontal line, then the same absolute deviation from the trend in a particular quarter will be identical to looking at the percentage change from the trend in that quarter; i.e. with a constant (or stationary) trend, both models will give the same result, as illustrated in Figure 20.3.

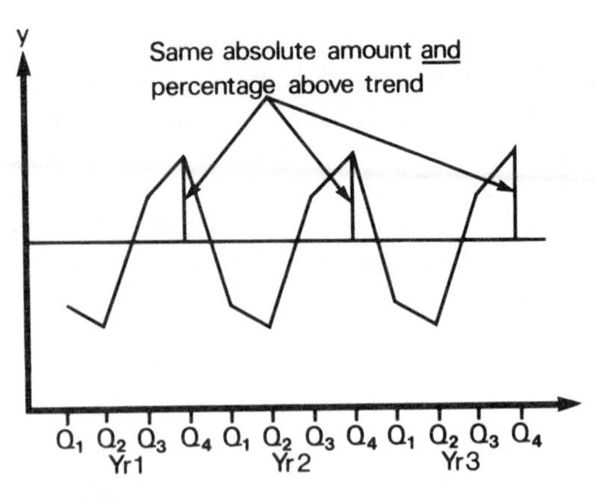

Figure 20.3

20.2 THE TREND

Most series follow some sort of long term movement, upwards, downwards or at some constant level. The first step in analyzing a time series is to construct a graph of the data, as in Figure 20.4, to see if there is any obvious underlying movement. In the diagram, we see that the data values are, generally, increasing with time. For some purposes, where all we need is an overall impression of how the data is behaving, such a graph may be all that is necessary. Where we need to go further and make predictions of the series into the future, then we will need to identify the trend as a series of values.

The graph will give some guidance on which method to use to identify the trend. If the broad underlying movement appears to be linear, then a regression trend would be appropriate; however, if there appears to be a curvilinear trend, then the moving average might provide better answers.

20.2.1 A linear regression trend

To find a linear trend we can return to the ideas of regression presented in Chapter 18, since there

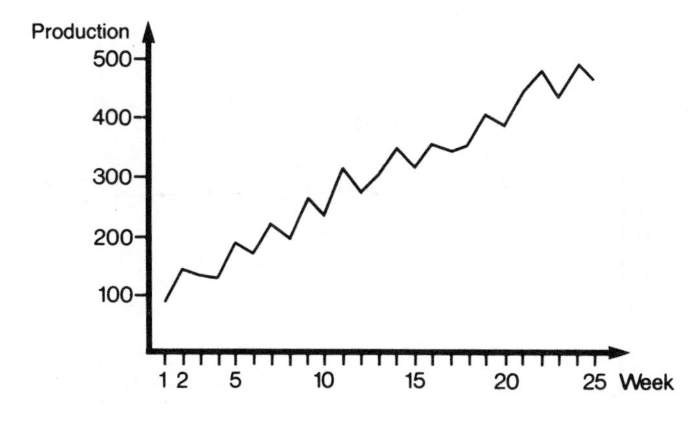

Figure 20.4
Note that points are joined by straight lines

we were trying to fit a straight line through a series of points. (In this case there can be no query about which to use as x and which to use as y, since it is the variable that we will eventually want to predict, and the passage of time that we will use to predict from.) The variable will form the y values, and x will represent time. Thus x will be a dummy variable, with 1 representing the earliest time period, 2 the next time period, and so on until the final time period (often the present day). Having established the values for the x and y variables, we may now use the formulae derived in Chapter 18 to identify a linear trend through the data. These were:

$$\hat{y} = a + bx$$

where

$$b = \frac{n\Sigma xy - \Sigma x\Sigma y}{n\Sigma x^2 - (\Sigma x)^2}$$

and

$$a = \bar{y} - b\bar{x}$$

Table 20.1 Sales of shirts

Date		Sales (000) y	x	xy	x^2
Year 1	Q_1	10	1	10	1
	Q_2	14	2	28	4
	Q_3	11	3	33	9
	Q_4	21	4	84	16
Year 2	Q_1	11	5	55	25
	Q_2	16	6	96	36
	Q_3	10	7	70	49
	Q_4	22	8	176	64
Year 3	Q_1	14	9	126	81
	Q_2	18	10	180	100
	Q_3	13	11	143	121
	Q_4	22	12	264	144
Year 4	Q_1	13	13	169	169
	Q_2	16	14	224	196
	Q_3	9	15	135	225
	Q_4	25	16	400	256
		245	136	2193	1496

For example, consider the figures given in Table 20.1.
Thus:

$n = 16; \Sigma x = 136;\ \Sigma y = 245;\ \Sigma xy = 2193;\ \Sigma x^2 = 1496.$

$$b = \frac{16 \times 2193 - 136 \times 245}{16 \times 1496 - (136)^2}$$
$$= \frac{35\,088 - 33\,320}{23\,936 - 18\,496}$$
$$= \frac{1768}{5440}$$
$$= 0.325$$

$$\bar{y} = \frac{245}{16} = 15.3125,$$

$$\bar{x} = \frac{136}{16} = 8.5$$

$$a = 15.3125 - 0.325(8.5) = 12.55$$

Thus the trend line is:

$$\hat{y} = 12.55 + 0.325x$$

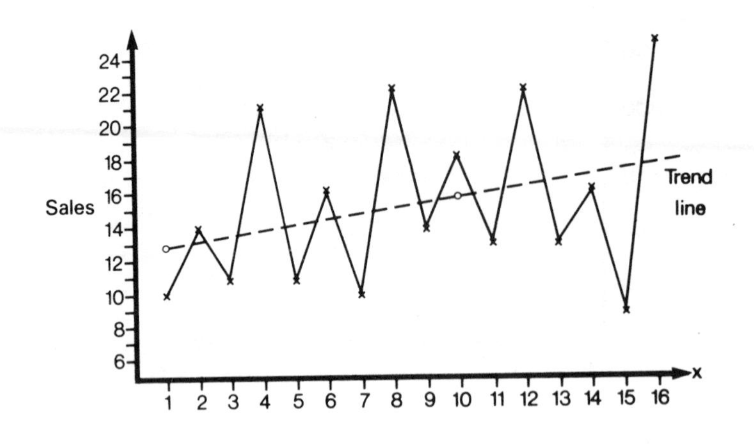

Figure 20.5

This may now be placed on the graph of the data in the usual way (Figure 20.5).

Numerical values for the trend may be found by substituting the values of x back into the trend equation.

Fox $x = 1$, we have $\hat{y} = 12.55 + 0.325(1) = 12.875$
Fox $x = 2$, we have $\hat{y} = 12.55 + 0.325(2) = 13.2$
etc.

Results are given in Table 20.2.

Table 20.2

Date		*Sales*	*Trend*
Year 1	Q_1	10	12.875
	Q_2	14	13.200
	Q_3	11	13.525
	Q_4	21	13.850
Year 2	Q_1	11	14.175
	Q_2	16	14.500
	Q_3	10	14.825
	Q_4	22	15.150
Year 3	Q_1	14	15.475
	Q_2	18	15.800
	Q_3	13	16.125
	Q_4	22	16.450
Year 4	Q_1	13	16.775
	Q_2	16	
	Q_3	9	
	Q_4	25	

This type of trend will be very easy to predict from, since the next time period will be represented by $x = 17$, and this can be substituted into the trend equation:

$$\hat{y} = 12.55 + 0.325(17) = 18.075$$

EXERCISES

1. Work out the trend for the three subsequent time periods (Year 4, quarters 2, 3 and 4). (Answer: 17.1, 17.425, 17.75.)
2. What would be the effect of using the method given above for data with a non-linear trend? How could this problem be overcome? (Hint: look back to Chapter 17.)

Using MICROSTATS we would place the time values into column C1 by using the GENERATE command, for example:

GENERATE from 1 to 16 in C1
or GENE 1 16 C1

and then place the data values into C2. A graph can be obtained by using the PLOT command. We can now find the linear trend of column C1 by using

LTRE C2 C3

when the trend values will be placed into column C3. To find the equation of the trend line we would need to use the command:

REGR C2 C1

EXERCISE

Use MICROSTATS to build the model shown in Table 20.1, find the trend values, and make predictions of the trend values for time periods 17 to 20 inclusive. (Answers: 18.075, 18.4, 18.725, 19.05)

20.2.2 A moving-average trend

This type of trend tries to smooth out the fluctuations in the original series by looking at relatively small sections, finding an average, and then moving on to another section. The size of the small section will often be related to the type of data that we are looking at; if it is quarterly data, we would use subsets of 4; if monthly data, subsets of 12; if daily data, subsets of 5 or 7.

Table 20.3 Breakdowns on a complex production line

Week	*Day*	*No. of breakdowns*	*Σ5 days*	*Average*
1	Monday	3		
	Tuesday	4		
	Wednesday	7	32	6.4
	Thursday	8	34	6.8
	Friday	10	34	6.8
2	Monday	5	34	6.8
	Tuesday	4	36	7.2
	Wednesday	7	38	7.6
	Thursday	10	39	
	Friday	12		
3	Monday	6		
	Tuesday	4		
	Wednesday	5		
	Thursday	8		
	Friday	9		
4	Monday	5		

In Table 20.3 we have daily data, and the appropriate subset size is 5, representing one complete cycle, which here is 1 week. Taking the first 5 days' breakdowns, we find that the total number of breakdowns is 32; dividing by 5 (the number of days involved) gives an average numner of breakdowns of 6.4. Since both of these figures relate to the first 5 days, it will be appropriate to record them opposite the middle day (i.e. Wednesday). If we now move the subset forward in line by 1 day, we will have another subset of 5 days (from week 1, Tuesday, Wednesday, Thursday, Friday, from week 2, Monday). For this group, the total number of breakdowns is 34, giving an average of 6.8. These two results will be recorded in the middle of this subset (i.e. opposite Thursday of week 1). This process is continued until we reach the last group of 5 days.

EXERCISE

Calculate the total number of breakdowns for each of the subsets of 5 days, and hence the averages for each subset.
(Answer: the averages will be 6.4, 6.8, 6.8, 6.8, 7.2, 7.6, 7.8, 7.8, 7.4, 7.0, 6.4, 6.2.)

Once each of the averages is recorded opposite the middle day of the subset to which it relates, we have found the moving-average trend. Two points should be noted about this trend, firstly that there are no trend figures for the *first* two data points, nor for the *last* two data points. Secondly, that the extension of this trend into the future to make predictions will be more difficult than from the linear trend calculated above: however, the trend is not limited to a small group of functional shapes and will thus be able to follow data where the trend does change direction.

A feature of using a subset with an odd number of data points is that there will be a middle item opposite which to record the answers; if we use subsets with an even number of data points, there will be no middle item. Does this matter? If the only thing that we want to do is to identify the trend then the answer is no! However, we usually want to go on and look at other aspects of the time series, and in this case we will want each trend value to be associated with a particular data point. To do this we need to use an extra step in the calculations to centre the average we are finding.

Table 20.4

		Sales	*Σ4 qtrs*	*Σ8 qtrs*	*Average*
Year 1	Q_1	100			
	Q_2	120			
			460		
	Q_3	150		930	116.25
			470		
	Q_4	90		930	116.25
			460		
Year 2	Q_1	110		910	113.75
			450		
	Q_2	110		890	111.25
			440		
	Q_3	140		870	108.75
			430		
	Q_4	80		850	106.25
			420		
Year 3	Q_1	100		830	
			410		
	Q_2	100		820	
			410		
	Q_3	130		810	
			400		
	Q_4	80		800	
			400		
Year 4	Q_1	90		790	
			390		
	Q_2	100		770	
			380		
	Q_3	120			
	Q_4	70			

The data in Table 20.4 are quarterly sales, and the appropriate subset size will be 4 (or 1 year). Summing sales for the first four quarters gives a total of 460 which is again recorded in the middle of the subset (here, between quarters 2 and 3 of year 1). Moving the subset forward by one quarter, and summing gives a total of 470 sales, recorded between quarter 3 and quarter 4 of year one. As above, this process continues up to and including the final subset of 4. None of these sums of four numbers is directly opposite any of the original data points. To bring a total number of sales (and hence the average too) opposite a data point, we may now add the first column of totals in *pairs*, putting the new total between them, and hence opposite a particular time period. (In the first case this will be opposite quarter 3 of year 1.) Each of these new totals is the sum of two sums of four numbers, i.e. a sum of 8 numbers, so we need to divide by 8 to obtain the average.

EXERCISE

Complete the final column of Table 20.4.
(Answer: the averages will be 103.75, 102.5, 101.25, 100, 98.75, 96.25.)

This set of figures is a centred four point moving-average trend. To graph this trend, we plot each of the trend values at the time period where it appears in our calculations, as in Figure 20.6. As with the previous example, there are no trend values for the first two data points, nor the last two. If the size of the subset had been larger, say 12s for monthly data, we would again need to add the Σ12s in pairs to give a sum of 24 and then divide by 24 to get the centred moving average trend.

EXERCISE

How may data points would not have an associated trend value in this case?

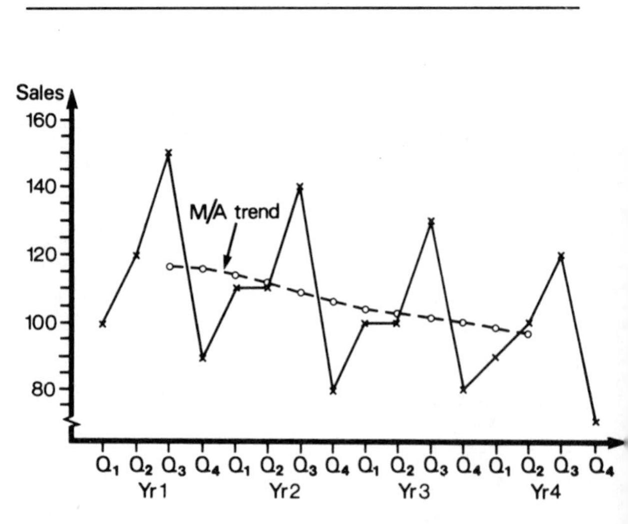

Figure 20.6

MICROSTATS will calculate a centred moving average trend from quarterly data which is placed into a column. If the data were in column C1, then we would use the command MAVE C1 C2, and the trend values would be placed into column C2. Note that MICROSTATS will only calculate moving averages for subsets of four.

EXERCISE

Put the sales data from Table 20.4 into column C1 of MICROSTATS and find the moving average values.

20.3 THE SEASONAL FACTORS

Seasonal effects are familiar to most businesses as sales are higher than average in one part of the year and lower than average in another period. Many government statistics are quoted as seasonally adjusted with statements such as 'the figure for unemployment increased last month but, taking into account seasonal factors, the underlying trend is downwards'. Our aim when analyzing a time series is to identify *when* the actual sales differ from the average (or trend) and to find the *magnitude* of the variations. Knowing such information on sales, for example, would be of considerable benefit in planning both production and stock holding policies for a company.

As we saw in section 20.1, there are two basic models used in analyzing time series data and the method of calculation of the seasonal factors will depend upon which model has been chosen.

20.3.1 Using an additive model

Looking first at an additive model, for a short run of data, we have:

$$A = T + S + R$$

where A is known, T has been isolated by using either the linear regression trend or the moving-average trend and R may be assumed to be zero. Therefore:

$$A = T + S$$

$$\text{or} \quad S = A - T$$

This subtraction may be applied to each data point for which we have a trend value and an actual value.

An example is given in Table 20.5 where we have used the data analyzed in section 20.2 and the linear trend values. Exactly the same process would be followed if we were to use the centred moving average trend except that there would be no values in the first two rows and the last two rows of the $S = A - T$ column.

Table 20.5

		Sales (000)(A)	*Trend (T)*	$S = A - T$
Year 1	Q_1	10	12.875	−2.875
	Q_2	14	13.200	0.800
	Q_3	11	13.525	−2.525
	Q_4	21	13.850	7.150
Year 2	Q_1	11	14.175	−3.175
	Q_2	16	14.500	1.500
	Q_3	10	14.825	−4.825
	Q_4	22	15.150	6.850
Year 3	Q_1	14	15.475	−1.475
	Q_2	18	15.800	2.200
	Q_3	13	16.125	−3.125
	Q_4	22	16.450	5.550
Year 4	Q_1	13	16.775	−3.775
	Q_2	16	17.100	−1.100
	Q_3	9	17.425	−8.425
	Q_4	25	17.75	7.250

Having completed this process we now have a considerable number of 'seasonal factors', but if we examine them a little more closely, we see that there are four different figures associated with quarter 1. A similar observation may be made for each of the other quarters. Each figure may be treated as an estimate of the seasonal factor for that particular quarter, and the average of these four estimates provides a single representative estimate of the seasonal variation for that quarter. The easiest approach to this procedure is to rewrite the final column of Table 20.5, the estimates, into a working table (Table 20.6) that gathers together the various estimates for each quarter.

Table 20.6

	Q_1	Q_2	Q_3	Q_4
Year 1	−2.875	−0.800	−2.525	7.150
Year 2	−3.175	1.500	−4.825	6.850
Year 3	−1.475	2.200	−3.125	5.550
Year 4	−3.775	−1.100	−8.425	7.250
	−11.300	3.400	−18.900	26.800
Average (/4)	−2.825	0.850	−4.725	6.700

The averages will be the additive seasonal factors.

EXERCISE

Work out the $S = A - T$ column from Table 20.3 (Answer: Table 20.7a).

Table 20.7a

Week	*Monday*	*Tuesday*	*Wednesday*	*Thursday*	*Friday*
1	—	—	0.6	1.2	3.2
2	−1.8	−3.2	−0.6	2.2	4.2
3	−1.4	−3.0	−1.4	1.8	—
	−3.2	−6.2	−1.4	5.2	7.4
Average	−1.6	−3.1	−0.467	1.733	3.7

(Note that in some cases we divided by 2 and in other cases by 3 since for this example we have used a moving-average trend which does not give a trend value for the first and last two values; we thus cannot calculate S for these observations.)

Above, we have specified that the seasonal factors are fluctuations *within* a year or a week; thus the effect over the whole period should be zero. The effects should cancel out over the year or week. Looking at the results of Table 20.7a we find that this is not the case! The sum of the averages is 0.266. To overcome this problem we may distribute this excess equally amongst the seasonal factors; i.e. subtract $(0.266/5) = 0.0532$ from each estimate so that the sum of the averages is zero (Table 20.7b).

Note that if the sum of the averages had been negative, we would have *added* a fifth to each estimate: with quarterly data, you divide the sum of averages by four before redistribution. For most practical purposes, this adjustment is not necessary!

EXERCISE

Why do the seasonal factors sum to zero in Table 20.6 so that no corrections are necessary?

Using MICROSTATS with the original data in column C1 and the trend values in column C2 we can find the additive seasonal factors by using the command, ADDM C1 C2 C3 C4 which will place the $(A - T)$ values into column C3 and the residuals into column C4. The four seasonal factors will be shown on the screen.

EXERCISE

Using the sales data in Table 20.5 and MICROSTATS find the additive seasonal factors when

Table 20.7b

Averages	−1.6000	−3.1000	−0.4670	1.7330	3.7000
Correction	−0.0532	−0.0532	−0.0532	−0.0532	−0.0532
	−1.6532	−3.1532	−0.5202	1.6798	3.6468

(a) a linear trend is used, and (b) when a centred moving average trend is used.

20.3.2 Using a multiplicative model

For short runs of data using the multiplicative model, we have:

$A = T \times S \times R$

As before, A and T are known and R may be assumed to be equal to 1. Therefore

$A = T \times S$

or

$S = A/T$

Taking the data from Table 20.1, we have Table 20.8.

Table 20.8

		Sales (A)	*Trend (T)*	$S = A/T$
Year 1	Q_1	10	12.875	0.777
	Q_2	14	13.200	1.061
	Q_3	11	13.525	0.813
	Q_4	21	13.850	1.516
Year 2	Q_1	11	14.175	0.776
	Q_2	16	14.500	1.103
	Q_3	10	14.825	0.675
	Q_4	22	15.150	1.452
Year 3	Q_1	14	15.475	0.905
	Q_2	18	15.800	1.139
	Q_3	13	16.125	0.806
	Q_4	22	16.450	1.337
Year 4	Q_1	13	16.775	0.775
	Q_2	16	17.100	0.936
	Q_3	9	17.425	0.517
	Q_4	25	17.750	1.408

Since there are four quarters in Table 20.9, the sum of these factors should be four (it is 3.999). These seasonal factors are often expressed as percentage figures or seasonal indices:

80.825 105.975 70.275 142.825

Table 20.9 Seasonals table

	Q_1	Q_2	Q_3	Q_4
Year 1	0.777	1.061	0.813	1.516
Year 2	0.776	1.103	0.675	1.452
Year 3	0.905	1.139	0.806	1.337
Year 4	0.775	0.936	0.517	1.408
Average (/4)	3.233	4.239	2.811	5.713
	0.808 25	1.059 75	0.702 75	1.428 25

Using MICROSTATS with the original data in column C1 and the trend values in column C2 we can find the multiplicative seasonal factors by using the command MULM C1 C2 C3 C4 which will place the (A/T) values into column C3 and the multiplicative residuals into column C4. The four seasonal factors will be shown on the screen.

EXERCISE

Using the sales data in Table 20.5 and MICROSTATS find the multiplicative seasonal factors when (a) a linear trend is used, and (b) when a centred moving average trend is used.

20.3.3 Seasonal adjustment of time series

Many published series are quoted as being 'seasonally adjusted'. The aim is to show how the trends in the figures are moving without the hindrance of seasonal variation, which may tend to obscure such trends. This is more, however, than just quoting a trend value since we still have the effects of the cyclical and random variations retained in the quoted figures.

Taking the additive model, we have:

$$A = T + C + S + R$$

and $A - S = T + C + R$

Constructing such an adjusted series relies heavily upon having correctly identified the seasonal factors from the historic data; that is having used the appropriate model over a sufficient time period. There is also a heroic assumption that seasonal factors identified from past data will still apply to current and future data. This may be a workable solution in the short term, but the seasonal factors should be re-calculated as new data becomes available.

By creating a new column in MICROSTATS which has the four seasonal factors repeated through time, and then subtracting this from the column containing the original data, we will create the seasonally adjusted series.

20.4 THE CYCLICAL FACTORS

Although we can talk about there being a cyclical factor in time series data and can try to identify it by using *annual* data, these cycles are rarely of consistent lengths. A further problem is that we would need six or seven full cycles of data to be sure that the cycle were there, and for some proposed cycles this would mean obtaining 140 years of data!

Several cycles have been proposed and a brief outline of a few of these is given below.

(a) Kondratieff Long Wave: 1920s, a 40–60 year cycle, there seems to be very little evidence to support the existence of this cycle.
(b) Kuznets Long Wave: a 20 year cycle, there seems to be evidence to support this from studies of GNP and migration.
(c) Building cycle: a 15–20 year cycle, some agreement that it exists in various countries.
(d) Major and minor cycles: Hansen 6–11 year major cycles, 2–4 year minor cycles; cf. Schumpeter inventory cycles. Schumpeter: change in rate of innovations leads to changes in the system.
(e) Business cycles: recurrent but not periodic, 1–12 years, cf. minor cycles, trade cycle.

At this stage we can construct graphs of the annual data and look for patterns which match one of these cycles. Since one cycle may be superimposed upon another, this identification is likely to prove difficult. Removing the trend from the data may help, so that we consider graphs of $(A - T) = C + R$ rather than graphs of the original time series. (Note that there is no seasonal factor since we are dealing with annual data.)

Producing such a series using either a spreadsheet or MICROSTATS is relatively easy since we have the original data in one column and can place the trend figures into another column, and then subtract one from the other.

20.5 THE RESIDUAL OR RANDOM FACTOR

We have made a series of assumptions about this element of a time series. These assumptions may be checked by now looking at the *actual* values (Table 20.10) for R at each data point; i.e.

$$R = A - T - S$$

where S is the average seasonal variation (Table 20.6)

or $R = A/T \times S$

where S is the average seasonal index (Table 20.9)

In the additive case, the average value of R was assumed to be 0; here the average is 0.106 25 which is not particularly dissimilar. For this type of model, we should also look for long runs of positive or negative values, as this may be evidence of a cyclical effect. With a multiplicative model, R was assumed to be 1; here the total is 16.001 giving an average of 1. A long run of values above or below 1 may again give evidence of a cyclical effect in the model.

20.6 PREDICTIONS

To make a prediction, we will recombine the time series elements that have been identified above. The first step is to extend the trend into the future. For a linear trend we have seen (section 20.2.1) that extension of the trend is achieved by substituting appropriate values for x; however, the problem is somewhat more difficult from a

Table 20.10

		Sales (A)	$R = A - T - S$ (additive)	$R = A/TS$ (multiplicative)
Year 1	Q_1	10	−0.05	0.961
	Q_2	14	−0.05	1.001
	Q_3	11	2.20	1.157
	Q_4	21	0.45	1.061
Year 2	Q_1	11	−0.35	0.960
	Q_2	16	0.65	1.041
	Q_3	10	−0.10	0.961
	Q_4	22	0.15	1.017
Year 3	Q_1	14	1.35	1.120
	Q_2	18	1.35	1.075
	Q_3	13	1.60	1.147
	Q_4	22	−1.15	0.936
Year 4	Q_1	13	−0.95	0.959
	Q_2	16	−0.25	0.883
	Q_3	9	−3.70	0.736
	Q_4	25	0.55	0.986

moving-average trend. Predictions of the trend are done by extending its graph in an appropriate direction, consistent with its past behaviour. There are two problems here: there is a considerable amount of judgement (or assumption) used in the process and, since the first few predictions are the best (being closest to the data), these are being used to find a trend for past data. (This is because the moving-average trend ends before the end of the data, section 20.2.2.)

Having obtained the appropriate number of trend values, the average seasonal factors (for the predicted quarter or day) are added to, or multiplied by, the trend value to give a prediction. Extending the sales of shirts example, using a linear trend purely for convenience, we have the results given in Table 20.11.

Table 20.11a Additive

		Trend (T)	Seasonal (S)	Prediction = $T + S$
Year 5	Q_1	18.075	−2.825	15.25
	Q_2	18.400	+0.850	19.25
	Q_3	18.725	−4.725	14.00
	Q_4	19.050	+6.700	25.75
Year 6	Q_1	19.375	−2.825	16.55

Table 20.11b Multiplicative

		Trend (T)	Seasonal (S)	Prediction = $T \times S$
Year 5	Q_1	18.075	0.808 25	14.609
	Q_2	18.400	1.059 75	19.499
	Q_3	18.725	0.702 75	13.159
	Q_4	19.050	1.428 25	27.208
Year 6	Q_1	19.375	0.808 25	15.660

EXERCISE

Make sure that you can get the answers in Table 20.11.

Time series models are often judged on the accuracy of the predictions.

Seasonal or other factors are often removed from past or present time series data in order to highlight salient features. For example:

deseasonalized data $= A - S$ (for appropriate period)
detrended data $= A - T$

20.7 EXPONENTIALLY-WEIGHTED MOVING AVERAGES

Where we require short term forecasts, for example in stock control situations, we may use a smoothing method known as **exponentially-weighted moving averages** (EWMA). This takes into account the movements in the data that have occurred in the past, but weights the more recent figures more heavily than those in the distant past. The prediction of the next value of y is shown in the equation given below:

$$y_{t+1} = y_t + (1-\alpha)y_{t-1} + (1-\alpha)^2 y_{t-2} + (1-\alpha)^3 y_{t-3} + \ldots$$

where α (alpha) is a **smoothing constant**. Further analysis of this equation shows that the right-hand side can be reduced to the sum of only two terms:

$$y_{t+1} = y_t + (1-\alpha)y_t$$

provided that we have an initial prediction for y_t. (In practice we often use the first actual value of y as the first predicted value ($\hat{y}_1 = y_1$) since the effects of any error in predicting the first actual value will very quickly be minimized.)

The choice of alpha will depend upon experience with the data and how quickly we wish the predictions to react to changes in the data. If we choose a *high* value for alpha, the predictions will be *very sensitive* to changes in the data. Where the data is subject to random shocks, then these will be passed on to the predictions. Where a *low* value of alpha is used, the predictions will be *slow to react* to changes in the data, but will be less affected by random shocks. Thus if we are dealing with data which rarely experiences random shocks, we would want to use a high value for alpha so that genuine changes of direction in the data are reflected in the predictions as soon as possible. In practice, values of alpha are usually chosen from within the range 0.3 to 0.6.

An example of the use of EWMA is shown in Table 20.12.

Table 20.12 EWMA using alpha = 0.3

Year (t)	*Sales (y_t)*	*$0.3y_t$*	*$0.7\hat{y}_t$*	$\hat{y}_{t+1}$
0				147
1	147	44.1	102.9	147.0
2	152	45.6	102.9	148.5
3	163	48.9	103.95	152.85
4	162	48.6	106.995	155.595
5	158	47.4	108.917	156.317
6	161	48.3	109.422	157.722
7	164	49.2	110.405	159.605
8	171	51.3	111.724	163.024
9	180	54.0	114.117	168.117
10	175	52.5	117.682	170.182

Table 20.13 EWMA using alpha = 0.6

Year (t)	*Sales (y_t)*	$\hat{y}_{t+1}$
0		147
1	147	147.0
2	152	150.0
3	163	157.8
4	162	160.32
5	158	158.928
6	161	160.171
7	164	162.469
8	171	167.587
9	180	175.035
10	175	175.014

As you can see, as a new data point becomes available, the model allows a prediction to be made for the next value. Using an alpha value of 0.3 has given reasonable predictions of the next period's sales, but given the fairly static nature of the data, a higher value of alpha could give better results. Using an alpha value of 0.6 gave the results shown in Table 20.13. This gives slightly better predictions for sales.

As you can see from the layout of Tables 20.12 and 20.13, the building of EWMA models is a natural candidate for the use of spreadsheets. We suggest that you use individual columns for each calculation (as in Table 20.12) when you first build such models, but that the whole formula can be put into a single cell (as in Table 20.13) once you see the way in which the model works. Building such spreadsheet models will allow you to experiment with different values of alpha to find the 'best' value for a particular set of data.

There are many extensions to this basic

EWMA model, and for those who are interested in extending their knowledge, we recommend the book by Lewis (1981). *Scientific Inventory Control*, Butterworths.

20.8 CONCLUSIONS

Any analysis of time series data must inevitably make the heroic assumption that the behaviour of the data in the past will be a good guide to its behaviour in the future. For many series, this assumption will be valid, but for others, no amount of analysis will predict the future behaviour of the data, for example, Stock Market prices. Wherever the data reacts quickly to external information (or random shocks) then time series analysis on its own, will not be able to predict what the new figure will be.

Even for more 'well-behaved' data, shocks to the system will lead to variations in the data which cannot be predicted, and thus the predictions should be treated as a guide to the future and not as a statement of exactly what will happen. For many businesses, the process of attempting to analyze the past behaviour of data may be more valuable than the actual predictions obtained from the analysis, since it will highlight both trends and seasonal effects which can then be taken into account in planning for the future.

20.9 PROBLEMS

1. Sales from company JCR's motor division have been monitored over the last four years and are presented below:

Year	*Quarter 1*	*Quarter 2*	*Quarter 3*	*Quarter 4*
1	20	30	39	60
2	40	51	62	81
3	50	64	74	95
4	55	68	77	96

   (a) Construct a graph of this data.
   (b) Find a centred four-point moving average trend and place it on your graph.
   (c) Calculate the four seasonal components using an additive model.
   (d) Use your model to predict sales for years 5 and 6.
   (e) Prepare a brief report to the company on future sales.
2. Use the data in question 1 to calculate a linear regression trend through the data, calculate the seasonal components, again using an additive model, and predict sales for years 5 and 6. Does this model represent an improvement upon the previous model?
3. The number of calls per day to a Local Authority department have been logged for a four-year period and are presented below:

Year	*Quarter 1*	*Quarter 2*	*Quarter 3*	*Quarter 4*
1	20	10	4	11
2	33	17	9	18
3	45	23	11	25
4	60	30	13	29

   (a) Construct a graph of this data.
   (b) Find a linear regression trend through the data.
   (c) Calculate the four seasonal components using a multiplicative model.
   (d) Predict the number of calls for the next two years.
4. Use the data and the trend from question 3 to find the average residual component for the time series using:
   (a) an additive model;
   (b) a multiplicative model.
5. A company's advertising expenditure has been monitored for 3 years, giving the following information:

Year	*Advertising expenditure*			
	Q1	*Q2*	*Q3*	*Q4*
1	10	15	18	20
2	14	16	19	23
3	16	18	20	25

   (a) Calculate a linear regression trend for this data.

(b) Graph the data and the trend.
(c) Find the additive seasonal components for each quarter.
(d) Predict the level of advertising expenditure for each quarter of year 4.

6. Calculate the multiplicative seasonal components for each quarter, using the data in Exercise 3. Does this represent an improvement on the previous model?
7. Using the additive model and the data in Exercise 3, calculate the average value of the residual component.
8. The level of economic activity in a county of England have been recorded for 15 years as follows:

Year	*Activity rate*	*Year*	*Activity rate*
1	52.7	8	52.6
2	54.4	9	50.7
3	54.7	10	49.8
4	55.4	11	48.3
5	53.8	12	43.8
6	53.5	13	40.3
7	53.4	14	37.8
		15	35.1

(a) Graph the data.
(d) Find a linear regression trend through this data and place it on your graph.
(c) Predict the activity rate for year 25.
(d) How confident are you of your prediction?

9. Sales over the last 24 months have been recorded for a certain company, and are shown in the following table. Find the exponentially-weighted moving averages for this data using the following values of alpha.
(a) 0.2
(b) 0.4
(c) 0.6

Month:	1	2	3	4	5	6	7	8	9	10
Sales:	225	230	226	240	245	260	280	310	320	280
Month:	11	12	13	14	15	16	17	18	19	20
Sales:	240	220	230	238	239	251	255	265	300	325
Month:	21	22	23	24						
Sales:	356	300	264	231						

Plot the series and the EWMAs and decide which of these values is most appropriate in this case.

10. A package holiday company sends out brochures to prospective clients. At certain times of the year it employs extra staff to cope with the demand. Monthly enquiries over the last three years have been recorded and are shown below.

Month :	1	2	3	4	5
Enquiries:	2000	2500	2000	1800	1400
Month :	6	7	8	9	10
Enquiries:	800	200	100	900	1400
Month :	11	12	13	14	15
Enquiries:	1600	1800	2300	2800	2400
Month :	16	17	18	19	20
Enquiries:	2000	1400	600	200	50
Month :	21	22	23	24	25
Enquiries:	1000	1600	1900	2000	2600
Month :	26	27	28	29	30
Enquiries:	3000	2800	1900	1100	400
Month :	31	32	33	34	35
Enquiries:	100	50	1100	1800	2000
Month :	36	37	38	39	40
Enquiries:	2300	2800	3100	2900	1800

Use this information to find the EWMA and hence predict the number of enquiries in the following month. (Use an alpha value of 0.3.) Attempt to find a better value of alpha to use with this data.

PART SIX
CONCLUDING EXERCISE

Collect quarterly data from *Economic Trends Annual Supplement for the Retail Prices Index, Average Wages, and the Money Supply* for a period of at least the last six years. Use the ideas contained in this section to analyze this data and prepare a report on the relationships that you discover.

The following list represents some of the relationships you might decide to investigate:

1. graphs of each series;
2. regression relationships against time;
3. seasonal variations;
4. dependence of RPI on Wages;
5. dependence of RPI on Money Supply;
6. dependence of RPI on Wages and Money Supply.

The size of the model you will be able to build will vary with the availability of data, but we suggest a minimum of 30 time periods. Put your data into MICROSTATS and experiment with the relationships suggested above. Are there any more meaningful relationships? Should any of the relationships be non-linear?

PART SEVEN MATHEMATICAL MODELS

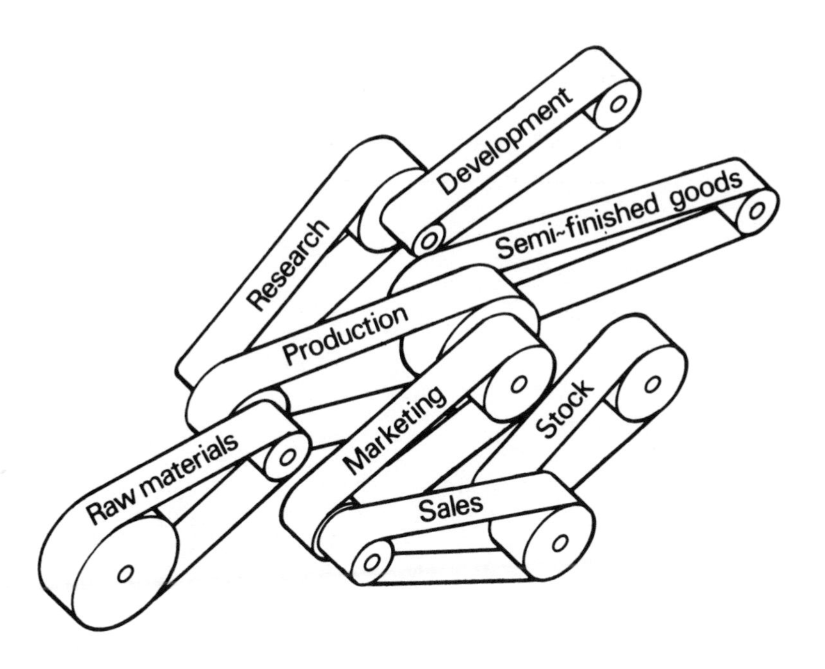

Business activity by its very nature is complex. Our need to understand the complexity will depend on our rôle within the business situation and on the problems we need to manage. In order to understand any problem, it is likely that we will make a number of simplifying assumptions. To understand the implications of any change proposed, it is necessary to explore how one factor impacts on one or more other factors. A mathematical model attempts to describe a real-world problem in a concise and manageable way using mathematical and statistical techniques. A mathematical model will usually involve the use of equations with some values kept fixed as parameters and other allowed to vary as input (independent) variables or output (dependent) variables.

Mathematical models are unlikely ever to give us perfect answers — who would want 'people' managers if they did? What these models do, is to allow us to work through our assumptions about the business situation and see the possible consequences. We can try out all sorts of ideas on a mathematical model at little or no cost. We can even use such models to plan for disasters. For example, what will happen to cash flow next month if we lose a major order? We can always test and improve a mathematical model by simulating realistic situations and making a comparison.

The models described in the following chapters have found a wide range of business applications. Indeed, the computer solutions that we read about are often based on the application of these models.

EXERCISES

1. Consider one aspect of one type of business and list the events which are certain to happen, and list the events which could possibly happen.
2. List a number of business situations which could be modelled mathematically or statistically.
3. List a number of factors which might limit the expansion of a business.

LINEAR PROGRAMMING 21

Linear programming describes graphical and mathematical procedures that seek the optimum allocation of scarce or limited resources to competing products or activities. It is one of the most powerful techniques available to the decision-maker and has found a range of applications in business, government and industry. The determination of an optimum production mix, media selection and portfolio selection are just a few possible examples. They all require definition and, for a numerical solution, mathematical formulation. Typically, the objective is either to maximize the benefits while using limited resources or to minimize costs while meeting certain requirements.

21.1 DEFINITION OF A FEASIBLE AREA

If a company needs to decide what to produce, as a matter of good management practice it would want to know all the possible options. In mathematical terms, it would want a definition of a **feasible space** or **feasible area**. Suppose that the company were involved with the production of two products, X and Y, and that each unit of X required 1 hour of labour and each unit of Y required 2 hours of labour. Labour hours, in this case, are resource requirements, X and Y the competing products. If all the labour required were available at *no* cost, there would be *no* scarcity and *no* production problem. However, if only 40 hours of labour were available each week then there would be an allocation problem. A decision would have to be taken as to whether only X, or only Y, or some combination of the two be produced. In mathematical terms the allocation problem could be written as an inequality:

$$x + 2y \leqslant 40$$

where $\leqslant$ is read as 'less than or equal to'. This is of course, properly written as $X + 2Y \leqslant 40$.

As we have already seen (Chapter 7) an equation can be represented graphically by a straight line. All the points on the line would provide a solution to the equation. If we were dealing only with the equation $X + 2Y = 40$, we would first find two points that provide solutions, plot these on a graph and finally join them with a straight line. Two possible solutions are:

$X = 0$, $Y = 20$;
$X = 40$, $Y = 0$.

An interpretation of these two solutions would be that if only Y is to be produced then 20 units can be made and if only X is to be produced then 40 units can be made. Another possible solution would be to produce 10 units of X and 15 units of Y. All three solutions are shown in Figure 21.1.

In the same way that an equation can be represented by a line, an inequality can be represented

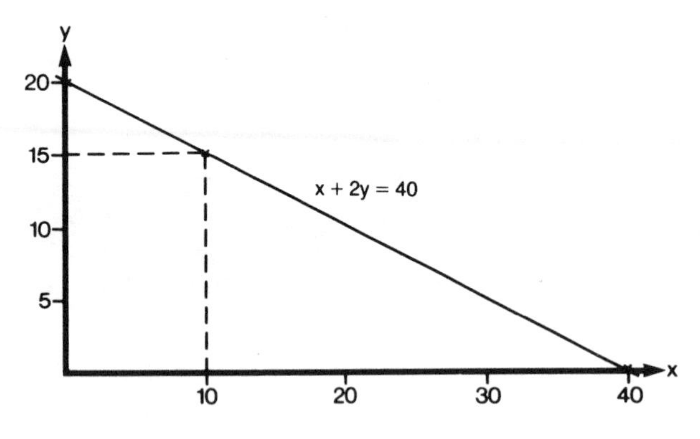

Figure 21.1 The definition of a line

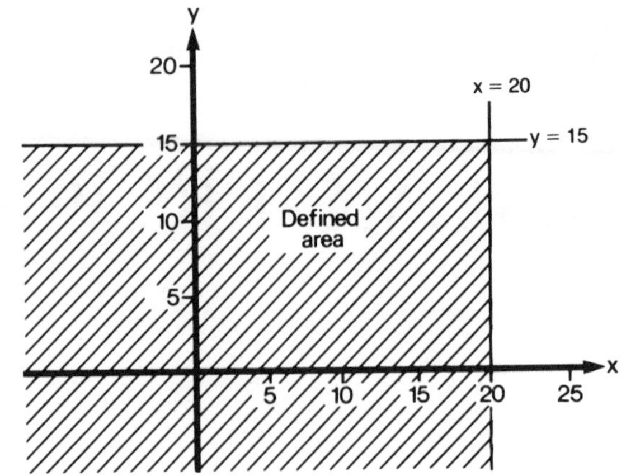

Figure 21.3 Area defined by two inequalities

by an *area*. If we consider the example $X + 2Y \leqslant 40$, the three points

$X = 0,\ Y = 20$
$X = 40,\ Y = 0$
$X = 10,\ Y = 15$

still provide solutions. In addition to these points, others that give an answer of less than 40 are also acceptable. The point $X = 20$, $Y = 5$ (answer 30) is acceptable whereas the point $X = 20$, $Y = 15$ (answer 50) is not. The 'less than or equal to' inequality defines an area that lies to the left of the line as shown in Figure 21.2.

Inequalities can also take the form

$X \leqslant 20$
$Y \leqslant 15$

The area jointly defined by these two inequalities is shown in Figure 21.3.

The definition of what is possible can be represented mathematically by a number of inequalities and together they can define a feasible area.

EXERCISES

1. Show graphically the following inequalities:
   (a) $5X + 2Y \leqslant 80$;
   (b) $X + Y = 30$;
   (c) $X \leqslant 10$;
   (d) $Y \geqslant 12$.

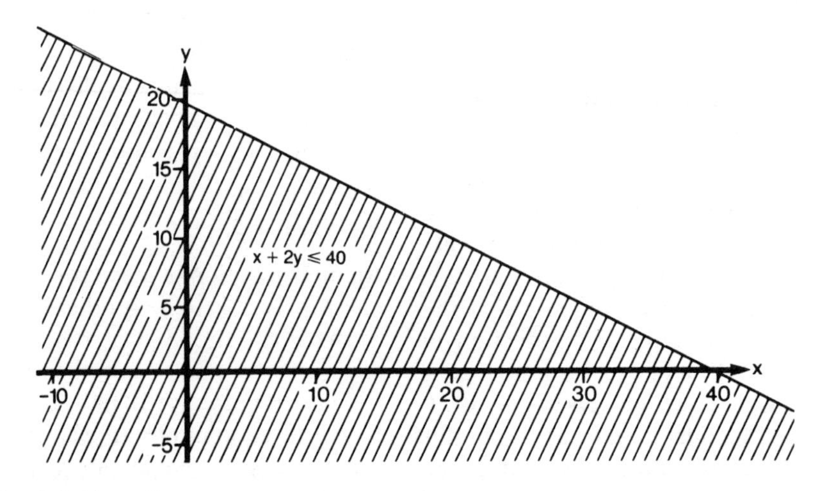

Figure 21.2 The definition of area

2. Show graphically a feasible area defined jointly by:

$$2Y + 5X \geqslant 140$$
$$X + Y \leqslant 40$$

21.2 THE SOLUTION OF A LINEAR PROGRAMMING PROBLEM

All linear programming problems have three common characteristics:

1. a linear objective function;
2. a set of linear structural constraints;
3. a set of non-negativity constraints.

The **objective function** is a mathematical statement of what management wishes to achieve. It could be the maximization of profit, minimization of cost or some other measurable objective. **Linearity** implies that the parameters of the objective function are *fixed*; for example, a constant cost per unit or constant contribution to profit per unit. The **structural constraints** are the physical limitations on the objective function. They could be constraints in terms of budgets, labour, raw materials, markets, social acceptability, legal requirements or contracts. **Linearity** means that all these constraints have fixed coefficients and can be represented by straight lines on a graph. Finally, the **non-negativity constraints** limit the solution to positive (and *meaningful*) answers only.

21.2.1 The maximization problem

Consider the following example where the objective is to maximize one function (profit).

EXAMPLE

A small company produces two products, X and Y. Suppose that each unit of product X requires 1 hour of labour and 6 tons of raw materials, whereas each unit of product Y requires 2 hours of labour and 5 tons of raw materials. Suppose also that the number of labour hours available each week is 40 and the available raw materials each week is 150 tons. If the contribution to profit is £2 per unit of X and £3 per unit of Y, determine the weekly production mix that will maximize profits.

This linear programming problem can be formulated mathematically as:

maximize

$z = 2X + 3Y$ **the linear objective function**

subject to

$$\left.\begin{aligned} X + 2Y &\leqslant 40 \\ 6X + 5Y &\leqslant 150 \end{aligned}\right\}$$ **the linear structural constraints**

and

$X \geqslant 0,\ Y \geqslant 0$ **the non-negativity constraints**

The objective function is the sum of the profit contributions from each product. If we were to decide to produce 5 units of X and 7 units of Y then the total contribution to profit would be $z = 2 \times 5 + 4 \times 7 = £38$. Linear programming provides a method to find what combination of X and Y will maximize the value of z subject to the given constraints.

In this example there are two structural constraints, one in terms of available labour and the other in terms of available raw materials. The usefulness of the solutions will depend on how realistically the constraints model the decision problem. If, for example, marketing considerations were ignored the optimum solution may suggest production levels that are incompatible with sales opportunities. Mathematically, we need a structural constraint to correspond to each limitation on the objective function. If we consider the labour constraint as an example, the coefficients represent the labour requirements of 1 hour for product X and 2 hours for product Y. A product mix of 5 units of X and 7 units of Y will require only 19 hours of labour ($1 \times 5 + 2 \times 7$), does not exceed the available labour time of 40 hours and satisfies the first constraint. A production mix of 18 units of X and 13 units of Y exceeds the available labour time, does not satisfy the constraint and therefore could not provide a possible solution. The product mix of 5 units of X and 7 units of Y also satisfies the remaining structural constraint, $6 \times 5 + 5 \times 7 \leqslant 150$, and is one of the possible or feasible solutions. Jointly the structural constraints define the feasible area. In this example, the feasible area is defined by the labour and raw material constraints. No account is taken of the many other factors that could affect the optimum production mix.

The feasible area is found graphically by treating each constraint as an *equation*, plotting the corresponding straight lines and defining an area bounded by the straight lines which satisfies all the *inequalities*. We proceed as follows.

The **labour constraint**: if we were to use all the labour time available then

$$X + 2Y = 40$$
$$(X = 0,\ Y = 20)(X = 40,\ Y = 0)$$

The two points shown in brackets are the 'one-product' solutions; we can use the 40 hours of labour to make 20 units of Y each week or 40 units of X. A line joining the two points will show combinations of X and Y that will require 40 hours of labour.

The **raw materials constraint**: if we were to use all the raw material available then

$$6X + 5Y = 150$$
$$(X = 0,\ Y = 30)(X = 25,\ Y = 0)$$

A line joining two possible solutions shown in brackets will show the combinations of X and Y that will require 150 tons of raw materials.

The **non-negativity constraints**: the constraints $x \geqslant 0$ and $y \geqslant 0$ exclude any possibility of negative production levels which have no physical counterpart. Together they include the x-axis and the y-axis as possible boundaries of the feasible area.

The feasible area defined by the two structural constraints and the two non-negativity constraints is shown in Figure 21.4. The feasible area is contained within the boundaries of $OABC$. It is now a matter of deciding which of the points in this area provides an optimum solution. The choice is determined by the objective function. In this example, the choice of whether to produce just X, or just Y or some combination of the two will depend on the **relative profitability** of the two products.

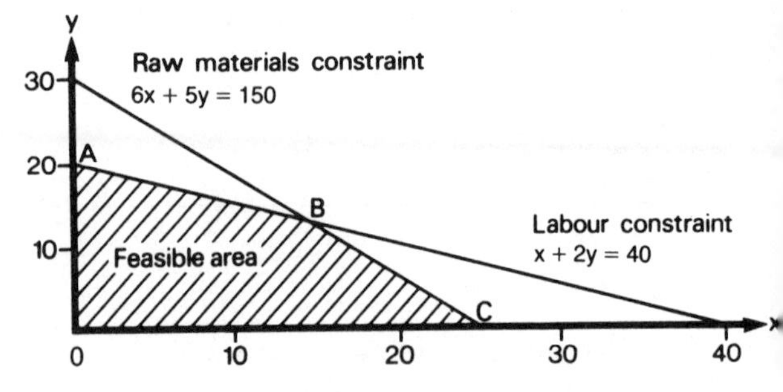

Figure 21.4 The definition of a feasible area

Profit has been expressed as a mathematical function $z = 2x + 3y$, where z is the profit level. If we were to fix the level of profits, z, the necessary combination of X and Y could be shown graphically as a straight line. This is referred to as a **trial profit** line. If we let $z =$ £30 then:

$$30 = 2X + 3Y$$
$$(X = 0,\ Y = 10)(X = 15,\ Y = 0)$$

A profit of £30 can be made by producing 15 units of X, or 10 units of Y or some combination of the two. This trial profit is shown in Figure 21.5. All the points on the trial profit line will produce a profit of £30. The gradient gives the 'trade-off' between the two products. In this case, to maintain a profit level you would need to trade 2 units of Y against 3 units of X (a loss of £6 against a gain of £6). The trial profit line shown violates none of the constraints so profit can be increased from £30.

Consider a second trial profit line where the profit level is fixed at £60.

If we let $z =$ £60 then

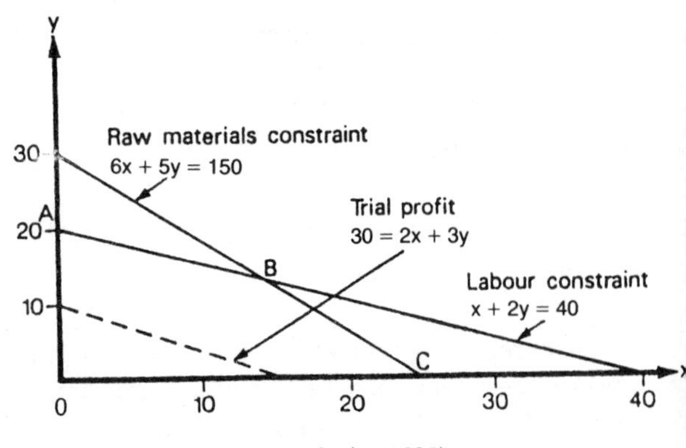

Figure 21.5 A trial profit ($z =$ £30)

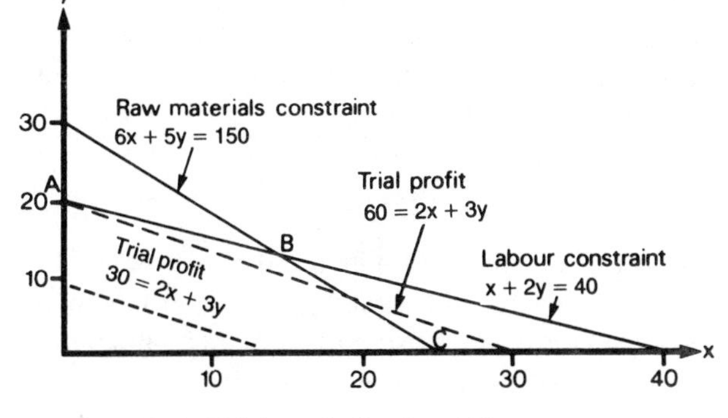

Figure 21.6 Trial profit fixed at £60

$60 = 2X + 3Y$
$(X = 0,\ Y = 20)(X = 30,\ Y = 0)$

This trial profit line is shown in Figure 21.6.

This second trial profit line is higher than and *parallel* to the first. It can be seen that a profit of £60 can be achieved at point *A*, and that some points on the trial profit line violate the raw materials constraint. To operate at point *A* would exhaust all the available labour hours but leave surplus raw materials. This solution can be improved upon by trading-off the more labour intensive product *Y* against the more raw material intensive product *X*. In terms of the graphical approach, we can note that any line that is **higher** and **parallel** to the existing trial profit line represents an improvement. By inspection we can see from Figure 21.6 that higher trial profit lines will eventually lead to the optimum point *B*. This point *B* ($x = 14\frac{2}{7}$, $Y = 12\frac{6}{7}$) can be determined directly from the graph, or by simultaneously solving the two equations that define lines crossing at point *B*. The resultant profit is found by substitution into the objective funcion:

$z = 2 \times 14\frac{2}{7} + 3 \times 12\frac{6}{7} = £67\frac{1}{7}$ per week

---

EXERCISE

A manufacturer of fitted kitchens produces two units, a base unit and a cabinet unit. The base unit requires 90 minutes in the production department and 30 minutes in the assembly department. The cabinet unit requires 30 minutes and 60 minutes respectively in these departments. Each day 21 hours are available in the production department and 18 hours are available in the assembly department. It has already been agreed that not more than 15 cabinet units are produced each day. If base units make a contribution to profit of £2 per unit and cabinet units £5 per unit, what product mix will maximize the contribution to profit and what is this maximum?

(Answer: 6 base units, 15 cabinet units, contribution to profit = £87.)

21.2.2 The minimization problem

Consider the following example where the objective is to minimize one function (cost).

EXAMPLE

A company operates two types of aircraft, the RS101 and the JC111. The RS101 is capable of carrying 40 passengers and 30 tons of cargo, whereas the JC111 is capable of carrying 60 passengers and 15 tons of cargo. The company is contracted to carry at least 480 passengers and 180 tons of cargo each day. If the cost per journey is £500 for a RS101 and £600 for a JC111, what choice of aircraft will minimize cost?

This linear programming problem can be formulated mathematically as:
Minimize

$z = 500X + 600Y$ **the linear objective function**

where *X* is the number of RS101 and *Y* is the number of JC111 subject to the constraints

$$\left.\begin{array}{l} 40X + 60Y \geqslant 480 \\ 30X + 15Y \geqslant 180 \end{array}\right\}$$ **the linear structural constraints**

and

$X \geqslant 0,\ Y \geqslant 0$ **the non-negativity constraints**

In this case we are attempting to minimize the cost of a service subject to the operational constraints. These structural constraints, the requirement to carry so many passengers and so many tons of cargo, are expressed as 'greater than or equal to'. The inequalities are again used to define the feasible area.

The **passenger constraint**: if we were to carry the minimum number of passengers then

$40X + 60Y = 480$
$(X = 0,\ Y = 8)(X = 12,\ Y = 0)$

We could use 8 JC111s to carry 480 passengers or 12 RS101s or some combination of the two as given by the above equation.

The **cargo constraint**: if were to carry the minimum amount of cargo then

$30X + 15Y = 180$
$(X = 0,\ Y = 12)(X = 6,\ Y = 0)$

We could use 12 JC111s to carry 180 tons of cargo or 6 RS101s or some combination of the two as given by the above equation.

The **non-negativity constraints**: to ensure that the solution excludes negative numbers of aircraft, $X \geqslant 0$ and $Y \geqslant 0$ are included as possible boundaries of the feasible area.

The resultant feasible area is shown in Figure 21.7. The objective is to locate the point of minimum cost within the feasible area.

We proceed as before, by giving a convenient value to z which defines a trial cost line, but in this case attempting to make the line as near to the origin as possible, while retaining at least one point within the feasible area so as to minimize cost. If we let $z =$ £6 000 then

$6\,000 = 500X + 600Y$
$(X = 0,\ Y = 10)(X = 12,\ Y = 0)$

A cost of £6 000 will be incurred by operating 12 RS101s, or 10 JC111s or a combination of the two as defined by the above equation, including combinations which are non-feasible. This trial cost line is shown in Figure 21.8.

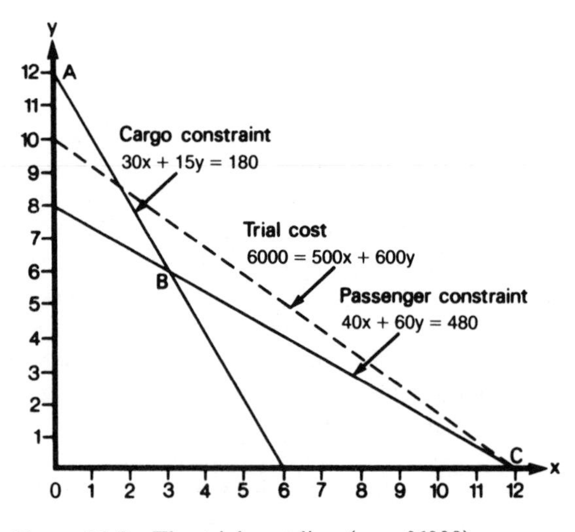

Figure 21.8 The trial cost line ($z =$ £6000)

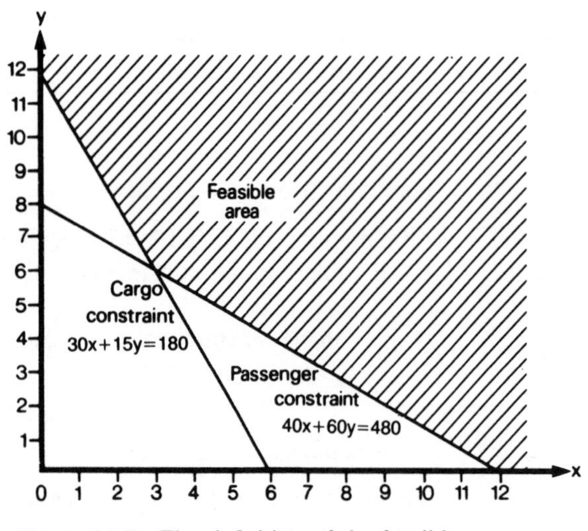

Figure 21.7 The definition of the feasible area

All lines *lower* than and *parallel* to the trial cost line show aircraft combinations that produce lower costs. By inspection we can see that point B is the point of lowest cost ($X = 3,\ Y = 6$). The level of cost corresponding to the use of 3 RS101s and 6 JC111s can be determined from the objective function:

$z = 500 \times 3 + 600 \times 6 =$ £5 100 per day

EXERCISE

A company currently operates two types of van, the 'Loader' and the 'Superloader', which it uses

for the transportation of pallets from its production plant to its distribution plant. The company fleet consists of four Loader vans and one Superloader which it needs to increase to meet increased demand. The Loader is capable of carrying 30 pallets a day and the Superloader 40 pallets a day with daily costs of £100 and £120 respectively. If demand estimates require the movement of 240 pallets each day, how should the fleet be increased (from the existing four Loaders and one Superloader) so as to minimize transportation costs?
(Answer: add two more Superloaders to fleet.)

21.3 SPECIAL CASES

Linear programming problems do not always yield a unique optimal solution. There are a number of special cases and we shall consider just two of them: no feasible solution and multiple optimum solutions.

21.3.1 No feasible solution

If the constraints are mutually exclusive, no feasible area can be defined and no optimum solution can exist. Consider again the maximization problem.

Maximize $z = 2X + 3Y$ **the linear objective function**

subject to

$$\left.\begin{aligned} X + 2Y &\leqslant 40 \\ 6X + 5Y &\leqslant 159 \end{aligned}\right\} \text{ the linear structural constraints}$$

and

$X \geqslant 0,\ Y \geqslant 0$ **the non-negativity constraints**

The feasible area has been defined by the constraints as shown earlier in Figure 21.4. Suppose that in addition to the existing constraints the company is contracted to produce at least 30 units each week. This additional constraint can be written as:

$$X + Y \geqslant 30$$

As a boundary solution the constraint would be:

$$X + Y = 30$$
$$(X = 0,\ Y = 30)(X = 30,\ Y = 0)$$

The three structural constraints are shown in Figure 21.9.

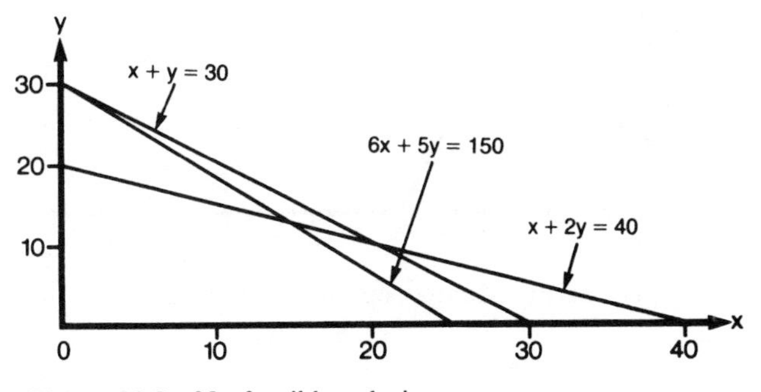

Figure 21.9 No feasible solution

This case presents the manager with demands which cannot *simultaneously* be satisfied.

21.3.2 Multiple optimum solutions

A multiple optimum solution results when the objective function is **parallel** to one of the boundary constraints. Consider the following problem: Minimize

$z = 600X + 900Y$ **the linear objective function**

subject to

$$\left.\begin{aligned} 40X + 60Y &\geqslant 480 \\ 30X + 15Y &\geqslant 180 \end{aligned}\right\} \text{ the linear structural constraints}$$

and

$X \geqslant 0,\ Y \geqslant 0$ **the non-negativity constraints**

This is the aircraft scheduling problem from section 21.2.2 with different cost parameters in the objective function. If we let $z =$ £8 100 then

$$8\,100 = 600X + 900Y$$
$$(X = 0,\ Y = 9)(X = 13.5,\ Y = 0)$$

The resultant trial cost line is shown in Figure 21.10.

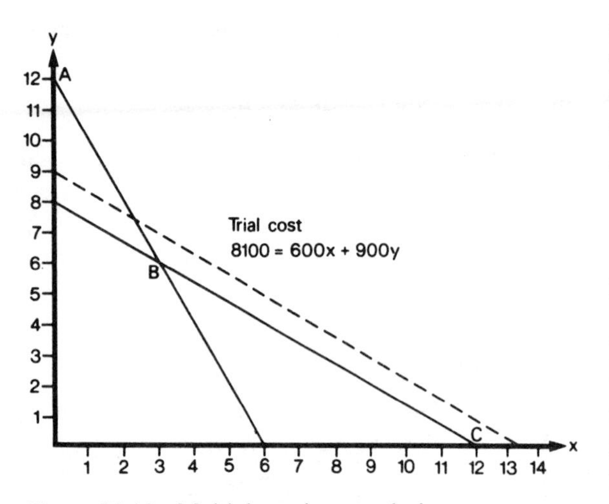

Figure 21.10 Multiple optimum solutions

This line is parallel to the boundary line BC. The lowest acceptable cost solution will be coincidental with the line BC making point B, point C and any other points on the line BC optimal. Multiple optimum solutions present the manager with choice and hence some flexibility.

EXERCISE

A textile firm makes two types of cloth, a Sutton Tweed and a Moseley Mohair. Each yard of tweed requires 3 oz of shoddy, 12 oz of mungo and 2 oz of flock, whilst the mohair requires 2 oz of shoddy, 5 oz of mungo and 7 oz of flock.

An agreement with the Dudley Mungo Manufacturers limits the amount of mungo available each day to 3 000 oz. The transport restrictions only enable 800 oz of shoddy to be delivered each day, while conservation restrictions on the hunting of flock limits this to 1 500 oz per day.

It has been estimated that the profit on tweed will be 20p per yard, and on mohair 15p per yard.

1. How many yards of each cloth should be produced and what is the maximum profit?
2. If a further constraint is added, that at least 300 yards of tweed be produced, what should the production mix be?

(Answer: (a) 170.6 yards of mohair, 152.9 yards of tweed, profit = £56.17; (b) constraints do not define a feasible area hence no optimum solution exists.)

21.4 CONCLUSIONS

Linear programming provides the decision-maker with a means of formulating and solving a wide range of problems. If these problems are defined in terms of two variables, then the solution can be found by graphical methods as shown. However, in practice, decision problems are likely to have three or more variables, one or more objective functions and a number of complex constraints. To solve these problems a number of methods have developed, such as the **Simplex Method**, but these are beyond the scope of this book.

There are a number of computer programs available for the solution of linear programming problems. Many are easy to use and their successful application will depend on your skills at formulating a realistic mathematical model.

21.5 PROBLEMS

1. (a) Maximize $z = 2x + 2y$

$$\begin{aligned} \text{subject to}\quad 3x + 6y &\leq 300 \\ 4x + 2y &\leq 160 \\ y &\leq 45 \\ x &\geq 0 \\ y &\geq 0 \end{aligned}$$

You should include in your answer the values of the optimum point and the corresponding value of z.

(b) Determine the optimum solution if the objective function were changed to

$$z = x + 3y.$$

(c) Determine the optimum solution if the objective function were changed to

$$z = 5x + 10y.$$

2. (a) Minimize $z = 2x + 3y$

$$\begin{aligned} \text{subject to} \quad 3x + y &\geqslant 15 \\ 0.5x + y &\geqslant 10 \\ x + y &\geqslant 13 \\ x &\geqslant 1 \\ y &\geqslant 4 \end{aligned}$$

   (b) Identify the acting constraints (those that affect the solution).
   (c) If the constraint $y \geqslant 4$ were changed to $y \geqslant 8$ how would this affect your solution?
3. A company produces a home computer and a small business computer. Each finished computer requires assembly, testing and packaging time. The home computer takes 10 hours to assemble, half an hour to test and one hour to pack. The business computer takes 24 hours to assemble, two hours to test and three hours to pack. Each week there are 600 hours available in the assembly department, 40 hours available in the testing department and 90 hours available in the packaging department. It has been estimated that weekly demand is for at most 70 home computers and at most 15 business computers of this make. The selling price for the business computer is £299 and £99 for the home computer.
   (a) Formulate a linear programming model for the maximization of sales revenue.
   (b) Solve your linear programming model by the graphical method.
   (c) Describe how each of the constraints affect the solution and what the implications are for management.
4. A confectionery company has decided to add two new lines to its product range: a chocolate fairy and a chocolate swan. It has been estimated that each case of chocolate fairies will cost £10.00 to produce and sell for £13.00, and that each case of chocolate swans will cost £11.40 to produce and sell for £15.00. Each case of chocolate fairies requires 30 oz of chocolate and each case of chocolate swans requires 50 oz. Only 450 oz of chocolate is available each day from existing facilities. It takes one hour of labour to produce either one case of chocolate fairies or chocolate swans and only 10 hours of labour are available each day.
   (a) Formulate a linear programming model to maximize the daily profit from chocolate fairies and chocolate swans.
   (b) Use the graphical linear programming method to solve the problem.
   (c) Describe how the company could achieve an output of five cases of chocolate fairies and six cases of chocolate swans.
5. Suppose you have been given the job of providing advance transport to a new exhibition hall and it is possible to use existing railway connections and bus routes.

   Each bus will cost £30 000 and carry 40 passengers. Each railway train will cost £45 000 and carry 50 passengers. A bus can make 15 trips a day and a railway train 12 trips a day.

   The system has a number of financial and design constraints. It must carry at least 48 000 passengers a day and cost no more than £2 700 000. There is also an agreement to use at least 10 trains, while no more than 66 buses are available.

   Formulate a linear programming model for the minimization of costs and solve using the graphical method.
6. A company has decided to produce a new cereal called 'Nuts and Bran' which contains only nuts and bran. It is to be sold in the standard size pack which must contain at least 375g. To provide an 'acceptable' nutritional balance each pack should contain at least 200g of bran. To satisfy the marketing manager, at least 20% of the cereal's weight should come from nuts. The production manager has advised you that nuts will cost 20 pence per 100g and that bran will cost 8 pence per 100g.
   (a) Formulate a linear programming model for the minimisation of costs.
   (b) Solve your linear programming model graphically.
   (c) Determine the cost of producing each packet of cereal.

NETWORKS 22

For any large project there are a host of smaller jobs that will need to be completed along the way. Some will have to await the completion of earlier tasks, while others can be started together. A **network** is a way of illustrating the various tasks, and the order in which they must be done, so that planning for the whole project may take place. It will also highlight those tasks which must not be delayed if the whole project is to be completed on time. These tasks collectively are known as the **critical path**. The technique of drawing up networks was developed during World War II and in the late 1950s in both the UK and the USA. In the UK it was developed by the Central Electricity Generating Board where its application reduced the overhaul time at a power station to 32% of the previous average. The US Navy independently developed the Programme Evaluation and Review Technique (PERT), whilst the Du Pont company developed the Critical Path Method, said to have saved the company $1 million in 1 year. All of these techniques are similar and have been applied in the building industry, in accountancy, and in the study of organizations, as well as their original uses.

Since for most projects only some of the tasks, or activities, will be **critical** to the overall timing of the project, the other activities can be seen as having **slack time**. This means that there will be some flexibility in when they need to be started or the length of time they may take. The objectives of the network analysis are:

1. to locate the critical activities;
2. to allocate the timing of other activities to obtain the most efficient use of manpower and other resources;
3. to look for ways to reduce the total project time by speeding up the activities on the critical path, while monitoring the other activities to ensure that they do not, themselves, become critical.

22.1 NOTATION AND CONSTRUCTION

A network consists of a series of **nodes** linked together by **arrows**. There are two ways of representing the activities within the project; these are 'activity between nodes' and 'activity on nodes': we shall be using the first of these alternatives, so that each arrow will represent an activity. Nodes are drawn with three sections, as shown in Figure 21.1.

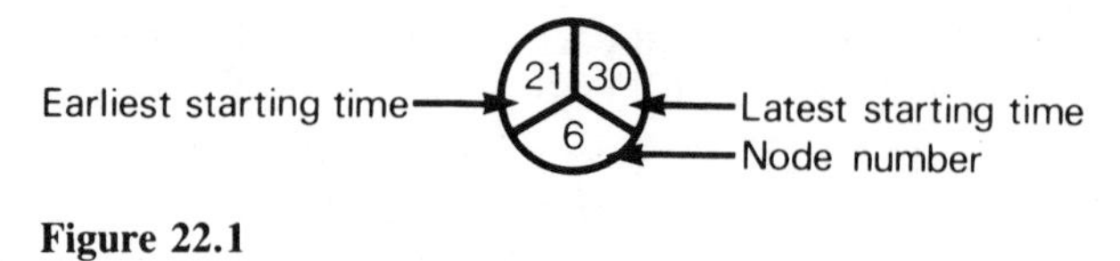

Figure 22.1

The **node number** will allow us to discuss with others where on the network we are, the **earliest starting time** (EST) will be the soonest that the next activity, or activities, can begin, while the **latest starting time** (LST) also relates to the next activity. Each arrow which joins a node to a later node will represent an activity, for example, selecting a sample of invoices in an audit check. **Dummy activities**, represented by dotted lines, are often used in networks where one activity can begin immediately a previous one is complete, but another activity has to await the completion of a second previous activity. These activities take no time.

To meet the objective of locating the critical activities, we must:

1. find the duration of the various activities and any activities which must be completed before a particular one may commence;
2. arrange these into a network;
3. find the earliest starting time by a forward pass through the network;
4. find the latest starting time by a backward pass through the network;
5. identify the critical path as consisting of those activities which have the same earliest and latest starting times.

22.1.1 Duration

Information about duration will often be available from previous projects, for example, the company will know how long it takes to obtain copies of plans from their head office. If the activity has not been undertaken before, then estimates can be obtained from those who have conducted this activity, or those most qualified to estimate the time needed. Activities which must be completed before a particular one may commence are usually a matter of technical information or common sense; for example, if you were constructing a new house, you would be unlikely to fit the carpets before you built the walls!

Table 22.1

Activity	*Duration*	*Any previous activities*
A	10	—
B	3	*A*
C	4	*A*
D	4	*A*
E	2	*B*
F	1	*B*
G	2	*C*
H	3	*D*
I	2	*E, F*
J	2	*I*
K	3	*G, H*

If we consider the example given in Table 22.1, we can develop the method of analyzing networks.

22.1.2 Construction

The speedy construction of a network is a matter of practice and experience, and, especially with large and complex projects, there may need to be several versions before a neat, legible network is drawn. Every effort should be made to avoid having arrows which cross each other, although this is sometimes inevitable.

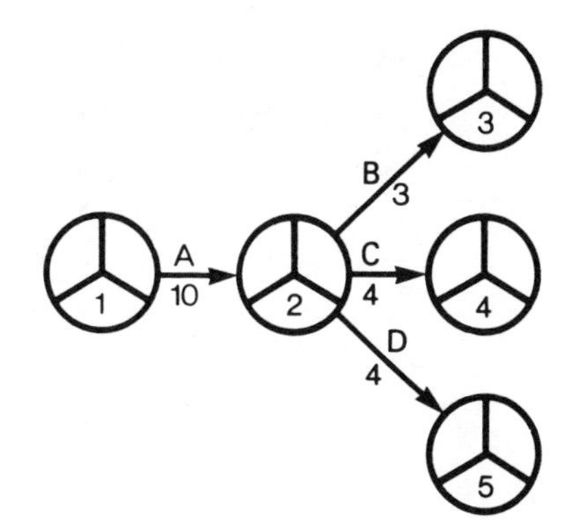

Figure 22.2

In Figure 22.2 we see the beginnings of the network to represent the project set out in Table 22.1. Activity *A* is the only activity which may begin immediately, since it has no prerequisites, and it is represented by the arrow between node 1 and node 2. Once this activity is completed, activities *B*, *C* and *D* may begin, and these are shown going to nodes 3, 4 and 5 respectively. Note that each arrow is labelled with its activity and duration at this stage, to avoid confusion at a later stage. Each branch of the network will eventually end at a common node.

EXERCISE

Draw up the whole network for the project represented in Table 22.1.

Your network should be similar to that shown in Figure 22.3. (Note the use of dummy activities from node 6 to node 7 and from node 8 to node 9.)

22.1.3 Earliest starting times (EST)

To find earliest starting times, place a *zero* in the EST section of node 1 and follow the arrows *forwards* through the network. The EST at node 2 will be 10, at 3 it will be 13, at 6 it will be 14 and so on.

Note that activity *I* can only begin when both *E* and *F* are complete; thus if your network has a dummy activity from node 6 to node 7 (as in Figure 22.4(a), *I* starts at time 15. If the dummy activity goes from node 7 to node 6, then the EST at node 6 is now 15, since activity *E* must be completed before *I* can begin (as shown in Figure 22.4b). It is possible to construct this network without using dummy activities, by making both *E* and *F* terminate at the *same* node. In this case the EST is the *highest* number of (EST at node

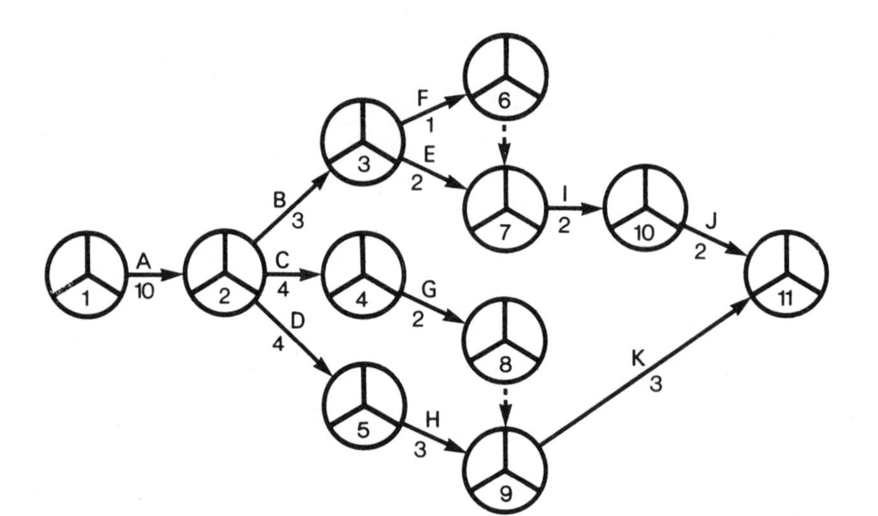

Figure 22.3

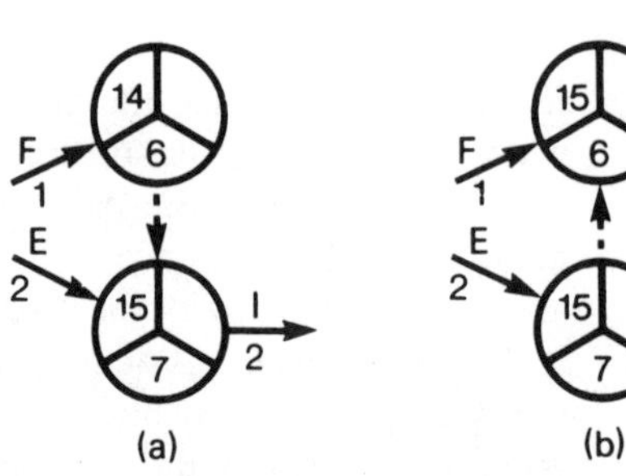

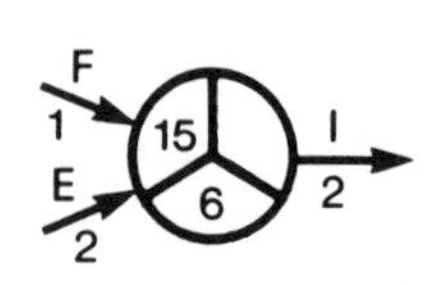

Figure 22.4

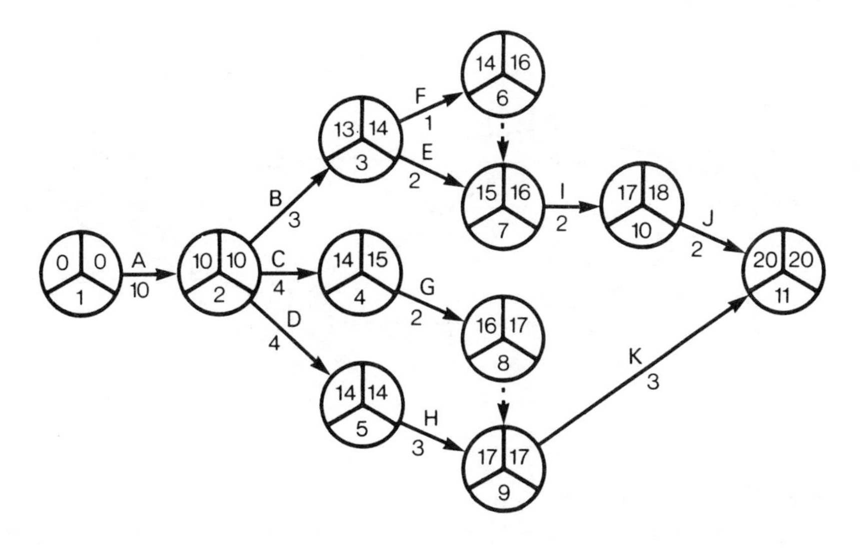

Figure 22.5

3 + duration of E) and (EST at node 3 + duration of F). This is illustrated in Figure 22.4(c). The total project time at node 11 will be 20.

22.1.4 Latest starting times (LST)

To find the latest starting times, the total project time is inserted in the LST section of the final node, and we go *backwards* through the network, *subtracting* the activity times at each stage. If two routes lead back to the same node, the *lowest* value is put into LST. This will eventually give 0 in the LST section of node 1 (Figure 22.5).

22.1.5 Critical path

The critical path consists of those activities which have the *same* earliest starting time and latest starting time; here A, D, H and K, and any delay in these activities will extend the total project time. There could be a delay of *one* time period elsewhere in the project without the total project time being increased.

22.2 PROJECT TIME REDUCTION

In the introduction to this chapter we specified three objectives for network analysis, the third of which was to look for ways to reduce the overall project time. Where an activity involves some chemical process it may be impossible to reduce the time taken successfully, but in most other activities the duration can be reduced *at some cost.* Time reductions may be achieved by using a (different) machine, adopting a new method of working, allocating extra personnel to the task or buying-in a part or a service. The minimum possible duration for an activity is known as the **crash duration**. Considerable care must be taken when reducing the times of activities on the network to make sure that the activity time is not reduced by so much that it is no longer critical. New critical paths will often arise as this time reduction exercise continues.

The project in Table 22.2 will have a critical path consisting of activities A and D, a project time of

Table 22.2

Activity	*Duration*	*Preceding activities*	*Cost*	*Crash duration*	*Crash cost*
A	8	—	100	6	200
B	4	—	150	2	350
C	2	*A*	50	1	90
D	10	*A*	100	5	400
E	5	*B*	100	1	200
F	3	*C, E*	80	1	100

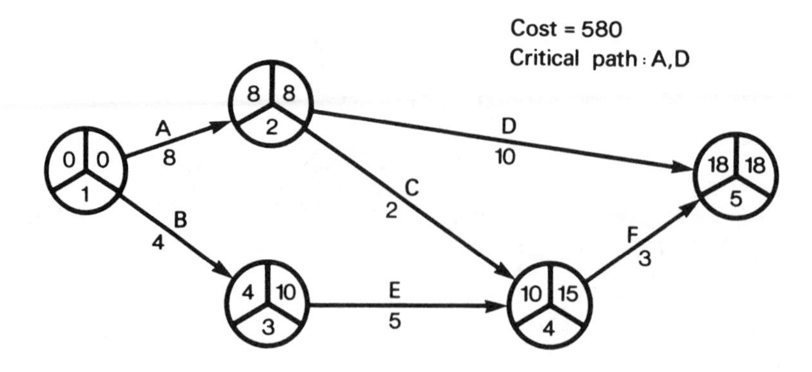

Figure 22.6

18 and a cost of 580 if the normal activity durations are used. This is illustrated in Figure 22.6.

Since cost is likely to be of prime importance in deciding whether or not to take advantage of any feasible time reductions, the first step is to calculate the cost increase per time period saved for each activity. This is known as the **slope** for each activity. For activity A, this would be:

$$\frac{\text{increase in cost}}{\text{decrease in time}} = \frac{100}{2} = 50$$

The slopes for each activity are shown in Table 22.3.

Table 22.3

Actvity	*A*	*B*	*C*	*D*	*E*	*F*
Slope	50	100	40	60	25	10

A second step is to find the **free float time** for each *non-critical* activity. This is the difference between the earliest time that the activity can finish and the earliest starting time of the next activity. By definition, activities on the critical path will have zero free float time. The initial free float times are shown in Table 22.4.

To reduce the project time, select that activity on the critical path with the lowest slope (here A) and note the difference between its normal duration and its crash duration (here, $8 - 6 = 2$). Look for the *smallest* (non-zero) free float time (here 1 for activity E), select the *minimum* of these two

Table 22.4

Activity	*EST*	*EFT*	*EST of next activity*	*Free float time*
B	0	4	4	0
C	8	10	10	0
E	4	9	10	1
F	10	13	18	5

numbers and reduce the chosen activity by this amount (here A now has a duration of 7). Costs will increase by the time reduction multiplied by the slope (1×50). It is now necessary to reconstruct the network (Figure 22.7).

Table 22.5

Activity	*EST*	*EFT*	*EST of next activity*	*Free float time*
B	0	4	4	0
C	7	9	9	0
E	4	9	9	0
F	9	12	17	5

The procedure may now be repeated. The activity on the critical path with the lowest slope is still A, but it can only be reduced by one time period. If this is done, we have the situation illustrated in Figure 22.8. Any further reduction in the project time must involve activity D, since A is now at the crash duration; the minimum non-zero free float

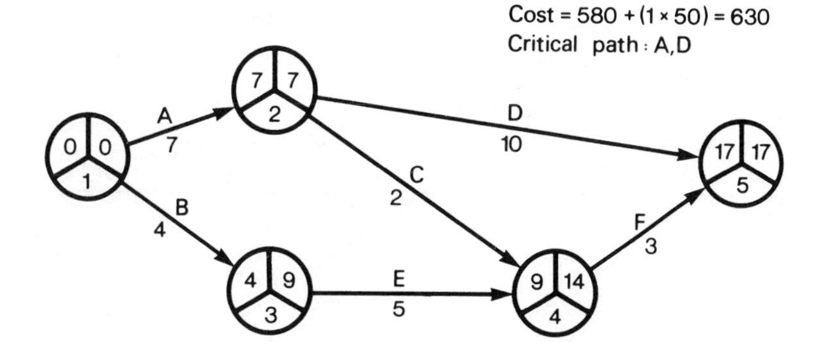

Figure 22.7

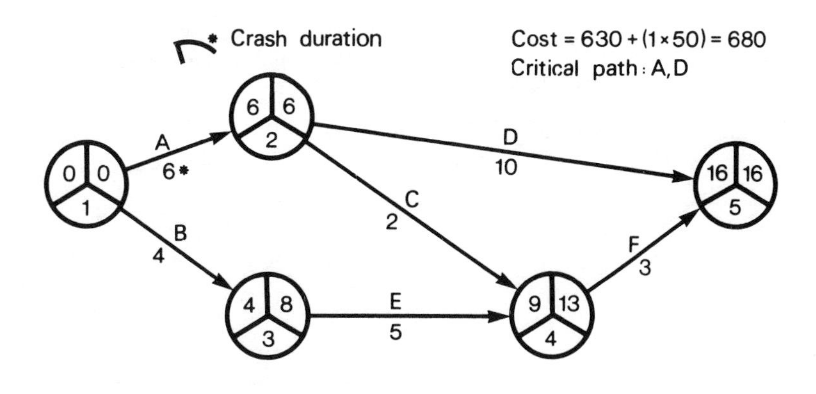

Figure 22.8

time is 1, but if this is used, we find that the free float time is reduced.

EXERCISE

Construct the network with activity D having a duration of 9 and recalculate the free float times for each non-critical activity.

Table 22.6

Activity	*EST*	*EFT*	*EST of next activity*	*Free float time*
B	0	4	4	0
C	6	8	9	1
E	4	9	9	0
F	9	12	16	4

Table 22.7

Activity	*EST*	*EFT*	*EST of next activity*	*Free float time*
C	6	8	9	1

Reducing activity D to a duration of 6 (i.e. $10 - 4$) we have the situation given in Figure 22.9.

There are now two critical paths through the network and thus for any further reduction in the project time it will be necessary to reduce *both* of these by the same amount. On the original critical path, only activity D can be reduced, and only by 1 time period at a cost of 60. For the second critical path, the activity with the lowest slope is F, at a cost of 10. If this is done, we have the situation in Figure 22.10.

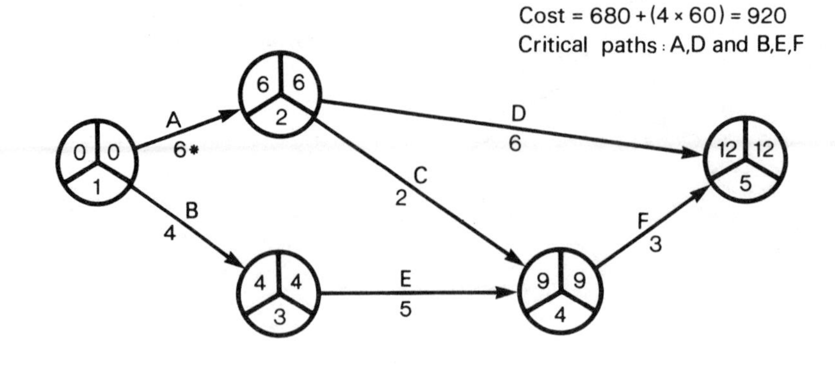

Figure 22.9

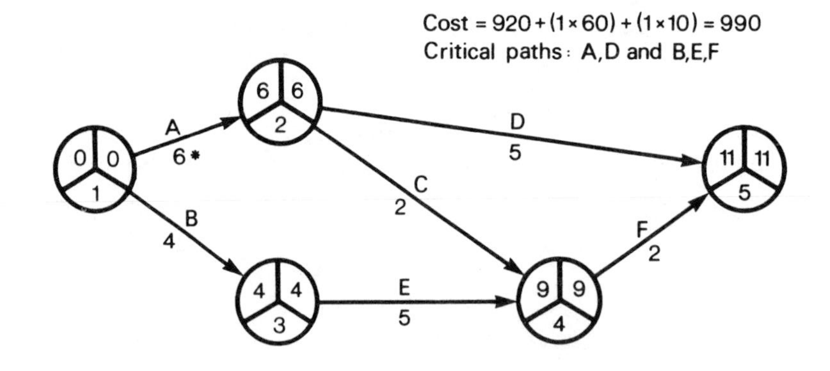

Figure 22.10

Since both activities on the original critical path are now at their crash durations, it will not be possible to reduce the total project time further.

22.3 UNCERTAINTY

In all of the calculations so far in this chapter we have assumed that the time taken for an activity will be the projected time as assessed at the planning stage, or some given reduced time due to the diversion of extra resources into that activity. In practice this may well not be true. Time may be saved if conditions are particularly favourable or the activity may take longer due to unforeseen difficulties. In general, delay is more likely than time saving, and thus if we wish to estimate the time for each activity, we cannot use the Normal distribution (since that was symmetric, see Chapter 12). It has been found that the **beta distribution** gives a good representation of the situation proposed above, and we shall use this to produce a confidence interval for the total project time. To produce this interval we shall need further information about the duration of the various activities, especially those on the critical path. The estimates needed are for the shortest duration and the longest duration of each activity, whilst the initial estimate of duration, which we have used in the earlier parts of this chapter, is seen as the most likely duration.

For the beta distribution, the mean and variance of the ith activity time are:

$$\text{mean} = \mu_i = \left(\frac{a + 4b + c}{6}\right)$$

Table 22.8

Activity	*Shortest (a)*	*Most likely (b)*	*Longest (c)*	*Mean*	*Variance*
A	8	10	14	10.333	1.000
D	3	4	6	4.167	0.250
H	2	3	6	3.333	0.444
K	1	3	7	3.333	1.000
				21.166	2.694

$$\text{variance} = \sigma_i^2 = \left(\frac{c-a}{6}\right)^2$$

where a = shortest time, b = most likely time and c = longest time.

The 95% confidence interval for the total project time will be:

$$\mu \pm 1.96\sigma \quad \text{where } \mu = \Sigma\mu_i \quad \text{and } \sigma = \sqrt{\Sigma\sigma_i^2}$$

(since we assume the sum of beta variables has approximately a Normal distribution).

If a network has been constructed for a particular project and the critical activities have been found to be A, D, H and K, then we may calculate the confidence interval as in Table 22.8.

Therefore, $\mu = 21.666$, $\sigma^2 = 2.695$, $\sigma = 1.641$.

Therefore, confidence interval is $21.166 \pm 1.96(1.641)$, i.e. 17.95 to 24.38.

Where the time taken for the critical activities is longer than the initial estimate, there need be no effect on the other activities, but if this time is shorter than the initial estimate it may be found that other activities become critical. Similarly, even if the activities on the critical path take their most likely durations, if some of the non-critical activities take their longest duration this may well extend the total project time. This is a further reason for calculating the free float time available on the non-critical activities.

22.4 CONCLUSIONS

The use of networks is a valuable aid to project planning and may result in considerable time-saving through the identification of those activities which are critical to the overall completion of the project within some target time. It will also help with resource planning and the phasing of activities, but it should be an ongoing process, especially where the project is large and complex. By monitoring the project in relation to the network, it is possible to identify where delays over the expected duration of an activity are occurring and what the implications will be for subsequent activities. Thus, if a non-critical activity is delayed through unforeseen circumstances, a new critical path may emerge through the network and the phasing of resource allocations, or the delivery of materials can be changed to meet the new timing of activities.

22.5 PROBLEMS

For each of the following projects, determine the earliest starting times and latest starting times for each activity, identify the critical path, and determine the total project time.

1.

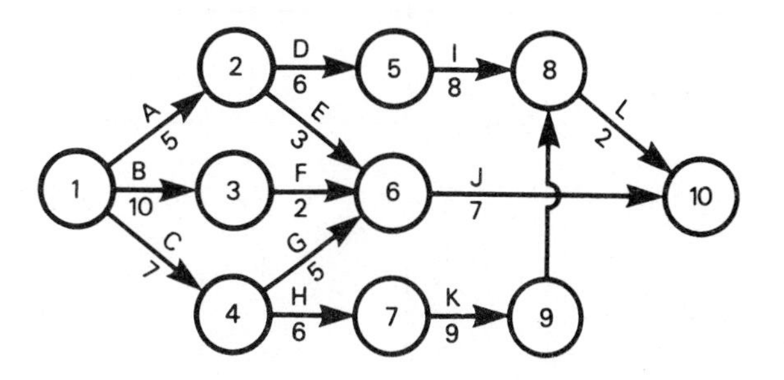

2.

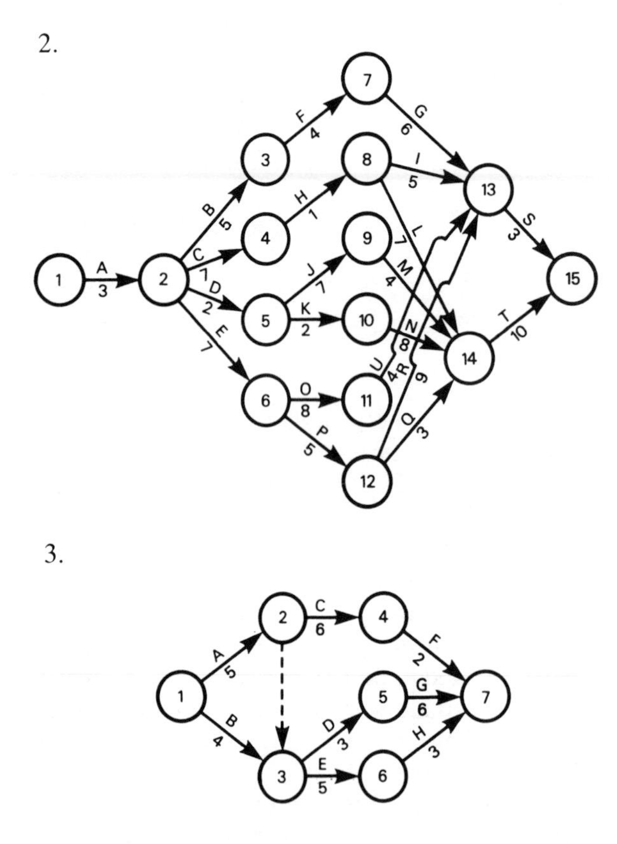

3.

For each of the following projects, construct a network, identify the critical path, and determine the total project time.

4.

Activity	*Duration*	*Preceding activities*
A	5	—
B	14	—
C	8	*A*
D	7	*A*
E	4	*B, C*
F	10	*B, C*
G	16	*B, C*
H	8	*D, E*
I	9	*F*
J	8	*G*
K	12	*H, I*

5.

Activity	*Duration*	*Preceding activities*
A	4	—
B	3	—
C	2	—
D	3	—
E	4	—
F	5	—
G	6	*A*
H	6	*A*
I	5	*B*
J	4	*B*
K	6	*C*
L	7	*C*
M	7	*C*
N	12	*D*
O	14	*D*
P	7	*E*
Q	8	*F*
R	4	*H, I, K*
S	6	*J, L, N*
T	8	*J, L, N*
U	5	*J, L, N*
V	6	*M, P, Q*
W	7	*G, R, S*
X	7	*O, U, V*

6. A person decides to move house and immediately starts to look for another property (this takes 40 working days.) She also decides to get three valuations on her own property; the first takes two working days, the second, four working days and the third, three working days. When all of the valuations are available she takes one further day to decide on the price to ask for her own property. The process of finding a buyer takes 40 working days. When she has found a house, she can apply for a mortgage (this takes 10 days); have two structural surveys completed, the first taking 10 days and the second 8 days; and instruct a solicitor which takes 35 days. When the surveys have been done on her new property and her old house, the removal firm can be booked. (NB. assume that her buyer's times for surveys, solicitors, mortgages, etc. are the same as her own.) When both solicitors have finished their work, and the removals are booked, she can finally move house, and this takes one day.

(a) Find the minimum total time to complete the move.

(b) What would be the critical activities in the move?

(c) What would be the effect of finding her new house after only twenty-five days?

7. Construct a network for the project outlined below and calculate the free float times on the non-critical activities and the critical path time. Find the cost and shortest possible duration for the whole project.

Activity	*Preceding activities*	*Duration*	*Cost*	*Crash duration*	*Crash cost*
A	—	4	10	2	60
B	—	6	20	3	110
C	*A*	5	15	4	50
D	*B, C*	4	25	3	70
E	*B*	3	15	1	55
F	*E*	5	25	1	65
G	*D, F*	10	20	4	50

8. (a) Construct a network for the project outlined below and calculate the free float time and slope for each activity.

(b) Identify the critical path and find the duration of the project using normal duration for each activity.

(c) Find the cost of reducing the duration for the whole project to 54 days.

(d) If there is a penalty of £500 per day over the contract time of 59 days and a bonus of £200 per day for each day less than the contract time, what will be the duration and cost of the project?

Activity	*Preceding activities*	*Duration*	*Cost (£)*	*Crash duration*	*Crash cost (£)*
A	—	5	200	4	300
B	*A*	7	500	3	1 000
C	*A*	6	800	4	1 400
D	—	6	500	5	700
E	*D*	6	700	3	850
F	*D*	8	900	5	1 050
G	*D*	9	1 000	5	1 240
H	—	8	1 000	4	1 320
I	*H*	7	600	4	900
J	*H*	7	800	6	1 000
K	—	5	1 000	4	1 200
L	*K*	9	500	5	700
M	*K*	10	1 200	8	1 240
N	*B*	8	600	4	760
O	*N, Q, S*	14	1 500	10	1 780
P	*R, U*	15	2 000	10	2 500
Q	*C, E, Y*	10	2 000	8	2 400
R	*C, E, Y*	15	1 500	7	1 900
S	*F, I*	20	3 000	15	3 750
T	*V, W*	10	2 000	7	3 200
U	*G, J, L*	14	1 800	9	2 250
V	*G, J, L*	22	5 000	13	7 700
W	*M*	18	4 000	10	5 280
X	*O, P, T*	11	3 000	9	4 000
Y	*H*	3	300	2	350

9. Construct a network for the project outlined below, and calculate the free float times on the non-critical activities, and the critical path time. Find the cost of the normal duration time. Find the shortest possible completion time, and its cost.

Activity	*Preceding activities*	*Duration*	*Cost*	*Crash duration*	*Crash cost*
A	—	5	100	4	200
B	—	4	120	2	160
C	*A*	10	400	4	1 000
D	*B*	7	300	3	700
E	*B*	11	200	10	250
F	*C*	8	400	6	800
G	*D, E, F*	4	300	4	300

MODELLING STOCK CONTROL AND QUEUES 23

Business involves many complex activities. In some cases these activities can be simplified to such an extent that the response becomes routine, such as invoicing customers; in other cases the number of unknowns involved make a routine response more difficult and we need to evaluate the different options e.g. investment appraisal. Managers, in general, deal with a reality that is complex, dynamic and incomplete. They are not in a position to understand every detail but need to be aware of the quantities involved and how one factor or variable is likely to relate to another.

It is useful to distinguish between the *system* and a *model of the system*. A system is a particular aspect of reality which can be described in terms of its inputs and its outputs. In business, for example, we refer to the production system, or the marketing system or the financial system. A manager's responsibilities may well be defined in terms of such a business system, and yet we cannot expect a manager to know a system, such as the production system, down to the 'last nut and bolt'. A model is a description of the system and as such, a simplification of reality. A model is specified in terms of relationships (equations) between variables and parameters. A **parameter** is a variable *fixed* for a particular situation. If price cannot change this month, it is a parameter of any problem specified in monthly terms. However, if price can change over the period of a year, it becomes a variable within this extended time frame. A model should be constructed, so that the system can be analyzed and the consequences of change (e.g. decisions) can be better understood. The Balance Sheet, Profit and Loss Account and other accounting statements, for example, model the financial position of a company.

In this chapter, we consider models that describe two aspects of business reality: the need to manage stock levels and the need to manage queues. The general principles of building a model can be extended to other aspects of the business reality e.g. machine loadings or product profitability. Models can be presented as a series of equations or in a spreadsheet format. The accuracy and complexity of the model will depend on how close it comes to reality. Given that a model is based on a series of assumptions, its applications may be limited. It is important that the model allows the decision-maker to change variables of interest and work through the possible consequences e.g. the consequences of a change in interest rate or number of service points available. How to use the output from a model will depend on the skill and judgement of the decision-maker.

23.1 INTRODUCTION TO THE ECONOMIC ORDER QUANTITY MODEL

The economic order quantity (EOQ) model, or the economic batch quantity (EBQ) model as it is often called, is one mathematical representation of the costs involved in stock control. It is one of the simplest mathematical stock control models, but has been found to provide reasonable solutions to a number of practical problems.

The derivation of this model involves two simplifying assumptions:

1. there is **no uncertainty** — demand is assumed to be known and constant;
2. the lead time is zero — the **lead time** is the time between placing an order and receiving the goods.

The economic order quantity model is developed from two types of cost associated with stock control: **ordering costs** and **holding costs**. Ordering costs are those costs incurred each time an order is placed. They can involve administrative work, telephone calls, postage, travel or a combination of two or more of these. Holding costs are the costs of keeping an item in stock. These can include the cost of capital, handling, storage, insurance, taxes, depreciation, deterioration and obsolescence.

EXAMPLE

A retailer, *A* to *Z* Limited, has a constant demand for 300 items each year. The cost of each item to the retailer is £20. The cost of ordering, handling and delivering is £18 per order regardless of the size of the order. The cost of holding items in stock amounts to 15% of the value of the stock held. If the lead time is zero, determine the order quantity that minimizes total inventory cost.

If Q is the quantity ordered on each occasion and D is the constant annual demand then the stock level as a function of time appears as in Figure 23.1.

We can start with an order of Q and since the lead time is zero the inventory rises from 0 units

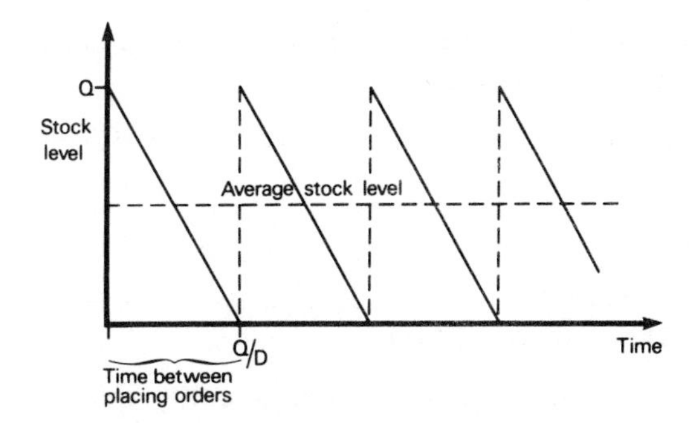

Figure 23.1 Stock levels as a function of time

to Q units. Thereafter, the stock level falls at the constant rate of D units per year. The time taken for the stock level to reach 0 is Q/D years. If we consider the annual demand of 300 items per year as given in our example, the stock level would fall to 0 in 2 years if we placed orders for 600 and would fall to 0 in 3 months ($\frac{1}{4}$ of a year) if we placed orders for 75.

It follows that the **number of orders** per year is D/Q. If orders of 75 units are placed to meet an annual demand of 300 then four deliveries are required each year. It also follows that the **average stock level** is $Q/2$ units, since the stock level varies uniformly between 0 and Q units. On this basis we are able to determine ordering costs and holding costs for this particular stock control model:

Total ordering costs (TOC) = cost per order × number of orders

$$= 18 \times \frac{D}{Q}$$

Total holding cost (THC) = cost of holding per item × average number of items in stock

$$= (15\% \text{ of } £20) \times \frac{Q}{2}$$

$$= 3 \times \frac{Q}{2}$$

We can note that actual holding cost will vary directly with the stock level from (15% of £20) × Q to 0 and that the above is an average. We can also note that it is typical for holding costs to be expressed as a percentage of value.

In seeking stock control policy we need to consider **total variable cost** (TVC) and **total annual cost** (TAC), where

total variable cost = total ordering costs + total holding costs

and

total annual cost = total ordering costs + total holding costs + purchase cost (PC)

We can determine an order quantity that will minimize the cost of stock control, in our example by 'trial and error' as shown in Table 23.1.

Table 23.1 The costs of stock

Ordering frequency p.a. f	Order size Q	Total ordering cost p.a. 18 × f	Average stock Q/2	Holding costs 3 × Q/2	TVC (TOC + THC)	TAC (TVC + PC) TVC + 6 000
1	300	18	150.0	450.0	468.0	6 468.0
2	150	36	75.0	225.0	261.0	6 261.0
3	100	54	50.0	150.0	204.0	6 204.0
4	75	72	37.5	112.5	184.5	6 184.5
5	**60**	**90**	**30.0**	**90.0**	**180.0**	**6 180.0**
6	50	108	25.0	75.0	183.0	6 183.0
10	30	180	15.0	45.0	225.0	6 225.0
12	25	216	12.5	37.5	253.5	6 253.5

This type of analysis would normally be undertaken using a spreadsheet with ordering cost and holding cost as parameters.

In this example both total variable cost and total annual cost are minimized if we place orders for 60 items, 5 times each year.

The structure of the table is as follows:

1. Demand is uniform at 300 items each year. It can be satisfied by one order of 300 each year, two orders of 150 every 6 months, three orders of 100 every four months and so on.
2. The cost of ordering remains fixed at £18 per order regardless of the size of an order. The total annual ordering cost is the ordering cost (C_0) multiplied by the ordering frequency. As the size of the order increases, the total annual order cost decreases.
3. Total annual holding cost is the cost of keeping an item in stock for a year (often expressed as a percentage of value) multiplied by the average stock level. At the time an order is received, holding cost will be high but will decrease to zero as the stock level falls to zero. Total annual holding cost is derived from the averaging process. As the size of the order increases the total annual holding cost increases.
4. Total variable cost is the sum of one cost that decreases with order quantity (TOC) and one cost that increases with order quantity (THC). These cost functions are shown in Figure 23.2.

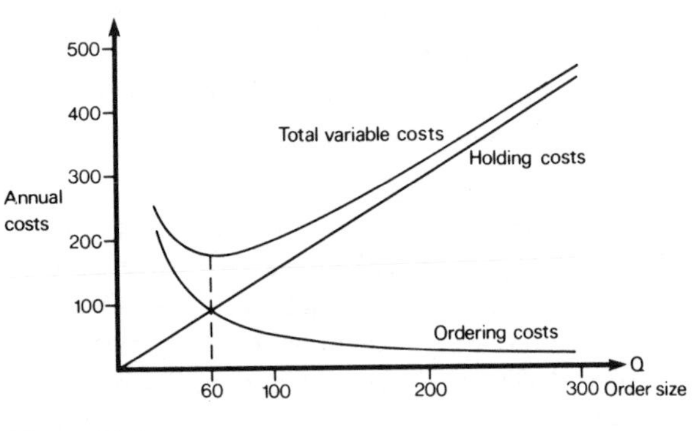

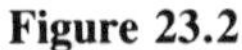

Figure 23.2

5. Total annual cost includes the annual purchase cost which in this case is £6 000 (£20 × 300). The order quantity that minimizes the total variable cost need not minimize the more important total annual cost. It is the total annual cost of the inventory that the retailer will need to pay and it is this cost which can be affected by such factors as price discounts.

The order quantity that minimizes the total variable cost could have been determined by substitution into the following equation:

$$Q = \sqrt{\left(\frac{2 \times C_0 \times D}{C_H}\right)}$$

where Q is the economic order quantity, D is the annual demand, C_0 is the ordering cost and C_H is the holding cost per item per annum.

By substitution we are able to obtain the solution found by 'trial and error'.

$$Q = \sqrt{\left(\frac{2 \times 18 \times 300}{3.00}\right)} = 60$$

The derivation of this formula is given in section 23.10.

EXAMPLE

A company uses components at the rate of 500 a month which are bought at the cost of £1.20 each from the supplier. It costs £20 each time to place an order, regardless of the quantity ordered.

The total holding cost is made up of the capital cost of 10% per annum of the value of stock plus 3p per item per annum for insurance plus 6p per item per annum for storage plus 3p per item for deterioration.

If the lead time is zero, determine the number of components the company should order, the frequency of ordering and the total annual cost of the inventory.

Annual demand = 500 × 12 = 6 000
Ordering cost = £20
Holding cost = £1.20 × 0.10 + £0.03 + £0.06 + £0.03
= £0.24

By substitution,

$$Q = \sqrt{\left(\frac{2 \times 20 \times 6\,000}{0.24}\right)} = 1\,000$$

To minimize total variable cost, and in this case also total annual cost, we would place an order for 1 000 components when the stock level is zero. The number of orders (or ordering frequency) is

$$\frac{D}{Q} = \frac{6\,000}{1\,000} = 6$$

We would therefore expect to place orders every 2 months.

$$\begin{aligned} \text{TAC} &= \text{TOC} + \text{THC} + \text{PC} \\ &= C_0 \times \frac{D}{Q} + C_H \times \frac{Q}{2} + \text{price} \times \text{quantity} \\ &= £20 \times \frac{6000}{1000} + £0.24 \times \frac{1000}{2} + £1.20 \times 6000 \\ &= £120 + £120 + £7\,200 \\ &= £7\,440 \end{aligned}$$

23.2 QUANTITY DISCOUNTS

In practice it is common for a supplier to offer discounts on items purchased in larger quantities. This reduces the purchase cost. The ordering cost is also reduced since fewer orders need to be placed each year. However, holding costs are increased with the larger average stock level. The economic order quantity formula cannot be used directly since unit cost and hence purchase cost is no longer fixed.

EXAMPLE

Suppose the retailer, *A* to *Z* Limited (see section 23.1), is offered a discount in price of $2\frac{1}{2}$% on purchases of 100 or more. If the lead time remains zero, determine the order that minimizes total inventory cost.

The total annual cost function, as shown in the last column of Table 23.1, now becomes discontinuous at the point where the discount is effective. Not only is the purchase cost reduced, but also the holding cost, as it is expressed as a percentage of value (15% of the value of the stock level).

If Q = 100, the minimum order quantity to qualify for the discount, then:

$$\text{TOC} = £18 \times \frac{D}{Q} = £18 \times \frac{300}{100} = £54$$

purchase price (with $2\frac{1}{2}$% discount) = £20 × 0.975 = £19.50

holding cost per item (15% of value) = £19.50 × 0.15 = £2.925

total holding cost = $£2.925 \times \frac{Q}{2} = £2.925 \times \frac{100}{2} =$ £146.25

total variable cost = £54 + £146.25 = £200.25

total annual cost = total variable cost + purchase cost
= £200.25 + £19.50 × 300 = £6 050.25

The discount on price of $2\frac{1}{2}$% will give the retailer (refer to Table 23.1) an annual saving of £6 204 − £6 050.25 = £153.75 if orders of 100 are placed.

The effects on costs of the discount are shown in Table 23.2.

Table 23.2 The costs of stock taking into account a price discount

Ordering frequency p.a. f	Order size Q	Total ordering cost p.a. 18 × f	Average stock Q/2	Holding costs	TVC (TOC + THC)	TAC (TVC + PC)
1	300	18	150.0	438.750	456.750	6 306.750
2	150	36	75.0	219.375	255.375	6 105.375
3	100	54	50.0	146.250	200.250	6 050.250
4	75	72	37.5	112.500	184.500	6 184.500
5	60	90	30.0	90.000	180.000	6 180.000
6	50	108	25.0	75.000	183.000	6 183.000
10	30	180	15.0	45.000	225.000	6 225.000
12	25	216	12.5	37.500	253.500	6 253.500

If this were being modelled on a spreadsheet, only the holding cost column and percentage cost component of Table 23.1 would need to be changed.

By inspection of total annual cost we can see that a minimum is achieved with an order size of 100. This function is shown as Figure 23.3.

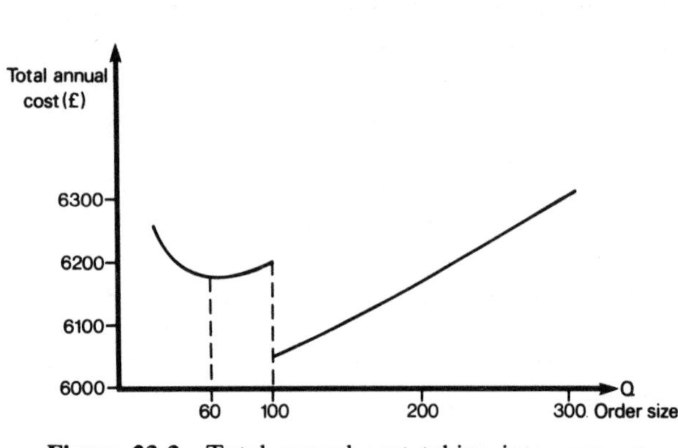

Figure 23.3 Total annual cost taking into account a price discount

The formula for the economic order quantity can be used within a price range to obtain a local minimum (up to 100 for example) but needs to be used with caution.

23.3 NON-ZERO LEAD TIME

One of the assumptions of the basic economic order quantity model (see section 23.1) was a lead time of zero. The model can be made more realistic if we allow a time lag between placing an order and receiving it. The calculation of the economic batch quantity does not change; it is the time at which the order is placed that changes. **A reorder point** must be calculated.

23.3.1 Non-zero lead time and constant demand

Suppose the retailer, A to Z Limited, retains a constant demand of 300 items each year but has a lead time of 1 month between placing an order and receiving the goods. The monthly demand for the items will be

$$300 \times \frac{1}{12} = 25$$

When the stock level falls to 25, the retailer will need to place the order (60 items in this case) to be able to meet demand in 1 month's time. Stock levels are shown in Figure 23.4.

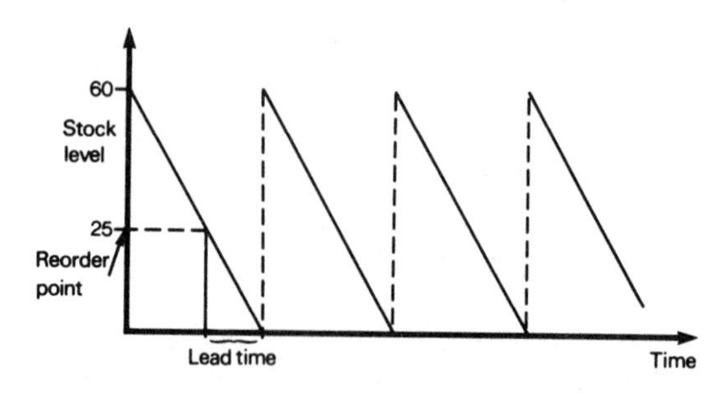

Figure 23.4 Stock levels with non-zero lead time

23.3.2 Non-zero lead time and probabilistic demand

If demand is probabilistic, as shown in Figure 23.5, a reorder level cannot be specified that will guarantee the existing stock level reaching zero at the time the new order is received.

To determine the reorder point we need to know the lead time (assumed constant), the risk a retailer or producer is willing to take of not being able to meet demand and the probabilistic demand function. The number of items held in stock

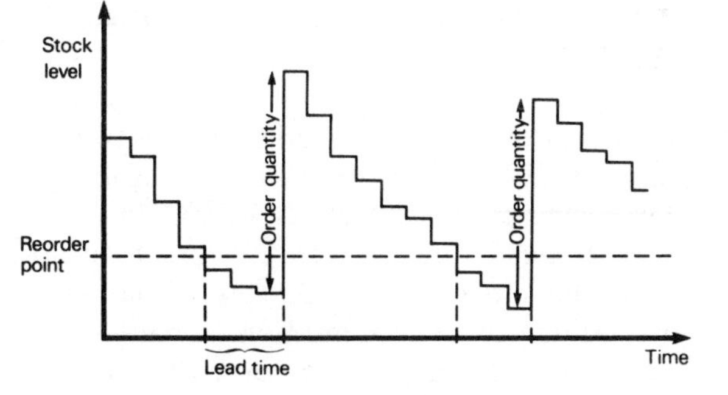

Figure 23.5 Stock levels with non-zero lead time and probabilistic demand

when an order is placed will depend on the risk the stockholder is prepared to take in not being able to meet demand during the lead time. The probability distribution of demand is often described by the Poisson or Normal distributions.

EXAMPLE

Suppose the retailer, A to Z Limited, now finds that the lead time is 1 month and that demand for items follows a Poisson distribution with a mean of 25 items per month. What should the reorder level be if the retailer is willing to take a risk of 5% of not being able to meet demand?

To determine the reorder level we need to refer to cumulative Poisson probabilities as shown in Table 23.3.

Table 23.3 Cumulative Poisson probabilities for $\lambda = 25$

r	$P(r \text{ or more})$
32	0.1001
33	0.0715
34	0.0498
35	0.0338

If the reorder level were 32, then

P (not being able to meet demand)
= P(33 or more items in 1 month)
= 0.0715 or 7.15%.

If the reorder level were 33, then

P(not being able to meet demand) = P(34 or more items in 1 month)
= 0.0498 or 4.98%.

Given that the risk the retailer is willing to take of not being able to meet demand is at most 5%, the reorder level would be set at 33 items. Once stock falls to this level a new order is placed.

EXAMPLE

You have been asked to develop a stock control policy for a company which sells office typewriters.

The cost to the company of each machine is £250 and the cost of placing an order is £50. The cost of holding stock has been estimated to be 10% per annum of the value of the stock held. If the company runs out of stock then the demand is met by special delivery, but the company is willing to take only a 5% risk of this service being required. From past records it has been found that demand is normally distributed with a mean of 12 machines per week and a standard deviation of 2 machines.

If there is an interval of a week between the time of placing an order and receiving it, advise the company on how many machines it should order at one time and what its reorder level should be.

By substitution, we are able to calculate the economic order quantity.

$$Q = \sqrt{\left(\frac{2 \times C_O \times D}{C_H}\right)}$$

Let

$C_O = £50$
$C_H = £250 \times 0.10 = £25$
$D = 12 \times 52 = 624$

We have assumed that demand can exist for 52 weeks each year although production is not likely to take place for all 52 weeks:

$$Q = \sqrt{\left(\frac{2 \times 50 \times 624}{25}\right)} = 50$$

To minimize costs, assuming no quantity discounts, orders for 50 office typewriters would be made at any one time. Given a lead time of a week, we need now to calculate a reorder level such that the probability of not being able to meet demand is no more than 5%.

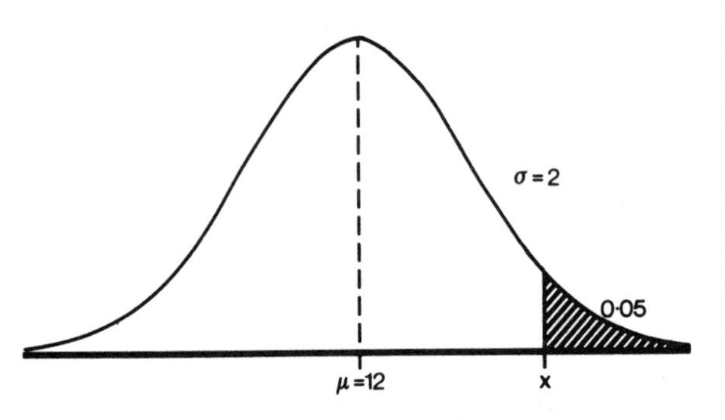

Figure 23.6 Weekly demand for office typewriters

Consider Figure 23.6. With reference to tables for the Normal distribution, a z-value of 1.645 will exclude 5% in the extreme right-hand tail area. By substitution, we are able to determine x:

$$z = \frac{x - \mu}{\sigma}$$

$$1.645 = \frac{x - 12}{2}$$

$$x = 12 + 1.645 \times 2$$
$$= 15.29.$$

To ensure that the probability of not being able to meet demand is no more than 5%, a reorder level of 16 would need to be specified.

It should be noted that any increase in the reorder level will increase the average stock level and as a result will increase holding costs. As we include more factors, the model becomes more and more complex. The inclusion of uncertainty, for example, makes the stock control model more realistic in most circumstances. As we have seen, the economic order quantity model can describe some stock control situations and can be developed to describe a number of others.

23.4 INTRODUCTION TO MODELLING QUEUES

Queues in one form or another are part of our daily lives. We may queue in the morning to buy a newspaper, or as a car driver we may queue to enter a multi-storey car park. Not all queues involve people so directly. Queues may consist of jobs waiting to be processed on a computer, or aeroplanes waiting to land, or components in batches waiting to be machined.

A queue forms when a customer cannot get immediate service. To the customer the queue may represent a lack of quality in the service. To the business, eliminating queues may increase costs for little benefit. If for example, an employee of an engineering company has to wait at a central store for a component, there is the opportunity cost associated with the possibility of lost production. However, to decrease waiting times at the central store, more service staff may need to be employed and the service arrangements changed at increased cost. Like other aspects of business, queues need to be managed and the relative costs balanced. If the service cost is high and the service has a high level of demand e.g. dentists, queues may be lengthy. In other cases, where the service cost may be relatively cheap and lost custom expensive, queues may be short e.g. supermarket checkouts and garages (you serve your own petrol).

Queues can take many forms and we can mathematically model only a few of these. Customers may form a single queue at a single service point or separate queues at multiple service points. Customers may arrive singly e.g. telephone calls through a switchboard, or in batches e.g. passengers leaving a train. Customers may be served in the order of their arrival e.g. football supporters, or on some other basis e.g. oil tankers may be given priority in the entry to a port. The way a queue is managed or served is known as the **queue discipline**. In general, the length of any queue will depend on the rate of customer arrival, the time taken to serve the customer, the number of service points and the queue discipline. It is worth noting the difference between the system and the queue. The queue refers to the customers waiting to be served, whereas the system refers to the customers waiting to be served plus the customer or customers being served.

Simple queueing can be reasonably described mathematically but as we attempt to make the queueing model more realistic, the mathematics quickly become complex. In this chapter, the mathematics of the single-channel and multichannel queues are presented. However as the queueing situation becomes more complex, simulation techniques (Chapter 24) offer a useful and practical alternative.

23.5 A MODEL FOR A SINGLE QUEUE

The derivation of a model for a single-server queue involves a number of simplifying assumptions.

1. There is a single queue with a single service point.
2. The queue discipline is 'first come first served'.
3. The arrivals are random and can be described by the Poisson distribution. Given the integer nature of arrivals, only a discrete probability distribution would have the necessary characteristics. The Poisson distribution is particularly suitable for modelling random arrivals in intervals of time (see section 11.3).
4. Service times follow a negative exponential distribution (service times are random). The properties of the negative exponential distribution need not concern us here. It is a continuous distribution and in many ways similar to the Poisson; it is defined by one parameter, the mean.
5. There is no limit to the number of customers waiting in the queue.
6. All customers wait long enough to be served.

To understand the characteristics of a queue, we need to determine the average arrival rate, λ, and the average service rate, μ. The ratio of the average arrival rate to the average service rate, ρ, provides a useful measure of how busy the queue is, and is known as the **traffic intensity** or **utilization factor**.

$$\rho = \frac{\lambda}{\mu} = \frac{\text{average arrival rate}}{\text{average service rate}}.$$

If, for example, the average arrival rate is 18 customers per hour and the average service rate is 60 customers per hour, the traffic intensity is given by

$$\rho = \frac{18}{60} = 0.3$$

If the average arrival rate is greater the average service rate, $\lambda > \mu$, then customers will be entering the queue more frequently than they leave it, and the queue will get longer and longer. An important assumption of queuing theory is that $\rho < 1$; in which case the queue can achieve a steady state or equilibrium.

The operating characteristics of the single queue model are as follows:

1. The probability of having to queue is given by the traffic intensity:

$$\rho = \frac{\lambda}{\mu}$$

   Clearly, as the average service rate increases relative to the average arrival rate, the probability of having to queue decreases.
2. The probability that the customer is served immediately, number one in the system, is given by

$$P_0 = 1 - \rho$$

3. The probability of n customers in the system is

$$P_n = (1 - \rho)\rho^n$$

4. The average number of customers in the system is

$$L_s = \frac{\rho}{1-\rho} = \frac{\lambda}{\mu - \lambda}$$

5. The average number of customers in the queue is

$$L_q = \frac{\rho^2}{1-\rho} = \frac{\lambda^2}{\mu(\mu - \lambda)}$$

6. The average time spent in the system is

$$T_s = \frac{1}{\mu(1 - \rho)} = \frac{1}{\mu - \lambda}$$

7. The average time spent in the queue is

$$T_q = \frac{\rho}{\mu(1 - \rho)} = \frac{\lambda}{\mu(\mu - \lambda)}$$

EXAMPLE

Given the average arrival rate of 18 customers per hour and the average service rate of 60 customers per hour, determine the probability of having to queue, the probability of not having to queue and the probability of 1, 2 and 3 customers in the system.

The probability of having to queue is equal to the traffic intensity,

$$\rho = \frac{18}{60} = 0.3$$

The probability of not having to queue is

$$P_0 = 1 - \rho = 1 - 0.3 = 0.7$$

The probabilities of there being 1, 2 and 3 customers in the system is given by

$$P_1 = (1 - 0.3) \times 0.3 = 0.21$$
$$P_2 = (1 - 0.3) \times 0.3^2 = 0.063$$
$$P_3 = (1 - 0.3) \times 0.3^3 = 0.0189$$

With such a favourable service rate relative to the arrival rate, the chance of having to join a queue of any significant length is quite small.

EXAMPLE

Customers arrive at a small Post Office with a single service point at an average rate of 12 per hour. Determine the average number of customers in the queue and the average time in the queue if service takes 4 minutes on average. If service times can be reduced to an average of 3 minutes, what would the impact be?

Given that $\lambda = 12$ and $\mu = \frac{60}{4} = 15$ then

$$L_q = \frac{\lambda^2}{\mu(\mu - \lambda)} = \frac{12^2}{15(15 - 12)} = 3.2$$

and

$$T_q = \frac{\lambda}{\mu(\mu - \lambda)} = \frac{12}{15(15 - 12)} = 0.27 \text{ hours}$$
$$= 16.2 \text{ mins}$$

If average service time can be reduced to 3 minutes, then $\lambda = 12$ and $\mu = 60/3 = 20$, and

$$L_q = \frac{\lambda^2}{\mu(\mu - \lambda)} = \frac{12^2}{20(20 - 12)} = 0.9$$

and

$$T_q = \frac{\lambda}{\mu(\mu - \lambda)} = \frac{12}{20(20 - 12)} = 0.075 \text{ hours}$$
$$= 4.5 \text{ mins}$$

The 25% reduction in average service time has made a considerable difference to the queue characteristics.

23.6 QUEUES — MODELLING COST

Whether a queue involves people or items there are associated costs. In the business situation it may be difficult to identify all the costs e.g. the opportunity cost of an employee waiting in a queue (there may be benefits if these employees in the queue discuss ways of improving the quality of their work), but it is important to achieve a balance between those costs increasing and those decreasing with queue length. Costs are generally of two types: **service costs** and **queueing costs**. Service costs include the labour necessary to provide the service and related equipment costs. Queuing cost is the opportunity cost of customers waiting and could include, for example, the lost contribution to profit incurred while an employee is waiting in a queue. The two following examples illustrate the possible methods of solution.

EXAMPLE

A spare parts department is manned by a receptionist with little technical knowledge of the product concerned. The receptionist is paid £3.50 per hour but can only deal with 12 requests per hour because of the need to refer to other staff for technical advice. A person with more technical knowledge would be paid £6.50 per hour but could deal with 15 requests per hour. On average, 10 requests per hour are made. If the opportunity cost of staff making the requests to the spare parts department is valued at £5 per hour, determine whether a change in staffing is worthwhile.

If a receptionist with more technical knowledge is employed, the traffic density will fall from

$$\rho = \frac{10}{12} = 0.83$$

to

$$\rho = \frac{10}{15} = 0.67.$$

If the receptionist has little technical knowledge, the average number of requests (or average number in the system)

$$L_s = \frac{\rho}{1-\rho} = \frac{0.83}{1-0.83} = \frac{0.83}{0.17} = 4.88$$

and the average cost per hour = 4.88 × £5 + £3.50 = £27.90

If the receptionist has more technical knowledge, the average number of requests is

$$L_s = \frac{\rho}{1-\rho} = \frac{0.67}{1-0.67} = \frac{0.67}{0.33} = 2.03$$

and the average cost per hour = 2.03 × £5 + £3.50 = £16.65

In this case there are cost benefits in paying the higher hourly rate to have a receptionist with more technical knowledge.

EXAMPLE

Photocopying is done in batches. On average, 6 employees per hour need to use the facility and their time is valued at £3.00 per hour. A decision needs to be made on whether to rent a type *A* or type *B* photocopying machine. The type *A* machine can complete an average job in 5 minutes and has a rental charge of £16 per hour, whereas the type *B* machine will take 8 minutes but has a rental charge of £12 per hour. Which machine is most cost effective?

For the type *A* machine, the service rate is 60/5 = 12 per hour and the arrival rate is 6 per hour.

The average time lost per worker per hour (average time spent in the system)

$$T_s = \frac{1}{\mu-\lambda} = \frac{1}{12-6} = \frac{1}{6}$$

The cost per hour = $6 \times \frac{1}{6} \times$ £3 + £16 = £19.

For the type *B* machine, the service rate is 60/8 = 7.5 per hour and the arrival rate is 6 per hour.

The average time lost per worker per hour

$$T_s = \frac{1}{\mu-\lambda} = \frac{1}{7.5-6} = \frac{1}{1.5} = \frac{2}{3}$$

The cost per hour = $6 \times \frac{2}{3} \times$ £3 + £12 = £24.

23.7 MODELLING MULTI-CHANNEL QUEUES

It is possible to reduce the average length of a queue and the average waiting times by increasing the number of service points. In this model a number of service points, S, are available and customers join a single queue. Whenever a service point becomes free, the next customer in line takes the place. This system is operated by a number of banks, building societies and car exhaust replacement centres. If the service rate at each channel is μ, then the traffic intensity is given by

$$\rho = \frac{\lambda}{\mu \times S}$$

To achieve a steady state, the traffic intensity, ρ, must be less than 1 or in other words, the total arrival rate must be less than the total service rate, $\lambda < \mu \times S$.

The derivation of the model for a multi-channel queue involves the following simplifying assumptions.

1. There is a single queue with S identical service points.
2. The queue discipline is 'first come first served'.
3. Arrivals follow a Poisson distribution.
4. Service times follow the negative exponential distribution.
5. There is no limit to the number of customers waiting in the queue.
6. All customers wait long enough to be served.

The operating characteristics of the multi-channel queue model are as follows:

1. The probability of there being no-one in the system is given by

$$P_0 = \frac{1}{\sum_{i=0}^{s-1} \frac{(\lambda/\mu)^i}{i!} + \frac{(\lambda/\mu)^S \times \mu}{(S-1)! \times (S \times \mu - \lambda)}}$$

2. The probability of there being n customers in the system is

$$P_n = \frac{(\lambda/\mu)^n}{S! \times S^{n-s}} \times P_0 \quad \text{for} \quad n > S$$

and

$$P_n = \frac{(\lambda/\mu)^n}{n!} \times P_0 \quad \text{for} \quad 0 \leqslant n \leqslant S$$

3. The average number of customers waiting for service is

$$L_q = \frac{(\lambda/\mu)^S \times \lambda \times \mu}{(S-1)! \times (S \times \mu - \lambda)^2} \times P_0$$

4. The average number of customers in the system is

$$L_s = L_q + \lambda/\mu$$

5. The average time spent in the queue is

$$T_q = \frac{L_q}{\lambda}$$

6. The average time spent in the system is

$$T_s = T_q + 1/\mu$$

EXAMPLE

A fast food outlet, McBurger, has customers arriving at the rate of 50 per hour. There are 3 identical service points and each can handle 20 customers per hour. Determine the operating characteristics of the queue.

The probability of there being no-one in the system is given by

$$P_0 = \frac{1}{\sum_{i=0}^{2} \frac{(50/20)^i}{i!} + \frac{(50/20)^3 \times 20}{(3-1)! \times (3 \times 20 - 50)}}$$
$$= 0.0449$$

The probabilities of 1, 2, 3, 4 or 5 customers in the system are

$$P_1 = 50/20 \times 0.0449 = 0.1123$$

$$P_2 = \frac{(50/20)^2}{2} \times 0.0449 = 0.1403$$

$$P_3 = \frac{(50/20)^3}{6} \times 0.0449 = 0.1169$$

$$P_4 = \frac{(50/20)^4}{6 \times 3} \times 0.0449 = 0.0974$$

$$P_5 = \frac{(50/20)^5}{6 \times 9} \times 0.0449 = 0.0812$$

The average number of customers waiting for service is

$$L_q = \frac{(50/20)^3 \times 50 \times 20}{2! \times (3 \times 20 - 50)^2} \times 0.0449 = 3.51$$

The average number of customers in the system is

$$L_s = 3.51 + 2.5 = 6.01$$

The average time spent in the queue is

$$T_q = \frac{3.51}{50} = 0.0702 \text{ hours} = 4.212 \text{ minutes}$$

The average time spent in the system is

$$T_s = 0.0702 + \frac{1}{20} = 0.1202 \text{ hours} = 7.212 \text{ minutes}$$

Whether the queue characteristics are acceptable will depend on the business context. In terms of understanding the options available, the exercise could be repeated using 4 or 5 or more service points (let $S = 4$, or 5 or more). To make the problem more realistic, costs could also be considered. However, as we increase the complexity of the model, we increase the complexity of the mathematics and a point can be reached where the mathematics cannot be extended usefully.

23.8 CONCLUSIONS

In this chapter, we have developed mathematical models to describe stock control and the formation of queues. As we attempt to make these models more realistic, e.g. by including cost or more uncertainty, the mathematical description becomes more complex. These models have found useful business applications but it is important to recognize their limitations. It is always worth checking, for example, how meaningful the assumptions are. If demand tends to cluster in a predictable way, for example, this has major implications for how we model the business situation.

Some computer software is available for stock control and queueing models. This software is based on the type of mathematics presented in this chapter and essentially allows quicker computation. These models can also be developed in a spreadsheet format. However, merely analyzing a model, even a very good model, does not necessarily provide the basis for the best business decision. A model tends to describe the business situation in one limited way. The economic order quantity model used in stock control, for example, is now seen as rather dated and management interest has turned to techniques like materials requirement planning (MRP) and just-in-time (JIT).

23.9 PROBLEMS

1. The demand for brackets of a particular type is 3 000 boxes per year. Each order, regardless of the size of order, incurs a cost of £4. The cost of holding a box of brackets for a year is reckoned to be 60p. Determine the economic order quantity and the frequency of ordering.
2. A company needs to import programming devices for numerically controlled machines it produces and services. The necessary arrangements for import include considerable paperwork and managerial time which together have been estimated to cost the company £175 per order. Each item is bought at a cost of £50 and is not subject to discount. The cost of holding each item is made up of a storage charge of £2 and a capital charge of 12% per annum of the value of stock. Demand is for 700 programming devices per year.
   (a) Determine the economic order quantity and the frequency of ordering.
   (b) Determine the total cost of the ordering policy.
   (c) If the lead time is one month and demand can be assumed constant, what should the reorder level be?
3. A company requires 28 125 components of a particular type each year. The ordering cost is £24 and the holding cost 10% of the value of stock held. The cost of a component is reduced from £6 to £5.80 if orders for 2 000 or more are placed and reduced to £5.70 if orders for 3 000 or more are placed. Determine the ordering policy that minimizes the total inventory cost.
4. Demand for a product follows a Normal distribution with a mean of 15 per week and standard deviation of four per week. If the lead time is one week, to what level should the reorder level be set to ensure that the probability of not meeting demand is no more than:
   (a) 5%;
   (b) 1%?
5. If demand for a product follows a Poisson

distribution with a mean of two items per week, to what level should the reorder level be set if:

(a) the probability of not meeting demand is to be no higher than 5% and the lead time is one week;

(b) the probability of not meeting demand is to be no higher than 1% and the lead time is two weeks?

6. Tourists arrive randomly at an Information Centre at an average rate of 24 per hour. There is only one receptionist and each enquiry takes 2 minutes on average. Determine
   (a) the probability of queuing,
   (b) the probability of not having to queue,
   (c) the probabilities of there being 1, 2 or 3 customers in the queue,
   (d) the average number of customers in the system,
   (e) the average number of customers in the queue,
   (f) the average time spent in the system, and
   (g) the average time spent in the queue.
7. Products arrive for inspection at the rate of 18 per hour. What difference would it make to
   (a) the average number of products in the system,
   (b) the average number of products in the queue,
   (c) the average time in the system, and
   (d) the average time in the queue

   if the service rate could be increased from 24 to 30 per hour?
8. A maintenance engineer is able to complete 5 jobs each day on average, at a cost to the company of £110 per day. The number of completed jobs could be increased to 7 each day if the engineer was given some unskilled assistance, at a total cost to the company of £150 per day. On average there are 4 new jobs each day. It has been estimated that the cost to the company of a job not done (an opportunity cost) is £200 per day. Should the engineer be given some unskilled assistance?
9. Lorries carrying animal feed can be unloaded manually or automatically. On average, 8 lorries arrive each hour. The manual system can unload lorries at the rate of 11 per hour whereas the automatic system can deal with 13 per hour. The hourly cost of the manual system is £80 and the automatic system £95. If the cost of keeping a lorry waiting is £12 per hour, which system is most cost effective?
10. A fast-food outlet, McBurger, has increased the number of service points from 3 to 4. Customers still arrive at the rate of 50 per hour and each service point can handle, on average, 20 customers per hour. Determine the characteristics of the queue for the 4-service-point system and compare with the 3-service-point system.
11. Cars approach a 3-tunnel system at the rate of 30 per hour. Each tunnel can accept 18 cars per hour. Determine the characteristics of the queue.

23.10 APPENDIX — PROOF OF EBQ

This appendix need only concern those interested in the proof of the economic order quantity formula:

$$\begin{aligned} \text{TVC} &= \text{THC} + \text{TOC} \\ &= C_H \times \frac{Q}{2} + C_0 \times \frac{D}{Q} \end{aligned}$$

We can differentiate this function with respect to Q (see Chapter 8) to obtain

$$\frac{\text{d(TVC)}}{\text{d}Q} = \frac{C_H}{2} - C_0 \frac{D}{Q^2}$$

To find a turning point we set this function for gradient equal to zero:

$$\frac{C_H}{2} - C_0 \frac{D}{Q^2} = 0$$

$$Q^2 = \frac{2 \times C_0 \times D}{C_H}$$

$$Q = \pm \sqrt{\left(\frac{2 \times C_0 \times D}{C_H}\right)}$$

To identify a maximum or a minimum we can differentiate a second time to obtain

$$\frac{d^2(\text{TVC})}{\text{d}Q^2} = \frac{2 \times C_0 \times D}{Q^3}$$

This second derivative is positive (and thus identifies a turning point which is a minimum) when Q is positive. Hence the order quantity that minimizes total variable cost is

$$Q = +\sqrt{\left(\frac{2 \times C_0 \times D}{C_H}\right)}$$

SIMULATION 24

Mathematical models can be classified as being of one of two types: analytical or simulation. **Analytical models** are those solved by mathematical techniques e.g. differentiation. We have seen in earlier chapters, how aspects of business can be described using equations and how a solution can be obtained using appropriate techniques e.g. breakeven analysis (Chapter 6) or linear programming (Chapter 21). The model can be relatively simple e.g. a linear equation relating total cost to output ($c = a + bx$) or fairly complex e.g. the characteristics of multi-channel queues. The application of analytical models is limited by the necessary assumptions e.g. linear constraints or arrivals following a Poisson distribution and the complexity of the mathematics. We have seen how intractable the mathematics can become as we attempt to make the model more realistic in business terms. When developing a model for queueing, we did not allow for the possibility of customers leaving very long queues, for example.

An alternative to solution by mathematical manipulation is simulation. **Simulation models** may also be specified in terms of equations and distributions but a solution is sought through *experimentation* rather than *derivation*. In a business application (there are many others), a simulation model attempts to imitate the reality of the business system in the same way that a flight simulator attempts to imitate aircraft flight. The flight simulator allows the experience of flight to be repeated any number of times, under varying conditions, and the consequences studied. A business simulation allows the business situation to be studied under a range of conditions. Managers, for example, may take the decisions in a simulation exercise and in a matter of hours be given the experience of running a business over a number of years. Simulation can be very quick and can effectively address the 'what if' type of question.

24.1 AN INTRODUCTION TO SIMULATION MODELS

Typically, a simulation model will attempt to describe a business system by a number of equations. These equations are characterized by four types of variable.

1. **Input variables** are determined outside of the model; they are exogenous, and are subject to change for a particular simulation. It is the input variables that create the business situation; they give the model circumstances e.g. increasing or decreasing demand. An input variable can be **controllable** e.g. the manager may decide the re-order level, or **uncontrollable** e.g. demand may be probabilistic and follow a known distribution.
2. **Parameters** (fixed variables) are input variables given a constant value for a particular simulation exercise. If, for example, a variable cost was allowed to increase during the

simulation it would be regarded as an input variable, however if its value were kept constant it would be a parameter.

3. A **status variable**, generally not included in the equations, gives definition to the simulation model. Status variables describe the state of the system. If, for example, the pattern of demand varies according to the month of the year, the status variable would specify the month. If a set of equations apply to the shoe industry and not the car components industry then this again is a matter of status.
4. The **output variables** provide the results of interest e.g. the economic batch quantity or the number of service points that minimize queueing costs.

As the input variables are allowed to change, perhaps follow a known distribution, the consequences on the output variables can be studied. The usefulness of any particular simulation model will depend on how well the equations relate the output variables to the input variables and parameters. A simulation may be 'run' through many times, even hundreds of times, to allow the output pattern to emerge.

24.2 DEVELOPING A SIMPLE SIMULATION MODEL

Suppose you were invited to play a game of chance. If a fair coin shows a head you win £1 and if it shows a tail you lose £2. In this case the input variable represents a head or tail, the value that can be won or lost a parameter, the property of the coin ($P(\text{Head}) = \frac{1}{2}$) a status variable and the actual amount won or lost each round the output variable. The system is represented in Figure 24.1

To understand the characteristics of the system we would need to 'run' the model a number of times and observe the outcomes. To simulate the fairness of play, the input values are based on random numbers e.g.

6 7 8 8 9 9 9 2 5 9 3 0

To obtain a sequence of heads and tails, an even number (including zero as even) is taken to represent a Head and an odd number a Tail:

H T H H T T T H T T T H

and the outcomes are then

£1 −£2 £1 £1 −£2 −£2 −£2
£1 −£2 −£2 −£2 £1

The average win in this short sequence is $-£9/12 = -£0.75$.

We have not achieved the expected result of $-£0.50$ (see section 10.5 on expected values) which is only likely to emerge with any stability over a longer run of plays.

EXERCISE

Repeat the above simulation with a set of 20, 30 and 50 Random numbers and compare the average win values.

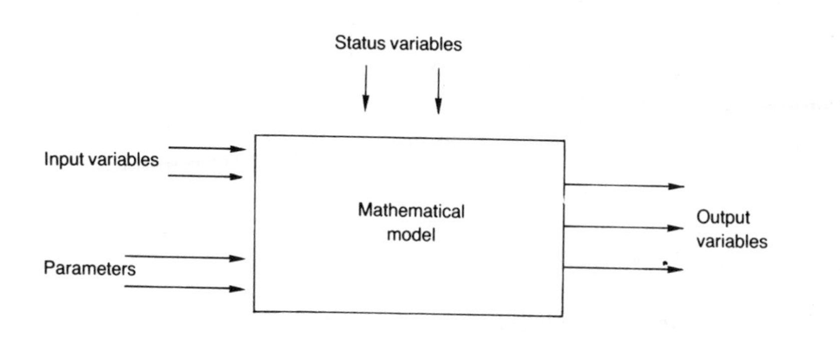

Figure 24.1

One major advantage of simulation is that we can consider changes to the system that can make the mathematics far more difficult but yet require only modest adaption to our simulation method.

Suppose the rules of the 'coin' game are changed to your advantage. If you get two heads in a row you win an extra £2, and if you get three heads in a row you win an extra £3 and so on.

Suppose the next set of random numbers give

6 1 5 0 0 4 3 9 8 2 1 8

the sequence of heads and tails would be

H T T H H H T T H H T H

and the outcome

£1 −£2 −£2 £1 £1 £1 −£2 −£2 £1 £1 −£2 £1
+£3 +£2

the average win in this case is £2/12 = £0.17.

To be sure of a reliable result, the number of trials would need to be increased.

EXERCISE

Outline how you would solve the above problem mathematically.

Simulation is a step by step approach that gives results by **iteration**. The procedure can be effectively described using a flowchart and is particularly suitable for computer solution. In practice, all simulation is computer based. Procedures can be programmed using a 'high-level' language such as FORTRAN or BASIC or using a specialised simulation language such as GPSS (General Purpose Simulation System). In a limited number of cases, a simulation exercise can be undertaken on a spreadsheet.

24.3 RANDOM EVENT GENERATION

Random numbers remove the chance of bias from any experimentation. In the previous example, we wanted to simulate the outcomes from a fair coin and used random even and odd numbers to do this. A sequence of alternating heads and tails e.g. H T H T H T H is possible but should be no more likely than the sequence shown. Typically, random number tables will give clusters of digits e.g.

14 86 50 97 03

or

46905 41003 84341 28752

How you use such numbers clearly depends on the type of input variable you need to simulate.

In MICROSTATS the commands IRAN and URAN give a number of options for the generation of random numbers e.g. a column of random number in the range 1 to 100 or a column of random numbers in the range 0 to 1 respectively.

If you are using random number tables, and always start from the top left-hand corner, the input values will be predictable and therefore not random. To ensure a random sequence from random number tables, you should always cross-off the used numbers. You should also check any random number generator to ensure a different sequence is generated for each simulation.

Random numbers can be used to generate all sorts of random events.

EXAMPLE

Given the random numbers 12 64 93 23 68 simulate the outcomes of sampling components where the probability of a defective is 0.3.

In this case, we could take the digits one at a time e.g. 1 2 6 4 etc. and let the digits 0 to 6 represent a non-defective component and the digits 7 to 9 a defective component. Using the random numbers given, the 5th and 10th components would be defective. In advance of the simulation, you could allocate digits differently but in the same proportions.

EXAMPLE

Given the random numbers 31246 50584 62791 83739 simulate the selection of voters for an opinion poll, if overall 25% vote for Party X, 30% for Party Y, 40% for Party Z with the remainder abstaining

In this case, digits could be taken in pairs (31 24 65 05 84 62 etc.) and allocated as follows:

00 to 24 Party X
25 to 54 Party Y
55 to 94 Party Z
95 to 99 Abstaining

The input values (simulated order of selection) would be:

Y, X, Z, X, Z etc.

Again, the digits could be assigned differently but in the same proportions. The following three examples show the generation of random events from the Binomial, Poisson and Normal distributions. In each case the following sequence of random numbers is used (to demonstrate method):

4 6 7 2 3 6 4 8 6 4
1 1 0 6 6 9 9 1 9

EXAMPLE

In a market research test, groups of 5 adults are selected and their views on beverages sought. If 60% of adults prefer tea to coffee, simulate the possible groupings in terms of this preference.

This distribution is Binomial with $n = 5$. To use the available tables (Appendix A), we will need to let $p = 0.4$ (tables only give values for p up to 0.5) and consider the number of individuals that prefer coffee to tea. The results are shown in Table 24.1.

Using this probability information, we are able to assign sets of 4 random digits to each of the possible outcomes as shown in Table 24.2.

Table 24.1

No. who prefer coffee (r)	*Cumulative prob. (r or more)*	*Exact prob (r)*	*No. who prefer tea*
0	1.0000	0.0778	5
1	0.9222	0.2592	4
2	0.6630	0.3456	3
3	0.3174	0.2304	2
4	0.0870	0.0768	1
5	0.0102	0.0102	0

Table 24.2

No. who prefer tea	*Probability 'weight'*	*Allocation of random numbers*
0	102	0000 to 0101
1	768	0102 to 0869
2	2304	0869 to 3137
3	3456	3174 to 6629
4	2592	6630 to 9221
5	778	9222 to 9999

Using the random numbers given, the selected groups of 5 would have the following number who prefer tea: 3, 3, 4, 2, 5. This set of values could then be input to the simulation model.

EXAMPLE

The demand for a particular component follows a Poisson distribution with a mean of 2 per day. Generate input data for a simulation model of inventory costs.

Given the Poisson parameter of $\lambda = 2$ (see Appendix B), we can allocate random numbers as shown in Table 24.3.

Taking the random digits in sets of 4, the number of components required over 5 simulated days is 2, 1, 4, 0, 6.

Table 24.3

Component demand	Cumulative probability	Exact probability	Allocation of random numbers
0	1.000	0.1353	0000 to 1352
1	0.8647	0.2707	1353 to 4059
2	0.5940	0.2707	4060 to 6766
3	0.3233	0.1804	6767 to 8570
4	0.1429	0.0902	8571 to 9472
5	0.0527	0.0361	9473 to 9833
6	0.0166	0.0155	9834 to 9988
7 or more	0.0011	0.0011	9989 to 9999

Even from these figures the variation in the input data is apparent. It is possible to use the simulation model to consider the effect of more control on input variation, or in the case of this example, managing demand more carefully.

EXAMPLE

The time taken to complete a task follows a Normal distribution with a mean of 30 minutes and standard deviation of 4 minutes. Simulate events from this distribution.

The Normal distribution is continuous, and we need to consider the probability of an event being within a given interval (see section 12.1). The determination of probability and the allocation of random numbers is shown in Table 24.4 below.

Table 24.4

Interval	z (lower boundary)	Probability	Allocation of random numbers
<22		0.02275	00000 to 02274
$22<26$	-2	0.13595	02275 to 15869
$26<30$	-1	0.3413	15870 to 49999
$30<34$	0	0.3413	50000 to 84129
$34<38$	1	0.13595	84130 to 97724
38 or more	2	0.02275	97725 to 99999

In this case, taking the random digits in sets of 5, the time intervals are 26 but less than 30 minutes, 30 but less than 34 minutes, 26 but less than 30 minutes and 30 but less than 34 minutes. If the simulation model requires a more precise time, then the size of the interval would need to be reduced e.g. 29 but less than 30 minutes.

The tables 24.2, 24.3 and 24.4 are often referred to as 'look-up tables, since they allow you to look up the effect on the model of a particular random.

24.4 THE CONSTRUCTION OF A SIMULATION MODEL

The generation of random event values for the input variables is only one aspect of simulation. The models developed for a business system need to make business sense. A model can only be as good as its specification, and the relationships between variables need to be investigated and understood. Suppose, for example, that a simulation model was developed to manage demand for an assembled product but did not take account of the need to supply replacement parts. The model may meet the requirements of production planning but not the needs of the purchasing department. The purchasing department would require an estimate that included components for assembly and replacement.

Simulation modelling should demonstate an understanding of the business system. It involves:

1. the formulation of the problem;
2. problem analysis;
3. model development;
4. implementation;
5. the questioning of results; and
6. the questioning of the model.

Any of these steps may be repeated a number of times in the light of experience. To understand the business problem a description will be required from those concerned; the problem owners.

Formulation of the problem requires an agreement on what the variables will be and to what extent they should be controllable. It also needs to be recognized that simulation is only one method of problem solution and there are likely to be others. Simulation may model the existing stock control system, for example, but should the problem owners consider alternative systems of stock management such as 'just-in-time'? Having agreed the type of model and the variables, the system will need further investigation and observation.

Problem analysis will involve the collection of data and other information on the variables included.

Model development involves the description of the business system by a set of equations. It may be necessary to make some simplifying assumptions e.g. that the relationships are linear, but the equations must reasonably relate the output variables to the input variables and parameters.

The **implementation** of a simulation model is likely to involve the repetition of a set of calculations using a computer. It is only when some stability emerges in the pattern of output that the number of repetitions is likely to be sufficient. We can then apply statistical method to these output results e.g. calculate averages. It is not for the problem owner, i.e. the manager, just to accept the output results; rather he/she should question their meaning.

It is only by the **questioning of results** that the implications for the business system are likely to be understood. If the output results do not match the experience of the business system, the model may need to be improved.

The **questioning of the model** is an important step in understanding the business system. It is useful to know, for example, that linear relationships do not adequately describe the problem.

EXAMPLE

Demand for a particular component is known to follow a Poisson distribution with a mean of 2 per day. The cost of holding this component is £0.30 per day and the cost of making an order is £2.50. Lead time is 2 days and the stock on hand is 7. The stock control policy is to order 6 when the stock level falls to 5 or less.

In terms of developing a simulation model, the above would have been agreed as a reasonable problem description at the formulation stage and the data collected at the problem analysis stage. Modelling stock control costs for this type of problem is well established (see Chapter 23) and can be presented using a spreadsheet format. Using the daily demand generated by an earlier example, we can develop a simulation of one week's stock control.

Table 24.5

Day	Stock b/fwd	Demand	Stock c/fwd	Holding cost	Order cost	Total cost
1	7	2	5	£1.50		£1.50
2	5	1	4	£1.20	£2.50	£3.70
3	4	4	0			
4	6 + 0	0	6	£1.80		£1.80
5	6	6	0			
						£7.00

To understand how weekly costs vary, this simulation would need to be repeated a number of times. Any of the values used e.g. order quantity or re-order level can be changed and the consequence on average weekly cost studied.

24.5 CONCLUSIONS

Simulation provides a useful alternative to other methods of problem solving. Equations that can be solved with relative ease for simple business situations can become increasingly difficult as we attempt to take account of the realities of business. In practice, all business simulation is likely to be computer based. It is not the calculations that require the investment of time but rather the development of an adequate mathematical model.

24.6 PROBLEMS

1. Why use simulation to solve business problems when most equations can be solved using other techniques provided a number of simplifying assumptions are made?
2. Why are random numbers important in simulation?
3. Lead time for a particular component can be 1, 2 or 3 weeks with respective probabilities of 0.3, 0.2 and 0.5. Using random numbers generate a sequence of 20 lead times.
4. Use random numbers to take 10 random samples from
   (a) a Binomial distribution where $n = 5$ and $p = 0.45$,
   (d) a Poisson distribution with a mean of 6, and
   (c) a Normal distribution with a mean of 70 and standard deviation of 5.)
5. The number of cars arriving each hour has the following probability distribution:

Arrivals	0	1	2	3	4	5
Probability	0.12	0.24	0.33	0.14	0.12	0.05

The service times vary according to the needs of the customer. There is a 40% chance that 2 cars are serviced in one hour and a 60% chance that 3 cars are serviced. Simulate a 35-hour working week and assess the maximum queue length.

PART SEVEN CONCLUDING EXERCISE

Draw a floor plan of a work place with at least two workers, two machines or two desks engaged in at least one activity with which you are familiar. It could look something like this:

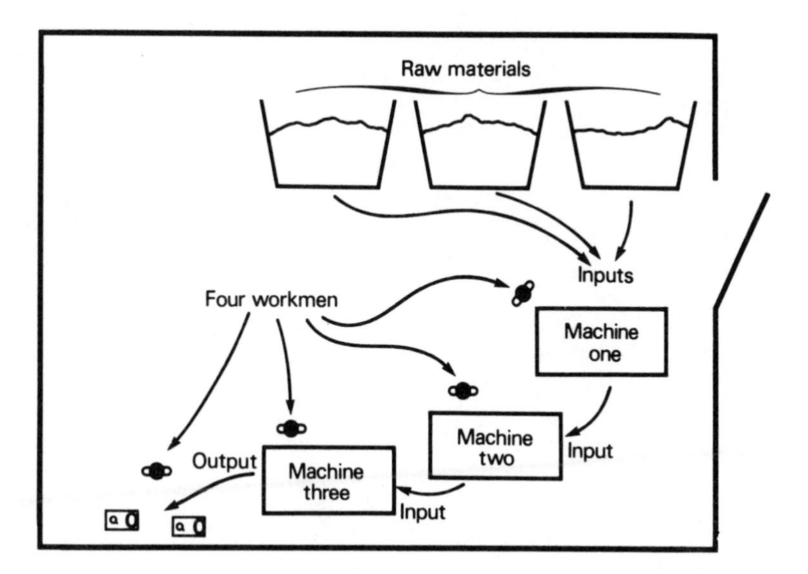

1. Identify the work processes and organization at this work place.
2. Identify where it would be appropriate to use mathematical models and the most realistic type of model in each case with special reference to any formulae and parameters needed.

PART EIGHT
REVIEW AND REVISION

In this section of the book we have brought together a selection of typical coursework assignments and examination questions in the Quantitative Methods area, some from first-year degree papers, and some from the professional bodies. We feel that in this subject it is only practice with such questions that will ensure a favourable result in end-of-year examinations. It is suggested that you might use these questions as an aid to revision, either by attempting a single examination question in 45 minutes, or by setting aside three hours and attempting to answer the required number of questions in that period. Whilst these questions will act as a guide to the type of questions you might meet, it is strongly suggested that you acquire copies of recent examination papers from your own course, and practise answering those too.

When answering these questions, and in examinations themselves, remember that it is usual for marks to be awarded for method as well as for getting the correct answer. Also, remember that the first 10 or 15 marks are usually the easiest to get, and thus you should answer the required number of questions, rather than spending 90 minutes on a single question.

Good luck in your examinations!

We would like to express our thanks to Mr. A.J. Mabbett and Mr. D. Deakins of Birmingham Polytechnic for permission to use questions from B.A. (Hons) Economics and to Mrs. D. Goodkin, also of Birmingham Polytechnic, for permission to use questions from the Accountancy Foundation Course. We would also like to thank Mr. G. O'Sullivan for permission to adapt coursework (given as Coursework 3) for use with MICROSTATS.

COURSEWORK 1

Since the Equal Pay Act of the 1970s it has been necessary for men and women to be paid the same wage for the same work, but men still earn more, on average, than women.

1. Use the figures from the latest NES on the 'distribution of gross weekly earnings of employees whose pay for the survey pay-period was not affected by absence' (Table B33) for calculate descriptive statistics of
   (a) males,
   (b) females on adult rates.

   Present your findings and suggest reasons for any differences between male and female earnings.
2. Use the figures from the NES to plot a graph showing the movement of average earnings for men and women since 1974. Select whichever figures you feel are most appropriate, justifiying your selection, and comment on the relationship shown by your graph.
3. Obtain figures from the Inland Revenue to show average (mean) earnings before tax; briefly compare this to the figures from the NES.

Notes

1. NES is New Earnings Survey; figures taken from Book B.
2. Inland Revenue data is taken from the *Annual Abstract of Statistics.*
3. Some further commentary is available in *Social Trends.*
4. It is advisable to check on the coverage and definition of surveys used in compiling statistics before comparisons are made.
5. You are expected to use MINITAB or MICROSTATS in section 1.
6. Keep your commentary brief and relevant, your whole report should be approximately 1000 words.

COURSEWORK 2

'New Car Registrations' forms part of the leading indicators used in government forecasting[1], and are said to turn approximately six months before other main series. Obtain quarterly data on New Car Registrations (unadjusted)[2] for the period 1968–1983Q_2 inclusive and present a graph of this data.

Analysis of this data using a centred four-point moving average trend has yielded the following results:

Trend figures

	Q_1	Q_2	Q_3	Q_4
1968	n/a	n/a	88.0250	83.7625
1969	84.9250	83.7875	83.1000	84.9750
1970	86.9875	89.6875	92.2750	93.4125
1971	97.0625	104.4375	112.3250	122.5875
1972	131.9875	136.7875	141.8375	143.3500
1973	142.0875	139.8500	130.7875	120.4375
1974	112.0000	105.2125	103.8250	103.1625
1975	101.4750	99.9750	98.5875	100.4625
1976	102.0625	104.2000	106.7375	106.5250
1977	108.0000	109.6625	113.1625	119.8625
1978	126.1875	130.5625	132.3500	139.3500
1979	142.6375	140.8750	143.3000	137.1375
1980	131.0750	129.5125	123.4250	120.3750
1981	120.6750	122.6875	124.6000	124.4625
1982	126.7875	130.6750	135.3500	140.9875

Additive seasonal factors

Q_1	Q_2	Q_3	Q_4
12.7059	6.2014	6.3884	−25.2957

Average random factor: 0.1809

Use an appropriate computer package to calculate a linear trend through the original data and find the seasonal factors.

Choose *one* of the two models to predict the average monthly level of the New Car Registrations for the period 1983Q_3–1984Q_4 inclusive justifying your choice.

Obtain more recent data and assess whether or not this remains consistent with your chosen model.

Critically assess the model and the predictions you have made in the light of the economic importance of the UK car industry. (Approx. 1200 words)

References

1. *Economic Progress Report* (1982) **149** September.
2. *Economic Trends Annual Supplement* (1984).

COURSEWORK 3

You are employed by an estate agent and need, as part of your duties, to estimate the value of houses. Your experience in doing this valuation has suggested to you that the statistical procedure of regression may be able to help you. House prices seem to depend primarily on a small number of variables or factors: number of bedrooms, approximate distance in miles from the city centre, score for house type, score for garage type and score for central heating facility.

Scores are allocated as follows:

Score for house type	*Score for garages*	*Score for central heating*
1 = terraced	0 = no garage	0 = no central heating
2 = semi-terraced	1 = garage	1 = part central heating
3 = detached	2 = double garage	2 = full central heating

You are required to undertake the following four tasks.

1. Collect relevant data from a local newspaper for a sample of 50 houses. You should explain briefly your sampling method.
2. Read your collected data values for each variable (house price, number of bedrooms etc.) into MICROSTATS, MINITAB or a similar package and obtain a plot of house price of y against each individual explanatory variable. Using appropriate statistics explain briefly how house prices relate to each explanatory variable.
3. To improve the accuracy of your prediction of house prices you decide to allocate to each house a number of points x, where:
   $x = \{6 \times \text{number of bedrooms}\}$
   $+ \{\text{approximate distance in miles from city centre}\}$
   $+ \{8 \times \text{score for house type}\}$
   $+ \{2 \times \text{score for garage}\}$
   $+ \{\text{score for central heating}\}$
   Again, using MICROSTATS, MINITAB or a similar package, determine the values of x for your 50 houses. Plot y against x and determine the regression line. Discuss your results.
4. Explain how the model of house prices developed in task 3 could be improved. Suggest other variables or factors that could be important in deciding the price of a house.

TYPICAL BUSINESS STUDIES DEGREE PAPER

Time allowed: 3 hours.

Answer any FOUR questions.

1. A company has monitored the amount of business generated by its sales representatives, and the results are recorded below:

Sales value (£000s)	*Number of sales reps*
Under 10	3
10 but under 20	20
20 but under 30	50
30 but under 50	20
50 and more	7

(a) Calculate: (i) the mean sales value;
(ii) the standard deviation of sales value;
(iii) the median sales value;
(iv) the quartile deviation of sales value. (*13 marks*)

(b) Construct a histogram for the above data. (*5 marks*)

(c) In the previous year, the mean was £25 200, the median was £20 150 and the standard deviation was £10 130. Comment on the changes, if any, from the previous year in the performance of the sales representatives. (*7 marks*)
(NB Assume limits of £8 000 and £80 000.)

2. A firm's demand curve passes through three known points:

if price = 100 demand = 2,
if price = 50 demand = 4,
if price = 46 demand = 10.

(a) Assume that $P = f(d)$ and that the demand function is quadratic and hence find the equation of the demand function. (*6 marks*)

(b) The supply function for the same firm is linear and passes through

$p = 20$ $s = 5$,
$p = 90$ $s = 9$.

If $p = f(s)$ find the supply function. (*3 marks*)

(c) Find the equilibrium point. (*6 marks*)

(d) Sketch the demand and supply functions. (*4 marks*)

(e) Calculate the price elasticity of demand at the equilibrium point and explain its significance. (*6 marks*)

3. The market demand function for a firm is

$P = 8 - 0.125Q$

and its average cost function is

$\mathrm{AC} = \frac{8}{Q} + 6 - 0.4Q + 0.08Q^2$

Find the level of Q to:

(a) maximize total revenue; (*4 marks*)
(b) minimize marginal cost; (*5 marks*)
(c) maximize profit. (*6 marks*)

Then

(d) find the level of profit at its maximum, and *(2 marks)*

(e) find the price level at maximum profit and the price elasticity of demand. Explain the situation faced by the firm. *(8 marks)*

4. Following a survey, men and women were classified by income group: the results are given below:

Sex	*Salary (£000s)*			
	Under 5	*5 but under 10*	*10 but under 20*	*20 or more*
Male	100	200	260	40
Female	100	200	90	10

(a) Calculate the value of the chi-squared statistic and hence test if there is an association between salary range and sex. *(15 marks)*

(b) Give likely reasons for the conclusion you have arrived at in part (a), noting any reservations you might have. *(10 marks)*

5. A company is cosidering undertaking one of two expansion projects. After preliminary anaylsis the following projected cash figures have been produced.

Year	*Cash flow I (£)* *In*	*Out*	*Cash flow II (£)* *In*	*Out*
1	8 000	10 000	5 000	—
2	15 000	10 000	12 000	10 000
3	25 000	10 000	24 000	—
4	55 000	—	35 000	20 000
5	50 000	—	65 000	—
6	40 000	—	70 000	—
7	40 000	—	30 000	—
8	20 000	—	20 000	—
Initial cost		30 000		30 000

(a) Use a discount rate of 18% to find the net present value of each project and hence recommend which should be undertaken. *(15 marks)*

(b) Discuss any reservations you have about your decision in part (a). *(10 marks)*

6. The size of a company's marketing department is thought to be related to the number of new products introduced. Data for ten companies is given below:

Size of marketing department	*Number of new products in last 5 years*
3	3
5	4
10	6
7	2
2	3
1	2
6	4
8	4
4	2
8	5

(a) Calculate the coefficient of determination and explain its meaning. *(10 marks)*

(b) Calculate the regression line of 'new products' *on* 'size of marketing department'. *(4 marks)*

(c) Construct a scatter diagram and place your regression line upon it. *(5 marks)*

(d) Predict the number of new products from departments of size

(i) 7, and

(ii) 15.

Discuss the applicability of your model. *(6 marks)*

7. A company's sales in the last three years are given below:

Year	*Q1*	*Q2*	*Q3*	*Q4*
1	20	15	8	32
2	21	17	8	36
3	23	20	9	40

(a) Find a centred 4-point moving averge trend. *(6 marks)*

(b) Construct a graph of the data and the trend. *(5 marks)*

(c) Calculate the 4 seasonal elements. (*6 marks*)

(d) Predict sales for Year 4 and comment on your model. (*8 marks*)

8. A company conducts an annual assessment of its employees, grading them on an appropriate scale. By comparing successive years' results, the following transition matrix was derived.

	Excellent	Good	Fair	Poor	Left the company
Excellent	0.8	0.1	0	0	0.1
Good	0.1	0.7	0.1	0	0.1
Fair	0	0.2	0.5	0.1	0.2
Poor	0	0.1	0.1	0.1	0.7
Left the company	0	0	0	0	1.0

(a) Construct a directed graph for this matrix. (*4 marks*)

(b) If at the end of last year, a cohort of 240 were split:

10 Excellent
100 Good
100 Fair
10 Poor
20 Left the Company

predict the expected numbers in each category at the end of this year. (*5 marks*)

(c) Predict the expected numbers in each category at the end of next year and the following. (*5 marks*)

(d) Discuss the likely reasons for the structure of the transition matrix and the implications for the company. (*11 marks*)

9. (a) State the assumptions of the Binomial model. (*4 marks*)

(b) A quality sampling scheme takes samples of 10 from batches which have 30% defective items. Calculate the following probabilities:

(i) of 3 defectives;
(ii) of 3 or less defectives;
(iii) of more than 3 defectives. (*9 marks*)

(c) Batches are also subject to a second sampling, where a sample of 10 is taken, but there are 5% defective in a different way. Find the probability of more than one defective in the sample. (*4 marks*)

(d) If batches are subjected to both samplings, find the probabilities that a batch:

(i) passes both tests (i.e. less than 3 in the first and less than 2 in the second);
(ii) fails both tests.

Discuss the implication of your results. (*8 marks*)

10. Either

(a) discuss the problem of achieving a representative sample of a known population or

(b) discuss the problems of designing an unbiased questionnaire. (*25 marks*)

TYPICAL ECONOMICS DEGREE PAPER

Time allowed: 3 hours

SECTION A

Answer TWO questions ONLY from this section.

1. A building society manager wishes to investigate the degree of association between the level of interest rates and the flow of building society deposits. The following historical data is available:

Date	*Interest rates**	*Building society deposits (inflows) £m*
1983 Jan	11	0.8
1983 July	9.5	1.1
1984 Jan	9	1.4
1984 July	12	1.2
1985 Jan	12	1.5
1985 July	12	1.3
1986 Jan	12.5	1.7
1986 July	10	1.1
1987 Jan	11	2.0
1987 July	10	1.1
1988 Jan	8.5	1.2
1988 July	10.5	1.9
1989 Jan	13	2.5
1989 July	14	2.2
1990 Jan	15	2.6

* Banks' base rates (Source: *Monthly Digest of Statistics*)

Investigate the degree of association by:

a) calculating Pearson's product moment correlation coefficient, r; *(5 marks)*

b) estimating a regression line of deposits on interest rates; *(5 marks)*

c) calculating the coefficient of determination. *(2 marks)*

Then:

d) draw a scatter diagram and place your regression line on the diagram; *(3 marks)*

e) comment on the degree of association in the light of your findings; *(5 marks)*

f) predict the level of building society deposits that the manager can expect if interest rates are:

(i) 17%, or

(ii) 10%,

and discuss the validity of these predictions. *(5 marks)*

2. The Plant Manager of a manufacturer of chemicals uses five inputs in the production process of fertilizers. He wishes to monitor accurately the changes in quantities and prices in the inputs that the company have been using. Information on prices and quantities used of the five inputs are given below:

Input	*1987*		*1988*		*1989*	
	Price	*Qty*	*Price*	*Qty*	*Price*	*Qty*
A	3.00	150	3.50	180	3.80	220
B	8.00	50	8.50	50	9.50	40
C	5.20	175	5.80	180	6.50	175
D	4.50	210	4.50	240	4.70	260
E	5.60	80	5.80	95	6.00	110

Setting 1987 as the base year calculate, for 1988 and 1989, calculate:

a) Laspeyre's and Paasche price indices; *(5 marks)*

b) Laspeyre's and Paasche quantity indices; *(5 marks)*

c) Fisher's ideal index for prices and quantities; *(3 marks)*

d) an expenditure index for 1988 and 1989; *(2 marks)*

Then;

e) carry out a factor reversal test for the indices you have calculated in parts (a) to (c); *(5 marks)*

f) comment on the differences in your results with reference to the relative advantages of Laspeyre's and Paasche indices. *(5 marks)*

3. Data from the *New Earnings Survey* (1989) reveals the following information about the distribution of gross weekly earnings for men and women:

Weekly earnings	*Percentage with gross weekly earnings* *Males*	*Females*
Under £120	5.3	22.6
£120 under £160	12.6	27.4
£160 under £200	16.9	18.1
£200 under £300	36.6	23.6
£300 under £450	20.4	7.1
£450 under £600	5.1	0.8
£600 and more	3.1	0.4

For both distributions:

(a) construct ogives (cumulative frequency polygons); *(5 marks)*

(b) estimate the median weekly incomes; *(2 marks)*

(c) estimate the uppe. and lower quartiles of weekly incomes and hence calculate the quartile deviations; *(3 marks)*

(d) calculate the arithmetic means and standard deviations; *(10 marks)*

(f) comment on the differences between the two distributions. *(5 marks)*

NB. Assume a lower limit of £100, and an upper limit of £750.

4. A company produces three products and divides its operations in the UK into 4 regions. The sales for 1989 are given below:

Region	*SALES (1989)* *Product A*	*Product B*	*Product C*
South East	55	65	40
West Midlands	40	55	45
South West	30	46	35
North and East	35	40	40

(a) Calculate the value of the chi-squared statistic. Use this to determine if there is an association between the product sales and the region of the UK at the 5% level of significance. *(14 marks)*

(d) Discuss the assumptions necessary to carry out this test of hypothesis. *(4 marks)*

(c) What reservations would you have when making recommendations to the company on the basis of your results? *(7 marks)*

5. *EITHER*

a) Training and Enterprise Councils (TEC) are a new agency which have been set up to coordinate training within a local area and in the West Midlands. One of their tasks is to establish the training needs of employers in their region.

Explain how you would design a *postal* questionnaire which is designed to discover the training needs, (skills, manpower requirements, qualifications, wage rates, ages) and requirements of local employers, for the TEC.

Advise the TEC on the relative advantages and disadvantages of a postal questionnaire compared to interviews for a survey of this nature. (*25 marks*)

Or

b) A review by economic commentators has pointed out that a 1% increase in mortgage rates makes a difference to the General Index of Retail Prices (RPI) of approximately 0.5 percentage points.

Explain *exactly* how the weighting system is produced for the RPI which results in this relatively high weighting for the mortgage rate in the calculation of the index. (*25 marks*)

SECTION B

Answer TWO questions ONLY from this section.

6. A retailer of electrical goods has two stores in a city; one in the centre of the city and one in the suburbs. This month, the retailer is concentrating the firm's sales efforts on a particular good. The quantity, Q_s, of the good that can be sold at the suburban store depends on the price per unit, P_s, charged at that store according to the demand function:

$$Q_s = 600 - 20P_s$$

The quantity, Q_c, that can be sold at the city centre store depends on the price per unit, P_c, charged at that store according to the demand function:

$$Q_c = 400 - 10P_c$$

a) If the retailer wishes to maximize the sales revenue from the two stores, determine how much should be sold at each store, the price and the total sales revenue. (*10 marks*)

b) If the retailer, on inspecting the firm's warehouse, finds that only 230 units of the good is available, determine the effect on the firm. (*12 marks*)

c) If one unit of the good costs the retailer £10 is it worthwhile trying to obtain another unit to sell? (*3 marks*)

7. The sole brewery on a small Atlantic Island can separate its consumers into two distinct markets: the islands's resident garrison (Market I) and the local inhabitants (Market II). The demand function in each of the two markets is given by:

Market I : $P_1 = 80 - 5Q_1$
Market II: $Q_2 = 90 - P_2/20$

where P_1 and P_2 are the price per gallon (in pence) in each market and Q_1 and Q_2 are the number of gallons of beer demanded (in '00 units) in each market. The monopolist's total cost function (in pounds) is given by:

$$TC - 50 - 20Q = 0,$$

where Q is the total output of beer from the brewery (in '00 units). If the brewery wishes to maximize profits, calculate:

(a) the price per gallon that should be charged in each market; (*13 marks*)

(b) the brewery's profit with and without price discrimination. (*12 marks*)

8. a) If

$$\mathbf{A} = \begin{bmatrix} 2 & 1 \\ 1 & 2 \end{bmatrix} \quad \mathbf{B} = \begin{bmatrix} 1 & 0 \\ 0 & 1 \end{bmatrix} \quad \mathbf{C} = \begin{bmatrix} 1 & 2 \\ 3 & 4 \\ 1 & 2 \end{bmatrix}$$

$$\mathbf{D} = \begin{bmatrix} 0 & 1 & 0 \\ 1 & 1 & 1 \end{bmatrix}$$

Find

(i) $\mathbf{A} + \mathbf{B} - 2\mathbf{DC}$

(ii) $\mathbf{A}^{-1}\mathbf{B}^{-1}$

(iii) $\mathbf{C}[\mathbf{A} + \mathbf{B}]\mathbf{B}^{-1}$

(iv) $\mathbf{CA}^{-1}\mathbf{B}^{-1}\mathbf{D}$ (*10 marks*)

b) A national-income model can be written as follows:

$$Y = C + I_0 + G_0$$
$$C = a + bY \ (a>0,\ 0<b)$$

where Y and C stand for the endogenous variables national income and consumption expenditure respectively, and I and G represent the exogenously determined investment and government expenditures.

Using Cramer's rule, find an expression for the equilibrium values of Y and C. *(5 marks)*

c) The equilibrium condition for three related markets is given by:

$$11P_1 - P_2 - P_3 = 31$$
$$-P_1 + 6P_2 - 2P_3 = 26$$
$$-P_1 - 2P_2 + 7P_3 = 24$$

Using matrix algebra, find the equilibrium price for each market. *(10 marks)*

9. A manufacturer of reproduction furniture faces the following demand function for its corner units:

$$Q_d = 308 - P,$$

where Q_d is the quantity demanded and P is the price (£) of a unit. The firm's total variable cost function is given by

$$\text{TVC} = 3Q^2 + 96Q,$$

where Q is the quantity produced. The fixed cost of production is £960.

If the firm only produces what it can sell, and no more, determine the breakeven point(s).

If the firm decides to produce 35 units, what is the profit?

Determine the output at which profit is a maximum, demonstrating your answer graphically. *(25 marks)*

10. A manufacturer of camera equipment wishes to build a mathematical model in order to explain the market forces which face the company. The business analyst in charge of the project has found that when the price of the camera is put at £50, then 50 cameras of a fixed type are available for sale. Also, when the price is raised to £75, 100 cameras are available for sale. Further investigation has shown that when the price was set at £150, no camera was sold; but when the price was lowered to £114, then 28 cameras were sold.

(a) Determine the firm's supply function. *(6 marks)*

(b) Determine the firm's demand function. *(5 marks)*

(c) Determine the market equilibrium for the camera. *(2 marks)*

(d) Suppose the government levies a tax of £t per camera sold. If the market is in equilibrium and the tax is increased, how will the price, quantity, and tax revenue change once the new equilibrium has been attained? *(12 marks)*

11. a) Define the following terms:
(i) simple interest,
(ii) compound interest,
(iii) present value,
(iv) net present value,
(v) internal rate of return. *(5 marks)*

b) A company is about to undertake a new project which requires investment in a specialist machine costing £75 000. The project will last for five years, after which it is estimated that the scrap value of the machine will be £1 250, to be received at the end of the sixth year. It is necessary to inject money into the project at the end of each year to ensure that the machine is kept in proper working order for the next year of the project. The amount needed at the end of the first year is £1 000 and it is estimated that this amount needs to be increased by 10% at the end of each succeeding year over the period of the project.

The revenue produced from the project through the use of the machine is estimated to be £20 000 at the end of the first year, and this will increase by 7.5% at the end of each succeeding year over the period of the project.

(i) Establish and tabulate the net cash flows for the project.
(ii) Establish the net present value of the project using discount rates of 10%, and 15%.
(iii) Establish the internal rate of return of the project and interpret your answer. *(20 marks)*

TYPICAL ACCOUNTANCY DEGREE PAPER

Time allowed: 3 hours.

Answer FIVE questions.

1. Information has been collected from a survey on the weekly income of a group of people who left school at 16 years of age although they were qualified to stay on into the sixth form. This information is presented below.

Weekly income (£)	*No. of people*
90 but less than 100	2
100 but less than 110	12
110 but less than 130	11
130 but less than 150	7
150 but less than 180	5
180 but less than 210	1

(a) Explain what is meant by the terms 'sample statistic' and 'population parameter'. (*4 marks*)

(b) Calculate the mean and median weekly income from the data given. Explain any difference in these two figures. (*5 marks*)

(c) Calculate the standard deviation of weekly income from the data given. (*5 marks*)

(d) A survey of a similar group of people who stayed on at school reported an average weekly income of £124. Formulate and test the hypothesis that staying on at school for this group of people has resulted in higher weekly income. (*6 marks*)

2. The age structure of a particular company's workforce has been presented as follows:

Age (in years)	*Male*	*Female*
under 30	610	244
30 but under 50	754	226
50 and over	666	174

(a) The management of this company has decided to select a sample from its workforce to obtain further information about their characteristics and opinions.
   (i) Describe how you would design the sample to ensure that it was representative.
   (ii) Describe the factors you would need to consider when deciding the sample size. (*6 marks*)

(b) The management of the company would like to use the survey results to assess the reaction of the workforce to merger plans. Determine the sample size required to estimate the percentage in favour, given that no prior information is available:

(i) to within ± 5% for the complete workforce with a 95% confidence interval;

(ii) to within ± 5% for male and female employees separately with 95% confidence intervals.

(6 marks)

(c) Suppose the results for weekly overtime earnings from a survey of this workforce were

	Male	*Female*
Sample size	50	45
Mean	£15.30	£9.80
Standard deviation	£6.30	£3.90

(i) Determine 95% confidence intervals for the mean weekly overtime earnings of males and females.

(ii) Determine the overall mean weekly overtime earnings (regardless of gender). Explain your use of weighting factors if applicable.

(8 marks)

3. A company which specializes in servicing household appliances in the home has found that the average time taken to complete a job is 2.7 hours with a standard deviation of 0.68 hours. It can be assumed that the time taken to complete a job will follow a Normal distribution.

(a) If the company assesses the cost of each job at £15 per hour, how much should it charge to cover:

(i) the cost of an average job?

(ii) the cost of 90% of jobs?

(8 marks)

(b) The company has now decided to charge for jobs as follows:

Time taken (hours)	*Charge (£)*
under 3	40
3 but under 4.5	50
4.5 and over	70

Determine the expected charge for a job and explain the consequences, if any, of the change in cost structure. *(12 marks)*

4. You have inherited building land and must decide whether to sell the plot for £58 000 or build houses for sale. To build houses will require an immediate investment of £40 000 with net returns subject to the state of the market. The market has been categorized as high, medium or low and forecasts of net returns made as follows

	Net returns (£)		
End of year	*high*	*medium*	*low*
1	60 000	40 000	20 000
2	42 000	42 000	42 000
3	22 000	44 000	44 000
4	23 000	23 000	46 000

Market research indicates that there is a 50% chance of a 'high' market, 30% of a 'medium' market and 20% chance of a 'low' market. You have been advised to use a discount rate of 20%.

(a) Explain what methods are available to evaluate the above decision problem.

(8 marks)

(b) Evaluate the above problem and suggest the best course of action. *(12 marks)*

5. A company is currently selling its product in its home market. The demand for the product is given by

$Q = 10\,000 - 100P$

where Q is quantity demand
and P is the price.
Total cost is given by

$\mathrm{TC} = 30\,000 + 60Q$

(a) Determine a revenue function in terms of Q and on the same graph plot revenue and total cost against Q for the range $Q = 0$ to $Q = 4\,000$. Indicate the breakeven points on your graph.

(6 marks)

(b) Using differentiation, determine the output that maximizes profit. State maximum profit and the price required to achieve this. *(6 marks)*

(c) The company has now decided to expand into the export market. The demand curve is the same in both markets. The distribution charges add an extra £2 per unit to the cost of selling in the export market and the fixed costs increase to £36 000. Determine the price to be charged in each market in order to maximize profits and state the resultant profit of the company. *(8 marks)*

6. Records of the unit sales of a particular item are given below:

Year	*Quarters* *I*	*II*	*III*	*IV*
1983	180	133	105	170
1984	203	150	102	193
1985	236	152	127	194
1986	241			

(a) Construct a graph of this data and state whether you would consider the additive or multiplicative model most appropriate, giving your reasons. *(5 marks)*

(b) Determine the trend and plot this on your graph. *(5 marks)*

(c) Calculate the seasonal adjustment factors using the time series model considered most appropriate in part (a). *(5 marks)*

(d) Predict unit sales for quarters III and IV of 1986 using your time series model. Describe briefly the difficulties of marking predictions of this kind. *(5 marks)*

7. The number of faults identified and the time taken to correct them was recorded for a sample of 10 new cars. The information was presented as follows:

Car	*Number of faults*	*Time taken to correct faults (min)*
1	16	26
2	8	13
3	10	18
4	5	16
5	1	2
6	8	15
7	13	23
8	21	27
9	12	22
10	3	4

(a) Graph this data and describe the relationship observed. *(4 marks)*

(b) Calculate Pearson's coefficient of correlation and test for significance. *(4 marks)*

(c) Determine the regression line to predict the time taken to correct faults and show this on your graph. *(4 marks)*

(d) Using your answer from part (c), predict the time taken to correct faults if a new car had:
(i) 4 faults;
(ii) 25 faults. *(4 marks)*

(e) Describe any reservations you have about the predictions made in part (d). *(4 marks)*

8. (a) Given the square matrices

$$\mathbf{A}=\begin{bmatrix}3 & 2\\ 4 & 1\end{bmatrix},\mathbf{B}=\begin{bmatrix}2 & -1\\ 0 & 1\end{bmatrix},\mathbf{C}=\begin{bmatrix}1 & 0\\ 0 & 1\end{bmatrix}$$

evaluate

(i) $\mathbf{A}+\mathbf{B}-\mathbf{C}$
(ii) $3(\mathbf{A}-\mathbf{B}+\mathbf{C})$
(iii) $\mathbf{AB}+\mathbf{BC}$ *(6 marks)*

(b) Verify that the matrix **E** is the inverse of matrix **D**

$$\mathbf{D}=\begin{bmatrix}1 & 1 & 0 & 2\\ 3 & 2 & 1 & 3\\ 2 & 1 & 3 & 0\\ 0 & 2 & 1 & 4\end{bmatrix}\quad \mathbf{E}=\begin{bmatrix}5 & -2 & 1 & -1\\ -34 & 16 & -7 & 5\\ 8 & -4 & 2 & -1\\ 15 & -7 & 3 & -2\end{bmatrix}$$

Explain your method. *(6 marks)*

(c) A company manufactures four products ($X1$, $X2$, $X3$, $X4$) which require time from four processes (alpha, beta, gamma and delta) as given below:

	Time required (hours)			
Process	*X1*	*X2*	*X3*	*X4*
alpha	1	1	0	2
beta	3	2	1	3
gamma	2	1	3	0
delta	0	2	1	4

Time available is 142 hours from process alpha, 348 hours from process beta, 245 hours from process gamma and 204 hours from process delta.

The company requires to know how many of each product to manufacture to use all the processing time available.

Express the problem as a set of simultaneous linear equations and use matrix **E** from part (b) to provide the solution.

(*8 marks*)

TYPICAL ACCOUNTING FOUNDATION PAPER

1. A firm uses three basic raw materials labelled A, B and C in its production process. The following data shows the average cost (in £ per unit) and quantity (in units) used over a period of three years:

		Year 1		*Year 2*		*Year 3*	
Material	*Unit*	*Cost*	*Qty*	*Cost*	*Qty*	*Cost*	*Qty*
A	kg	2.50	2250	2.80	2500	3.10	3200
B	litre	12.15	12000	15.00	14500	16.10	15200
C	item	5.25	1250	5.30	1400	5.55	1650

(a) Construct a Laspeyre price index for raw materials for years 2 and 3 using year 1 as base year and interpret your results. *(6 marks)*

(b) Construct a Paasche quantity index for raw materials for years 2 and 3 using year 1 as base year and interpret your results. *(6 marks)*

(c) Compare and contrast the relative advantages and disadvantages of Laspeyre and Paasche indices. *(4 marks)*

(d) Describe some of the possible uses of price and quantity index numbers to a manufacturing firm. *(4 marks)*

2. A manufacturing company has two service departments (maintenance and administration), and two production departments, (assembly and finishing.) The direct overheads of each department are as follows:

Maintenance	*Administration*	*Assembly*	*Finishing*
£2 500	£3 250	£22 500	£11 400

The labour hours of each service department's time used by the other departments is as follows:

	Maintenance	Administration	Assembly	Finishing
Maintenance		200	1 000	800
Administration	150		400	350

Using matrix algebra, calculate the total cost of each production department including the allocation of service department costs. *(20 marks)*

4. (a) The Polytechnic have decided to build a new sports centre. An area of 11 250 square feet is required for the swimming pool. To one side of the pool, an area 40 feet wide is needed for squash courts and changing rooms. On the other side of the pool, an area 30 feet wide is required for offices. A car park 50 feet wide is to be built at the front and an area 10 feet wide is needed at the back for boiler room, filtration system and storage.

What is the minimum area that will be required for the sports centre and what are the dimensions of the pool for this site area? (*12 marks*)

(b) A firm has estimated that its total cost of production is given by:

$C = 2068 + 20x + 0.08x^2$

where

C is total production cost in £000s
x is number produced/sold in thousands

and that the selling price is related to the quantity sold by the following:

$P = 314 - 6x$

where £P is the selling price.

Find the price and quantity that will maximize profit, and the maximum profit. (*8 marks*)

5. A firm has up to £100 000 to invest in stocks and bonds. Stock cost £50 each and each stock is expected to produce an annual return in the form of a £10 dividend. Each bond costs £120 and will provide a return of 10% p.a. As stocks are riskier than bonds, no more than 25% of the total amount invested can be in stocks.
   (a) Formulate the linear programming problem to maximize the total annual return from these two types of investment. (*5 marks*)
   (b) Solve the problem using graphical linear programming (indicating the feasible region on your graph) and state the solution. (*13 marks*)
   (c) Briefly describe how you would deal with a non-integer solution to a problem of the type described above. (*2 marks*)

6. A timber merchant estimates that its annual demand for a certain type of oak plank is 1600. The planks are kept in a storage bin which has capacity for 150 planks. Each plank costs £17.50. The clerical costs involved with placing an order are £3, the delivery charge per order is £45 and the cost of inspecting and unloading the order are £2. The firm's cost of capital is 12%.
   (a) Calculate the optimum order quantity and total costs incurred. (*4 marks*)
   (b) If the storage bin could be expanded at a cost of £75 to hold 500 planks, would it be worthwhile expanding the bin? (*3 marks*)
   (c) Given that the storage bin has been expanded to hold 500 planks, would it be worthwhile for the firm to take advantage of a 10% discount offered by the suppliers for batches of 500 planks? (*5 marks*)
   (d) What is the minimum discount on batches of 500 planks for it to be worthwhile to the firm to take advantage of the discount (assuming that it has storage facilities for 500 planks). (*5 marks*)
   (e) Given that it is very difficult to forecast demand accurately, do you think that EOQ stock control model is valid? (*3 marks*)

7. The following data show a firm's quarterly sales for a three year period:

		Sales (000s)		
Quarter	*1*	*2*	*3*	*4*
Year 1	10.1	11.9	12.8	11.6
Year 2	11.1	12.4	13.6	12.6
Year 3	12.0	12.9	14.4	13.5

   (a) Using the method of moving averages, calculate the trend and hence the seasonal variation for the data. (*13 marks*)
   (b) Forecast sales for the first two quarters of year 4. (*3 marks*)
   (c) Specify the movements which may be identified when analyzing economic and business statistics over a long time span. (*4 marks*)

8. A firm wishes to test consumer reactions to a revolutionary type of instant food. It has selected an area, containing 19 000 households, close to the factory and Head Office, to assess initial reactions to the product. As the product is expensive, the firm believes it will appeal mostly to high income groups and has already ascertained that out of the 19 000 households, 5 000 are in the high income group and the remainder in the low income

group. In an initial survey, of a 1 in 20 sample of the high income group, 40% expressed an interest in the product: of a 1 in 50 sample of the low income group, 35% expressed an interest.

(a) Estimate the percentage in the entire population of 19 000 households who are interested in the product. (*3 marks*)

(b) Calculate a 95% confidence interval for the proportion of high income group households interested in the product, and interpret your result. (*4 marks*)

(c) Test the hypothesis that there is no difference between the proportions in the two samples interested in the product. (*5 marks*)

(d) Outline the main advantages and disadvantages of mail questionnaires and personal interviews for market research surveys. (*4 marks*)

(e) Describe what is meant by stratification in sampling of data and explain some of its uses in market research surveys. (*4 marks*)

APPENDIX A CUMULATIVE BINOMIAL PROBABILITIES

		$p = 0.01$	0.05	0.10	0.20	0.30	0.40	0.45	0.50
$n = 5$	$r =$ 0	1.0000	1.0000	1.0000	1.0000	1.0000	1.0000	1.0000	1.0000
	1	0.0490	0.2262	0.4095	0.6723	0.8319	0.9222	0.9497	0.9688
	2	0.0010	0.0226	0.0815	0.2627	0.4718	0.6630	0.7438	0.8125
	3		0.0012	0.0086	0.0579	0.1631	0.3174	0.4069	0.5000
	4			0.0005	0.0067	0.0308	0.0870	0.1312	0.1875
	5				0.0003	0.0024	0.0102	0.0185	0.0313
$n = 10$	$r =$ 0	1.0000	1.0000	1.0000	1.0000	1.0000	1.0000	1.0000	1.0000
	1	0.0956	0.4013	0.6513	0.8926	0.9718	0.9940	0.9975	0.9990
	2	0.0043	0.0861	0.2639	0.6242	0.8507	0.9536	0.9767	0.9893
	3	0.0001	0.0115	0.0702	0.3222	0.6172	0.8327	0.9004	0.9453
	4		0.0010	0.0128	0.1209	0.3504	0.6177	0.7430	0.8281
	5		0.0001	0.0016	0.0328	0.1503	0.3669	0.4956	0.6230
	6			0.0001	0.0064	0.0473	0.1662	0.2616	0.3770
	7				0.0009	0.0106	0.0548	0.1020	0.1719
	8				0.0001	0.0016	0.0123	0.0274	0.0547
	9					0.0001	0.0017	0.0045	0.0107
	10						0.0001	0.0003	0.0010

where
p is the probability of a characteristic (e.g. a defective item),
n is the sample size and
r is the number with that characteristic.

APPENDIX B CUMULATIVE POISSON PROBABILITIES

	$\lambda = 1.0$	2.0	3.0	4.0	5.0	6.0	7.0
$x =$ 0	1.0000	1.0000	1.0000	1.0000	1.0000	1.0000	1.0000
1	0.6321	0.8647	0.9502	0.9817	0.9933	0.9975	0.9991
2	0.2642	0.5940	0.8009	0.9084	0.9596	0.9826	0.9927
3	0.0803	0.3233	0.5768	0.7619	0.8753	0.9380	0.9704
4	0.0190	0.1429	0.3528	0.5665	0.7350	0.8488	0.9182
5	0.0037	0.0527	0.1847	0.3712	0.5595	0.7149	0.8270
6	0.0006	0.0166	0.0839	0.2149	0.3840	0.5543	0.6993
7	0.0001	0.0011	0.0335	0.1107	0.2378	0.3937	0.5503
8		0.0002	0.0119	0.0511	0.1334	0.2560	0.4013
9			0.0038	0.0214	0.0681	0.1528	0.2709
10			0.0011	0.0081	0.0318	0.0839	0.1695
11			0.0003	0.0028	0.0137	0.0426	0.0985
12			0.0001	0.0009	0.0055	0.0201	0.0534
13				0.0003	0.0020	0.0088	0.0270
14				0.0001	0.0007	0.0036	0.0128
15					0.0002	0.0014	0.0057
16					0.0001	0.0005	0.0024
17						0.0002	0.0010
18						0.0001	0.0004
19							0.0001

where
λ is the average number of times a characteristic occurs and
x is the number of occurrences.

APPENDIX C
AREAS IN THE RIGHT-HAND TAIL OF THE NORMAL DISTRIBUTION

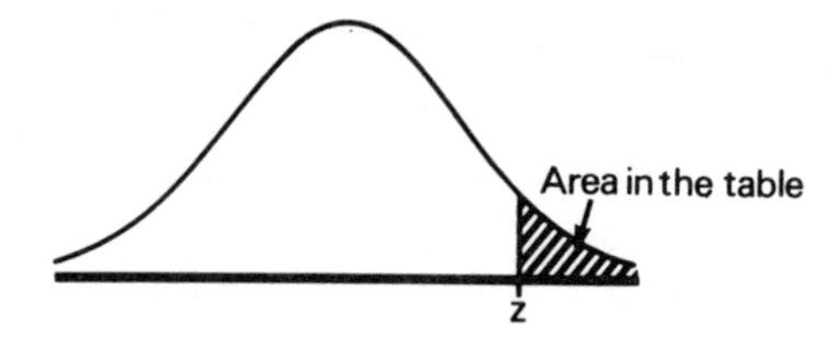

z	.00	.01	.02	.03	.04	.05	.06	.07	.08	.09
0.0	.5000	.4960	.4920	.4880	.4840	.4801	.4761	.4721	.4681	.4641
0.1	.4602	.4562	.4522	.4483	.4443	.4404	.4364	.4325	.4286	.4247
0.2	.4207	.4168	.4129	.4090	.4052	.4013	.3974	.3936	.3897	.3859
0.3	.3821	.3783	.3745	.3707	.3669	.3632	.3594	.3557	.3520	.3483
0.4	.3446	.3409	.3372	.3336	.3300	.3264	.3228	.3192	.3156	.3121
0.5	.3085	.3050	.3015	.2981	.2946	.2912	.2877	.2843	.2810	.2776
0.6	.2743	.2709	.2676	.2643	.2611	.2578	.2546	.2514	.2483	.2451
0.7	.2420	.2389	.2358	.2327	.2296	.2266	.2236	.2206	.2177	.2148
0.8	.2119	.2090	.2061	.2033	.2005	.1977	.1949	.1922	.1894	.1867
0.9	.1841	.1814	.1788	.1762	.1736	.1711	.1685	.1660	.1635	.1611
1.0	.1587	.1562	.1539	.1515	.1492	.1496	.1446	.1423	.1401	.1379
1.1	.1357	.1335	.1314	.1292	.1271	.1251	.1230	.1210	.1190	.1170
1.2	.1151	.1131	.1112	.1093	.1075	.1056	.1038	.1020	.1003	.0985
1.3	.0968	.0951	.0934	.0918	.0901	.0885	.0869	.0853	.0838	.0823
1.4	.0808	.0793	.0778	.0764	.0749	.0735	.0721	.0708	.0694	.0681

z	.00	.01	.02	.03	.04	.05	.06	.07	.08	.09
1.5	.0668	.0655	.0643	.0630	.0618	.0606	.0594	.0582	.0571	.0559
1.6	.0548	.0537	.0526	.0516	.0505	.0495	.0485	.0475	.0465	.0455
1.7	.0446	.0436	.0427	.0418	.0409	.0401	.0392	.0384	.0375	.0367
1.8	.0359	.0351	.0344	.0336	.0329	.0322	.0314	.0307	.0301	.0294
1.9	.0287	.0281	.0274	.0268	.0262	.0256	.0250	.0244	.0239	.0233
2.0	.02275	.02222	.02169	.02118	.02068	.02018	.01970	.01923	.01876	.01831
2.1	.01786	.01743	.01700	.01659	.01618	.01578	.01539	.01500	.01463	.01426
2.2	.01390	.01355	.01321	.01287	.01255	.01222	.01191	.01160	.01130	.01101
2.3	.01072	.01044	.01017	.00990	.00964	.00939	.00914	.00889	.00866	.00842
2.4	.00820	.00798	.00776	.00755	.00734	.00714	.00695	.00676	.00657	.00639
2.5	.00621	.00604	.00587	.00570	.00554	.00539	.00523	.00508	.00494	.00480
2.6	.00466	.00453	.00440	.00427	.00415	.00402	.00391	.00379	.00368	.00357
2.7	.00347	.00336	.00326	.00317	.00307	.00298	.00289	.00280	.00272	.00264
2.8	.00256	.00248	.00240	.00233	.00226	.00219	.00212	.00205	.00199	.00193
2.9	.00187	.00181	.00175	.00169	.00164	.00159	.00154	.00149	.00144	.00139
3.0	.00135									
3.1	.00097									
3.2	.00069									
3.3	.00048									
3.4	.00034									
3.5	.00023									
3.6	.00016									
3.7	.00011									
3.8	.00007									
3.9	.00005									
4.0	.00003									

APPENDIX D STUDENT'S t CRITICAL POINTS

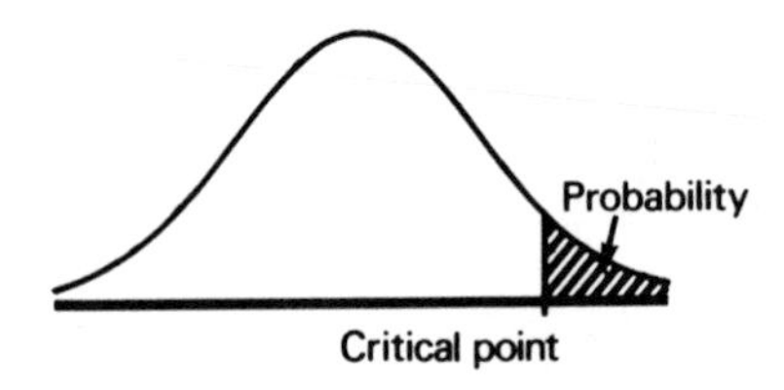

Probability	0.10	0.05	0.025	0.01	0.005
$\nu =$ 1	3.078	6.314	12.706	31.821	63.657
2	1.886	2.920	4.303	6.965	9.925
3	1.638	2.353	3.182	4.541	5.841
4	1.533	2.132	2.776	3.747	4.604
5	1.476	2.015	2.571	3.365	4.032
6	1.440	1.943	2.447	3.143	3.707
7	1.415	1.895	2.365	2.998	3.499
8	1.397	1.860	2.306	2.896	3.355
9	1.383	1.833	2.262	2.821	3.250
10	1.372	1.812	2.228	2.764	3.169
11	1.363	1.796	2.201	2.718	3.106
12	1.356	1.782	2.179	2.681	3.055
13	1.350	1.771	2.160	2.650	3.012
14	1.345	1.761	2.145	2.624	2.977
15	1.341	1.753	2.131	2.602	2.947

Probability	0.10	0.05	0.025	0.01	0.005
16	1.337	1.746	2.120	2.583	2.921
17	1.333	1.740	2.110	2.567	2.898
18	1.330	1.734	2.101	2.552	2.878
19	1.328	1.729	2.093	2.539	2.861
20	1.325	1.725	2.086	2.528	2.845
21	1.323	1.721	2.080	2.518	2.831
22	1.321	1.717	2.074	2.508	2.819
23	1.319	1.714	2.069	2.500	2.807
24	1.318	1.711	2.064	2.492	2.797
25	1.316	1.708	2.060	2.485	2.787
26	1.315	1.706	2.056	2.479	2.779
27	1.314	1.703	2.052	2.473	2.771
28	1.313	1.701	2.048	2.467	2.763
29	1.311	1.699	2.045	2.462	2.756
30	1.310	1.697	2.042	2.457	2.750
40	1.303	1.684	2.021	2.423	2.704
60	1.296	1.671	2.000	2.390	2.660
120	1.289	1.658	1.980	2.358	2.617
∞	1.282	1.645	1.960	2.326	2.576

where v is the number of degrees of freedom.

APPENDIX E
χ^2 CRITICAL VALUES

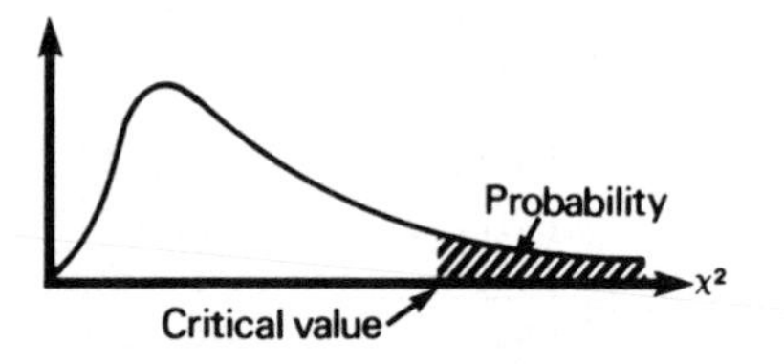

Probability		0.250	0.100	0.050	0.025	0.010	0.005	0.001
$\nu =$	1	1.32	2.71	3.84	5.02	6.63	7.88	10.8
	2	2.77	4.61	5.99	7.38	9.21	10.6	13.8
	3	4.11	6.25	7.81	9.35	11.3	12.8	16.3
	4	5.39	7.78	9.49	11.1	13.3	14.9	18.5
	5	6.63	9.24	11.1	12.8	15.1	16.7	20.5
	6	7.84	10.6	12.6	14.4	16.8	18.5	22.5
	7	9.04	12.0	14.1	16.0	18.5	20.3	24.3
	8	10.2	13.4	15.5	17.5	20.3	22.0	26.1
	9	11.4	14.7	16.9	19.0	21.7	23.6	27.9
	10	12.5	16.0	18.3	20.5	23.2	25.2	29.6
	11	13.7	17.3	19.7	21.9	24.7	26.8	31.3
	12	14.8	18.5	21.0	23.3	26.2	28.3	32.9
	13	16.0	19.8	22.4	24.7	27.7	29.8	34.5
	14	17.1	21.1	23.7	26.1	29.1	31.3	36.1
	15	18.2	22.3	25.0	27.5	30.6	32.8	37.7
	16	19.4	23.5	26.3	28.8	32.0	34.3	39.3
	17	20.5	24.8	27.6	30.2	33.4	35.7	40.8
	18	21.6	26.0	28.9	31.5	34.8	37.2	42.3
	19	22.7	27.2	30.1	32.9	36.2	38.6	43.8

Probability	0.250	0.100	0.050	0.025	0.010	0.005	0.001
20	23.8	28.4	31.4	34.2	37.6	40.0	45.3
21	24.9	29.6	32.7	35.5	38.9	41.4	46.8
22	26.0	30.8	33.9	36.8	40.3	42.8	48.3
23	27.1	32.0	35.2	38.1	41.6	44.2	49.7
24	28.2	33.2	36.4	39.4	43.0	45.6	51.2
25	29.3	34.4	37.7	40.6	44.3	46.9	52.6
26	30.4	35.6	38.9	41.9	45.6	48.3	54.1
27	31.5	36.7	40.1	43.2	47.0	49.6	55.5
28	32.6	37.9	41.3	44.5	48.3	51.0	56.9
29	33.7	39.1	42.6	45.7	49.6	52.3	58.3
30	34.8	40.3	43.8	47.0	50.9	53.7	59.7
40	45.6	51.8	55.8	59.3	63.7	66.8	73.4
50	56.3	63.2	67.5	71.4	76.2	79.5	86.7
60	67.0	74.4	79.1	83.3	88.4	92.0	99.6
70	77.6	85.5	90.5	95.0	100	104	112
80	88.1	96.6	102	107	112	116	125
90	98.6	108	113	118	124	128	137
100	109	118	124	130	136	140	149

where ν is the number of degrees of freedom.

APPENDIX F
PRESENT VALUE FACTORS

Years	1%	2%	3%	4%	5%	6%	7%	8%	9%	10%
1	.9901	.9804	.9709	.9615	.9524	.9434	.9346	.9259	.9174	.9091
2	.9803	.9612	.9426	.9426	.9070	.8900	.8734	.8573	.8417	.8264
3	.9706	.9423	.9151	.8890	.8638	.8396	.8163	.7938	.7722	.7513
4	.9610	.9238	.8885	.8548	.8227	.7921	.7629	.7350	.7084	.6830
5	.9515	.9057	.8626	.8219	.7835	.7473	.7130	.6806	.6499	.6209
6	.9420	.8880	.8375	.7903	.7462	.7050	.6663	.6302	.5963	.5645
7	.9327	.8706	.8131	.7599	.7107	.6651	.6227	.5835	.5470	.5132
8	.9235	.8535	.7894	.7307	.6768	.6274	.5820	.5403	.5019	.4665
9	.9143	.8368	.7664	.7026	.6446	.5919	.5439	.5002	.4604	.4241
10	.9053	.8203	.7441	.6756	.6139	.5584	.5083	.4632	.4224	.3855
11	.8963	.8043	.7224	.6496	.5847	.5268	.4751	.4289	.3875	.3505
12	.8874	.7885	.7014	.6246	.5568	.4970	.4440	.3971	.3555	.3186
13	.8787	.7730	.6810	.6006	.5303	.4688	.4150	.3677	.3262	.2897
14	.8700	.7579	.6611	.5775	.5051	.4423	.3878	.3405	.2992	.2633
15	.8613	.7430	.6419	.5553	.4810	.4173	.3624	.3152	.2745	.2394
16	.8528	.7284	.6232	.5339	.4581	.3936	.3387	.2919	.2519	.2176
17	.8444	.7142	.6050	.5134	.4363	.3714	.3166	.2703	.2311	.1978
18	.8360	.7002	.5874	.4936	.4155	.3503	.2959	.2502	.2120	.1799
19	.8277	.6864	.5703	.4746	.3957	.3305	.2765	.2317	.1945	.1635
20	.8195	.6730	.5537	.4564	.3769	.3118	.2584	.2145	.1784	.1486
21	.8114	.6598	.5375	.4388	.3589	.2942	.2415	.1987	.1637	.1351
22	.8034	.6468	.5219	.4220	.3418	.2775	.2257	.1839	.1502	.1228
23	.7954	.6342	.5067	.4057	.3256	.2618	.2109	.1703	.1378	.1117
24	.7876	.6217	.4919	.3901	.3101	.2470	.1971	.1577	.1264	.1015
25	.7798	.6095	.4776	.3751	.2953	.2330	.1842	.1460	.1160	.0923

Years	11%	12%	13%	14%	15%	16%	17%	18%	19%	20%
1	.9009	.8929	.8850	.8772	.8696	.8621	.8547	.8475	.8403	.8333
2	.8116	.7972	.7831	.7695	.7561	.7432	.7305	.7182	.7062	.6944
3	.7312	.7118	.6931	.6750	.6575	.6407	.6244	.6086	.5934	.5787
4	.6587	.6355	.6133	.5921	.5718	.5523	.5337	.5158	.4987	.4823
5	.5935	.5674	.5428	.5194	.4972	.4761	.4561	.4371	.4190	.4019
6	.5346	.5066	.4803	.4556	.4323	.4104	.3898	.3704	.3521	.3349
7	.4817	.4523	.4251	.3996	.3759	.3538	.3332	.3139	.2959	.2791
8	.4339	.4039	.3762	.3506	.3269	.3050	.2848	.2660	.2487	.2326
9	.3909	.3606	.3329	.3075	.2843	.2630	.2434	.2255	.2090	.1938
10	.3522	.3220	.2946	.2697	.2472	.2267	.2080	.1911	.1756	.1615
11	.3173	.2875	.2607	.2366	.2149	.1954	.1778	.1619	.1476	.1346
12	.2858	.2567	.2307	.2076	.1869	.1685	.1520	.1372	.1240	.1122
13	.2575	.2292	.2042	.1821	.1625	.1452	.1299	.1163	.1042	.0935
14	.2320	.2046	.1807	.1597	.1413	.1252	.1110	.0985	.0876	.0779
15	.2090	.1827	.1599	.1401	.1229	.1079	.0949	.0835	.0736	.0649
16	.1883	.1631	.1415	.1229	.1069	.0930	.0811	.0708	.0618	.0541
17	.1696	.1456	.1252	.1078	.0929	.0802	.0693	.0600	.0520	.0451
18	.1528	.1300	.1108	.0946	.0808	.0691	.0592	.0508	.0437	.0376
19	.1377	.1161	.0981	.0826	.0703	.0596	.0506	.0431	.0367	.0313
20	.1240	.1037	.0868	.0728	.0611	.0514	.0433	.0365	.0308	.0261
21	.1117	.0926	.0768	.0638	.0531	.0443	.0370	.0309	.0259	.0217
22	.1007	.0826	.0680	.0560	.0462	.0382	.0316	.0262	.0218	.0181
23	.0907	.0738	.0601	.0491	.0402	.0329	.0270	.0222	.0183	.0151
24	.0817	.0659	.0532	.0431	.0349	.0284	.0231	.0188	.0154	.0126
25	.0736	.0588	.0471	.0378	.0304	.0245	.0197	.0160	.0129	.0105

APPENDIX G
MANN–WHITNEY TEST STATISTIC

n_1	p	$n_2=2$	3	4	5	6	7	8	9	10	11	12	13	14	15	16	17	18	19	20
	.001	0	0	0	0	0	0	0	0	0	0	0	0	0	0	0	0	0	0	0
	.005	0	0	0	0	0	0	0	0	0	0	0	0	0	0	0	0	0	1	1
2	.01	0	0	0	0	0	0	0	0	0	0	0	1	1	1	1	1	1	2	2
	.025	0	0	0	0	0	0	1	1	1	1	2	2	2	2	2	3	3	3	3
	.05	0	0	0	1	1	1	2	2	2	2	3	3	4	4	4	4	5	5	5
	.10	0	1	1	2	2	2	3	3	4	4	5	5	5	6	6	7	7	8	8
	.001	0	0	0	0	0	0	0	0	0	0	0	0	0	0	0	1	1	1	1
	.005	0	0	0	0	0	0	0	1	1	1	2	2	2	3	3	3	3	4	4
3	.01	0	0	0	0	0	1	1	2	2	2	3	3	3	4	4	5	5	5	6
	.025	0	0	0	1	2	2	3	3	4	4	5	5	6	6	7	7	8	8	9
	.05	0	1	1	2	3	3	4	5	5	6	6	7	8	8	9	10	10	11	12
	.10	1	2	2	3	4	5	6	6	7	8	9	10	11	11	12	13	14	15	16
	.001	0	0	0	0	0	0	0	0	1	1	1	2	2	2	3	3	4	4	4
	.005	0	0	0	0	1	1	2	2	3	3	4	4	5	6	6	7	7	8	9
4	.01	0	0	0	1	2	2	3	4	4	5	6	6	7	9	8	9	10	10	11
	.025	0	0	1	2	3	4	5	5	6	7	8	9	10	11	12	12	13	14	15
	.05	0	1	2	3	4	5	6	7	8	9	10	11	12	13	15	16	17	18	19
	.10	1	2	4	5	6	7	8	10	11	12	13	14	16	17	18	19	21	22	23

Source: Conover W. J. (1971), *Practical Nonparametric Statistics*, New York: Wiley, 384–8.

n_1	p	$n_2=2$	3	4	5	6	7	8	9	10	11	12	13	14	15	16	17	18	19	20
	.001	0	0	0	0	0	0	1	2	2	3	3	4	4	5	6	6	7	8	8
	.005	0	0	0	1	2	2	3	4	5	6	7	8	8	9	10	11	12	13	14
5	.01	0	0	1	2	3	4	5	6	7	8	9	10	11	12	13	14	15	16	17
	.025	0	1	2	3	4	6	7	8	9	10	12	13	14	15	16	18	19	20	21
	.05	1	2	3	5	6	7	9	10	12	13	14	16	17	19	20	21	23	24	26
	.10	2	3	5	6	8	9	11	13	14	16	18	19	21	23	24	26	28	29	31
	.001	0	0	0	0	0	0	2	3	4	5	5	6	7	8	9	10	11	12	13
	.005	0	0	1	2	3	4	5	6	7	8	10	11	12	13	14	16	17	18	19
6	.01	0	0	2	3	4	5	7	8	9	10	12	13	14	16	17	19	20	21	23
	.025	0	2	3	4	6	7	9	11	12	14	15	17	18	20	22	23	25	26	28
	.05	1	3	4	6	8	9	11	13	15	17	18	20	22	24	26	27	29	31	33
	.10	2	4	6	8	10	12	14	16	18	20	22	24	26	28	30	32	35	37	39
	.001	0	0	0	0	1	2	3	4	6	7	8	9	10	11	12	14	15	16	17
	.005	0	0	1	2	4	5	7	8	10	11	13	14	16	17	19	20	22	23	25
7	.01	0	1	2	4	5	7	8	10	12	13	15	17	18	20	22	24	25	27	29
	.025	0	2	4	6	7	9	11	13	15	17	19	21	23	25	27	29	31	33	35
	.05	1	3	5	7	9	12	14	16	18	20	22	25	27	29	31	34	36	38	40
	.10	2	5	7	9	12	14	17	19	22	24	27	29	32	34	37	39	42	44	47
	.001	0	0	0	1	2	3	5	6	7	9	10	12	13	15	16	18	19	21	22
	.005	0	0	2	3	5	7	8	10	12	14	16	18	19	21	23	25	27	29	31
8	.01	0	1	3	5	7	8	10	12	14	16	18	21	23	25	27	29	31	33	35
	.025	1	3	5	7	9	11	14	16	18	20	23	25	27	30	32	35	37	39	42
	.05	2	4	6	9	11	14	16	19	21	24	27	29	32	34	37	40	42	45	48
	.10	3	6	8	11	14	17	20	23	25	28	31	34	37	40	43	46	49	52	55
	.001	0	0	0	2	3	4	6	8	9	11	13	15	16	18	20	22	24	26	27
	.005	0	1	2	4	6	8	10	12	14	17	19	21	23	25	28	30	32	34	37
9	.01	0	2	4	6	8	10	12	15	17	19	22	24	27	29	32	34	37	39	41
	.025	1	3	5	8	11	13	16	18	21	24	27	29	32	35	38	40	43	46	49
	.05	2	5	7	10	13	16	19	22	25	28	31	34	37	40	43	46	49	52	55
	.10	3	6	10	13	16	19	23	26	29	32	36	39	42	46	49	53	56	59	63
	.001	0	0	1	2	4	6	7	9	11	13	15	18	20	22	24	26	28	30	33
	.005	0	1	3	5	7	10	12	14	17	19	22	25	27	30	32	35	38	40	43
10	.01	0	2	4	7	9	12	14	17	20	23	25	28	31	34	37	39	42	45	48
	.025	1	4	6	9	12	15	18	21	24	27	30	34	37	40	43	46	49	53	56
	.05	2	5	8	12	15	18	21	25	28	32	35	38	42	45	49	52	56	59	63
	.10	4	7	11	14	18	22	25	29	33	37	40	44	48	52	55	59	63	67	71
	.001	0	0	1	3	5	7	9	11	13	16	18	21	23	25	28	30	33	35	38
	.005	0	1	3	6	8	11	14	17	19	22	25	28	31	34	37	40	43	46	49
11	.01	0	2	5	8	10	13	16	19	23	26	29	32	35	38	42	45	48	51	54
	.025	1	4	7	10	14	17	20	24	27	31	34	38	41	45	48	52	56	59	63
	.05	2	6	9	13	17	20	24	28	32	35	39	43	47	51	55	58	62	66	70
	.10	4	8	12	16	20	24	28	32	37	41	45	49	53	58	62	66	70	74	79

n_1	p	$n_2=2$	3	4	5	6	7	8	9	10	11	12	13	14	15	16	17	18	19	20
	.001	0	0	1	3	5	8	10	13	15	18	21	24	26	29	32	35	38	41	43
	.005	0	2	4	7	10	13	16	19	22	25	28	32	35	38	42	45	48	52	55
12	.01	0	3	6	9	12	15	18	22	25	29	32	36	39	43	47	50	54	57	61
	.025	2	5	8	12	15	19	23	27	30	34	38	42	46	50	54	58	62	66	70
	.05	3	6	10	14	18	22	27	31	35	39	43	48	52	56	61	65	69	73	78
	.10	5	9	13	18	22	27	31	36	40	45	50	54	59	64	68	73	78	82	87
	.001	0	0	2	4	6	9	12	15	18	21	24	27	30	33	36	39	43	46	49
	.005	0	2	4	8	11	14	18	21	25	28	32	35	39	43	46	50	54	58	61
13	.01	1	3	6	10	13	17	21	24	28	32	36	40	44	48	52	56	60	64	68
	.025	2	5	9	13	17	21	25	29	34	38	42	46	51	55	60	64	68	73	77
	.05	3	7	11	16	20	25	29	34	38	43	48	52	57	62	66	71	76	81	85
	.10	5	10	14	19	24	29	34	39	44	49	54	59	64	69	75	80	85	90	95
	.001	0	0	2	4	7	10	13	16	20	23	26	30	33	37	40	44	47	51	55
	.005	0	2	5	8	12	16	19	23	27	31	35	39	43	47	51	55	59	64	68
14	.01	1	3	7	11	14	18	23	27	31	35	39	44	48	52	57	61	66	70	74
	.025	2	6	10	14	18	23	27	32	37	41	46	51	56	60	65	70	75	79	84
	.05	4	8	12	17	22	27	32	37	42	47	52	57	62	67	72	78	83	88	93
	.10	5	11	16	21	26	32	37	42	48	53	59	64	70	75	81	86	92	98	103
	.001	0	0	2	5	8	11	15	18	22	25	29	33	37	41	44	48	52	56	60
	.005	0	3	6	9	13	17	21	25	30	34	38	43	47	52	56	61	65	70	74
15	.01	1	4	8	12	16	20	25	29	34	38	43	48	52	57	62	67	71	76	81
	.025	2	6	11	15	20	25	30	35	40	45	50	55	60	65	71	76	81	86	91
	.05	4	8	13	19	24	29	34	40	45	51	56	62	67	73	78	84	89	95	101
	.10	6	11	17	23	28	34	40	46	52	58	64	69	75	81	87	93	99	105	111
	.001	0	0	3	6	9	12	16	20	24	28	32	36	40	44	49	53	57	61	66
	.005	0	3	6	10	14	19	23	28	32	37	42	46	51	56	61	66	71	75	80
16	.01	1	4	8	13	17	22	27	32	37	42	47	52	57	62	67	72	77	83	88
	.025	2	7	12	16	22	27	32	38	43	48	54	60	65	71	76	82	87	93	99
	.05	4	9	15	20	26	31	37	43	49	55	61	66	72	78	84	90	96	102	108
	.10	6	12	18	24	30	37	43	49	55	62	68	75	81	87	94	100	107	113	120
	.001	0	1	3	6	10	14	18	22	26	30	35	39	44	48	53	58	62	67	71
	.005	0	3	7	11	16	20	25	30	35	40	45	50	55	61	66	71	76	82	87
17	.01	1	5	9	14	19	24	29	34	39	45	50	56	61	67	72	78	83	89	94
	.025	3	7	12	18	23	29	35	40	46	52	58	64	70	76	82	88	94	100	106
	.05	4	10	16	21	27	34	40	46	52	58	65	71	78	84	90	97	103	110	116
	.10	7	13	19	26	32	39	46	53	59	66	73	80	86	93	100	107	114	121	128
	.001	0	1	4	7	11	15	19	24	28	33	38	43	47	52	57	62	67	72	77
	.005	0	3	7	12	17	22	27	32	38	43	48	54	59	65	71	76	82	88	93
18	.01	1	5	10	15	20	25	31	37	42	48	54	60	66	71	77	83	89	95	101
	.025	3	8	13	19	25	31	37	43	49	56	62	68	75	81	87	94	100	107	113
	.05	5	10	17	23	29	36	42	49	56	62	69	76	83	89	96	103	110	117	124
	.10	7	14	21	28	35	42	49	56	63	70	78	85	92	99	107	114	121	129	136

n_1	p	$n_2=2$	3	4	5	6	7	8	9	10	11	12	13	14	15	16	17	18	19	20
	.001	0	1	4	8	12	16	21	26	30	35	41	46	51	56	61	67	72	78	83
	.005	1	4	8	13	18	23	29	34	40	46	52	58	64	70	75	82	88	94	100
19	.01	2	5	10	16	21	27	33	39	45	51	57	64	70	76	83	89	95	102	108
	.025	3	8	14	20	26	33	39	46	53	59	66	73	79	86	93	100	107	114	120
	.05	5	11	18	24	31	38	45	52	59	66	73	81	88	95	102	110	117	124	131
	.10	8	15	22	29	37	44	52	59	67	74	82	90	98	105	113	121	129	136	144
	.001	0	1	4	8	13	17	22	27	33	38	43	49	55	60	66	71	77	83	89
	.005	1	4	9	14	19	25	31	37	43	49	55	61	68	74	80	87	93	100	106
20	.01	2	6	11	17	23	29	35	41	48	54	61	68	74	81	88	94	101	108	115
	.025	3	9	15	21	28	35	42	49	56	63	70	77	84	91	99	106	113	120	128
	.05	5	12	19	26	33	40	48	55	63	70	78	85	93	101	108	116	124	131	139
	.10	8	16	23	31	39	47	55	63	71	79	87	95	103	111	120	128	136	144	152

APPENDIX H WILCOXON TEST STATISTIC

Table of critical values of T in the Wilcoxon matched-pairs signed-ranks test

N	Level of significance, direction predicted		
	.025	.01	.005
	Level of significance, direction not predicted		
	.05	.02	.01
6	0	—	—
7	2	0	—
8	4	2	0
9	6	3	2
10	8	5	3
11	11	7	5
12	14	10	7
13	17	13	10
14	21	16	13
15	25	20	16
16	30	24	20
17	35	28	23
18	40	33	28
19	46	38	32
20	52	43	38
21	59	49	43
22	66	56	49
23	73	62	55
24	81	69	61
25	89	77	68

Source: S. Siegel, *Nonparametric Statistics*. McGraw-Hill Book Company, New York, 1956, table G.

n	d	*Confidence coefficient*	α''	α'
22	49	.991	.009	.005
	50	.990	.010	.005
	66	.954	.046	.023
	67	.950	.050	.025
	76	.902	.098	.049
	77	.895	.105	.053
23	55	.991	.009	.005
	56	.990	.010	.005
	74	.952	.048	.024
	75	.948	.052	.026
	84	.902	.098	.049
	85	.895	.105	.052
24	62	.990	.010	.005
	63	.989	.011	.005
	82	.951	.049	.025
	83	.947	.053	.026
	92	.905	.095	.048
	93	.899	.101	.051
25	69	.990	.010	.005
	70	.989	.011	.005
	90	.952	.048	.024
	91	.948	.052	.026
	101	.904	.096	.048
	102	.899	.101	.051

APPENDIX I
RUNS TEST

Lower critical values of r in the runs test

n_1 \ n_2	2	3	4	5	6	7	8	9	10	11	12	13	14	15	16	17	18	19	20
2											2	2	2	2	2	2	2	2	2
3					2	2	2	2	2	2	2	2	2	3	3	3	3	3	3
4				2	2	2	3	3	3	3	3	3	3	3	4	4	4	4	4
5			2	2	3	3	3	3	3	4	4	4	4	4	4	4	5	5	5
6		2	2	3	3	3	3	4	4	4	4	5	5	5	5	5	5	6	6
7		2	2	3	3	3	4	4	5	5	5	5	5	6	6	6	6	6	6
8		2	3	3	3	4	4	5	5	5	6	6	6	6	6	7	7	7	7
9		2	3	3	4	4	5	5	5	6	6	6	7	7	7	7	8	8	8
10		2	3	3	4	5	5	5	6	6	7	7	7	7	8	8	8	8	9
11		2	3	4	4	5	5	6	6	7	7	7	8	8	8	9	9	9	9
12	2	2	3	4	4	5	6	6	7	7	7	8	8	8	9	9	9	10	10
13	2	2	3	4	5	5	6	6	7	7	8	8	9	9	9	10	10	10	10
14	2	2	3	4	5	5	6	7	7	8	8	9	9	9	10	10	10	11	11
15	2	3	3	4	5	6	6	7	7	8	8	9	9	10	10	11	11	11	12
16	2	3	4	4	5	6	6	7	8	8	9	9	10	10	11	11	11	12	12
17	2	3	4	4	5	6	7	7	8	9	9	10	10	11	11	11	12	12	13
18	2	3	4	5	5	6	7	8	8	9	9	10	10	11	11	12	12	13	13
19	2	3	4	5	6	6	7	8	8	9	10	10	11	11	12	12	13	13	13
20	2	3	4	5	6	6	7	8	9	9	10	10	11	12	12	13	13	13	14

Source: Swed Frieda S. and Eisenhart C. (1943) Tables for testing randomness of grouping in a sequence of alternatives, *Ann. Math. Statist.*, **14**, 66–87

Note: For the one-sample runs test, any value of r that is is equal to *or* smaller than that shown in the body of this table for given value of n_1 and n_2 is significant at the 0.05 level.

Upper critical values of r in the runs test

n_1 \ n_2	2	3	4	5	6	7	8	9	10	11	12	13	14	15	16	17	18	19	20
2																			
3																			
4				9	9														
5			9	10	10	11	11												
6			9	10	11	12	12	13	13	13	13								
7				11	12	13	13	14	14	14	14	15	15	15					
8				11	12	13	14	14	15	15	16	16	16	16	17	17	17	17	17
9					13	14	14	15	16	16	16	17	17	18	18	18	18	18	18
10					13	14	15	16	16	17	17	18	18	18	19	19	19	20	20
11					13	14	15	16	17	17	18	19	19	19	20	20	20	21	21
12					13	14	16	16	17	18	19	19	20	20	21	21	21	22	22
13						15	16	17	18	19	19	20	20	21	21	22	22	23	23
14						15	16	17	18	19	20	20	21	22	22	23	23	23	24
15						15	16	18	18	19	20	21	22	22	23	23	24	24	25
16							17	18	19	20	21	21	22	23	23	24	25	25	25
17							17	18	19	20	21	22	23	23	24	25	25	26	26
18							17	18	19	20	21	22	23	24	25	25	26	26	27
19							17	18	20	21	22	23	23	24	25	26	26	27	27
20							17	18	20	21	22	23	24	25	25	26	27	27	28

APPENDIX J DURBIN–WATSON STATISTIC

To test H_0: no positive serial correlation,

if $d < d_L$, reject H_0;
if $d > d_U$, accept H_0;
if $d_L < d < d_U$, the test is inconclusive.

To test H_0: no negative serial correlation, use $d = U - d$.

Level of significance $\alpha = 0.05$

	$p=2$		$p=3$		$p=4$		$p=5$		$p=6$	
n	d_L	d_U	d_L	d_U	d_L	d_U	d_L	d_U	d_L	d_U
15	1.08	1.36	0.95	1.54	0.82	1.75	0.69	1.97	0.56	2.21
16	1.10	1.37	0.98	1.54	0.86	1.73	0.74	1.93	0.62	2.15
17	1.13	1.38	1.02	1.54	0.90	1.71	0.78	1.90	0.67	2.10
18	1.16	1.39	1.05	1.53	0.93	1.69	0.82	1.87	0.71	2.06
19	1.18	1.40	1.08	1.53	0.97	1.68	0.86	1.85	0.75	2.02
20	1.20	1.41	1.10	1.54	1.00	1.68	0.90	1.83	0.79	1.99
21	1.22	1.42	1.13	1.54	1.03	1.67	0.93	1.81	0.83	1.96
22	1.24	1.43	1.15	1.54	1.05	1.66	0.96	1.80	0.86	1.94
23	1.26	1.44	1.17	1.54	1.08	1.66	0.99	1.79	0.90	1.92
24	1.27	1.45	1.19	1.55	1.10	1.66	1.01	1.78	0.93	1.90
25	1.29	1.45	1.21	1.55	1.12	1.66	1.04	1.77	0.95	1.89
26	1.30	1.46	1.22	1.55	1.14	1.65	1.06	1.76	0.98	1.88
27	1.32	1.47	1.24	1.56	1.16	1.65	1.08	1.76	1.01	1.86
28	1.33	1.48	1.26	1.56	1.18	1.65	1.10	1.75	1.03	1.85
29	1.34	1.48	1.27	1.56	1.20	1.65	1.12	1.74	1.05	1.84

	$p=2$		$p=3$		$p=4$		$p=5$		$p=6$	
n	d_L	d_U	d_L	d_U	d_L	d_U	d_L	d_U	d_L	d_U
30	1.35	1.49	1.28	1.57	1.21	1.65	1.14	1.74	1.07	1.83
31	1.36	1.50	1.30	1.57	1.23	1.65	1.16	1.74	1.09	1.83
32	1.37	1.50	1.31	1.57	1.24	1.65	1.18	1.73	1.11	1.82
33	1.38	1.51	1.32	1.58	1.26	1.65	1.19	1.73	1.13	1.81
34	1.39	1.51	1.33	1.58	1.27	1.65	1.21	1.73	1.15	1.81
35	1.40	1.52	1.34	1.58	1.28	1.65	1.22	1.73	1.16	1.80
36	1.41	1.52	1.35	1.59	1.29	1.65	1.24	1.73	1.18	1.80
37	1.42	1.53	1.36	1.59	1.31	1.66	1.25	1.72	1.19	1.80
38	1.43	1.54	1.37	1.59	1.32	1.66	1.26	1.72	1.21	1.79
39	1.43	1.54	1.38	1.60	1.33	1.66	1.27	1.72	1.22	1.79
40	1.44	1.54	1.39	1.60	1.34	1.66	1.29	1.72	1.23	1.79
45	1.48	1.57	1.43	1.62	1.38	1.67	1.34	1.72	1.29	1.78
50	1.50	1.59	1.46	1.63	1.42	1.67	1.38	1.72	1.34	1.77
55	1.53	1.60	1.49	1.64	1.45	1.68	1.41	1.72	1.38	1.77
60	1.55	1.62	1.51	1.65	1.48	1.69	1.44	1.73	1.41	1.77
65	1.57	1.63	1.54	1.66	1.50	1.70	1.47	1.73	1.44	1.77
70	1.58	1.64	1.55	1.67	1.52	1.70	1.49	1.74	1.46	1.77
75	1.60	1.64	1.57	1.68	1.54	1.71	1.51	1.74	1.49	1.77
80	1.61	1.66	1.59	1.69	1.56	1.72	1.53	1.74	1.51	1.77
85	1.62	1.67	1.60	1.70	1.57	1.72	1.55	1.75	1.52	1.77
90	1.63	1.68	1.61	1.70	1.59	1.73	1.57	1.75	1.54	1.78
95	1.64	1.69	1.62	1.71	1.60	1.73	1.58	1.75	1.56	1.78
100	1.65	1.69	1.63	1.72	1.61	1.74	1.59	1.76	1.57	1.78

APPENDIX K
MICROSTATS Manual

Author: Mike Hart, © 1991

Department of Public Policy and Managerial Studies
LEICESTER POLYTECHNIC
Scraptoft Campus
LEICESTER
LE7 9SU

CONTENTS

1 INTRODUCTION AND TUTORIAL

1.1 What is MICROSTATS?

MICROSTATS has been designed as a general purpose, easy-to-use statistics package for use on IBM and IBM compatible machines with at least 256K of memory. It closely resembles the MINITAB statistical package written by the University of Pennsylvania which is to be found on many main-frame installations. However, although the overall design and philosophy are consistent with those of MINITAB, there are significant differences between the two packages, not least in size, functions and price. MICROSTATS may be seen as an integrated program which fulfils the function of an introduction to more specialized statistical software such as MINITAB or SPSS (Statistical Package for the Social Sciences)

1.2 Capacities of MICROSTATS

MICROSTATS is written in Turbo Pascal which makes it both compact and fast in operation. The program consists of the principal MICROSTATS module which is kept in memory at all times and a series of overlay files which are switched in and out of memory automatically when required. On a floppy disk system, this process rarely takes more than a second or so and provides no noticeable delay whilst on a hard-disk system, it appears almost instantaneous. It is obvious, therefore, that all of the MICROSTATS modules must be kept together on the same disk or sub-directory for the correct operation of the program.

The MICROSTATS worksheet (explained in 1.3 below) allows for 200 rows of some 45 columns of data which constitute 9000 elements of data. (MINITAB, Version 5, by contrast, allows a storage capacity of 325 rows of some 50 columns of data which is practically double.) If memory constraints are a real problem then it is time probably to graduate onto a more powerful package in any case.

To install MICROSTATS, follow the instructions in the small text file named READ.ME! (At the MS–DOS prompt, type READ.ME!) The installation process is simple but illicit or uninstalled programs will not run.

If you need a statistical package in order to analyze a social survey, then a better application program may well be TURBOSTATS, details of which may be obtained from M.C. Hart. TURBOSTATS allows for the naming and labelling of both variables and values in a manner similar to SPSS and the output is, in fact, modelled upon the FREQUENCIES and the CROSSTABS facilities of that particular package.

1.3 How does MICROSTATS work?

MICROSTATS assumes that the user has several columns of data organized in a worksheet and is organized to reproduce (and eventually manipulate) that worksheet. The fundamental concept, which it is important to appreciate early on, is that MICROSTATS primarily manipulates *columns* of data and expects that data will be numerical, and not textual, data. In this, it differs from a spreadsheet which typically handles text and formulas in addition to numbers.

Columns of data, which are the basic unit of analysis used in the MICROSTATS system, may be called by a number which *must* be preceded by the letter C (or c) and no intervening space e.g. C1. Columns of data may also be allocated a name up to eight characters long and this name may also be used to refer to data provided that the name is always enclosed in (the same!) single quotation marks e.g. 'MARKS'. (On some IBM keyboards, this is the quotation mark found on the same key as the @ symbol or the shifted 7.)

MICROSTATS works by scanning a line which is input to the computer after the user has responded to the MICROSTATS prompt (Command ?) with a command and then pressed ⟨ENTER⟩. It picks out the keyword that it needs (or more specifically, the first four letters of a keyword) and then checks it against an internal dictionary of some 109 words. If the keyword is found, then MICROSTATS will activate the subroutine which performs the analysis suggested by the user. Notice that MICROSTATS does not necessarily display anything on the screen after certain commands, so if you get the prompt

(Command ?) back and no error message, you can generally assume that the system has obeyed your request! If you want to see the results of your commands, you can generally PRINT out the relevant columns to satisfy yourself that your command has worked!

If you use a command that MICROSTATS does not recognize (e.g. you may have mis-typed it) then you will get a warning beep and a 'Word not found' message.

MICROSTATS will take the whole of an input line and *parse* it which means that the relevant information will be extracted and the rest of the line will be ignored. Generally, everything that is not a keyword, a column number (preceded by C) or a value will be ignored. This means that you can address MICROSTATS with English-type commands, provided that your first word is always a keyword. Later on, as your familiarity with the system increases, you can progress to the shortened form of the command if you prefer.

EXAMPLE

Assume that we have a column of data in C1 and we wish to multiply this column by 10 and put the results into C2. We can use the long form of the command:

MULTiply data in C1 by 10 and put the result into C2

or we could use the short form of the command (which is what MICROSTATS actually acts upon):

MULTI C1 10 C2

Notice in the long form of the command that only the first four letters are significant. If we made a mistake in any of the letters from the fifth onwards then MICROSTATS would ignore this.

If you forget any of the words in the MICROSTATS dictionary, then you can either press ? or type the keyword HELP which will give you access to one of eight screens where the commands are displayed in logical groups.

1.4 Getting started

Make a backup copy of the distribution disk using the DISKCOPY A:A: (or DISKCOPY B:B:) command before running MICROSTATS for the first time. Use only this copy. Once the computer is "booted up" under MS-DOS, merely insert the working copy of the MICROSTATS disk and type MS-START. MICROSTATS should load, give a title page and invite you to press (RETURN) to start. You will be presented with information of the size of your worksheet and the prompt: Command?

Try the following, not forgetting to press ⟨RETURN⟩ (or ⟨ENTER⟩) at the end of each line:

SET data into C1

At this stage, a new prompt will appear as follows:

DATA>

and you should then type in some data with at least one space as a separator before each number, e.g.

DATA> 3 4 5 6 7 8 9

As soon as you press ⟨RETURN⟩, the prompt (DATA >) will reappear and you could enter as many more lines of data as you wish.

You should *not* attempt to enter more than a one full line of data (i.e. 80 characters of data) at a time, since it is always possible to enter additional data on subsequent lines.

To finish entering your data, make sure that you have pressed ⟨ENTER⟩ at the end of the previous line and the prompt (DATA >) is still in front of you. Now type END to signify that you have finished and the data you have just entered will be displayed to you on the screen with an indication of the number of items that you have input.

The whole transaction will look like this:

```
Command? SET data into C1
DATA >3 4 5 6 7
DATA >END
1.  3.000
2.  4.000
3.  5.000
4.  6.000
5.  7.000
  5 data items entered in C1
Command?
```

When the prompt (Command ?) re-appears, you may now issue another command. Print out the data that you have just entered with the following command:

Command ? PRINT data in C1

	C1
	(n = 5)
1.	3.000
2.	4.000
3.	5.000
4.	6.000
5.	7.000

and the data that you have just entered will appear on the screen. Now that you have some data in the computer, try the command:

Command? DESCribe C1

and you should see the following output:

Count of	C1 =	5.000
Minimum of	C1 =	3.000
Maximum of	C1 =	7.000
Sum of	C1 =	25.000
Sum of squares of	C1 =	135.000
Mean of	C1 =	5.000
Median of	C1 =	5.000
Mode of	C1 =	(No mode)
Standard dev-n [pop'n] of	C1 =	1.414
Standard dev-n [sample] of	C1 =	1.581
Skew of	C1 =	0.000
Standard error of mean of	C1 =	0.707
Quartiles [Q1] [Q3] of	C1 =	3.500
	C1 =	and 6.500
Percentiles [10] [90] of	C1 =	3.000
		and 7.000

(The meaning of these statistical terms are explained in the main text.)

You may now like to try some of the other commands that will perform statistics upon a single column of data. Før example, try giving the command:

Command? SUM data in C1

and then try extending the command even further by trying:

Command? SUM data in C1 — put data into C2

In this case, the sum of the data in C1 will be put into as many rows of C2 as there are rows in C1. However, note that

SUM C1 put into 1 row of C2

will only use row 1 of C2.

To give names to your two data columns, then use the NAME command as follows:

Command? NAME C1 'DATA' and C2 'SUM'

Now type INFO, or the full command INFOrmation, and you will see the status of your worksheet displayed as follows:

C1 'DATA' n = 5
C2 'SUM' n = 5

43 columns of length 200 still available for use

You should grow used to the following two commands as they can often help you keep track of your worksheet:

INFO	checks on the number and length of your columns
PRINT Cx	where x is a column number e.g. C1, or a range e.g. PRINT C1–C5

1.5 Going further

Let us assume that you still have the numbers

3 4 5 6 7

contained in C1. You can PRINt C1 to check whether this is still the case. Now we are going to create a second column of data to explore some of the other possibilities offered by MICROSTATS. Try the following:

Command? MULTiply C1 by C1 — put result into C2

If you prefer, you may use the short form of the command which merely gives MICROSTATS the bare minimum that it requires i.e. the first four letters of the keyword and the columns that it must use to operate upon. The short form of the command is:

Command? MULT C1 C1 C2

MICROSTATS will respond equally well to either the short or the long form of the command, so feel free to use whichever seems more natural to you.

If the prompt (Command ?) returns and there has been no error message, we can assume that the command has been successfully performed. To check this out, we can ask MICROSTATS to perform the following:

Command? PRINt C1 and C2

and we should find that C2 contains the square of C1 i.e. each number in C1 has been multiplied by itself and then put into C2.

In the print-out on the screen, you will notice that MICROSTATS has printed out each number to three places of decimals. It does this *by default* but it is possible to alter this by using a command such as

Command? DISPlay C1 to 2 dp

If you then PRINt C1 and C2 once again, you will see that you have output to two places of decimals in C1. You can experiment with this if you like.

Now we are going to do some more serious work. If a statistician wishes to see how much of one variable is associated with another, (s)he generally wishes to generate a statistical measure known as a **correlation coefficient**. If you have ever done this long-hand, you will know what a lot of effort is involved. MICROSTATS will do this for you quickly and easily by using the command:

Command? CORRelate C1 and C2

You should obtain a result that informs you that the correlation of C1 and C2 is equal to 0.9931, as well as other statistical information giving you the probability that you could have obtained this result by chance alone (in this case, about 3 in every 10,000 cases!).

This indicates a very high positive correlation. The correlation coefficient can take a value of +1 for a perfect positive correlation, −1 for a perfect negative correlation or any value in between. A value very close to +1 indicates a very high degree of association between the two sets of values, which is not surprising considering that one is the square of the other.

If we wish to derive a mathematical equation to link together the two sets of observations, we use what statisticians call a **regression equation**. This is an equation that informs you on the basis of the data that the computer has what would be a predicted value for one variable once we are a given a specific value for the other. Try the command:

Command? REGRess C2 upon C1

and your output should inform you that the regression equation of C2 upon C1 is equal to:

y = −23.0000 + 10.0000 * X

The asterisk is a standard method in computing of saying 'multiplied by' and this regression equation is telling us that y will take a value of 10.000 times a particular value of X minus 23.0

Now give names of the two variables created by using the commands:

Command? NAME C1 'Number' and C2 'Square'
Command? INFO

Finally, we are going to do something which is visually more exciting and also shows that you can use names in commands as well as column numbers. Once we have allocated names to columns we are quite free to use them instead of column numbers. We must ensure that the names are exactly those which have been allocated. For example the names 'NUMBER' and 'NUMBER ' look exactly alike to a human reader but are regarded as quite different names by the computer, where the space would be regarded as an additional character.

If we now use the PLOT command as follows:

Command? PLOT 'Number' v 'Square'

then we will find a plot of the two variables in which the maximum and minimum of each variable are displayed, as well as the names of the variables and the correlation coefficient. If you perform the PLOT in reverse order i.e. PLOT 'Square' v 'Number' you will find the axes are now reversed.

To clear the system of data, we can now type

Command? ERASe C1 C2

and the two columns of data will be 'rubbed out'. You may check that this is so by using INFO which will inform you that all of the columns are empty and unnamed and give you the total available on the system.

This concludes the tutorial on the use of the MICROSTATS system. The following sections will describe how to use the more sophisticated features contained in the package. This also assumes that you have a certain degree of statistical knowledge so that you can understand what type of analysis is being performed, and also that you are aware of the underlying assumptions.

If your statistical knowledge is limited, then you can try to use some of the other simple commands that are listed in the complete list of MICROSTATS commands and experiment to discover what they do.

2 LIBRARY OF MICROSTATS COMMANDS

2.1 Entering data

SET data into C1
  DATA > 3 4 5 6 7 8 9
  DATA > 10 11 12 13 14
  DATA > END

Use the SET command for entering data one column at a time. A long column of data may be split over several MICROSTATS input lines but your input line should not exceed about 100 characters. As a rule of thumb then whenever the cursor goes onto the second 'screen line' then you should press ⟨ENTER⟩ to register that line of data.

The SET command will overwrite any data already contained in that column. If you wish to append data to a column, then use the APPEnd command which has the same format as SET.

To exit, type END, or X (or nothing at all) on a BLANK data line and not at the end of a series of numbers.

READ data into C1–C3
  DATA Row 1 > 1 4 7
  DATA Row 2 > 2 5 8 3
Too many data items — re-enter line
  DATA Row 2 > 2 5
Too few data items — re-enter line
  DATA Row 2 > 2 5 8
  DATA Row 3 > 3 6 9
  DATA Row 4 > END

3 rows of data entered into C1–C3

Use the READ command for entering data into several columns simultaneously. This is best used when you are more experienced in data entry. Note that the number of columns indicated and the number of data items per line must tally exactly. Do not leave spaces on either side of the hyphen mark.

In the example, exactly three data items per line are expected. If you supply more or less then the system will warn you and prompt you to re-enter the correct number of data items.

NAME data in C1 'name1' and C2 'name2'

Names the data — check that you use the correct apostrophe!

2.2 Column statistics

SUM data in C1
SUM data in C1 put into C2
SUM data in C1, put into 1 row in C2

If a second column is specified but with no other value, then the sum of C1 will be put into as many rows of C2 as there are rows in C1. If you specify a value, e.g. 1, then this value will indicate how many rows of C2 should be filled.

Commands which work in an identical fashion are:

COUNt	Number of data items in the column
MAXImum	The maximum value in a column
MINImum	The minimum value in a column
AVERage MEAN	Both commands give the arithmetic mean
MEDIan	The median measure of

	central tendency i.e. the value which occupies the central position once the data has been sorted into ascending order.
MODE	The mode or the most frequently occurring value. If there are multiple modes, then only the first identified will be shown.
STANdard dev-n	Standard deviation calculated with a divisor of N (population)
STDE	Standard deviation calculated with a divisor of N-1 (sample)
SSQ (Sum of Squares)	The sum of x-squared, i.e. each value is squared and then the x-squareds are summed.
SEMEan	Standard error of the mean
SKEW	The skew of the distribution. A positive skew indicates that the median is to the left (i.e. smaller) than the mean whilst a negative skew indicates that median is to the right (i.e. larger) than the mean.
QUARtiles	Similar to the median except the data is divided into four quarters and the first and third quartile are shown
PERCentiles	Similar to quartiles except the n'th (and 100-n'th) percentiles are shown. If n is not specified the 10th and 90th percentiles are shown
DESCribe	This command will give summary statistics of any one column:

Count	Sum
Minimum	Mean
Maximum	Median
Sum of Squares	Mode

Standard deviation (population)
Standard deviation (sample)
Standard error of the mean

Quartiles	Percentiles
Skew	(10th) and (90th)

2.3 Printing out data

PRINT C1
PRINT C1-C7

You may print out any one column or up to seven adjacent columns of data. Any columns in excess of seven will be ignored. Do not leave a space on either side of the hyphen.

After a screenful of data (20 items) you will be prompted whether to view more data or exit the printing of the data.

2.4 Arithmetic on columns

You may perform simple arithmetic on your columns using either another column or a number. The results may be put into another column or even stored straight back into the original column, in which case they overwrite the previous contents. Examples include:

ADD C1 to C2 put into C3
SUBT C2 from C1 put into C3
MULTiply C1 by 10 put into C3
DIVIde C4 by 10 put into C4
RAISe C1 to the power of 3 put into C5
RECIprocal of C1 put into C11
INT ger of C1 put into C2

Once you have performed these manipulations, the MICROSTATS prompt (Command ?) will return. Print out the relevant columns if you want to satisfy yourself that the results are as you intended them to be.

2.5 Manipulations upon columns

These manipulations allow you to transform the data in some way e.g. by taking a log or a square root. You may then put the transformed data into another or even back into the same column. The manipulations include the following:

SQRT of C1 put into C2	SQRT ~ Square root
ABSolute of C1 put into C2	ABS ~ Absolute
ROUNd C1 to x decimal places	Rounds to specified no. of decimal places
LOGE C1 put into C2	LOGE ~ natural log
EXPOnent C2 put into C3	EXPO ~ exponent

LOGTen C1 put into C2	LOGTen = Log to base 10
ANTIlog C1 put into C2	ANTIlog = antilogarithm
SORT C1 put into C2	Sorts into ascending order

To sort data in C1 into a *descending* order, use the following sequence: (the −1 indicates a descending sort)

SORT C1 −1 C2

RANK data in C1, put into C2

RANK gives a ranking number i.e. when the numbers are are arranged in ascending order from lowest to highest, the lowest is awarded a rank of 1, the next lowest a rank of 2 and so on. When values are equal, the relevant ranking numbers are 'shared out'. If −1 is specified, the rank order will be from highest to lowest.

Two copying options are available:

COPY data in C1 to C3

which copies from C1 to C3, and

COPY data from C1–C5 to block starting at C11

which copies the block of data from C1–C5 into a new block starting at C11.
(NB no spaces on either side of the hyphen)

2.6 Plots

MICROSTATS will perform a scatter plot or scattergram of data on two matching variables:

The typical plotting command is:

PLOT C1 C2

but before you issue the command, you should take some elementary precautions. These include the following:

1. Ensure that the columns are of equal length.
2. Try not to plot data in which all of the values of one or other variable are identical.

If you observe numbers rather than asterisks (*) in your plot, this is because MICROSTATS is informing you that two data points are 'mapped' onto the same data co-ordinates. Notice that the plots will be scaled and the minimum, mid-point and maximum of each given as well as the column names (if any).

The first column that you specify will be the vertical axis and this is usually called the dependent variable (as it depends in causal terms upon the second variable). The second column specified will be the horizontal axis and this is usually called the independent variable (which you might think of, loosely, as the cause). The dependent variable is usually designated by the letter y and the independent variable by the letter x. For example, if we had two variables in which one was student grant ('GRANT') whilst the other was level of parental income ('INCOME'), then 'GRANT' would be the dependent variable (i.e. y) and would be entered first whilst 'INCOME' would be the independent variable (i.e. x) and would be entered second:

PLOT 'GRANT' v. 'INCOME'

Notice also that the correlation coefficient is calculated and displayed in the top right-hand corner of the graph.

Plots are useful to see if there is a tendency for the data to cluster and form one of the following patterns.

1. A line sloping upwards from bottom left to top right. The more closely the data clusters around such a line, the more it suggests a positive correlation in which as one variable increases, so does the other.
2. A line sloping downwards from top left to bottom right. The more closely the data clusters around such a line, the more it suggests a negative correlation i.e. one variable increases as the other decreases.
3. No apparent pattern at all. This suggests the absence of association which would be no correlation (or only a very small one).

2.7 Histograms, Tally

The HISTogram command requests a plot of a single variable so that you can examine its shape. A typical histogram of random numbers from 1–100 would show the following:

Command? HISTogram of data in C1

Choose first midpoint, interval (y/n?) y
First mid-point? 5.5 Interval? 10

Middle of Interval	Number of Observations	
5.5	10	**********
15.5	8	********
25.5	13	*************
35.5	8	********
45.5	8	********
55.5	12	************
65.5	6	******
75.5	9	*********
85.5	9	*********
95.5	17	*****************

There are several things to note about the HISTogram command.

1. You are given the option to choose the first midpoint and the interval. If you press 'n' or just ⟨RETURN⟩ then the command will choose what appears to be sensible defaults depending upon the shape of the data but which may appear strange to you. If you choose to select the midpoint and the interval, then you should have at least a rough idea of what the data looks like before you start.
2. Notice that statistically the midpoints may not be just where you expect them to be. For example, in the example given above, then all of the data lying in the range 0.5 upwards to 10.49999 downwards would be regarded as lying within the first block. The mid-position of a range which extends from 0.5 to 10.4999 is (10.4999 − 0.5)/2 (i.e. 5) + 0.5 which is 5.5 (to the nearest one place of decimals) and not 5.0!
3. The HISTogram command cannot deal very sensibly with very small or very large numbers. Under such circumstances, it is probably sensible to scale them up (or down) yourself and put the data into a new column and then try the effects of HIST on the scaled column. For example, a range of 100 numbers in the range 0–1 are best scaled up to 0–10 or even 0–100. After all, the HISTogram analysis is only intended to give you a visual representation of the actual 'shape' of the data rather than a precise mathematical result and therefore such scaling up or down is quite legitimate.

Tally

If you have 'discrete' data i.e. data which can take one of a range of values (e.g. numbers such as 1,2,3 which may be answers in a questionnaire) then TALLY will be superior to a histogram. Up to 20 consecutive integers will be accepted and the columns displayed are:

VALUE	(i.e. the number itself)
N	(the number of occurrences)
CUM N	(cumulative total of N's)
PERCENT	(the percentage)
CUMPCT	(the cumulative percentage)
Barchart	(a simple barchart)

You may LABEL your TALLY commands, thus LABEL 1 'Male' 'Female' where 1 is the *row* from which you want to start. LABELS can also used in a CHISquare analysis (page 399).

2.8 Bi-variate statistics

Bi-variate statistics is the name given to the statistical analysis of pairs of data, such as that dealt with already in the PLOT command. Several bivariate statistics are provided.

CORRelate C1 and C2

In this case, the **Pearson product-moment correlation coefficient** between the two stated variables will be performed. Before you issue the command, then check the following two points.

1. Via INFO make sure that you have equal numbers of data in each column.
2. Via PRINt make sure that all of the values of one or other column are not identical. If so, the command will fail and computer may well abort.

In order to interpret the value of the correlation coefficient, refer to the entry under PLOT above.

It is particularly important to remember the following two points.

1. A high positive (or negative) correlation coefficient cannot be taken to imply *causation*.
2. A high (or low) correlation may be significant in purely statistical terms but not be significant in social scientific terms. For example, a high correlation between heights and weights of a general sample of the population is not surprising, as taller people are generally heavier. Conversely, the absence of a correlation may not achieve statistical significance but may be highly significant in terms of a social scientific model. The absence of a relationship where we might be led to expect one (for example between unemployment and mental illness) might prove to be highly significant in terms of social scientific theory, even though the result does not achieve a degree of statistical significance.

A **regression line**, sometimes known also as a **least squares line** is a line that best fits a series of data pairs and which can be used to predict one variable once we know:

1. the regression equation itself;
2. the value of the independent variable.

We can use:

REGRess C1 upon C2
REGRess C1 upon C2 with intercept in C3, slope in C4, value of x, put predicted y in C5

REGR C1 C2 C3 C4 10 C5

Or, in more compact form, REGR C1 C2 10

The general form of a regression equation is:

y = a + b * (x)

where

y = dependent variable (that we wish to discover);
x = independent variable (which may be given);
a = intercept;
b = slope.

MICROSTATS will also calculate R−sq (i.e. R^2) which is the **coefficient of determination**. The correlation coefficient r is the square root of the coefficient of determination.

EXAMPLE

Put the following data pairs into C1 and C2 where:

C1 = Salary
C2 = Years of education since age 15

C1	5000	3000	6000	4000	7000	6000	9000
C2	2	4	5	6	7	8	9

Then obtain the regression equation, as follows:

```
Command? REGRess C1 upon C2
Regression of C1 upon C2 =
y = 2401.6393 + 565.5738 * X      (R - sq = 0.47591)
```

Now try the longer form of the command, but this time we wish to know what salary that can be expected from an individual with 3 years of 15+ education.

```
Command? REGRess C1 on C2, a in C3, b in C4, x = 3,
result in C5
y = 2401.6393 + 565.5738 * X      (R - sq = 0.47591)
Command? PRINt C3-C5
        C3          C4         C5

     (n = 1)     (n = 1)    (n = 1)
1.  2401.6393   565.574   4098.361
```

Here the critical result is in C5 that tells us that with 3 years of post 15+ education (x = 3) the predicted level of salary will be:

y = 2401.6393 + (565.574 * 3)
= 4098.361

Whereas in correlation the result does not depend upon which variable is C1 and C2, the same is *not* true of regression.

In regression, the dependent variable is regressed onto the independent variable. In terms of our example, the salary (the dependent variable) will be regressed upon years (the independent variable). If you experiment by trying to regress C2 on C1 then you will see a very different result, so it is important that the order of variables is understood before you use this command.

MICROSTATS also supports multiple regression, but the output requires a degree of statistical knowledge for its correct interpretation (See the command MREGress, page 404)

If you suspect that the data is curvilinear (e.g. the kind of relationship that is met when one number is the square or higher power of the other number) then try a logarithmic transformation of the data before you start to perform the regression command.

CONTingency table of data in C1 and C2

The CONTingency table command is designed to 'table' those cases where we have integer numbers in two columns which represent 'coding' numbers e.g. in C1 we might have the numbers 1–2 which represent female and male whilst in C2 we might have numbers 1–5 representing five categories of political identification. Such data is often known as categorical data. If we wish to see how many of one category are represented in the other (e.g. how many female Conservatives) then we would use the CONTingency table command.

EXAMPLE

Use the following commands which put 100 random cases of either 1,2 in C1 and either 1,2,3,4,5 in C2.

Command? IRAN 100 cases from 1 to 2 put into C1
Command? IRAN 100 cases from 1 to 5 put into C2

Now table the result:

Command? CONTingency C1 with C2

You should get a result similar in appearance to the following. It will probably not be identical because the random number generator may well have produced a different pattern of data to fill C1 and C2:

	C2>	1	2	3	4	5	
C1	1	8	12	5	13	13	51
	2	11	9	10	9	10	49
		19	21	15	29	23	100

There are two points to note about this command.

1. Only try to table consecutive integers up to a maximum of 10 in each direction of the table.
2. If you wish to put the cell results into another column for later analysis by chi-square then specify a third column as the starting point e.g. CONT C1 with C2, results from C10 onwards. In this case, MICROSTATS responds with a message:

   Data fed into C10–C14

   and you can confirm this result by PRINting out the relevant columns of data.

   Note that you can put 5 or less cell contents into the new block. If you attempt to put more than five, then the command will be ignored. (This is because the CHI-SQUARE command which uses these results will only accept a table 5 cells wide by 5 cells deep.)

CHISquare of data in C3–C5

The CHISquare command will accept a block of up to five rows/columns and perform a **chisquare test** upon the data. The underlying statistical assumption is that data is measured at the nominal or categorical level (e.g. code numbers representing a sex or a political party). To demonstrate CHISquare, then put in the following data using the random number generator. We are going to generate a sex coding (1,2) in C1 and a political party coding in C2 (1,2 or 3):

Command ? IRAN 100 nos from 1 to 2 put into C1
Command ? IRAN 100 nos from 1 to 3 put into C2
Command ? CONTingency C1 and C2 put cells into C3

	C2 >	1	2	3	
C1	1	18	14	14	46
C2	2	19	19	16	54
		37	33	30	100

Data fed into C3–C5

A 'one-sample' CHISquare is also allowed i.e. specification of one column of data (in which case, the number of 'expected' cells is inferred by splitting the total number of cases equally between the number of cells in the column). Up to ten rows are allowed in a one-sample test.

CHISquare of data in C3–C5
Expected frequencies are printed below observed frequencies

	C3	C4	C5	Totals
1	18 17.02	14 15.18	14 13.80	46
2	19 19.98	19 17.82	16 16.20	54
Totals	37	33	30	100

0.06 + 0.09 + 0.00 +
0.05 + 0.08 + 0.00 +
Total chi-square = 0.280 df = 2 p = 0.8695

(You will not get exactly these results because the random number generator will have generated a different pattern of initial data but it should not be too dissimilar).

CHISquare takes the initial sets of data in each cell (the observed data) and then works out the expected data in each cell on the assumption that one variable is exactly proportionately represented within the other.

In this case, we are trying to see if there are sex differences in the way in which people vote. For each cell, the expected differences are worked out according to the formula:

$$\frac{(\text{Observed} - \text{Expected})^2}{\text{Expected}}$$

and this is the chi-square for that cell. Finally all of the chi-squares are summed, the degrees of freedom (df) calculated according to the rule:

(number of rows − 1) × (number of columns − 1)

and the probability is worked out. The probability means the liklihood that we could have achieved a chi-square value as large as this by chance (where 1 ~ certainty or 100% whilst 0 ~ impossible or 0%) Any probability which is greater than 0.05 (or 5%) means that there is not a statistically significant difference in the proportions of C2 represented in C1 (in terms of the example above, a sex difference in voting behaviour) The level of 5% is only a convention in statistics; we could choose 1% to be even more sure of our results should we so wish.

Points to note about chisquare are listed below.

1. The data should be measured at the nominal level i.e. categories such as male/female.
2. Cells with an expected frequency of less than 5 can generate chi-squares that give a misleading result. If this is the case, then a warning message will be given. It is generally best to combine categories to make the numbers in each cell so much larger.
3. Zero cells will abort the analysis, with a division by zero error! Make sure that you have no zero cells in the analysis before you start by combining categories if necessary.
   NB. You may label your rows if you like. See LABEL, page 397.

2.9 Data generation

There are times when it is useful to generate displays of data for demonstration purposes. Several commands are provided in MICROSTATS as detailed below.

GENErate values from 1 to 100 in C1

This will generate data from the first value to the second value stated in the relevant column. It could be used to provide an index number for a series of data.

DEFIne the value of 10 into the first 5 rows of C1

This allows a constant to be put in as many rows of the column as you desire.

IRANdom 100 random integers from 1 to 100, put into C1
IRANdom 100 random integers from 1 to 100, put into C1–C5

This is an **integer random number generator**. You should remember to state the numbers of integers required, followed by the lower limit, the upper limit and the destination column.

URANdom 100 random numbers and put into C2
URANdom 100 random numbers and put into C2–C5

The URANdom random number generator generates floating point numbers in the range 0 to 1 and puts the required number in the destination column. You can multiply them up if required, to give numbers in the required range.

2.10 Edit commands

It is often necessary to edit data because it may have to be manipulated or sifted to meet particular needs. Several editing commands are supplied as detailed below.

PICK the rows from 1 to 2 in C1 and put into C2

This command enables the user to 'top' or 'tail' a column to ensure it is generally correct. If you had entered one too many items in a column in error, then the PICK command could be used to put the correct number of items back into the same or a different column.

RECOde the values from 3 to 5 in C1 to 1 and put into C2

This command enables the user to 'degrade' the data. For example, if there were several political parties coded 1 to 6 then they could be reduced to 2 groups by recoding all the values from 1–3 to a 1 and all the values from 4–6 to a 2.

CHOOse values 1 to 5 in C1 (and corresponding C2) and put into C3 and C4

This is one of the most powerful editing commands, as it enables us to make sub-groups for further analysis. For example, if males/females were coded as 1,2 in C1 then the 'male' data could be separated from the 'female' data.

EXAMPLE

```
SET C1
  DATA > 1 2 1 2 1 2 1 2 1 2
  DATA > END
SET C2
  DATA > 3 2 7 4 2 6 4 2 1 4
  DATA > END
CHOOSE 1 in C1 (corr C2) and put into C3 and C4
PRINT C3 C4
```

	C3 (n = 5)	C4 (n = 5)
1.	1.000	3.000
2.	1.000	7.000
3.	1.000	2.000
4.	1.000	4.000
5.	1.000	1.000

As you can see, the coding number in C1 i.e. 1 and the corresponding data from C2 have been sifted out and put into C3 and C4.

OMIT data from 5 to 7 in C3 put into C4

This data is the obverse of the CHOOse command. Whereas CHOOse will select the data that you request and transfer that data over to the destination columns, the OMIT command will transfer over all of the data except that which you wish to omit.

JOIN the data in C2 to the end of C1 and put back into C1

Notice here that the data you wish to join to the end of the other column is specified first. You have the opportunity to put the newly augmented column in a new column or back into one of the original ones.

SUBStitute (or PUT) the value of 10 into row 9 of C2

This command would be used if you had made an error (e.g. in data input) that you wish to correct after having entered the data. Remember that the value that you wish to substitute is specified first, and the row of the column for which it is destined is specified second.

LET Cx(y) = Value

This performs the same function as SUBStitute but in a somewhat more intuitive manner. For example, if you had entered the series of numbers 10 20 3 40 50 in C1 and had intended the third number to be 30 and not 3, you may correct your error at the end of data entry by:

LET C1(3) = 30 (i.e. put 30 into the 3rd row of C1)

DELETE row 2 from C1
DELETE row 2 from C1–C5

This command will delete a ROW of data from a

column (or a block of columns). Be careful with any DELETE or ERASE command as, once deleted, the data cannot be recovered.

ERASE data in COLUMNS C3–C5

This command erases single COLUMNS or blocks of columns.

COPY C1 into C2
COPY the block from C3–C6 into a new block starting at C13

The simple version of copy performs a straight copy of one column into another. The more advanced version will copy a block of data but the user should specify the source columns using a hyphen (no spaces!) and the start of the new block.

DISPlay the data in C3 to 1 decimal place.
DISPlay the data in unchanged cols to 2 decimal places.

This command alters the displayed value to the required number of decimal places but not the value which MICROSTATS holds internally which will be about 10 places of decimals.

You may either change one column specifically or the rest of the unchanged columns by specifying no column number. To display *no* decimal places then use the command DISPlay 0 (rather than the command DISPlay with no parameters).

2.11 ROW commands

The ROW commands exactly parallel the column statistics except that they operate upon *rows across columns* rather than individual columns which is the usual method of analysis. When any of the ROW commands are issued, a table is given from which users may select the value(s) in which they are interested. The ROW commands are listed below.

RDEScribe the data in row 1 of C1–C5
RSUM of data in row 1 of C1–C5, put results into C6
RSSQ (Sum of Squares) row 1 of C1–C5, put into C6
RMEAn of data in row 1 of C1–C5, put results into C6
RMEDian of data in row 1 of C1–C5, put results into C6
RMODe of data in row 1 of C1–C5, put results into C6
RMINimum of data in row 1 of C1–C5, put results into C6
RMAXimum of data in row 1 of C1–C5, put results into C6
RSTAndard dev-n [pop'n] data in row 1 of C1–C5, results in C6
RSTD dev-n [sample] data in row 1 of C1–C5, results in C6
RSEMean stand. error of mean of row 1 of C1–C5 results in C6
RSKEw (skew) of data in row 1 of C1–C5, results in C6
RQUArtiles of data in row 1 of C1–C5, results in C6
RPERcentiles of data in row 1 of C1–C5, results in C6

You may specify one row only for the relevant statistics.

If you do NOT specify a row but instead specify an extra column, then the results of each row will be put into the relevant row of the 'extra' column:

e.g. RMEAN C1–C5 C6 will put the mean of *each row* of C1–C5 into C6.

2.12 Statistical testing

To perform a statistical test of data which has been measured with a ratio or interval level of measurement, we can use either of the TWOSample or the POOLed commands.

The TWOSample command assumes that we wish to test the hypothesis that the means of two samples differ statistically from each other. The underlying assumption is that the population variances need not be approximately equal.

EXAMPLE

IRAN 100 values from 1 to 100, put into C1 and C2
TWOS C1 and C2
Twosample t C1 vs. C2

	n	mean	stdev	se mean
C1	100	52.9200	29.9504	2.9950
C2	100	48.4400	27.2528	2.7253

```
95.00 PCT C.I. for mu C1 – C2: (–3.508, 12.468)
ttest mu C1      = mu C2     (vs. n.e.):
T = 1.106    p = 0.2699 approx.    d.f. = 196
```

The output requires some interpretation. For information, the means, standard deviations, standard errors and 95% confidence intervals (= C.I.) of the mean are displayed. The critical results come on the last line of the display where the critical values are those for T and p. The letter T indicates the number of standard errors by which the two means differ and p, the probability of getting results like this by chance factors (where p varies from 0–1). As a rough rule of thumb, we would expect a significant result for T to be anything in excess of the value of 2.00. The value of p gives us the probability that the means differ by amount that they do under the influence of chance factors alone. A significant result is achieved when p is equal or less to the value of 0.05 (i.e. there is a 5% chance or less that the observed differences could have occurred by chance alone). The inference is, therefore, that non-chance factors are operating in which case we reject the null hypothesis that the population means (mu) are equal and accept the alternative hypothesis that the populations means (mu) are not, in fact, equal.

The d.f. (degrees of freedom) figure is used by MICROSTATS internally to calculate the values for T and p. Usually it is a figure approximately equal to $(n_1 + n_2 - 2)$ where n_1 is the number of rows of data in the first column and n_2 is the number of rows in the second column.

POOLed test for C1 vs. C2

The output and interpretation of the POOLed test is almost identical to that of the TWOSample command. However, this command may be used if the user is confident that the two populations have approximately equal variances. If in doubt, the user should generally use the TWOSample test.

MANN-Whitney test.

This is a non-parametric test which is generally regarded as almost as powerful as its parametric analogue (TWOSample test or t-test). The data may be measured at the ordinal level (as internally, the calculations are performed upon ranked data). For a full interpretation of 'u', the user should consult a statistical source. The Mann-Whitney test is regarded as closely related to the Wilcoxon rank-sum test. Strictly speaking, the test is used to evaluate the difference between population distributions, not population means but when the distributions of the groups are similar the test does in fact measure differences in central tendency. Column lengths are limited to 100 in this command, as the test requires the columns to be joined together.

As with the TWOSample and POOLed test, the critical value are those for T and the probability. As a rule of thumb, one is looking for a T value of approximately 2.00 or greater and a probability of equal or less than 0.05 (2-tailed test) in order to achieve evidence that the distributions differ from each other significantly. (Small samples (i.e. < 11) will require reference to Mann-Whitney tables to ascertain the significance level.)

KOLMogorov-Smirnov tests (one sample and two sample)

The Kolmogorov-Smirnov (or K-S) test is a non-parametric test which is much less sensitive to small numbers than the traditional chi-square test. It tests the cumulative frequency distributions (observed against expected) and reports a figure which needs to be interpreted by reference to K-S tables. It is most suited to the analysis, for example, of questionnaire items and is used in the following way. If there are four potential answers to a question and any answer is as likely as any other then the 'expected' cumulative frequencies will be 25%, 50%, 75% and 100%. The test will take the actual distribution of answers, test them against 'expected' and report the result.

If there are prior expectations of the results (e.g. testing against previous data) then these may be placed in the second column for a two-sample test.

For example, out of 80 answers to a question such as 'Should the death penalty be retained' scaled as

1 — Strongly opposed 2 — Opposed
3 — In favour 4 — Strongly in favour

we collect the data 1 (7 people) 2 (15 people)
3 (20 people) 4 (38 people)

and then analyse the data as follows:

(First set the data 7, 15, 20, 38 in C1)

KOLM C1

Kolmogorov-Smirnov [D] = 0.225
(Consult tables for significance)

If we have the proportions from a previous questionnaire, then we can put these into C2 and perform the analysis

KOLM C1 C2

(In this case, the 'expected' frequencies are those taken from C2 instead of the assumption of an equal spread which would be the case in a one-sample test)

TTEST data in C1 against a value of 50

This test is used to test a sample mean against a known value or population mean. The critical value is to observe whether or not the value is equal to or less than $p = 0.05$ in which case we conclude that there is a statistically significant difference between the sample mean and the value.

TDIST 1.9603 at 2500 df

This calculation will give the user the proportion of a distribution (one- and two-tailed) that corresponds to the value for the degrees of freedom specified. It may be thought of as an alternative to look-up tables.

TINT for data in C1 at 95%

This command gives the user the confidence intervals for the data at the confidence level requested.

CHID for value 3.84 at 1 df

This command is another alternative to a look-up table. It provides the user with the probability of achieving the specified value at specific degrees of freedom. In the above example, we would get the response

Probability = 0.0500

which informs us that with a normal chi-square table of 2 × 2, i.e. two rows and two columns which is 1 d.f. then a value of 3.84 would be achieved only 0.05 (5% of the time) by chance factors alone.

2.13 Normal distribution

To generate a sample that has the shape of a 'normal' i.e. bell-shaped distribution, use the following command:

NORMal 200 values into C1

By default, the mean will be 0.00 and the standard deviation (sigma) will be 0.15. To specify your own mean and standard deviation, then specify them as the second and third values on the command line e.g.

1st value 2nd value 3rd value
↓ ↓ ↓

NORMal 200 values, mean 100, sigma 15 in C1

To specify a seed (to help generate identical distributions) then make the seed (= 3125 by default) an odd integer between 0–32767 thus:

NORMal 200 values, mean 100, sigma 15 in C1, seed = 625

To draw a sample (from any column of data) then use:

SAMPle 20 values from C1, put into C2–C3

A seed may be specified to draw the same sample, if needed:

SAMPLE 10 values from C1 put into C2, seed = 625

The seed may be any odd integer from 0–32767 (6125 by default).

2.14 Time series analysis

LTREnd C1 C2
(Linear TREnd of data in C1, put into C2)

Performs a **linear trend** (i.e. regression) of the

data (assumed to be y in C1) against an imaginary column of data (assumed to be x with values from 1 . . .n). The results are then placed in C2.

MAVE C1 C2
Moving AVErage of data in C1, put into C2

performs a four-quarter **moving average** for the data in C1, putting the results in C2. The user should ensure that the number of rows is *exactly* divisible by 4 to give accurate results. Note that the first two rows and last two rows of a moving average column are 0 (representing no data)

ADDM (ADDitive Model of data in C1)

The form of the command is:

ADDM C1 C2 C3 C4

where all four columns are *essential*! (NB Hyphens are *not* allowed here)

C1 contains the [A]ctual data
C2 contains the [T]rend line
C3 contains the [S]easonal data
C4 contains the [R]esidual data

A table of quarterly deviations is printed together with the average quarterly deviation. The average of the residual data is also computed and this should take a value close to zero.

MULM (MULtiplicative Model of data in C1)

The form of this command, and the output from it, are almost identical to ADDM (above). The syntax is the same and the only difference is that the average of the Residual data should approximate to a value of 1.0 and not zero.

2.15 Multiple regression

The form of the command is:

MREGression of data in C1 on 'x' predictors in Cy–Cz, {optionally} put residuals in (Cr) and fitted Y's in (Cf).

For example, consider the two commands:

1. MREGress data in C1 on 3 predictors in C2–C4, residuals in (Cr) and fitted in (Cf)
2. MREGress data in C1 on 4 predictors in C2 C3 C4 C5

The output and interpretation of multiple regression is by its nature quite complex and attention will be drawn here only to the most salient points. The reader should consult a relevant textbook of statistics to enhance understanding of these techniques. Note particularly the following points.

1. First the regression equation is given.
2. This is followed by a table giving, *inter alia*, the constant term and the coefficients for each of the specified independent variables. Note particularly the t-ratios here because a large t (e.g. greater than 2) and a correspondingly small p indicates that that particular variable contributes significantly to the prediction of the dependent variable.
3. The value of R–sq (R–squared) is the coefficient of determination and is a measure of the variation in y accounted for by the combination of multiple factors of x. Generally speaking, we are concerned with finding those factors that best predict y by raising the R–sq value significantly and particularly the R–squared value (adjusted for the degrees of freedom)
4. In the analysis of variance table, the SS (sum of squares) figure for the regression shows the variation in y which is explained by the independent factors whilst the SS (sum of squares) for the residual shows the unexplained. The F statistic can be used to indicate whether the independent variables collectively can be said to have a significant effect upon the dependent variable. (Consult statistical tables here with degrees of freedom for regression [numerator] and degrees of freedom for residual [denominator].)
5. The Durbin–Watson statistic is a measure of autocorrelation in the residuals. The ideal value for DW is 2.0 in the range from 0.0–4.0: a low value indicates a significant positive autocorrelation and a high value indicates a significant negative autocorrelation.
6. The output then gives a table of observed, fitted and residual values.
7. The output concludes with a correlation matrix. Note that this is symmetrical i.e. a cor-

relation of C2 with C4 is the same value as the correlation of C4 with C2.

2.16 RETRieve and SAVE files

RETRieve (and then follow instructions)
RETRieve a:myfile

This command RETRieves files that have been previously saved under MICROSTATS. It will *not* retrieve other files which might be accessed with FREAD (q.v.)

SAVE (and then follow instructions)
SAVE a:myfile

This command will SAVE the workfile for the user. No extension should be used as MICROSTATS actually saves two files, one of which is in specially coded numerical format (for fast access and compact disk storage) and the other of which is a text file in which names, if allocated, are stored. The user does need to be concerned with such details but it does explain why two files will appear on the disk for every worksheet saved, one with a .MCS extension and the other with a .NAM extension.

FREAD
FREAD a:datafile

The FREAd command will read, or attempt to read, any file in which data has been saved in a straight ASCII format. As FREAD can only read in completely 'rectangular' blocks of data, it is *important* that any data that is exported by another package should be absolutely rectangular. For example, to export a spreadsheet of 2 × 10 columns and 2 × 5 columns, then pad the last columns to 10 with zeros to make a 'rectangle' that is 2 × 10. Adjustments could be made once the data is successfully imported into MICROSTATS.

Remember that MICROSTATS will only read numerical and not textual data. The user will be prompted for the start column of the data, which will then be read into consecutive columns.

If data is prepared using a text or word processor for input into MICROSTATS as well as other packages, then any legitimate data separator (; or, or ⟨space⟩) may be used.

FWRITE

FWRIte will write out data as a straight ASCII file with a choice of delimiters allowed. Before using this command, remind yourself of the start and end columns by using INFO as FWRIte will request your start and end columns.

FERAse
FERAse myfile

FERAse will erase any type of file whether saved under MICROSTATS or not. If there are non-MICROSTATS files that the user wishes to erase, then the full file name with drive letter, name and extension should be given.

2.17 Directory management

DIREctory [A:] {or DIR}

DIREctory will list all files on the logged drive (by default), or the files on the specified drive.

Note that DIRE does not change the logged disk drive. In addition to the normal directory display, a separate list of MICROSTATS files is given. Wildcard characters such as * or ? are not implemented in this command.

LOGD
LOGD A:

LOGD with no parameter will remind you of the drive upon which you are currently logged, and at the same time issue a directory of files.

LOGD with a legitimate file disk-drive parameter will *both* log the user onto the specified drive and also issue a directory.

CHDIR [Dr:] Subdirectory

Changes the logged subdirectory (and also gives a list of files) If a drive is specified, then the user is logged on to that drive also.

DISK [Dr:]

Gives system information on the drive including percentage used as a percentage of total disk space, and the free bytes available.

2.18 General commands

HELP (or ?) {The F1 key may also be used}

HELP gives access to eight help screens and an index page. At the bottom of each page the user may access the [N]ext Page, [L]ast Page, [E]xit or specify a page-number to access directly.

STOP

STOP completes the work-session. The user is prompted to save the work-sheet and also asked to confirm exit to ensure that an accidental exit does not occur.

NAME

This names columns of data
e.g. NAME C1 'Heights' C2 'Weights'

Take care that the same, single, apostrophe is used. (See also page 391)

INFOrmation

This command informs the user of the numbers of columns (and their length) still available for use. The column numbers, names allocated (if any) and number of data items in each column will be notified.

MICROSTATS users should use this command *frequently* to check on the status of their work-sheet and to confirm that the data that they have in their work-sheet conforms with their expectations. Similarly, PRINT should be used in conjunction with INFO to check on the data in columns.

PRINT C1
PRINT C1–C5
PRINT C10 C2 C5 C8

PRINT is a command which *always* requires information as to which columns of data to print. If a range of columns is requested, then it should be specified with a hyphen but with no spaces on either side of the hyphen. Only seven consecutive columns may be printed if the hyphen form of the command is used and it is not generally sensible to attempt to print out more than seven columns if the user wishes to preserve a 'clean' screen display.

Long columns will stop after twenty items (a screenful) and prompt the user to view the next screenful or to exit to the next command. Names are displayed together with column contents.

PRON

The PRON command stands for PRinter ON. Output normally directed to the screen will now appear on the printer. In some cases, this may mean performing 'blind' so the user should have rehearsed a particular sequence of commands first to verify their effect, taken a note of the same and then repeated the same with the PRON switch toggled on.

PROF

The PROF command stands for PRinter OFf. Output will be redirected back to the screen for a normal 'dialogue'.

NOTE

NOTE displays a comment for documentation purposes. MICROSTATS will ignore any data on a note line and in this respect it resembles REM in a BASIC program.

FKEY

This gives two diagrams of the distribution of commands on the function keys. It may also be activated with SHIFT-F1. The diagrams may be dumped to a printer to provide templates if desired.

PRTSC

This command enables the user to obtain a screen-dump (as the normal PrtSc key is often used to 'SNAP' screens) This function may also be activated by either CTRL-PRTSC or CTRL-Pg Dn (one or other will almost certainly work on your computer)

CLEAR

This command clears the screen (before a screen-snap for example, or a screen-dump to the printer). If the printer is engaged with PRON then a Form-Feed will also be sent to the printer.

2.19 Avoiding crashes!

Despite the warnings built in at various points, MICROSTATS will occasionally crash or 'abort' when it cannot cope with certain error conditions e.g. a calculation which involves a division by zero.

Here are some tips and hints to minimize the occasions upon which MICROSTATS will abort, or at least to make sure that the consequences are not too dire!

— *Do* make sure that you do not enter more than a line-full (or a little over) of data in the SET command.
— *Do* save your precious data after a fair amount of typing or column manipulation. A SAVE every 15–20 minutes only takes a few seconds and ensures that the potential loss of time is limited to this amount.
— *Do* make use of INFO, PRINT and the HELP screens in order to keep a check on the status of the worksheet.
— *Do* keep a note of events that caused the system to crash and avoid them in the future!

3.0 Saving output with 'SNAP'

Although not technically a part of the MICROSTATS package, SNAP is a specialized 'Terminate-and-Stay-Resident' program that allows you to take up to thirty 'snapshots' of your screen. When you conclude running MICROSTATS within the batch file MS.BAT, then the 'snapshots' you have taken will be 'developed' under the names SNAPSHOT.01 . . .SNAPSHOT.30 They will be written onto the drive you have specified when invoking the batch file e.g.

MS b:

will 'develop' your snapshots onto Drive B: These files may then be incorporated into a word-processor or text-editor of your choice for incorporation into other documents. If you wish to collate several snapshot files together into one working file, then this is easily accomplished by typing the following command from DOS:

COPY SNAPSHOT.* OUTPUT.DOC

with the result that all of the 'snapshot' files will be stitched together into one output file, named OUTPUT.DOC
The following five files are essential for the 'clean' operation of SNAP. These are:

MARK.COM	which marks the position of SNAP.EXE in memory
SNAP.EXE	the 'snapshot' Terminate-and-Stay-Resident program
DEVELOP.EXE	the 'developer' program
TSR.COM RELEASE.COM }	two files that between them release the memory used by SNAP back to DOS.

The Commands PRON and PROF may still be used to retain a measure of compatibility with MINITAB. However the SNAP system may still be preferred on the grounds that the output files produced allow for a degree of editing (e.g. to eliminate error messages) that the PRON/PROF routines do not.

4.0 DATA SETS

There are three data sets on your disk.

DATA1 contains the data of Sellmore plc from Table 2.1.
DATA2 contains the survey responses from the Shopping survey in the Concluding Exercise to Part V.
DATA3 contains the data from Table 19.2.

5.0 ALPHABETICAL INDEX OF COMMANDS

? (= HELP)

ABS
ADD
ADDModel
ANTIlog
APPEnd
AVERage

CHID
CHDIrectory
CHISquare
CHOOse
CLEAR
CONTingency
COPY
CORRelate
COUNt

DEFIne
DELEte
DESCribe
DIREctory (or DIR)
DISK
DISPlay
DIVIde

ERASe
EXPO

FERAse
FKEY
FREAd
FWRIte

GENErate

HELP
HISTogram

INFOrmation
INTeger
IRANdom

JOIN

KOLMogorov-Smirnov

LABEL
LET
LOGD
LOGE
LOGTen
LTREnd

MANN-Whitney
MAVErage
MAXImum
MEAN
MEDIan
MINImum
MODE
MREGess
MULModel
MULTiply

NAME
NORMal distribution
NOTE

OMIT

PERCentile
PICK
PLOT
POOL
PRINt
PROF
PRON
PRTScreen
PUT

QUARtiles

RAISe
RANK
RDEScribe
READ
RECIprocal
RECOde
REGRess
RETRieve
RMAXimum
RMEAn
RMEDian
RMINimum
RMODe
ROUNd
RPERcentiles
RQUArtiles
RSKEw

RSSQ (Row Sum of Squares)
RSTAndard dev-n
RSTD dev-n
RSEMean (Row Std. Error of Mean)
RSUM

SAMPLE
SAVE
SEMEan
SET
SKEW
SORT
SQRT
SSQ (Sum of Squares)

STANdard deviation
STDEv-n
STOP
SUBStitute
SUBTract
SUM

TALLY
TDIStribution
TINTerval
TTESt
TWOSample

URANdom

ANSWERS TO SELECTED PROBLEMS

CHAPTER 3

3. Mean = 2.84
   median = 3
   mode = 4
4. Mean = 1.3611
   median = 1
   mode = 0
5. Mean = 146.8
   median = 146
   mode = 135
   New mean = 162.22
   New median = 149.5
   New mode = 135
6. Mean = £9.2365
   median = £8.86
7. Arithmetic mean = 7.26
   Geometric mean = 6.5849
8. Mean = 1.14
   median = 1
   mode = 0
9. Mean = £111.93
   median = £95.93
   mode = £34
10. Mean = 5.818
    median = 3.66
    mode = 0.9
11. Mean = £0.21
    median = −£1.41
13. Mean = £108.26
14. Total value = £818 000
    mean = £454.44

CHAPTER 4

3. Range = 6
   quartile deviation = 5
   standard deviation = 1.67
4. Range = 4
   quartile deviation = 1
   standard deviation = 1.3775
5. Range = 57
   quartile deviation = 4.5
   Standard deviation = 12.89
   New range = 171
   New quartile deviation = 15.25
   New Standard deviation = 36.44
6. Mean = 9.2365
   standard deviation = 2.825

7. Range = 5
   quartile deviation = 1
   standard deviation = 1.2492
8. First quartile = 53.1
   third quartile = 151.227
   quartile deviation = 49.0635
   standard deviation = 74.0944
9. First quartile = 1.5833
   third quartile = 8.34459
   quartile deviation = 3.3806
   standard deviation = 5.532
10. Mean = £0.21
    standard deviation = £7.053
11. (a) Mean = £95.33
        standard deviation = £55.28
    (b) 55.8%
    (c) median = £85.21
        first quartile = £46.38
        third quartile = £123.66
        quartile deviation = £38.64
12. (a) A £12,208.33
          £1,946.77
        B £12,375
          £1,235.33
    (b) A var = 15.95
          skew = 0.03
        B var = 9.98
          skew = −0.06
13. (a) 6.25 minutes
    (b) 20.5 minutes
    (c) 27.22 minutes

CHAPTER 5

1. (a) 80, 92, 96, 100, 104, 116
   (b) 4.167%, 11.538%
2. (a) 52.63, 63.16, 84.21
   (b) 247, 266, 285, 313.5
3. (a) 100, 175, 475
   (c) 100, 168, 420.75
   (e) 100, 167, 386
   (g) 100, 97, 100.5
4. (a) 100, 108, 108
   (b) 100, 117.86, 139.29
   (c) 100, 122.5, 150
   (d) 100, 110.2, 117.1
5. 100, 110.4, 120.3
8. (a) 100, 102.2, 106.47,
       112.59, 118.58, 133.59,
       146.28, 153.49, 159.5,
       167.78, 174.46, 182.48,
       192.63, 200.78, 212.27,
       228.83
   (b) 100, 103.2, 107.5,
       112.4, 123.4, 136.3,
       145.1, 146.3, 150.5,
       158.1, 166.64, 176.91,
       191.46, 205.53, 223.4,
       244.42
9. (a) 103.81, 111.3
   (b) 108.34, 109.98
   (c) 104.35, 113.9
   (d) 108.92, 112.56
   (e) 113.06, 125.27
   (f) 103.81, 111.3
       108.34, 112.21
       104.35, 114.04
       108.92, 114.97
       113.06, 127.96
   (g) 103.81, 111.9
       108.34, 112.21
       104.35, 115.9
       108.92, 116.21
       113.06, 130.05
   (h) 103.81, 111.9
       108.34, 107.14
       104.35, 119.14
       108.92, 114.07
       113.06, 127.65

N.B. all answers (f) to (h) are cumulative.

CHAPTER 6

1. $x = 9y$
2. $x = -12y$
3. $x = -y$
4. $x = 3y$
5. $y = 0.8$
6. a^5
7. 1
8. $a^2 + a^3/b$
9. a^2
10. $a^3 + a^6 - a^8$
11. 1/8
12. $x = 22$
13. $2a^2 + ab$
14. $a^2 + ab + b^2$
15. $a^2 + 4ab + 4b$
16. $a^2 - b^2$
17. $6a^2 + 2b^3$
18. $12x^2 + 18xy - 12y^2$
19. (a) 32 (b) 258
20. (a) 238 (b) 2268
21. (a) −15 (b) 156
22. 35
23. (a) 2916 (b) 118096
24. (a) 384 (b) −2046
25. (a) 0.8387 (b) 26.78
26. 28125
30. (a) $x = 2.6$ $y = 0.3$
    (b) $x = 13$ $y = 2$
    (c) $x = 4$ $y = -2$
    (d) $x = 1.75$ $y = 3.15$
31. (a) $x = 1$ or 2
    (b) $x = -2$ or 4
    (c) $x = -1$ or -12
    (d) $x = -3.217$ or 6.217
    (e) $x = 0.607$ or 4.393
    (f) $x = -1$, $+1$ or 6
    (g) $x = -3.3935$ or 5.8935
32. $P = 70 - 2Q$
33. $P = 20$ $Q = 25$
34. $P = 20$ $Q = 25$
35. $P = 47.5 - 4.5Q$
    $P = 9.5 + 0.1Q$
    $P = 10.815$
    $Q = 13.152$
36. (a) demand: $P = 2Q^2 - 100Q + 2000$
    supply: $P = 5Q^2 + 4Q$
    (b) $Q = 13.765$
    $P = 1002.45$

CHAPTER 7

1. $\begin{bmatrix} 7 & 8 \\ 12 & 15 \end{bmatrix}$
2. $[16 \ \ 8 \ \ 4]$
3. 50
4. $\begin{bmatrix} 18 & 24 \\ 60 & 78 \end{bmatrix}$
5. $\begin{bmatrix} 38 & 40 \\ 56 & 58 \end{bmatrix}$
6. impossible
7. $\begin{bmatrix} 5 & 14 & 18 \\ 8 & 2 & 7 \\ 10 & 8 & 4 \end{bmatrix}$
8. $[39 \ \ 53 \ \ 45]$
9. $\begin{bmatrix} 24 & 64 & 72 \\ 14 & 49 & 79 \\ 13 & 39 & 95 \end{bmatrix}$
10. $\begin{bmatrix} 122 \\ 26 \\ 72 \end{bmatrix}$
11. $\frac{1}{6}\begin{bmatrix} 5 & -1 \\ -4 & 2 \end{bmatrix}$
12. $\frac{1}{204}\begin{bmatrix} 1 & -28 & 44 \\ 24 & -60 & 36 \\ -11 & 104 & -76 \end{bmatrix}$
13. 1/50
14. $\begin{bmatrix} 24 & 12 & 6 \\ 32 & 16 & 8 \\ 40 & 20 & 10 \end{bmatrix}$
15. $\begin{bmatrix} 1 & 0 & 0 \\ 0 & 1 & 0 \\ 0 & 0 & 1 \end{bmatrix}$

16. $\begin{bmatrix} 1 & 0 & 0 \\ 0 & 1 & 0 \\ 0 & 0 & 1 \end{bmatrix}$
17. impossible
18. $\begin{bmatrix} 60 & 30 & 48 \\ 80 & 40 & 64 \\ 100 & 50 & 80 \end{bmatrix}$
19. 180
20. $\begin{bmatrix} 138 & 456 & 708 \\ 184 & 608 & 944 \\ 230 & 760 & 1180 \end{bmatrix}$
21. $x = 2.6, y = 0.3$
22. $x = 4, y = -2$
23. $x = 4, y = -2$
24. $x = -1.7, y = 2.3$
25. $a = 1, b = 2, c = 5$
26. $a = 0.1, b = -1, c = 2.5$
27. $x = 8, y = 10, z = -5$
28. $a = 1, b = 3, c = 2$
29. $a = 1, b = -1, c = 2,$
$d = 2$
30. $x = 5, x = 3$
$x = -2, x = 1$
31. $X_1 = 220, X_2 = 170$
32. $A = 178.779, B = 704.942$
33. $A = 810, B = 1620,$
$C = 2250$
34. $X_1 = 1103.49, X_2 = 1151.81,$
$X_3 = 1952.24$

CHAPTER 8

1. $f'(x) = 0$
2. 2
3. $f'(x) = 3$
4. $10x^4$
5. $f'(x) = 42x^2 - 12x^3$
6. $34x + 14$
7. $f'(y) = 7 + 4y - 12y^2$
8. $4q - 4q^{-2}$
9. $f'(x) = e^x$
10. $2 + 8s + 0.5s^{-1/2}$
11. $f'(x) = 140x^6 + 24x^5 - 15x^4 - 28x^3 + 12x^2 - 22x + 2 + x^{-2} - 18x^{-3}$
12. $-2x^{-3} + 6x^{-4} + (16/3)x^{-5}$
13. $MR = 100 - 4x$; $P = 50$
14. $AC = (1/3)x^2 - 5x + 30$
$MC = x^2 - 10x + 30$
15. (a) $x = 19.67399$ (b) $x = 11.888$
16. $MC = MR$ at $x = 3.765$
$AC = AR$ at $x = 6.437$
17. $E_0 = -1$ at $x = 10$
18. (a) $P = 1000 - 5Q$ (b) $P = 1500 - 55Q$
(c) $MR = 1000 - 10Q$: $MR = 1500 - 110Q$
(e) $E_D = -19$: $E_D = -1.727$
(f) $400 < MC < 900$
19. Min at $x = 5, y = 1$
20. Max. at $x = 10, y = 1200$
21. Min at $x = -1, y = 2$
22. Linear, so no max. or min
23. Min at $x = 1.6667, y = -4.3333$
24. Max. at $x = 1, y = 12.33$
Min. at $x = 5, y = 1.33$
25. Max at $b = 0.25, a = 10.25$
26. Max. at $x = 0.5858, y = 13.3137$
Min. at $x = 3.414, y = -9.3137$
27. Neither, a constant cannot have max. or min.
28. Min. at $x = 1, y = 37.75$
Max. at $x = 2, y = 38$
Min. at $x = 3, y = 37.75$
29. Output $(x) = 10$: Profit $= 210$
30. Max. profit at $x = 11.888$ when profit $=$ £696.111
31. Output $(x) = 3.765$: Profit $= 92.58$
32. Min. AC at $x = 1$, when $AC = MC = 98$
33. $f'(x) = 6x^2 + 10x + 12$
34. $(2x^2 + 6x + 5)(6x + 4) + (3x^2 + 4x + 10)(4x + 6)$
35. $f'(x) = 3(4x^3 + 6x^2)^2 (12x^2 + 12x)$
$= 576x^8 + 2304x^7 + 3024x^6 + 1296x^5$
36. $xe^{x^2/2}$
37. $f'(x) = [(x^3 + 10)(4 + 12x) - (4x + 6x^2)(3x^2 + 6)]/[(x^3 + 6x)^2] = [-6x^4 - 8x^3 + 36x^2]/[x^6 + 12x^4 + 36x^2]$

38. $(20 - 18x^4)/(3x^4 + 10)^2$
39. $2x^2 + c$
40. $x^3 + 2x^2 + 10x + c$
41. $0.5x^4 + 2x^3 - 3.3333x^{-3} + c$
42. $10x + c$
43. $0.3333x^3 + 0.5x^2 + 5x + c$
44. 26
45. 200
46. 1433.33
47. (a) $x = 6.4934$
    (b) Profit = £526.5008; Price = £137.013
    (c) $E_0 = -10.55$
    (e) Output falls to $x = 6.2243$
48. (a) $TC = 8 + 16Q - Q^2$
    (b) $TR = 40Q - 8Q^2$
    (c) $Q = 2.5$
    (d) $Q = 1.7143$
    (e) $0.4286 \leqslant Q \leqslant 3.$
49. $\partial y/\partial x = 4z$; $\partial y/\partial z = 4x$
50. $6x^2 + 8xz + 2z^2$; $4x^2 + 4xz - 9z^2$
51. $\partial y/\partial x = 7 + 6xz - 3x^2z^2 - 2z + 5z^2 - 7z^3$
    $\partial y/\partial z = 3x^2 - 2x^3z - 2x + 10xz - 21xz^2$
52. $40x^3$; $45z^2$
53. $\partial q/\partial p_1 = 4p_1 - 3 + 4p_2$
    $\partial q/\partial p_2 = 4p_1 - 5 + 2p_2$
54. $4st + 14t - 6t^2 - 21$; $2s^2 + 14s + 10 - 12st$
55. Min. at $x = 2, z = 3, y = 1.5$
56. Min. at $x = 1, z = 5, y = 7$
57. Max. at $x = 5, z = 10, y = 400$
58. Max. at $x = 2, y = 1.5, z = 31$
59. Min. at $x = 0, z = 0, y = 0$
60. No max. or min. as a linear function
61. $L = 2, K = 3$, TC = 1900
62. $x = 2, y = 2.2$, profit = 52.6
63. (a) $X = 5, Y = 1$, profit = 130
    (b) $X = 5.2$, profit = 125.2
    (c) $Y = 2$, profit = 10
64. Max. at $x = 2.605, z = 0.868$
65. Min. at $z = 0.10177, x = 0.5088,$
    $y = -1.1687$
66. Max. at $x = -163, y = 142$
67. $x = 10.291, y = 2.233, z = -91.65$
68. Min. at $x = 0.5, y = 0.5$
69. $l = 33.333, k = 20, q = 103.7956$
70. $L = 250, K = 93.75, Q = 958.4147$
71. (a) $x = 1, y = 5$, profit = 718
    (b) $x = 0.185, y = 4.629,$
    profit = 715.924
    (c) $x = 1.685, y = 4.629,$
    profit = 715.394

CHAPTER 9

1. (a) £367.04
   (b) £390.23
2. (a) £16 105.10
   (b) £16 145.04
   (c) £15 529.70
3. £641.09
4. £3 519.61
5. £3 485.10
6. £353.91
7. Project 1: £18 708.70
   Project 2: £17 021.20
8. Large Process £4 297 560
   Two med processes £3 960 220
9. £1 221.02
10. £1 343.12
11. £12 646.93
12. £298.31
13. £996.26
14. 27%
15. (a) 23.1%
    (b) 21.6%
    (c) 16.6%
16. 1.94%
17. 9.70%

CHAPTER 10

1. 0.25
2. 0.25.
3. 0.125
4. 0.375.
5. 0.64, 0.16, 0.032, 0.096
6. 25.
7. 640, 160, 32, 96
8. (a) 1/18
   (b) 1/9
   (c) 1/6
   (d) 1/6.
9. 1/36
10. 9/16.
11. 1/12
12. 7/12.
13. 1/444
14. (a) 1/13
    (b) 1/14
    (c) 16/52
    (d) 1/52
    (e) 3/13
    (f) 1/2
    (g) 1/26
    (h) 3/26.
15. (a) 1/16
    (b) 0.14932
16. 116/221.
17. (a) 0.25
    (b) 0.5
18. $p(A) = 0.1$; $p(B) = 0.03333$; $p(C) = 0.025$.
19. (a) 140
    (b) 180
    (c) 402
    (d) Expect 120, so yes
20. (i) 0.075
    (ii) 0.08
    (iii) 0.001828.
21. (b) 0.9
22. (a) 0.25
    (b) 0.28.
23. (a) 0.168
    (b) 0.4667
    (c) 0.46659
24. (a) 0.729
    (b) 0.009
    (c) 0.0009.
25. £210
26. Project A: £5 600
    Project B: £4 350
27. EMV(ad) = £6 300
    EMV(no ad) = £5 750

30. $\mathbf{P}^2 = \begin{bmatrix} 0.04 & 0.96 \\ 0 & 1 \end{bmatrix}$

    $\mathbf{P}^4 = \begin{bmatrix} 0.0016 & 0.9984 \\ 0 & 1 \end{bmatrix}$

    $\mathbf{P}^8 = \begin{bmatrix} 0.00000256 & 0.99999744 \\ 0 & 1 \end{bmatrix}$

31. $\mathbf{P}^2 = \begin{bmatrix} 0.04 & 0.96 \\ 0 & 1 \end{bmatrix}$

    $\mathbf{P}^4 = \begin{bmatrix} 0.0016 & 0.9984 \\ 0 & 1 \end{bmatrix}$

    $\mathbf{P}^8 = \begin{bmatrix} 0.00000256 & 0.99999744 \\ 0 & 1 \end{bmatrix}$

32. (a) [85 48 11 16]
    (b) [75.5 44.3 11.4 28.8]
    (c) [69.02 40.67 11.27 39.04]
    (d) [64.419 37.52 10.829 47.232].
33. (a) [60 105 110 145 160 70]
    (b) [1.62 7.695 20.63 66.245 265.83 287.98]
    (c) £47 000
    (d) £40 717.20
34. (a) [74.5 20 10.5]
    (b) [70.15 14.9 19.95]
    (c) [62.515 14.03 28.455]
    (d) [54.9845 21.5605]
    (e) [55.73755 20.80745]
    (f) [55.662245 20.882755].

CHAPTER 11

1. 3
2. 120; 120.
3. 38 760 and 1
4. 45; 10; 1.
5. 635 013 560 000
6. 0.000 000 000 006 299 078.
7. $P(0) = 0.0256$, $P(1) = 0.1536$, $P(2) = 0.3456$, $P(3) = 0.3456$, $P(4) = 0.1296$
8. (a) 0.004096
   (b) 0.27648
   (c) 0.54432.
9. (a) 0.00065536
   (b) 0.27869184
   (c) 0.31539456
10. (i) 0.142625
    (ii) 0.0203419.
11. 0.6590022
12. $P(0) = 0.13533$, $P(1) = 0.27067$, $P(2) = 0.27067$, $P(3) = 0.180447$, $P(4) = 0.0902235$, $P(5) = 0.0360894$.
13. $P(>4) = 0.0529$; $P(>5) = 0.01682$.
14. $P(0) = 0.22313$, $P(1) = 0.334695$, $P(2) = 0.251021$, $P(3) = 0.125511$, $P(4) = 0.0470665$
15. $P(0) = 0.3679$; $P(1) = 0.3679$; $P(2) = 0.1839$; $P(3) = 0.0613$; $P(4) = 0.0153$; $P(5+) = 0.0037$
    Days = 134.28, 134.28, 67.14, 22.38, 5.59, 1.35
    Cost = £365 000.
16. $P(>3) = 0.5665276$, expected no. = 2719.33248
    Total cost = 4 × 4 800 × £7.84 = £150 528
17. Demand > supply for approximately 16 days.
18. (a) 0.0803
    (b) 8 030
    (c) £1 204 500
    (d) TR = £10 000 000:
    TC = £6 000 000 + £1 204 500:,
    expected profit = £2 795 500

CHAPTER 12

1. (a) 1
   (b) −0.6
   (c) −1.54
   (d) 2.14
   (e) 0
2. (a) 0.1587
   (b) 0.97725
   (c) 0.97725
   (d) 0.5398
   (e) 0.0446
   (f) 0.3821
   (g) 0.8023
   (h) 0.00494
   (i) 0.04482
   (j) 0.1747
   (k) 0.1337
   (l) 0.04556
   (m) 0.9258
   (n) 0.95.
3. (b) −0.52
   (e) 1.77
   (g) −1.645 and +1.645
4. (a) 3.01%
   (c) 93.39%
   (e) £89.21.
5. (a) 0.9973
   (b) 0.99379
   (c) 0.3085
6. (a) 0.0655
   (b) 0.2266
   (c) 0.98809
7. (a) 0.2266
   (b) 0.1056.
8. (a) 0.1446
   (b) 0.3632
9. (a) 0.834
   (b) 0.9292
   (c) 0.90531.
10. 0.2327
11. (a) 0.8643

(b) cannot assume $p = 0.3$, so cannot estimate.
12. 0.0526
13. (a) 0.648
    (b) 0.123.
14. (a) 0.6708
    (b) 0.4499
15. 0.6844.
16. (a) 0.00248
    (b) 0.1867
17. (a) 0.3446
    (b) 0.01426
18. 0.1587

CHAPTER 13

1. £2 400 ± £221.75
2. (a) $\bar{x} = 146.8$, $s = 13.05$
   146.8 ± 4.044
   (b) 73
3. $\bar{x} = 1.14$, $s = 1.25$
   1.14 ± 0.15
4. (a) 3 ± 0.65
   (b) 136.46
5. (a) 20% ± 2.5%
   (b) 6 147
6. (a) 374.33
   (b) 384
8. 7.333 and 22.667
9. 96.32% and 99.68%
10. (a) 5.40 ± 0.61
    (b) 45.25
11. 14, 27

CHAPTER 14

1. Greater than 19.65 cubic yards
2. Less than £13.81
3. Greater than £0.78
4. Greater than 94.48%
5. Less than 12.45
6. £0.50 ± £0.91
7. −0.564 ± 0.227
8. 4% ± 5%
9. 1% ± 7.7%
10. £1 259 ± £102.78
11. 7.9 minutes ± 0.83 minutes
12. 2.245 and 5.755
13. −0.899 and 3.699

CHAPTER 15

1. $z = 2.36$, reject H_0
2. $z = 12.39$, reject H_0
3. $z = 2.816$, reject H_0
4. $z = -13.622$, reject H_0
5. $z = 1.52$, cannot reject H_0
6. $z = -1.353$, cannot reject H_0
7. $z = -1.67$, cannot reject H_0
8. $z = 0.2886$, cannot reject H_0
9. $z = -1.3066$, cannot reject H_0
10. $z = 2.449$, reject H_0
11. $z = 1.01$, cannot reject H_0
12. reject H_0
13. $t = -0.9035$, cannot reject H_0
14. $t = -2.68$, reject H_0
15. $t = -0.535$, cannot reject H_0
16. $t = 1.018$, cannot reject H_0
17. $t = -1.6978$, cannot reject H_0
18. $t = 34.75$, reject H_0
19. (a) 29.5%
    (b) 19.6%

CHAPTER 16

1. Chi-sq. = 4.254, cannot reject H_0
2. Chi-sq. = 14.6, reject H_0
3. Chi-sq. = 140.555, reject H_0
4. Chi-sq. = 2.517, cannot reject H_0
5. Chi-sq. = 8.67, cannot reject H_0
6. Chi-sq. = 21.719, reject H_0
7. Chi-sq. = 4.4211, cannot reject H_0
8. Chi-sq. = 1.281, cannot reject H_0
9. Chi-sq. = 49.987, reject H_0
10. Chi-sq. = 177.007, reject H_0
11. Cannot reject H_0
12. Reject H_0
13. Cannot reject H_0
14. Cannot reject H_0
15. Cannot reject H_0
16. Cannot reject H_0
17. Cannot reject H_0
18. Cannot reject H_0
19. Cannot reject H_0
20. Reject H_0

CHAPTER 17

1. $r_s = 0.6$
2. $r_s = -0.47619$
3. $r_s = -0.3333$
4. $r_s = 0.5515$
6. $r = 0.989949$
7. $r = 0.9904$, $r^2 = 0.98089$
8. $r = 0.7028$
9. $r = 0$
10. $r = 0.9284$
11. $r = 0.88713$, $r = 0.98$
12. (b) 0.6898

CHAPTER 18

1. $y = 0.6 + 2.8x$
2. $y = 34.5152 - 0.90606x$
3. $y = 1.24 + 0.607x$
   $y_{(x=7)} = 5.489$
4. $y = 4$
5. $y = -15.9 + 5.21x$
6. (a) $y = 833.8028 + 65.7356x$
   (b) 3003.08
   (c) 4777.94
7. (a) ATC = 56.5619 − 3.304x
   (c) 28.48, 3.705, −9.5095

CHAPTER 19

1. (a) $Y = 7.52 + 3.93\ X_2$
   (b) $Y = 33.267 - 2.32\ X_3$
   (c) $Y = 31.18 + 0.323\ X_2 - 2.136\ X_3$
   (e) $R^2 = 0.9957$
2. (a) $Y = -4.1 + 1.76\ X_2 + 0.986\ X_3$
3. $Y = 4.705 + 1.728\ X_2 + 1.133\ X_3 + 0.986\ X_4$

CHAPTER 20

1. (b) 39.75, 44.875, 50.375, 55.875, 59.75, 62.625, 65.75, 69.0, 71.375, 73.375, 73.875
   (c) $Q_1 = -14.82$, $Q_2 = -5.24$, $Q_3 = 1.39$, $Q_4 = 18.68$
2. Trend = 28.88 + 3.67647 X
   $Q_1 = -13.365$, $Q_2 = -5.0425$, $Q_3 = 1.03$, $Q_4 = 17.355$
3. (b) Trend = 11.33 + 0.42 X
   (c) $Q_1 = 1.88$, $Q_2 = 0.89$, $Q_3 = 0.39$, $Q_4 = 0.83$
4. (a) Average $R = 0$
   (b) Average $R = 1$
5. (a) Trend = 12.83 + 0.72 X
   (c) $Q_1 = -3.35$, $Q_2 = -1.12$, $Q_3 = 0.78$, $Q_4 = 3.68$
6. $Q_1 = 79.5\%$, $Q_2 = 94.6\%$, $Q_3 = 105.7\%$, $Q_4 = 120\%$
7. Average = 0
8. (b) Trend = 59.595 − 1.3136 X
   (c) 26.756
9. Alpha = 0.6

CHAPTER 21

1. (a) $x = 20$, $y = 40$, $z = 120$
   (b) $x = 10$, $y = 45$, $z = 145$
   (c) Multiple solution including $x = 20$, $y = 40$ and $x = 10$, $y = 45$, $z = 500$
2. (a) $x = 6$, $y = 7$, $z = 33$.
   (b) Acting constraints $0.5x + y \geqslant 10$
   $x + y \geqslant 13$.
   (c) $x = 5$, $y = 8$, $z = 34$.
3. Business computer = 12.5:
4. (b) $F = 2.5$ and $S = 7.5$.
   (c) Increase labour available to 11 hours.
5. 10 trains; 70 buses
6. (b) $N = 75$ and $B = 300$.
   (c) 39p.

CHAPTER 22

1. Critical path: C, H, K, M, L: time = 26
2. Two critical paths: A, C, H, L, T and A, E, P, Q, T
   Total time = 28.
3. Critical path: A, D, G: time = 14
4. Critical Path: B, F, I, K; Time = 45.
5. Critical path: D, N, S, W: time = 28
6. (a) 80 working days
   (b) Critical activities are:
   Valuation 2, Decide on price, Find a buyer, Instruct solicitor, Move.
   (c) No effect, as not a critical activity.
7. Critical path: B, E, F, G: time = 24; cost = 130
   Can reduce to time = 13; cost = 390
8. (b) Critical path: H, I, S, O, X: Time = 60 days
   (c) £38 170
   (d) £35 840 in 58 days.
9. Critical path: A, C, F, G: time = 27; cost = 1 820
   Can reduce to time = 18; cost = 2 940

CHAPTER 23

1. 200, 15
2. (a) 174, 5
   (b) £36 400
   (c) 58.33.
3. Order 3 000, cost £161 392.50
4. (a) 21.58.
   (b) 24.304.
5. (a) 6
   (b) 10
6. (a) 0.8
   (b) 0.2
   (c) $P_1 = 0.16$
   $P_2 = 0.128$
   $P_3 = 0.1024$
   (d) 4
   (e) 3.2
   (f) 0.167 hrs or 10 mins
   (g) 0.133 hrs or 8 mins
7. (a) reduced by 1.5
   (b) reduced by 1.35
   (c) reduced by 5 mins
   (d) reduced by 4.5 mins
8. Cost without ass. = £910/day
   Cost with ass. = £416/day
9. Manual system £112.40/hr
   Auto system £114.56
10. 4 service point system:
    $P_0 = 0.0737$
    $P_1 = 0.1843$
    $P_2 = 0.2303$
    $P_3 = 0.1919$
    $P_4 = 0.1200$
    $P_5 = 0.0750$
    $L_q = 0.53$
    $L_s = 3.03$
    $T_q = 0.636$ mins
    $T_s = 3.636$ mins
11. $P_0 = 0.1727$
    $P_1 = 0.2878$
    $P_2 = 0.2399$
    $P_3 = 0.1333$
    $P_4 = 0.0740$
    $P_5 = 0.0411$
    $L_q = 8.995$
    $L_s = 10.662$
    $T_q = 17.988$ mins
    $T_s = 21.324$ mins

INDEX

Installing MICROSTATS on a 5¼″ floppy

IMPORTANT!

Before you run MICROSTATS for the first time, you should make a back-up copy by using the following command:

DISKCOPY A: A: (or DISKCOPY B: B:)

using the copy of DISKCOPY that was supplied with your version of MS-DOS. Then put your original away and work ONLY from your working copy of MICROSTATS.

Some production copies of the MICROSTATS diskette may come pre-installed i.e. with the authentification information already in place. You can check that this is so by typing SD/A on the original MICROSTATS distribution disk and you should then see the authentification file (which has no name and is 0 bytes long) at the end of the file-list. If this file does not exist, it will be created automatically the first time MICROSTATS is run when the batch-file MS-RUN is activated. However, you should take a back-up copy of the distribution disk BEFORE you run MICROSTATS.

You can make a backup copy ONLY by using the DISKCOPY command described above.

Running MICROSTATS from a 3½″ floppy

If your machine will only take 3½″ floppy disks, then you need to find a machine equipped with both a 5¼″ and a 3½″ disk drive in order to make yourself a working copy of MICROSTATS.

Make sure that your floppy in the 3½″ disk drive is formatted. Then copy over all of the files with the command:

COPY A:*.*B:

(assuming that Drive A: is your 5¼″ drive and Drive B: is your 3½″ drive. If it is the other way round, then you will need to reverse the position of the drive letters with COPY B:*.*A:)

If your version of MICROSTATS was already preinstalled on your distribution disk on the 5¼″ floppy, then you may need to create a special installation file on your 3½″ floppy if this is Drive B. You can check out whether you need to do this by typing.

SD/A

on your 5¼″ floppy. If you do NOT see the authentification file (which has no name and is 0 bytes long) at the end of your file list, then you will need to create an authentification file on Drive B: in the following manner.

(1) First log onto your 3½″ floppy drive.

Copy the file MSINST-C.COM to MSINST-B.COM with the command COPY MSINST-C.COM MSINST-B.COM

(2) Use the DEBUG utility program supplied with MS-DOS to type the following:

DEBUG MSINST-B.COM (followed by RETURN)

(3) When you see the hyphen prompt (-), type

E 11F 42

and press RETURN.

(4) When you see the hyphen prompt (-) again, type W (You should see the message 'Writing 00041 bytes' or similar appear)

(5) When you see the hyphen prompt (-) again, type Q

MSINST-B.COM should now be written on your 3½″ floppy.

If your 3½″ floppy is Drive B: then run the file MSINST-B by typing MSINST-B and then press RETURN. If your 3½″ floppy is Drive A: then run the file MSINST-A instead of MSINST-B.

Type SD/A and you should see the authentification file (no name, 0 bytes in length!) appear at the end of your file list.

Put your original distribution disk away in a safe place!

TECHNICAL SUPPORT

If you have questions or problems with your software which are not attributable to a careful reading of the manual, then you are entitled to a period of technical support for 30 days after the date of purchase. To take advantage of this support you must register your purchase (see the registration card).

If you do have problems with the software, please write to the address on the registration card and enclose the following details:

- The description of the computer upon which the problem arose and its configuration, e.g. number and type of drives, available memory, etc.
- The contents of the AUTOEXEC.BAT or CONFIG.SYS files, particularly if you feel that these might have contributed to the problem.
- The contents of any worksheets or data sets upon which you were working when the problem arose. Copies of any files, screen dumps and printouts should also be included, as well as the sequence of events which led to the problem arising, including where possible the sequence of keystrokes.
- Any other information which could lead to the problem being reproduced and therefore identified.

However, it is fair to warn users that problems can generally be averted by:

- correct installation of the software;
- careful reading of the manual.

There are some circumstances under which almost any statistical software will fail (for example, operations which cause the computer to 'divide by zero') but these are usually due to the fact that inappropriate techniques are being used on the data. A careful reading of the manual will generally help avoid these difficulties.

LICENSE STATEMENT